MFPT 2017

50 Years of Failure Prevention Technology Innovation

Virginia Beach, Virginia, USA
16-18 May 2017

ISBN: 978-1-5108-8839-5

Printed from e-media with permission by:

Curran Associates, Inc.
57 Morehouse Lane
Red Hook, NY 12571

proceedings.com

Some format issues inherent in the e-media version may also appear in this print version.

Printed by Curran Associates, Inc. (2019)

For permission requests, please contact Society for Machinery Failure Prevention Technology at the address below.

MFPT Society
715 Twining Road
Suite 135
Dresher PA 19025

Phone: (215) 346-2870
Fax: (215) 352-6902

https://mfpt.org

Additional copies of this publication are available from:

Curran Associates, Inc.
57 Morehouse Lane
Red Hook, NY 12571 USA
Phone: 845-758-0400
Fax: 845-758-2633
Email: curran@proceedings.com
Web: www.proceedings.com

TABLE OF CONTENTS

A COMPARATIVE STUDY ON ANOMALY DETECTION OF THE COMBUSTION SYSTEM IN GAS TURBINES

Jiao Liu
School of Energy Science and Engineering
Harbin Institute of Technology
Harbin, Heilongjiang, China
Telephone: (240)825-7497
liujiaohit@hotmail.com

Myeongsu Kang
Center for Advanced Life Cycle Engineering (CALCE)
University of Maryland
College Park, MD, USA
mskang@umd.edu

Jinfu Liu, Zhongqi Wang, Daren Yu*
School of Energy Science and Engineering
Harbin Institute of Technology
Harbin, Heilongjiang, China
jinfuliu@hit.edu.cn, wangzhongqi@hit.edu.cn, yudaren@hit.edu.cn

Michael G. Pecht
Center for Advanced Life Cycle Engineering (CALCE)
University of Maryland
College Park, MD, USA
pecht@umd.edu

Abstract: Due to the fact that the combustion system, the core component of gas turbines, works in the highly adverse environmental conditions of high temperature and high pressure, it frequently faces malfunctions, causing catastrophic accidents. Hence, anomaly detection plays an important role in helping the combustion system run safely and economically. In recent decades, some methods have been published on anomaly detection of the gas turbine combustion system. However, there are few studies that compare these methods. The aim of this paper is to review and provide analytical results of the anomaly detection methods. An overall assessment of the merits or weaknesses of the generic methods is provided by testing the methods with actual gas turbine operating data. Additionally, some possible research development of the anomaly detection of the gas turbine combustion system is presented in this comparative study.

Key words: Anomaly detection; gas turbine; combustion system; health management

---

* Corresponding author.

Introduction: The technology of gas turbines has improved significantly over the past 50 years. As the efficiency of gas turbines increases, the combustion system, the core component of the gas turbine, is subjected to severe thermal stress [1] and requires costly maintenance [2]. Prognostics and health management (PHM) can be used to solve these problems. Implementation of PHM can help gas turbines run safely and economically by improving their reliability, safety, and availability, while reducing the operational costs.

The gas turbine combustion system works in the highly adverse environmental conditions of high temperature, high pressure, and high speed. Malfunction of the combustion system is often followed by catastrophic accidents. Anomaly detection plays an important role in PHM systems, and usually focuses on detecting abnormal deviation from its nominal behavior and detecting fault timely. Therefore, anomaly detection of combustion system in gas turbines is particularly significant.

Typical faults in gas turbine combustion system include fuel uneven distribution, fuel supply channel damage or leak, nozzle carbon deposition, nozzle jam, components burn through, and so on. These faults may cause abnormal combustion chamber outlet temperature. A gas turbine combustion system is usually installed with several combustion chambers of the same structure. These chambers are equally arranged in the circumferential direction. Outlet temperature of the normal combustion chamber is almost same, while the outlet temperature of the malfunction combustion chamber is different from the others. However, the combustor outlet temperature is very high, and conventional sensors cannot work in such harsh environments for a long timc.

The owners usually take advantage of the exhaustive gas temperature (EGT) to monitor the performance of the combustion system. The EGT provides the most relevant information about the state of the hot components, which is measured by several thermocouples distributed equally at the outlet of the gas turbine. See Figure 1. All exhaust thermocouple readings are almost identical when the unit is in normal operation, while some exhaust thermocouple readings are different from the others when the combustion system malfunctions. The EGT spread, the difference between different thermocouples, can be treated as an indicator in anomaly detection of combustion system. But there are many problems to be considered. The malfunction of the combustion system is not the unique factor causing uneven temperature spreads. In normal operation, some other factors can also cause uneven temperature spreads [3], including operation conditions, ambient conditions, thermocouple position error, and manufacturing tolerances. At the same time, the hot gas rotates following by blades and the degree of rotation will vary with operation conditions. The hot gas from different combustion chambers mixes in the turbine, which reduces the amplitude of the EGT spreads compared with the spreads at the exit from the combustion chambers. Therefore, anomaly detection of gas turbine combustion systems is complex.

Many studies have been published on anomaly detection of gas turbine combustion systems. The research can be divided into two categories, including spread-based method and model-based method. In this paper, these methods are reviewed. The merits or weaknesses of the generic methods are discussed by testing the methods with actual gas turbine operating data. The paper presents some possible research developments in the anomaly detection of gas turbine combustion systems.

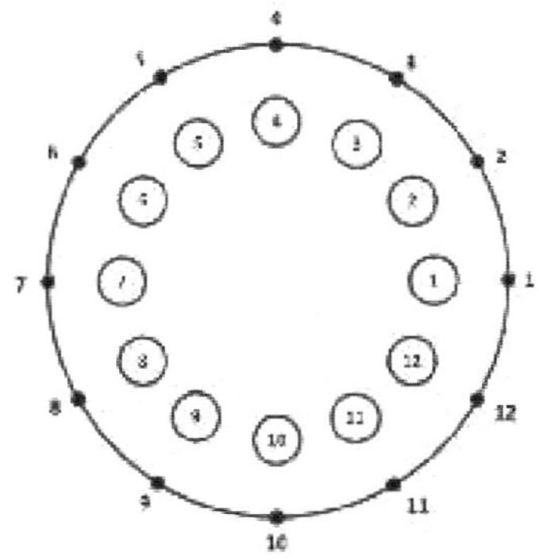

Figure 1: Combustion Chambers and Thermocouples Distribution

Method Review

Spread-based Method: The EGT profile is almost equal when the combustion system is in normal operation, while the EGT profile is uneven when there is a fault. This is an important feature of the EGT. Based on this feature, the spread-based method monitors the spread, the difference between different thermocouples, to identify the status of the system. The spread is small when the combustion system is in normal operation, while the spread is big when there is a fault. The spreads-based method is intuitive, so it is commonly applied in industry. The monitoring system introduced in Ref [4, 5] shows that the system monitors not only direct temperature but also variation in the exhaust gas distribution from the average value. Large variation can indicate combustion problems. In Ref [6], in addition to detecting the variation, combustion anomaly detection method also monitors the maximum EGT, median EGT and average EGT as supplements. Gulen et al. [7] introduce a heavy-duty industrial gas turbine real-time online performance diagnostic system. In the system, excessive and sudden changes or consistent upward trends in the ratio of the maximum EGT to the average value indicate the potential combustor problems.

The spread-based method utilizes unevenness of the fault EGT profile to monitor the performance of the combustion system. As mentioned above, some factors, operation conditions, ambient conditions, thermocouple position error, manufacturing tolerances, can also cause uneven temperature spreads. To remove the influence of the factors, the average EGT value is usually treated as the representative of operation conditions and ambient conditions in these methods. The value after subtracting or dividing by the average EGT value is taken as the monitoring indicator. Obviously, this kind of method is not very accurate. The average EGT value is roughly representative, and the relationship between the average EGT value and operation conditions and ambient condition is more complex. As one of the most widely used monitor systems, the MARK VI system [8] produced by GE makes a correction. It defines a parameter S, allowable exhaust temperature dispersion, which is the function of the average EGT value T_4^* and the compressor outlet temperature T_2^* , as shown in Eq. (1), where (100) means the equation should be added to 100 °F under the condition of variable operation condition.

$$S = \left(60 + 0.145T_4^* - 0.08T_2^* \Big|_{300}^{725}\right)\Big|_{50}^{170} + (100) \qquad (1)$$

The method calculates temperature differences between the maximum EGT and the three minimum EGTs. The higher the ratio of the temperature differences to the allowable exhaust temperature dispersion, the greater the possibility of the failure. However, some factors are still not considered, such as thermocouple position error, manufacturing tolerances, and the gas rotation. In practice, the operators have found the solution to lack adequate sensitivity, i.e., the hot components have been damaged seriously once the alarm is generated [9].

Model-based Method: The model-based method take advantage of an EGT model to detect anomaly in the gas turbine combustion system. The model may be built either from prior physical knowledge of the gas turbine (physical model) or from some prior identification based on a set of training data (data-driven model). The model-based method mainly contains two approaches.

The first one can be described as Figure 2, and the residuals are usually an indicator. The EGT model is built in normal operation. The output of model is the estimated EGT value. If estimated EGT value and actual EGT value are almost same, their residuals will be stable. Contrarily, if estimated EGT value and actual EGT value are totally different, their residuals will increase. The abnormalities are then identified by corresponding high residuals. The factors affecting EGT value in normal operation are taken into account by the model. Therefore, building an accurate model is the key problem in the model-based method.

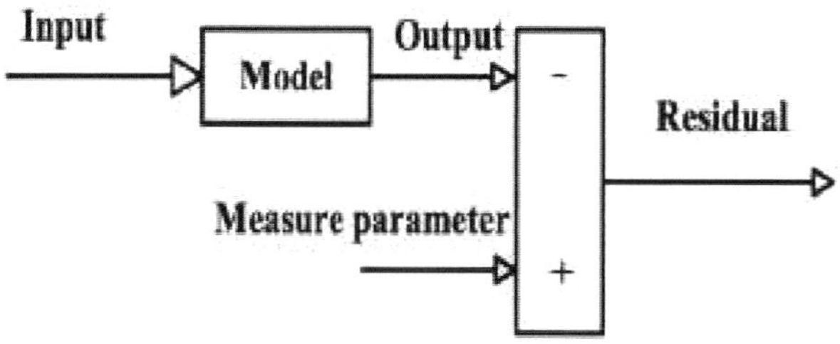

Figure 2: Diagram of Model-based Method

In combustion system anomaly detection methods, artificial neural network (ANN) is usually used to build the EGT model [10-12]. Artificial neural networks [13] are computational models inspired by the neural structure of the brain that are capable of machine learning and pattern recognition. Due to their high connectivity and parallelism, ANNs can link, in a non-linear way, a multi-dimensional input space with a multi-dimensional output space, allowing very high computational speed. Based on data, ANN does not need to consider a complex physics principle. Tarassenko et al. [10] present an EGT model of normality based on ANN and make use of the spatial correlations in the EGT profile. When a fault develops in one of the combustion chambers, there is a local effect on the temperature profile; only a small number of contiguous thermocouples are significantly affected whilst the rest of the profile is largely unchanged. The model of normality is therefore constructed by learning the function relating the temperature values from the four thermocouples opposite it. Song et al. [14] and Yılmaz [15] evaluate the relationship between EGT and operational parameters based on a multiple linear regression. The operational parameters Song et al. consider include low-pressure turbine outlet pressure, high-pressure rotator speed, high-pressure compressor outlet temperature and outlet pressure, and low-pressure rotator speed.

Yılmaz proposes that the fuel flow, thrust-specific fuel consumption, and take-off margin temperature are determined to be the most significant operational parameters. Kumar et al. [16] compare the statistical and neural network approach. As one of the breakthrough technologies in machine learning, deep learning has begun to be applied in anomaly detection of combustors. Stacked denoising autoencoder, one of the deep network architectures, has been used to improve combustor anomaly detection [17].

There are some EGT physical models to be proposed. Korczewski [18] describes an EGT physical model based on thermal and flow processes in gas turbines. Zhang [19] considers the gas rotating and mixing in the gas turbine and presents an EGT physical model. Medinaet al. [22] present an EGT model based on both basis function expansion and the Brayton cycle.

The second model-based method utilizes the parameters of EGT model as an indicator. The system can be described by the following equation

$$y_t = F(\theta_t,\ X_t) + v_t \tag{2}$$

where y_t and the variables X_t are observed, v_t is white noise independent of X_t, and the parameter vector is θ_t.

The parameter vector θ_t characterizes the system, which can be identified based on measurements. We can have a nominal model denoted by θ_t for such a system. Slight deviations of the actual system from its nominal behavior θ_t indicate early fault of the system [20].

Based on this opinion, a new indicator is presented in Ref [21]. Two main factors affecting the EGT value in normal operation are considered, including the operating and ambient conditions, and the structure deviation of different combustion chambers caused by processing and installation errors. Moreover, some studies focus on the parameter estimation and detection. Medina, P. et al. [22] analyzes the parameters with statistical techniques and with wavelets to detect abrupt changes in combustion chambers. The estimation is known to be biased because of the noisy input. M. Basseville et al. [23] present an asymptotic local approach for change detection to solve the problem.

Experiment: To assess the merits and weaknesses of the generic methods, two cases are presented in this section. The operating data come from the Taurus 70 gas turbine produced by Solar Turbines. The gas turbine has 12 combustion chambers and 12 thermocouples to measure the EGT. The gas turbine operated 14,407 min in normal operation, as can be seen in Figure 3. The normal operation data are used as the training set. In the first case, the fault data comes from actual operation. Figure 4 shows the fault EGT data in the first case. In the second case, the fault data is constructed by normal data. The 7th exhaust thermocouple reading downs 10 ºC slowly at 500 min, as shown in Figure 5. The training set is used to determine the threshold of the indicator. The other data is used to detect if the indicator exceeds the threshold. If the threshold is not exceeded, it means the system is normal. Otherwise, the system has an anomaly.

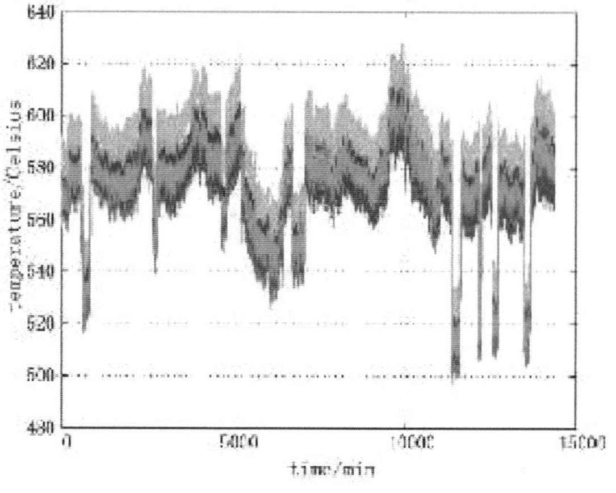

Figure 3: EGT in Normal Operation Condition

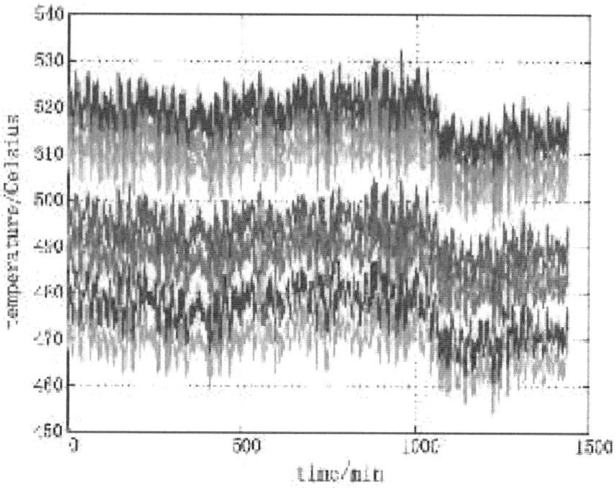

Figure 4: Fault Data in Case 1

In this section, the method introduced in Ref [5] is applied as the typical spread-based method. The detection indicator is variation in the exhaust gas distribution from the average value. The method introduced in Ref [10] is applied as the typical model-based method. The two models are, respectively, an ANN model and a physical model. In this paper, a three-layer ANN model is built. The output signal is an EGT value, and the input signals are EGT values of the four thermocouples opposite it. The neural node number in the hidden layer is 20. The detection indicator is the model's residuals.

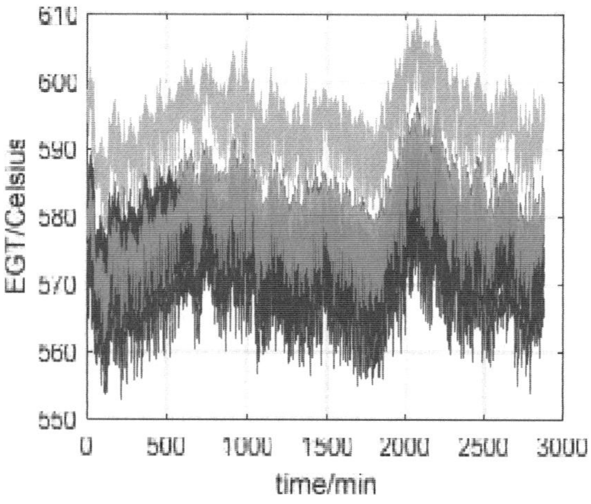

Figure 5: Fault Data in Case 2

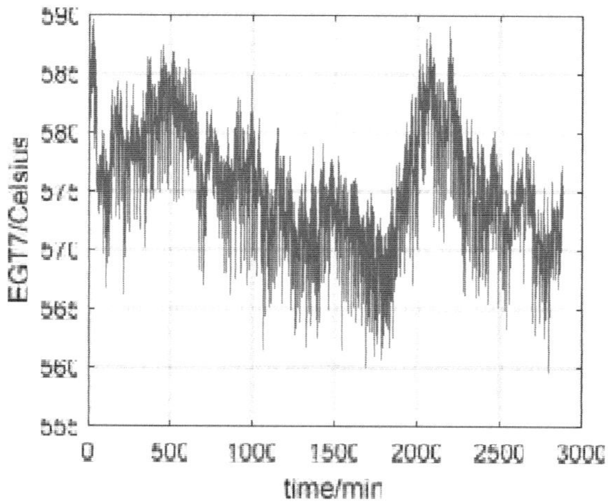

Figure 6: 7th Thermocouple Reading in Case 2

The detection results are shown in Figure 7 to Figure 11. As we can see in Figure 7 and Figure 10, the spread-based method can detect the fault in the combustion system, but it is powerless to detect slight faults. In comparison, the model-based method performs much better. It can detect both serious and slight faults. However, it still has problems.

Figure 8 reflects the model accuracy. The residual of training data has some sudden change because of operation condition changes. It causes a wider threshold and then affects the detection sensitivity. As shown in Figure 12, if the fault also happens at 500 min, but at a much slower deterioration rate, the model-based method is not sensitive as expected. It means the model could be modified more accurately to further improve the detection sensitivity.

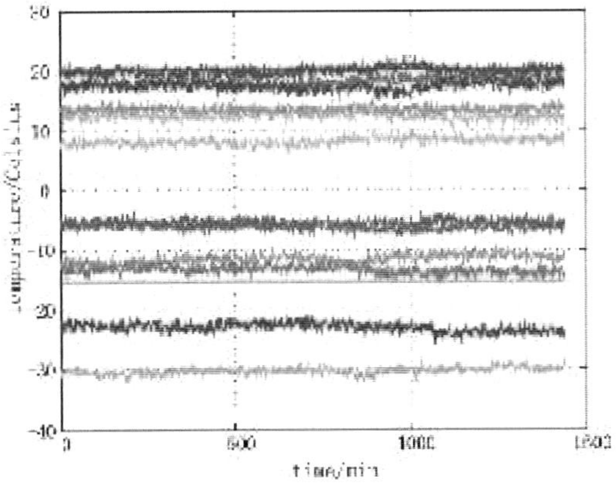

Figure 7: Spread-based Method Detection Result in Case 1

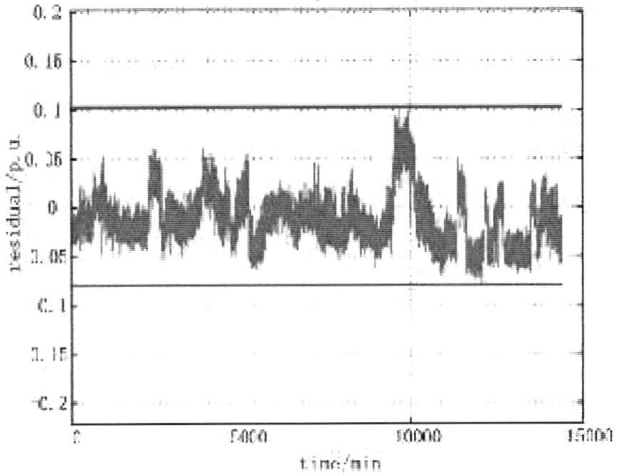

Figure 8: Model-based Method Training Data Detection Result in Case 1 (4th thermocouple)

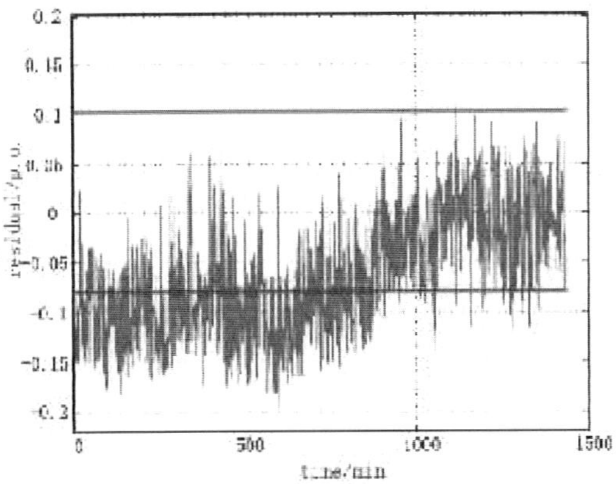

Figure 9: Model-based Method Detection Result in Case 1 (4th thermocouple)

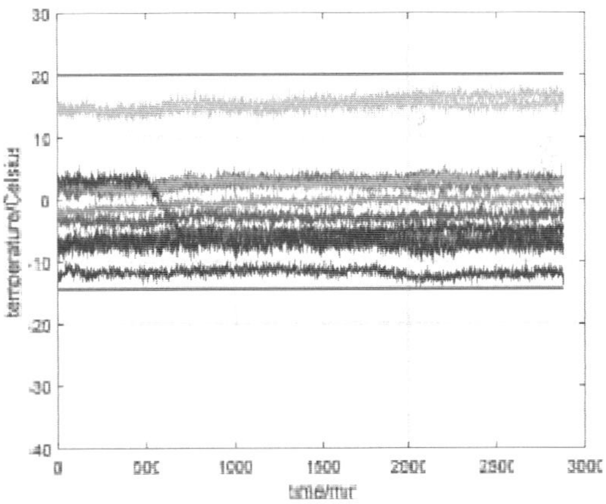

Figure 10: Spread-based Method Detection Result in Case 2

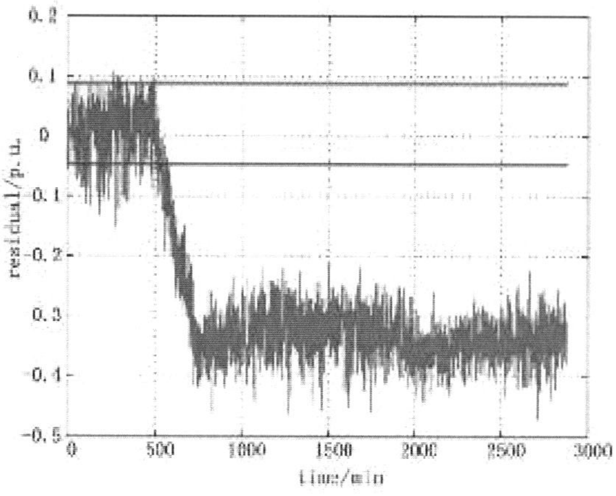

Figure 11: Model-based Method Detection Result in Case 2 (7th thermocouple)

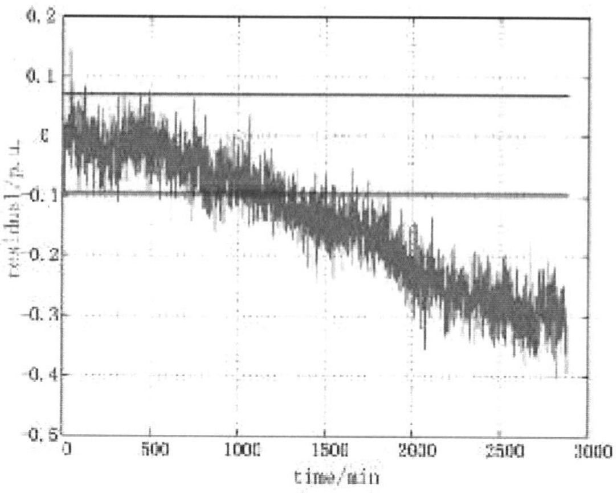

Figure 12: Model-based Method Detection Result in Case 3 (7th thermocouple)

Conclusions: The vulnerability of combustion chambers in gas turbines motivated researchers to develop and advance various anomaly detection methods and tools. In this paper, a review of spread- and model-based methods was presented with a description of their pros and cons. Although the spread-based method was easy to realize in real applications, it was not suitable to detect anomalies in gas turbines operating under non-stationary conditions. Instead, the spread-based method was generally suitable for an over-temperature alarm. The model-based method has become an alternative for the detection of anomalous behavior of thermodynamics in combustion chambers. This was mainly because the model-based method has been more effective for facilitating anomaly detection under non-stationary conditions rather than the spread-based method. However, the performance of the model-based method depends on the accuracy of the model that represents the normal behavior of gas turbine combustors. The model-based method can be improved in some aspects. For instance, the fundamental understanding of the failure mechanisms of combustion chambers can improve the accuracy of the model. Likewise, the inclusion of prior physical knowledge in the data-driven model can be a potential research direction that can enhance the accuracy of the model and be used for anomaly detection.

Acknowledgment: The authors would like to thank China Scholarship Council for supporting the first author of this paper while studying abroad. Likewise, this work was supported in part by the over 100 CALCE members of the CALCE Consortium.

References:

[1] Jiang H, Ren J, Xueying L I, et al. Status and Development Trend of the Heavy Duty Gas Turbine. Proceedings of the Chinese Society of Electrical Engineering, 2014, 34(29): 5096-5102.

[2] IATA, Airline Maintenance Cost Executive Commentary, 2011, International Air Transport Association, Montreal, Canada.

[3] Knowles, Martin, Gas Turbine Temperature Spread Monitoring Detection of Combustion System Deterioration, ASME 1994 International Gas Turbine and Aeroengine Congress and Exposition. American Society of Mechanical Engineers, 1994.
[4] Hamilton, PE Patrick, and D. Ha. Exhaust gas temperature capabilities now in system 1 software. Product Update 25.1, 2005: 88-89.
[5] General Electric Company, Heavy Duty Gas Turbine Monitoring & Protection, 2015, General Electric Company, Nevada, United States.
[6] Miller, K. W., Iasillo, R. J., & Lemmon, M. F. 2015. U.S. Patent No. 20,150,267,591. Washington, DC: U.S. Patent and Trademark Office.
[7] Gulen, S. Can, Patrick R. Griffin, and Sal Paolucci. Real-time on-line performance diagnostics of heavy-duty industrial gas turbines. Journal of engineering for gas turbines and power, 2002, 124(4): 910-921.
[8] YU L, JIN J. Study on Deflection Laws of 9FA Gas-Turbine Exhaust Temperature Field. Electric Power Construction, 2009, 3: 026.
[9] Liu, J., The Research for Exhaust Temperature Anomaly Detection in Gas Turbine, 2014, M.S. thesis, Harbin Institute of Technology, Harbin, China.
[10] Tarassenko, L., Nairac, A., Townsend, N., Buxton, I., and Cowley, Z., Novelty Detection for the Identification of Abnormalities, Int. J. Syst. Sci., 2000, 31(11), pp. 1427–1439.
[11] Patsimo Systems, Case Study of Anomaly Detection on a GE LM2500+ Gas turbine done by Patsimo Systems.
[12] Kenyon, A. D., V. M. Catterson, and S. D. J. McArthur. Development of an intelligent system for detection of exhaust gas temperature anomalies in gas turbines. Insight-Non-Destructive Testing and Condition Monitoring, 2010, 52(8): 419-423.
[13] Haykin, S., Neural Networks - A Comprehensive Foundation, 2nd Ed., 1999, NJ, USA: Prentice Hall International.
[14] Song, Y. X., Zhang, K. X., and Shi, Y. S., Research on Aeroengine Performance Parameters Forecast Based on Multiple Linear Regression Forecasting Method, J. Aerosp. Power, 2009, 24(2), pp. 427–431.
[15] Yılmaz, I, Evaluation of the Relationship Between Exhaust Gas Temperature and Operational Parameters in CFM56-7B Engines, Proc. Inst. Mech. Eng., Part G, 2009, 223(4), pp. 433–440.
[16] Kumar, Amar, et al. Gas turbine engine operational data analysis for anomaly detection: Statistical vs. neural network approach, Electrical and Computer Engineering (CCECE), 2013 26th Annual IEEE Canadian Conference on. IEEE, 2013.
[17] Yan W, Yu L. On accurate and reliable anomaly detection for gas turbine combustors: A deep learning approach, Proceedings of the Annual Conference of the Prognostics and Health Management Society. 2015.
[18] Korczewski, Z., Exhaust Gas Temperature Measurements in Diagnostic Examination of Naval Gas Turbine Engines, Pol. Marit. Res., 2011, 18(2), pp. 37–43.
[19] Q.ZHANG. Contribution a` la surveillance de proce´de´s industriels. Thesis, Univ. Rennes I, Dec. 1991.
[20] Zhang, Qinghua, Michèle Basseville, and Albert Benveniste. Early warning of slight changes in systems and plants with application to condition based maintenance. Diss. INRIA, 1992.

[21] Jinfu, Liu, et al. Early Fault Detection of Hot Components in Gas Turbines. Journal of Engineering for Gas Turbines and Power, 2017, 139(2): 021201.
[22] Medina, P., Saez, D., and Roman, R., On Line Fault Detection and Isolation in Gas Turbine Combustion Chambers, ASME Turbo Expo 2008: Power for Land, Sea, and Air, ASME Paper No. GT2008-51316.
[23] Basseville, M., Benveniste, A., Mathis, G., & Zhang, Q, Monitoring the combustion set of a gas turbine. In Proc. Safeprocess, 1994, 94, pp. 397-402.

A FAILURE POLYMORPHISM THEORY FOR SYSTEM RELIABILITY MODELING CONSIDERING FAILURE MECHANISMS CORRELATIVITY

Chaojie Qi, Yufeng Sun* and Yaqiu Li
School of Reliability and System Engineering, Beihang University
37 Xueyuan Road
Room 537
Haidian District, Beijing
Telephone: (86) 18801273795
1334225486@qq.com

Abstract: There is a 'polymorphism' phenomenon in biology, which means individuals evolve to two or more clearly different phenotypes in the same population of a species, as a result of internal species gene and external environment condition. A similar 'failure polymorphism' phenomenon exists in complex systems. The failure modes in a complex system originate from different failure mechanisms. Failure mechanism is the root cause of product failure, and experts are used to paying attention to failure dependence problem by studying the failure behavior, few researches is conducted from the failure mechanisms correlation perspective. In this paper, polymorphism related concepts are incorporated into reliability field, new definitions of failure polymorphism are discussed, and Bayesian framework is introduced to tackle the problems of uncertainty factors and dynamic evaluation. Based on physics of failure (PoF), the paper presents a system-level reliability evaluation method considering the mechanisms correlativity.

Key words: failure polymorphism; failure mechanisms correlativity; physics of failure; Bayesian framework

1 Introduction: Failure mechanism is the root cause of product failure. It describes the internal physical and chemical process during products' lifetime and reflects the relationship between internal variables and product life under certain circumstances. Plenty of engineering practices indicates that all failure scenarios of complicated system, which consists of numerous units (electronic component, material and structural part, etc.), could be finally traced back to several dependent failure mechanisms of certain units. One single unit may have different failure mechanisms in variant environment conditions. Moreover, different units may have the same failure mechanism in equivalent environment and units of the same sort may fail differently due to individual differences caused by processing technics [1]. One mechanism of certain unit may affect other mechanisms in the same unit, further one unit fails may have an influence on other dependent units in the same system, all these cases will aggravate the degradation of system until it fails [2]. Experts are used to paying attention to fault dependence problem by studying the failure behavior (common caused failure and failure propagation), few researches is conducted from the failure mechanisms correlation perspective.

There is a 'polymorphism' phenomenon in biology, which means individuals evolve to two or more clearly different phenotypes in the same population of a species. This phenomenon occurs frequently in hierarchical and social insect species (such as termits, aphid, bee, etc.), as a result of internal species gene and external environment condition. Differentiated individuals focus on their specified work, which is called polytheism, to improve work efficiency and strength the dependences among individuals, help the species adapt to the environment changes ultimately. The essence of polymorphism phenomenon is a nature selection process from microcosmic to macrocosmic, in which the species produces adaptive differentiation on individuals to keep surviving in dynamic environment, and the differentiated representations of individuals promote the evolution of the whole species in turn.

A similar 'failure polymorphism' phenomenon exists in complex systems. Hierarchical system is composed of various units in different levels, and different unit failure modes originated from diverse failure mechanisms may lead to different failure mode of system [3]. Assumptions in biologic theory are introduced for failure polymorphism, the internal factors of which are the uncertainty of units' properties (material, structural design, process level, etc.), while the external factors are the uncertainty of environment and load conditions. Failure mechanisms of unit that are susceptible to these factors may have intense reciprocal effects, and result in the unit failure polymorphism, which means the same unit may have different kinds of failure modes. The interactions among different units' failure modes generated by their functional or structural dependences are supposed as the failure polytheism. As a consequence of the small changes in system configuration, the reliability of system begins to decline, which can be regarded as the evolution of system failure state. In this paper, polymorphism related concepts are incorporated into reliability field, new definitions of failure polymorphism are discussed, and Bayesian framework is introduced to tackle the problems of uncertainty factors and dynamic evaluation. Based on physics of failure (PoF), this paper presents a system level reliability evaluation method considering the mechanisms correlativity.

2 Method for System Level Reliability Evaluation Considering the Mechanisms Correlativity: In a complex system, parts, components and subsystems have its own property and behavior, which can affect the final state of the system [4]. In order to study the similar 'failure polymorphism' phenomenon in complex systems, biological polymorphism phenomenon is introduced to reliability field. As shown in Figure 1, taking the micro electro mechanical system (MEMS) as an example, property and behavior of the parts, components and subsystems are redefined.

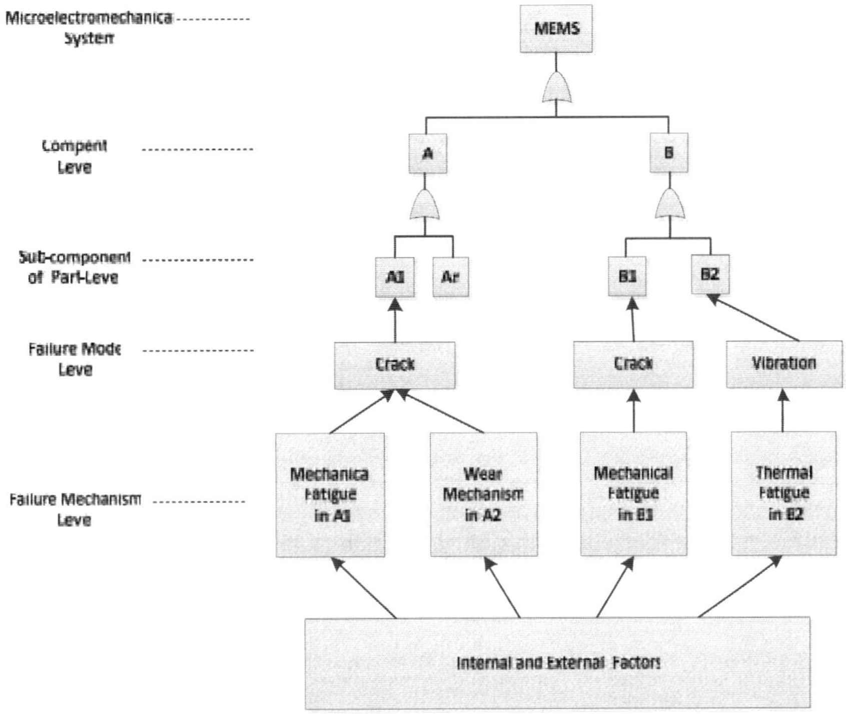

Figure 1: The Structure Diagram of MEMS

2.1 New Definitions of Failure Polymorphism: There is a ‘polymorphism’ phenomenon in biology, which means individuals evolve to two or more clearly different phenotypes in the same population of a species. Similarly, in a complex system, parts, components and subsystems will show a variety of failure modes, which is called ‘failure polymorphism’ phenomenon. For example, in the MEMS, both the sub-component A1 and B1 may cause crack failure, while the sub-component B2 may generate mechanical vibration [5].

The ‘polymorphism’ phenomenon in biology is mainly due to the differentiation of biotic population. And the causes of the differentiation include hereditary and non-hereditary factors. Analogously, the ‘failure polymorphism’ phenomenon is resulted from the differentiation of parts, components and subsystems. The reasons for differentiation including the differences among individuals and the effects of the environment, which is called failure differentiation factor.

There is biological adaptation in ‘polymorphism’ phenomenon. Biological adaptations means the organism can adjust themselves to the environment. In ‘failure polymorphism’ phenomenon, failure can also adapt to the environment. For example, in the MEMS (Fig. 1), the sub- component A1 will initiate crack (or crack propagation) to fit the combined effects of alternating mechanical stress and corrosion. The MEMS will eventually degeneration or failure to fit the harsh external environment.

In the MEMS, there is correlativity in the failure adaptation, such as the correlativity between the two failure mechanisms in sub-component A1. The failure of A1 is due to the combined effects of mechanical fatigue and corrosion. The failure between sub-components B1 and B2 are both affected by external alternating mechanical stresses. The correlativity is due to the same environment and working conditions. In addition, the same design, material, position or other factors will result failure correlativity. Correlativity also means the degradation of a part may be affected by the failure mechanism of other parts. It is great difficulties to evaluate the reliability of system level by considering the uncertainty factors and failure mechanism correlativity. In this paper, Bayesian framework is introduced to tackle the problems of uncertainty factors and dynamic evaluation [6]. Based on physics of failure (PoF), this paper presents agent oriented modeling method to evaluate the system reliability considering the mechanisms correlativity.

2.2 Single Mechanism Modeling Under Bayesian Framework: Based on PoF, a component has certainty parameters which mean the component has a determined life. However, for a number of components, their processing quality, structural size and material properties are uncertainty. Meanwhile, environmental stress and other parameters of components in service are also uncertainty, which will result the life of components becoming uncertain. Here, based on PoF, the uncertainty of these parameters is taken into account by the Bayesian updating, which makes the life evaluation result closer to the actual engineering situation.

Take the sub-component B1 as an example. As a mechanical unit of the MEMS, B1 has the main failure mode of crack propagation and the corresponding failure mechanism is mechanical fatigue failure. The physics of failure model in B1 is shown in Equation (1).

$$N_{B} = \frac{1}{K_{B} S_{B}^{n}} \tag{1}$$

Where N_{B1} is the number of cycles before failure; S_{B1} is the stress amplitude; K_{B1} is the proportional constant and n is the mechanical parameter.

By fitting the accelerated test data, the lifetime N of the unit B1, is given a logarithmic normal distribution: N_{B1} ~LN ($_{B1}$, σ_{B1}). So the mean value μ_{B1}can be represented by Equation (2).

$$\mu_B = \log\left(\frac{1}{K_B S_B^n}\right) \quad (2)$$

Therefore, the life-stress multivariate function of unit B1 is shown in Equation (3).

$$f(N_{B1} \mid S_{B1}, K_{B1}, n, \sigma_{B1}) = \frac{1}{\sigma_{B1} N_{B1}\sqrt{2\pi}} e^{-\frac{1}{2}\left(\frac{\ln(N_{B1})+\ln(K_{B1})+n\ln(S_{B1})}{\sigma_{B1}}\right)^2} \quad (3)$$

The parameter vector of the model is shown in Equation (4).

$$\theta = (K_B, n, \sigma_B)^T \quad (4)$$

The uncertainty parameter σ_{B1} has no prior information, and assume the parameter σ_{B1} follows a uniform distribution from -3 to 3 in the Bayesian analysis. By investigating the literature, the prior information matrix of the parameters K_{B1}, n are shown in Equation (5).

$$\theta_{K,n} = \begin{bmatrix} K & n \\ 2.06E-11 & 4.59 \\ 6.11E-14 & 4.42 \\ 3.00E-10 & 2 \\ 5.13E-10 & 3.61 \\ 1.13E-10 & 2 \\ 4.53E-10 & 2.09 \end{bmatrix} \quad (5)$$

By fitting the two parameters in the priori information matrix with a normal distribution, we can get the prior distribution of the parameters, K_{B1} ~ Normal (2.33E-10, 2.41E-10), n ~ Normal (3.10, 1.19), σ_{B1} ~ Uniform (-3, 3).

The data of Bayesian updating is derived from the accelerated fatigue tests where the stress amplitude S_{B1} is a constant. The test data D_{B1} contains F complete failure observations t_i and H right delete measured value T_j, namely $D_{B1}\{(t_i, T_j), i = 1, \dots, F, j = 1, \dots, H\}$.

The Bayesian update process uses the test data $D_{B1}\{(t_i, T_j), i = 1, \dots, F, j = 1, \dots, H\}$. The likelihood function of the life-stress model is shown in Equation (6).

$$\begin{aligned} L(D_B \mid K_B, n, \sigma_B) &= L(t_1, \dots, t_F, T_1, \dots, T_H \mid K_B, n, \sigma_B) \\ &= \prod_{i=1}^{F} f(t_i \mid K_B, n, \sigma_B) \prod_{j=1}^{H} R(T_j \mid K_B, n, \sigma_B) \end{aligned} \quad (6)$$

Where $R(T_j \mid K, n, \sigma)$ is the reliability equation, which is expressed by Equation (7).

$$R_B(T_j \mid K_B, n, \sigma_B) = 1 - \int_0^{T_j} f(x \mid K_B, n, \sigma_B)\,dx \quad (7)$$

According to the Bayesian equation, the posterior distribution of the model parameters is shown in Equation (8).

$$\pi(K_{B1}, n, \sigma_{B1} | D_{B1}) = \frac{L(D_{B1} | K_{B1}, n, \sigma_{B1})\pi_0(K_{B1}, n, \sigma_{B1})}{\iiint L(D_{B1} | K_{B1}, n, \sigma_{B1})\pi_0(K_{B1}, n, \sigma_{B1}) dK_{B1}\, d\ \sigma_{B1}} \quad (8)$$

The probability distribution of the life-stress model of sub-component B1 is shown in Equation (9).

$$f_{B1}(N_{B1}, S_{B1}) = \iiint f(N_{B1} | S_{B1}, K_{B1}, n, \sigma_{B1})\, \pi(K_{B1}, n, \sigma_{B1} | D_{B1}) dK_{B1}\, dn d\sigma_{B1} \quad (9)$$

The Markov chain Monte Carlo (MCMC) method is used to obtain the analytic solution of Equations (8) and 9 by WinBUGS [7]. After running the WinBUGS code, the posteriori distribution of the uncertainty parameters K_{B1}, n, and σ_{B1} are shown in Figure 2.

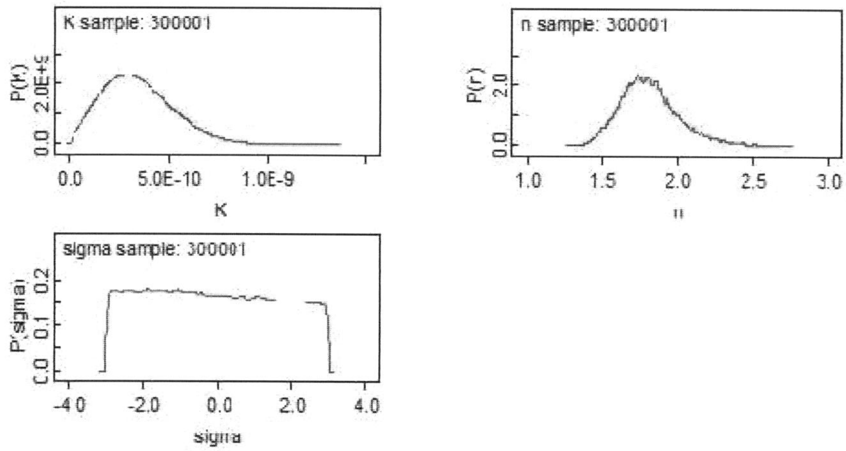

Figure 2: The Posteriori Distribution of Uncertainty parameters K_{B1}, n, and σ_{B1}

Table 1 shows the final results of the Bayesian updating process for the uncertain parameters K_{B1}, n, and σ_{B1}.

Table 1: The Final Result of K_{B1}, n, and σ_{B1} after 300000 Bayesian Parameters Update

Node	μ	σ	MC error	2.5%	median	97.5%
K_{B1}	3.445E-10	1.986E-10	4.894E-12	6.758E-11	3.271E-10	7.299E-10
n	1.815	0.198	0.006776	1.49	1.794	2.278
σ_{B1}	0.017	1.613	0.02243	-2.858	0.023	2.88

According to the stress-life model, the 300000 groups of posterior data K_{B1}, n, and σ_{B1} are simulated by Monte Carlo method. And the probability density distribution of the life distribution in sub-component B1 can be obtained. There is no need to know the life distribution of the sub-component B1. Each underlying failure data N_{B1} will be used as input of the upper level. And then we can use the data to simulate the degradation process of the system. It can be seen that the probability physical model obtained under the Bayesian framework can take full account of the material and environmental uncertainty [8]. So it is more objective and closer to the actual situation. Obviously, the modeling processes and steps of the single failure mechanism in sub-component B2 are similar to sub-component B1.

2.3 Considering the Mechanisms Correlativity for System Reliability Modeling Based on Multi-Agents: At present, only a few deterministic models involve the interaction of the failure mechanisms. And there is no probabilistic failure physical model considering the interaction between failure mechanisms. System-level modeling will be difficult because of the diversity of its components and the interaction between their failure mechanisms. Here we propose a hybrid modeling method based on multi-agents and system dynamics to solve the problems. In order to ensure that there is a failure mechanism correlativity in the studied system and meanwhile the system is simplified as much as possible, only the sub-components A1 and B1 in the MEMS are considered.

In the MEMS, each component, part, and subsystem is treated as an agent [9]. In response to internal and external changes, each agent is created with a single output variable whose value is a function of one or more input variables. Each input variable represents another agent, pointing to the output variable. The generic formula for the output variables is shown in Equation (10).

$$Y = f(X_1, X_2, \ldots, X_n) \tag{10}$$

Where Y represents the function of the output variable in the agent, consisting of the input variables $X_1, X_2, \ldots, X_n$. The functional equation $f(X)$ is a probabilistic failure physical model of the output in the agent, which has been determined in Section 2.2. Such agents are also called "multi-agents" because their evolution relies on input variables from external stress and the interaction among internal agents. There may be such agents that cannot get any input from other agents, which refers to the 'self-contained agents'. The development of the self-contained agents is independent of other agents, so they have independent output variables in the form of probability density functions.

As the uncertainty of $f(X)$ and X_i, the output in each agent is always stochastic. The output in each self-contained agent is also stochastic and consistent with the appropriate probability density function (modified or indeterminate parameter). Any feature or parameter given by a constant value should be treated as a constant instead of an agent [10]. A determined value, which can change in a known trend as the system evolves (such as time variables) or according to the time equation defined, need to be assigned as an agent.

For example, a case study of agent B1, the output of the agent B1 (the number of cycles) is related to the parameters K_{B1}, n, and σ_{B1}. The posterior probability distribution of K_{B1}, n, and σ_{B1} is acquired by the Bayesian update in Section 2.2. So the parameters K_{B1}, n, and σ_{B1} belong to agents. The parameter S_{B1} is not an agent as the parameter is a constant. According to Equation (1), the relationship between the output and each input in B1 can be determined. As shown in Figure 2, the result of the output is uncertainty with the form of probability density.

The output of the single failure mechanism in the sub-component A1 is similar to the sub-component B1. Besides, the interaction between the fatigue failure mechanism and the wear failure mechanism in A1 can be expressed by the interaction among the multi-agents. Meanwhile, the output of the agent A1 is the length of the crack. In this case, it is necessary to add a monitoring agent M_{A1} to monitor the crack length in real time. Once the crack length reaches the failure threshold, the monitoring agent M_{A1} is activated and output the total failure time N_A. Then the value of M_{A1} becomes 0 and continue to monitor the next time.

The failure of sub-components A1 and B1 can cause degradation in the MEMS. Therefore, sub-components A1 and B1 need monitoring agents MD_A and MD_B to represent the system degradation caused by the failure of A1 or B1. Besides, a monitoring agent M_S is added to monitor the current degradation of the MEMS. When the system reaches the degradation threshold, the monitoring agent M_S will be activated and output the failure time N_S of the MEMS. The specific relationships are shown in Figure 3.

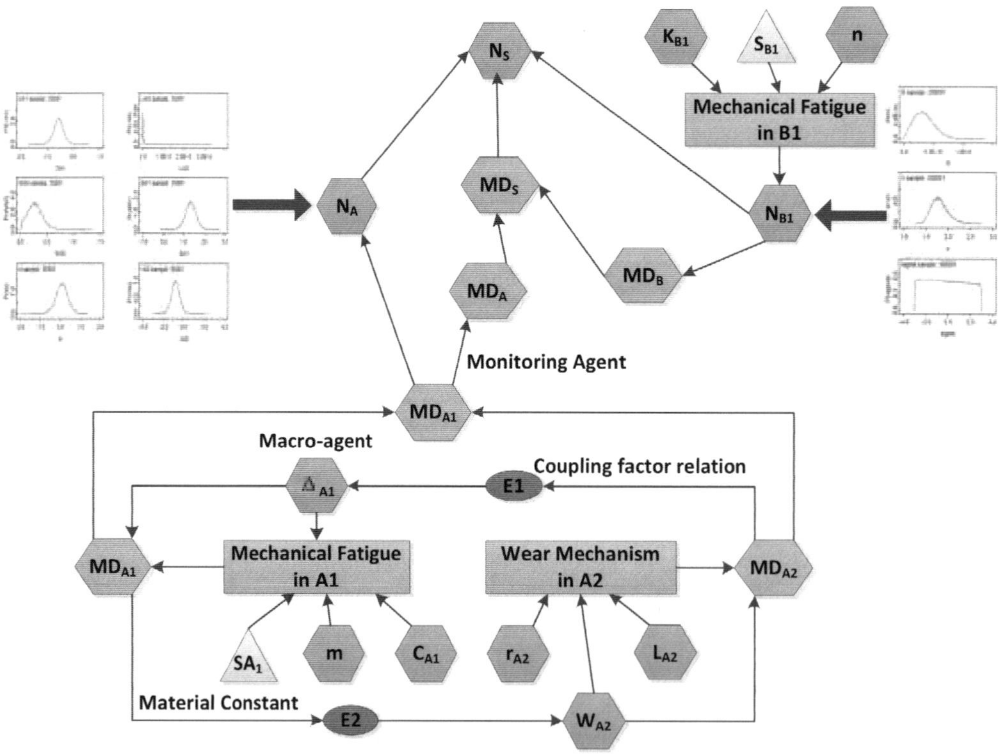

Figure 3: MEMS Model based on Multi-Agents

In Figure 3, there is a coupling relationship between the fatigue failure mechanism and the wear failure mechanism in sub-component A1. On the one hand, the damage caused by wear will affect the stress intensity factor Δ_{A1} in the fatigue failure mechanism and the agent E1 represents the coupling relation. On the other hand, the generation of fatigue cracks will affect the wear rate W_{A2} in the wear failure mechanism and the agent E2 represents the coupling relation. Thus the modeling of the MEMS can deal with the correlativity among different failure mechanisms. According to the system dynamics and characteristics in the MEMS, the system degradation amount M_{A} caused by the failure of the sub-component A1 and the system degradation amount M_{B} caused by the failure of the sub-component B1 can be expressed by the Equation (11).

$$MD_S = g(MD_A, MD_B) \tag{11}$$

When the system degradation amount MD_S reaches the failure threshold, the total failure time N_S of the MEMS is output. Depending on the total failure time, the remaining life of the MEMS can be evaluated. This modeling method considers not only the uncertainty of the parameters such as the process and structure in the underlying materials, but also the correlativity between different failure mechanisms, which will be very promising.

3 Case Study: The paper presents probability physics of failure model in single failure mechanism about sub-component B1. And meanwhile the paper presents a system-level reliability evaluation method considering the mechanisms correlativity based on multi-agents. The following are the simulation results.

3.1 Life Evaluation of the Single Failure Mechanism Under Bayesian Framework: The probabilistic failure physical model construction method of a single failure mechanism is analyzed in Section 2.2. After updating the Bayesian parameters in Section 2.2, Monte Carlo simulation is used to evaluate the life of sub-component B1 through the posteriori data parameters K_{B1}, n and σ_{B1}. According to Equation (1) and the posterior parameters, the probability density distribution histogram of the lifetime N_B can be obtained, and then connecting the midpoint in each rectangle block, we can get the probability density curve of the lifetime distribution about sub-component B1. Monte Carlo simulation results are shown in Figure 4. It can be seen that the distribution of the lifetime is closest to the posterior distribution of the uncertainty parameter K_{B1}, so the dispersion parameter K_{B1} is the main parameter that causes the lifetime to be dispersed.

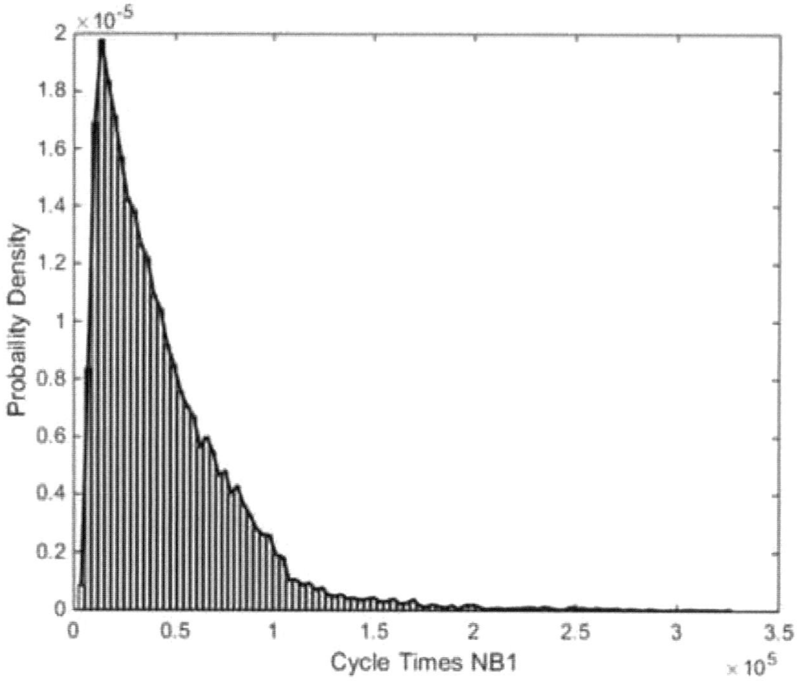

Figure 4: The Lifetime Probability Density Histogram and Probability Density of Sub-component B1

This method of modeling under the Bayesian framework combines the priori information with the test data. Meanwhile, the method considers the dispersion of factors, such as the material properties, dimensions and stress, adding these dispersions to the existing failure physical model. Under Bayesian framework, the probability of failing physical model will be achieved, and thus the result will be more accurate and closer to the actual situation.

3.2 Remaining Life Evaluation of MEMS Based on Multi-Agents: In Section 2.3, the MEMS modeling method based on Multi-agents is described in detail. When we measure the degradation of the sub-components and the number of load cycles, the remaining life of the MEMS can be evaluated based on multi-agents under Bayesian framework [11]. In the test, we get three sets of data, including the degradation of the sub-components and the number of load cycles N_t. The data D_a can be expressed in Equation (12).

$$Da = \{MDA_i, MDB_i, N_i, i = 1,2,3\} \quad (12)$$

The remaining life of the MEMS is represented by NL, and then NL can be represented by Equation (13).

$$NL_i = N_s - N_i \quad (13)$$

According to the three sets of data, the multi-agents model will be updated under Bayesian framework, and the obtained data will be fitted to evaluate the remaining life of the MEMS. The results are shown in Figure 5.

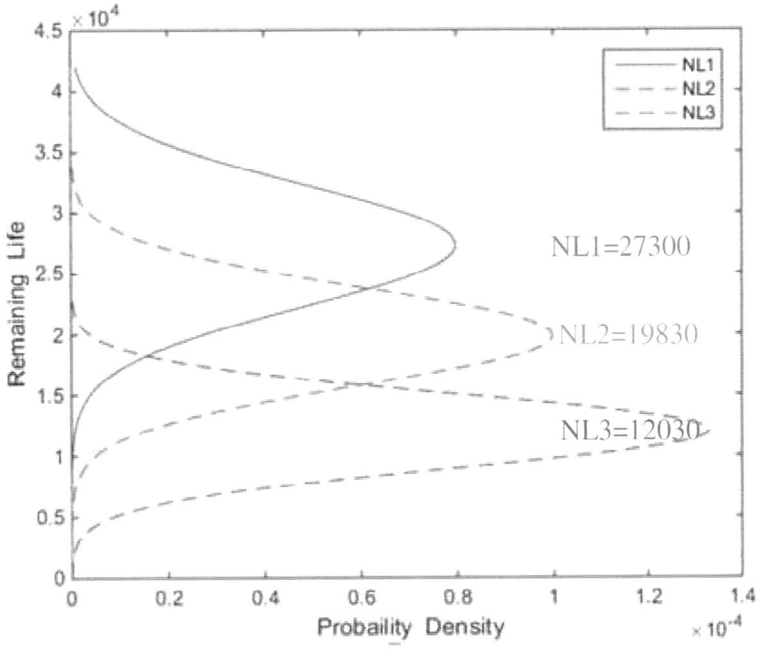

Figure 5: The Remaining Life of MEMS

The data obtained from Bayesian updating are fitted by MATLAB, and we can find that the lifetime distribution of MEMS is normal distribution. The type of life distribution in B1 is different from MEMS, which can be shown that the life distribution type of the sub-component is not directly related to the life distribution type of the whole system.

In Figure 5, NL1, NL2 and NL3 respectively represent the remaining life of the MEMS system under the first, second and third measurements. Since the lifetime distribution of the MEMS system conforms to the normal distribution, the expected value of the remaining life can be expressed by the mean of the normal distribution. The mean of NL1, NL2 and NL3 are 27300, 19830 and 12030 respectively. It can be seen that the expected value of the remaining life in MEMS decreases with the increase of the load cycle number.

Since the lifetime distribution of the MEMS system conforms to the normal distribution, the variance represents the life uncertainty. The three sets of data are used to update the Bayesian parameters, and then NL1, NL2 and NL3 are becoming narrower, which means that their variance is getting smaller. That indicates the uncertainty of remaining life NL1, NL2 and NL3 in the MEMS are getting smaller. After each Bayesian update, the multi-agents modeling will integrate the test information, and thus the uncertainty of system remaining life assessment is declining. The more data are measured, the more information will be integrated in multi-agents modeling and the remaining life assessment will be more accurate.

4 Conclusions: In this paper, polymorphism related concepts are incorporated into reliability field, new definitions of failure polymorphism are discussed, and Bayesian framework is introduced to tackle the problems of uncertainty factors and dynamic evaluation. Under the Bayesian framework, based on multi-agents, the reliability evaluation method of system-level is realized considering the correlativity between the failure mechanisms. Compared with other traditional methods, the method in the present paper will be very promising and has the following advantages:

(1) The posterior parameters under the Bayesian framework can fully consider the influence of the lower units including material, technology and environmental load uncertainties. Based on probability physics of failure, the life assessment method is more accurate and close to the actual situation.

(2) With the help of biological polymorphism, the reliability evaluation method of system-level can be realized considering the failure mechanisms correlativity. At the same time, this evaluation method can transfer the uncertainty of the bottom units to the system-level through the real-time information exchange among the agents, and finally realize the dynamic simulation of the system.

(3) The measured data can be used to update the Bayesian parameters, so that the uncertainty of the system will be reduced in the simulation process, which is conducive to the more accurate assessment to the remaining life of the system.

References:

[1] Pecht M, Gu L. Physics-of-failure-based prognostics for electronic products. IEEE Transactions of the Institute of Measurement and Control, 2009, 31(3-4):309-322.
[2] Song S, Coint D W, Feng Q, et al. Reliability analysis for multi-component systems subject to multiple dependent competing failure processes. IEEE Transactions on Reliability, 63(1):331-345,2014.
[3] Zeng Z, Kang R. A physics-of-failure-based approach for failure behavior modeling: With a focus on failure collaborations. Annual European Safety and Reliability Conference (ESREL). Wrocław, Poland.2014.
[4] Zeng Z, Chen YX. Failure Behavior Modeling: Towards a Better Characterization of Product Failures. CHEMICAL ENGINEERING. 33:571-576,2013.
[5] Zahra M, Mohammad M, Aris C. Proceeding of the ASME 2011 Power Conference. Denver, Colorado, USA. 2011.
[6] Martz H F, Waller R A. Bayesian Reliability Analysis, John Wiley & Sons, New York. 1982.
[7] Brooks S P. Markov Chain Monte Carlo Method and its Application. Royal Statistical Society. 47(1):69-100, 1998.

[8] Chookah M, Modarres M. A probabilistic Physics-of-Failure model for prognostic health management of structures subject to pitting and corrosion-fatigue. Reliability Engineering and System Safety. 96:1601-1610, 2011.
[9] Lyell M J, Krueger W, Zhang G. An agent-based approach to health monitoring systems applied to the ISS power system. The 45th AIAA Aerospace Sciences Meeting and Exhibit. Reno, Nevada, USA, 2008.
[10] Azarkhail M. Agent Autonomy Approach to Physics-Based Reliability Modeling of Structures and Mechanical Systems. University of Maryland, College Park, 2007.
[11] Azarkhail M. The Evolution and History of Reliability Engineering: Rise of Mechanistic Reliability Modeling. International Journal of Performability Engineering. 2012.

A Novel Approach for Machinery Health Prognostics Using Statistical Tools

Murtaza Hussain
PNEC, National University of Sciences and Technology
Habib Ibrahim Rehmatullah Road
Karachi, Pakistan 75350
Telephone: (0092) 21-48503021
murtaza@pnec.nust.edu.pk

Asif Mansoor
Bahria University, Karachi Campus
Karachi, Pakistan

Qaisar Ali
8179 Tributary Lane
Reynoldsburg, Ohio 43068

Abstract: Condition based maintenance of machinery is being much talked about in the engineering sector of defense and commercial industry. A lot of expenditure is generally incurred on condition monitoring of machinery to avoid unexpected downtimes and failures vis-à-vis optimizing machinery operation. The concept is ever evolving due to technological advancements as well as with the emergence of unique nature of defects. Vibration Analysis is a potent tool of condition monitoring for prediction and diagnostics of machinery failures. Presently, time and frequency spectra are being widely used for defect diagnostics of machinery. However, they require signal conditioning to eliminate noise and to enhance resolution of spectrum. Extensive research in the area of signal processing has been undertaken to refine time and frequency spectra. Notwithstanding application of statistical tools for analysis of various defects in machinery using condition monitoring data can be a viable option. Research in this area, where statistical models have been applied, revealed encouraging results. In this paper, we have modeled bearing vibration data by applying time varying Markov Switching Auto Regressive method which was found very helpful in estimating RUL of machinery.

Key words: Autoregressive; Defect frequency; Forecasting; Markov switching; Prognostics; Regime, Time variant

Introduction: In real life, machinery operates in diversified conditions. These conditions include environment, adopted maintenance philosophy, operators/maintainer skills, operating conditions etc. These operating conditions govern the machinery operation and health; machines have similar released life by the manufacturer tend to fail at different operational hours.

Prognostic on the basis of physical models is theoretically fine but its practical manifestation in real time environment is difficult, as machinery life is governed by various known and unknown effects. The effects generated on machinery during operations, at times, deprive the implementation of physical laws due to multifarious physical changes being encountered.

On the other hand, prognosis using data driven models, carries better practical manifestation in real life. Condition monitoring data provides a comprehensive understanding of machinery condition under prevailing operating conditions. This data while providing various condition indications of machinery can be manipulated using different tools for prognosis. Taking lead from huge usage in the fields of finance and medical, data driven techniques have become popular approaches for prognosis in engineering sector also. A comprehensive review [1] shows that many of the prognostics designs are based on data driven models.

Data driven methods are broadly divided in to two approaches i.e. Artificial Intelligence (AI) and Statistical approaches. AI approach has its unique computational power; however, it requires an extensive expertise. Comparatively, statistical methods are relatively easier to adopt and apply. Furthermore, it can be deduced from comparative analysis [2] that statistical tools offer an easier and reliable prognostics framework. Similar reviews for different scenarios already presented also support the above argument [3], [4].

Statistical data driven methods include many linear and nonlinear methodologies. These broadly include regression models, wiener processes, gamma processes, markovian models, stochastic filtering based models, covariate based hazard models etc. Comprehensive discussion on various statistical models with their suitability of application is discussed in [5].

The nature of machinery operation involving nonlinearity and randomness is challenging and carries modeling complexities for the estimation of machinery life. Nonlinearity in case of machinery can be a defect, whose propagation will be highly random under prevailing operating and environmental conditions. In term of statistics, the nature of this damage propagation will be stochastic. The situation becomes more complex when there are several nonlinearities (defects), evolving randomly with time under varying operating conditions [4].

Researchers have made enormous endeavors to estimate the machinery remaining useful life focusing on nonlinearities and randomness discretely or in combination. Particle filter method has been adopted by various researchers. This method provides more accuracy as compared to other algorithms. It has a good capability to model nonlinear dynamical systems with multivariate data [6],[7],[8] . Hidden Markov Models [9],[10] is a good tool for performing fault and degradation diagnosis on random dynamic systems with multi failure modes specially when failure state is unobserved. It can model different stages of failures and can distinguish fault types in a component.

Wiener process also referred to as Brownian motion apply concept of random walk of particle relating with machinery performance to measure the degradation. However, when we talk about the RUL estimation especially in case of a nonlinear phenomena then the Wiener method has to assume the mean degradation path as linear; hence the dynamics of nonlinearity cannot be adjudged efficiently [5], [11], [12]. Regression based models are common tools used for time series analysis as well as for RUL estimation due to their simple application [5]. These models utilize various condition monitoring independent variables regressed against some dependent variable to estimate the trend and subsequent calculation of RUL. The application of this tool is enough matured in the field of economics, finance etc [13]. Other techniques including Principle Component Analysis (PCA), Gaussian Mixture Model, Logistic Regression, Statistical pattern recognition, Kalman filter, Match matrix, Support vector machine have been applied and discussed for the machinery prognostics with detail in [1], [2], [3], [4].

In this paper we will discuss the application of Time Variant Markov Switching Autoregressive Model to estimate remaining useful life considering Roller element bearing (REB) as a case.

Bearing Prognosis: Roller Element Bearings (REB) are the most critical component of machinery and a determinant of machinery health. The loading of machinery either internal or external is supported by the bearings. Most of the condition monitoring tools primarily focused on bearing health; especially vibration technology which evolved on bearing health dynamics.

REBs are meant to eliminate sliding contacts between surfaces by use of rolling elements. REBs may have balls or rollers and accordingly they are classified. The REBs are normally classified on the basis of No of balls/rollers, ball/rollers diameter, inner/outer race diameter, pitch diameter etc. The four parts of REBs are balls/rollers, inner race, outer race and cage as illustrated in 'Figure 1'.

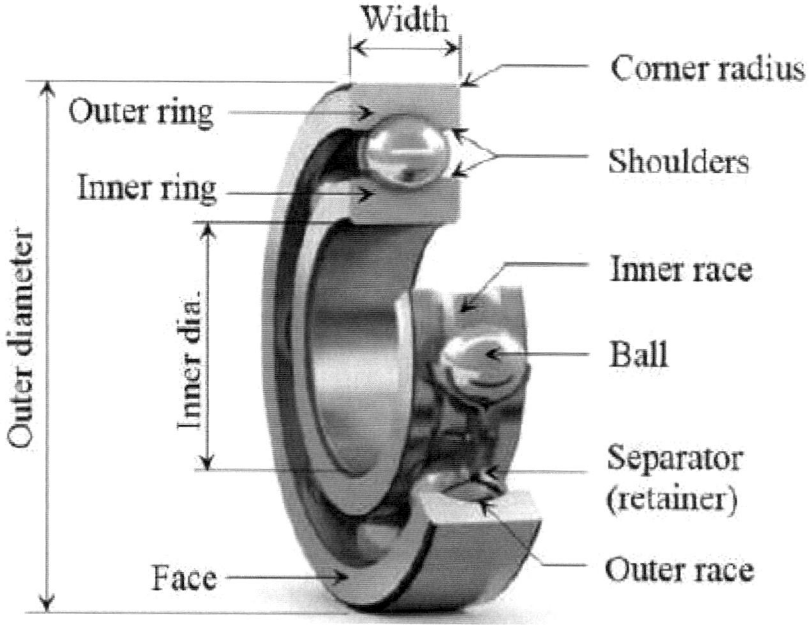

Figure 1: Deep Grooved Ball Bearing

Wear in bearings arises due to stresses in rolling contact surfaces. Under normal operating conditions with no defect, the wear process is linear. However, when there is some abnormality induced like abnormal operating conditions, defects (misalignment, foreign particle ingress, lubrication problems etc), then the stress level rises to enhanced metal to metal contact causing various parts of bearings to deteriorate. This deterioration results in exponential trend of bearing wear. The defects of bearings are seen as defect frequencies in the vibration spectrum as Fundamental train frequency or cage frequency (FTF), Ball pass frequency outer race (BPFO), Ball pass frequency inner race (BPFI) and Ball spin frequency (BSF).

Dynamic models for bearing wear evolution have been studied by various researchers. These include the modeling to study the effects of Elasto-Hydrodynamic lubrication, excessive clearances, brinelling phenomena, non-uniform loading conditions, modulation effects etc. Moreover, frequency response in multi degree of freedom systems using FEM and other mathematical tools has been under focused in various studies [14]. Beside physical models, data driven models are also adopted by various researchers [15]. The pros and cons of physics based and data driven based models have already been discussed in preceding lines under introduction; same are also relevant for bearing application.

Experimental setup: Test rig is a simple arrangement driven by a 2 Hp variable speed induction motor. A customized designed shaft is then supported by two bearings. The test bearing UC-203 (deep groove ball bearing) follows the support bearings which will be subjected to various radial and axial loading conditions, under modified housing. A screw type loading mechanism is installed for static load in radial direction and for axial loading a load is mounted on shaft see 'Figure 2'

Testing is done on the basis of varying loading and variable speed conditions. The concept of accelerated testing is adopted in order to perform the failure analysis with varying static load from 0N-75N and a constant axial load of 100N. The analysis is confined to the bearing only, in view of its criticality. The data collection was conducted using NI DAQ system.

Figure 2: Experimental Setup

Model and Method: Markov switching model explained by [16] is studied to account for nonlinearities in financial time series data and the auto regressive component to cater for the non stationary process in the data. The model provides a appropriate methodology to describe changes in the dynamic behaviour of machinery health data. It assumes the presence of a finite number of states, whereby the parameters are allowed to take on different values with regard to the regime prevailing at certain point of time. The regime shifts arise from the outcome of an unobserved random variable which is assumed to evolve following Markov chain. Consideration of time varying regime probabilities also make this model practically suitable for scenarios where condition are not constant and varying with time.

Model specification: Let y_t represents machinery health variable that undergo changes over time. Considering this series as a subset of stochastic process whose probability distribution depends on the realization of a hidden discrete stochastic process z_t . The stochastic process y_t is directly observable, whereas z_t is a latent random variable that is visible only indirectly through its effect on the identification of y_t . The changes in health variable across different regimes are modeled by the regime switching "Equation (1)":

$$z_t = \beta_{s_t} + \sum_j^k \beta_{j,s_t} z_{t-j} + \epsilon_t \qquad (1)$$

here ϵ_t is normally distributed term, so that $\epsilon_t \sim N(0, \sigma^2_{s_t})$ and 's_t' unobserved discrete state variable. Whereas parameter (β_{j,s_t}) depends on state variable $S_t = i$, $i = 1,2,\ldots,N$ which represents regime in process at time t.

Statistical inference and estimation procedure: The MARKOV Switching Regression model extends the simple exogenous probability framework by specifying the first order Markov process for the regime probabilities followed by likelihood computation, filtering, and smoothing.

Regime probabilities: The first order Markov assumption necessitates that the likelihood of being in a regime depends on the preceding state as in "Equation (2)", so that:

$$P(s_t = j | s_{t-1} = i) = p_i\ (t) \qquad (2)$$

Normally, these probabilities are assumed to be time-invariant in order that $p_i\ (t) = p_i\ \forall$ t. These probabilities in transition matrix form are shown in "Equation (3)":

$$p(t) = \begin{bmatrix} p_1\ (t) & \cdots & p_{1M}(t) \\ \vdots & \ddots & \vdots \\ p_{M1}(t) & \cdots & p_M\ (t) \end{bmatrix} \qquad (3)$$

here ij-th component signify the probability of transitioning from regime 'i' in period t-1 to regime 'j' in period t. Since each row of the transition matrix indicates a full set of conditional probabilities so a different multinomial specification for each row' i' is required which is defined in "Equation (4)":

$$p_i\,(z_{t-1},\delta_i) = \frac{e\quad (z_{t-1}^{\prime}\delta_t\,)}{\sum_{j=1}^{M} e\quad (z_{t-1}^{\prime}\delta_t\,)} \tag{4}$$

$f \quad j = 1,\ldots.,M\ a \quad i = 1,\ldots,M\ w\ \ h\ n \qquad\qquad \delta_{i,} \;= 0$

In Hamilton model [16], the regime probabilities are time invariant; however, in real life the probabilities are not constant in time rather they are time variant. In case of defect evolution in machinery it may happen in case of non stationary conditions that the defect growth may enhance due various loading/operating conditions in certain point of time. In contrast to that let $[\boldsymbol{S_t}]_{t=1}^{T}$ illustrate the first order sample path then its two state Markov process with time varying transition probabilities are explained by [17],[18] and represented in "Equation (5)":

Time *t*

State 0	**State 1**

$$\left[\begin{array}{c|c} \begin{array}{c} P_t^{0} \\ P(s_t=0\,|\,s_{t-1}=0, z_{t-1}, \beta_0) \\ \dfrac{e\quad (z_{t-1}^{\prime}\beta_0)}{1+e\quad (z_{t-1}^{\prime}\beta_0)} \end{array} & \begin{array}{c} P_t^{0} = (1-P_t^{0}\,) \\ P(s_t=1\,|\,s_{t-1}=0, z_{t-1}, \beta_0) \\ 1-\dfrac{e\quad (z_{t-1}^{\prime}\beta_0)}{1+e\quad (z_{t-1}^{\prime}\beta_0)} \end{array} \\ \hline \begin{array}{c} P_t^{0} = (1-P_t^{1}\,) \\ P(s_t=0\,|\,s_{t-1}=1, z_{t-1}; \beta_1) \\ 1-\dfrac{e\quad (z_{t-1}^{\prime}\beta_1)}{1+e\quad (z_{t-1}^{\prime}\beta_1)} \end{array} & \begin{array}{c} P_t^{1} \\ P(s_t=1\,|\,s_{t-1}=1, z_{t-1}; \beta_1) \\ \dfrac{e\quad (z_{t-1}^{\prime}\beta_1)}{1+e\quad (z_{t-1}^{\prime}\beta_1)} \end{array} \end{array}\right] t-1 \tag{5}$$

Here

$$z_{t-1} = (1, z_{1.t-1}, \ldots, z_{(k-1).t-1})'$$

$$\beta_i = (\beta_{i0}, \beta_{i1}, \ldots, \beta_{i(k-1)})' \text{ and } i = 0, 1$$

Likelihood Evaluation and Filtering: The likelihood contribution for given observation is computed by utilizing density function in each regime by the one step ahead probability of being in that regime as shown in "Equation (6)":

$$L_t(\beta,\gamma,\sigma,\delta) = \sum_{m=1}^{M} \frac{1}{\sigma_m}\emptyset\left(\frac{z_t-\mu_t(m)}{\sigma(m)}\right).\,P(s_t = m|\Im_{t-1},\delta) \tag{6}$$

$\beta = (\beta_1, \ldots, \beta_M), \sigma = (\sigma_1, \ldots, \sigma_M), \delta$ are parameters that determines the regime probabilities $\emptyset(.)$ is the standard normal density function and $\Im_{t-1}$ is the information set in period lag one. Besides in undemanding form δ represents the regime probabilities themselves.

The full log-likelihood is normal combination of

$$l(\beta,\gamma,\sigma,\delta) = \sum_{t=1}^{T} l_l \quad \left\{\sum_{m=1}^{M} \frac{1}{\sigma_m}\emptyset\left(\frac{z_t-\mu_t(m)}{\sigma(m)}\right).\,P(s_t = m|\Im_{t-1},\delta)\right\} \tag{7}$$

"Equation (7)" can be maximized w.r.t $(\beta,\gamma,\sigma,\delta)$

Filtering process estimates the updated probabilities which are computed by utilizing Bayes' theorem and the laws of conditional probability which can be expressed as "Equation (8)":

$$P(s_t = m|\Im_t) = P(s_t = m|z_t, \Im_{t-1}) = \frac{f(z_t|s_t = m, \ _{t-1})P(s_t = m|z_t, \ _{t-1})}{f(z_t| \ _{t-1})} \quad (8)$$

The right side of the expression (8) is obtained as a by-product of densities obtained during likelihood estimation. On substitution we have "Equation (9)":

$$P(s_t = m| \ _t) = \frac{\frac{1}{\sigma_m} \left(\frac{z_t-\mu_t(m)}{\sigma(m)}\right).p_m(G_{t-1},\delta)}{\sum_{j-1}^{M}\frac{1}{\sigma_j} \left(\frac{z_t-\mu_t(j)}{\sigma(m)}\right)p_j(G_{t-1},\delta)} \quad (9)$$

where G_{t-1} represents vector of exogenous variables.

Smoothing: Estimates of the regime probabilities may be improved by using all of the information given in the sample. The smoothed estimates for the regime probabilities in period t use the information set in the final period, $_T$.,in contrast to the filtered estimates which employ only contemporary information, $_T$. Intuitively, using information about future realizations of the dependent variable $y_s(s > t)$ improves estimates of being in regime m in period t, because the Markov transition probabilities link together the likelihood of the observed data in different periods. Another approach for smoothing requires only a single backward recursion through the data. Under the Markov assumption, the joint probability is given by "Equation (10)" and "Equation (11)":

$$P(s_t = i, s_{t+1} = j| \ _T) = P(s_t = i| \ s_{t+1} = j, | \ _T).P(s_{t+1} = j| \ _T) \quad (10)$$

$$= \frac{P(s_t=i, s_{t+1}=j| \ _t)}{P(s_{t+1}=j| \ _t)} P(s_{t+1} = j| \ _T) \quad (11)$$

"Equation (12)" shows the smoothed probability in period t obtained by marginalizing the joint probability with respect to s_{t-1}:

$$P(s_t = i| \ _T = \sum_{j=1}^{M} P(s_t = i, \ s_{t+1} = j| \ _T) \quad (12)$$

Initial Probabilities: The Markov Switching Filter requires initialization of the filtered regime probabilities in period zero, $P(s_0 = m| \ _0)$. There are few ways to proceed with initial probabilities. Most commonly, the initial regime probabilities are set to the steady state values implied by the Markov transition matrix. The values are thus treated as functions of the parameters that determine the transition matrix. Alternately, prior knowledge may be used to specify regime probability values, and lastly for treating of the initial probabilities as parameters which are to be estimated.

Application of Model: The dynamic behavior of Machinery health variables (z_t) observed over time is modeled by utilizing MASR with time varying transition probabilities. In this regards two state Markov switching model is estimated in which z_t is regressed over switching coefficients z_{t-1} and constant keeping AR(3) as non-varying regressor. Maximum likelihood estimated equations for different regime with standard deviation in parenthesis are shown "Equation (13)":

$$z_t = \begin{cases} 0.578 + 1.315\, z_{t-3} + [AR(3) = 0.343] + \varepsilon_{t_1} & \text{Healty State} \\ \quad (0.261) \quad (0.359) \qquad (0.057) & \\ 0.0821 + 0.813\, z_{t-3} + [AR(3) = 0.343] + \varepsilon_{t_2} & \text{Unhealty State} \\ \quad (0.013) \quad (0.027) \qquad (0.057) & \end{cases} \tag{13}$$

with $\varepsilon_{t_1} \sim N(0, 0.085)$ and $\varepsilon_{t_2} \sim N(0, 0.057)$

'Equation (13)' demonstrates that differences in the health state specific mean indicate the growth of defects in machinery for the period under study. The probabilities values for both regimes are significant ($p < 0.05$) which confirm that dynamics in both regimes are substantial.

The time varying transition probability matrix of the predicted 'Equation (13)' is presented in 'Table 1' which indicates that the probability of switching from healthy state to unhealthy state is 0.4066, while remaining in healthy state is 0.5933.

Table 1: Time Varying Transition Probability Matrix.

	Healthy State	**Unhealthy State**
Healthy state	0.593398	0.406602
Unhealthy State	0.048505	0.951495

The variations in the estimates of time varying transition probabilities for the constructed model in different states is presented in 'Figure-3' which reflect that machinery remain in its healthy for more hours.

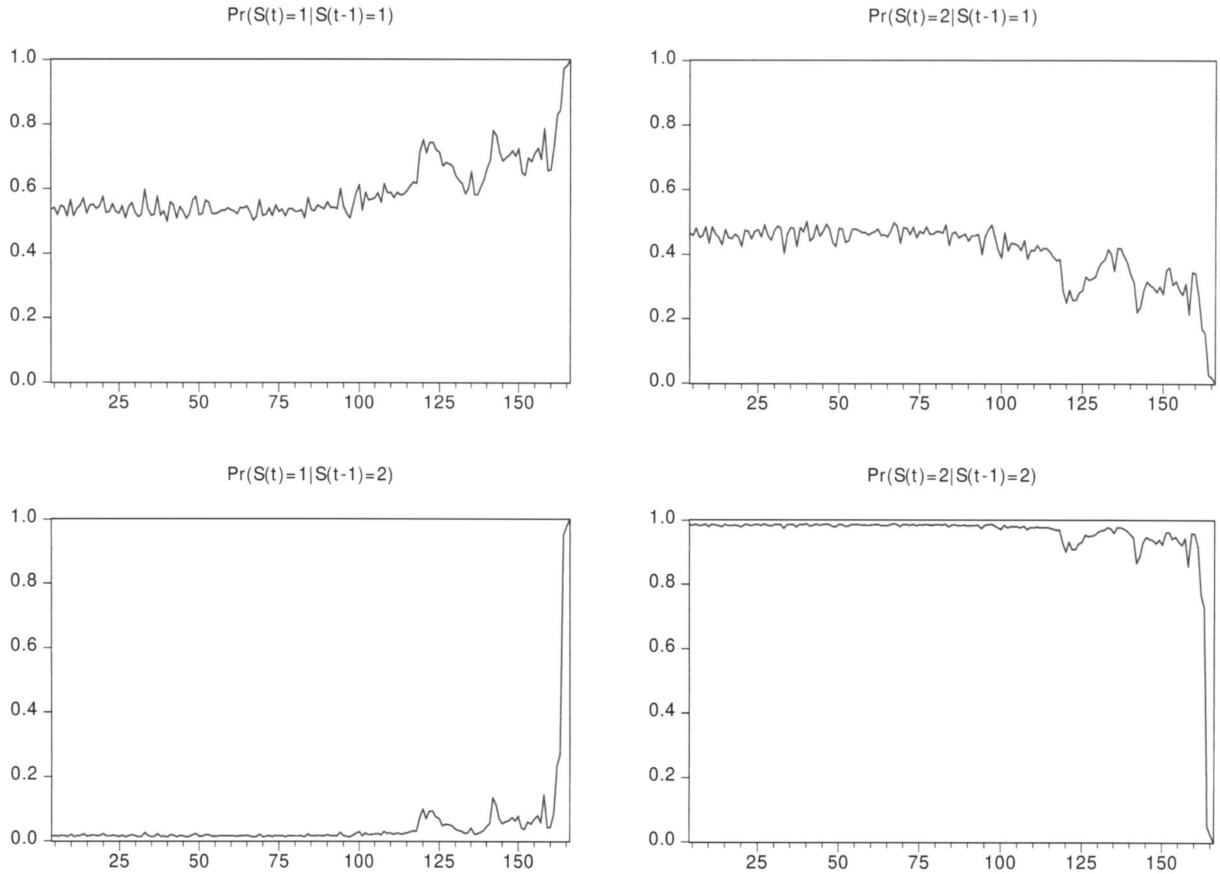

Figure 3. Fluctuation in Time-varying Markov Transition Probabilities

The subsequent time varying expected durations depicted under 'Table 2' imply that machinery remain in the healthy state for a longer duration before moving to unhealthy state.

Table 2: Time Varying Expected Duration Matrix.

	Healthy State	Unhealthy State
Mean	76.08617	49.90062
Std. Dev.	931.9972	23.58307

Based on above expected duration, the expected variation point between healthy & unhealthy states is shown in 'Figure 4'.

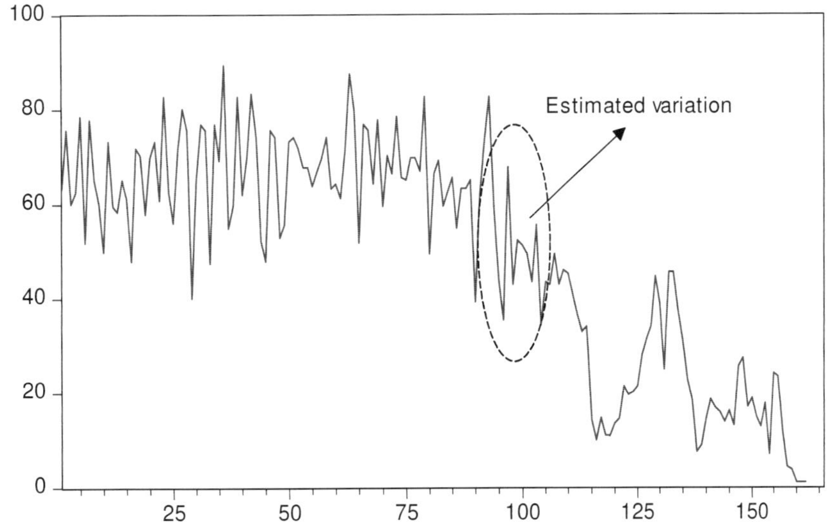

Figure 4. Expected variation between two states

'Figure 5' exhibit the filtered high & low variance probabilities of being in the two health states. The filtering process keeps on updating the estimated probability in order to determine the likelihood of machinery health from one regime to the other. It is evident from graph that machinery unhealthy states begin after 120 running hours.

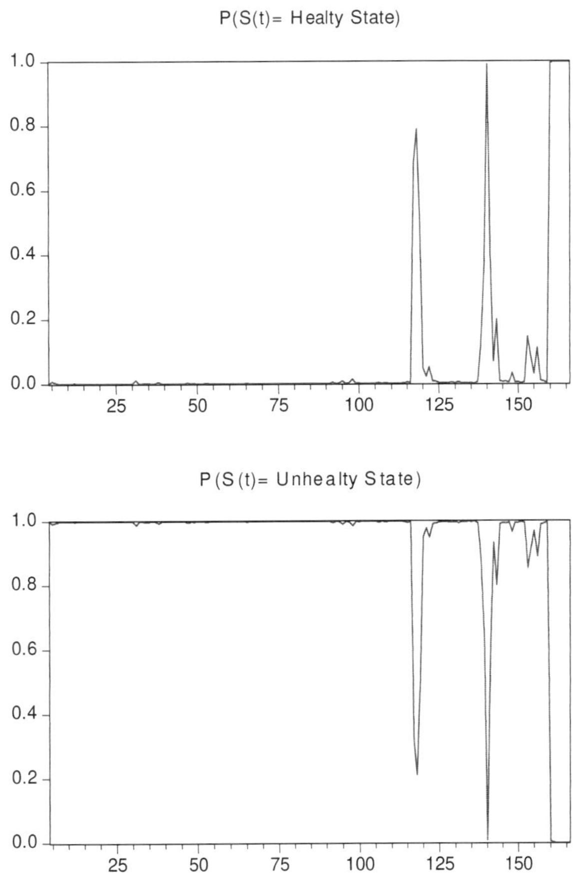

Figure 5. Filtered high & low variance probability plot.

'Figure 6' shows the real time instability in actual and forecasted health indicator which demonstrates inconsistency in machinery health after its usage of about 120 running hours and it keeps on growing afterwards. Its continuous handling without maintenance will distort the mechanism after 164 running hours.

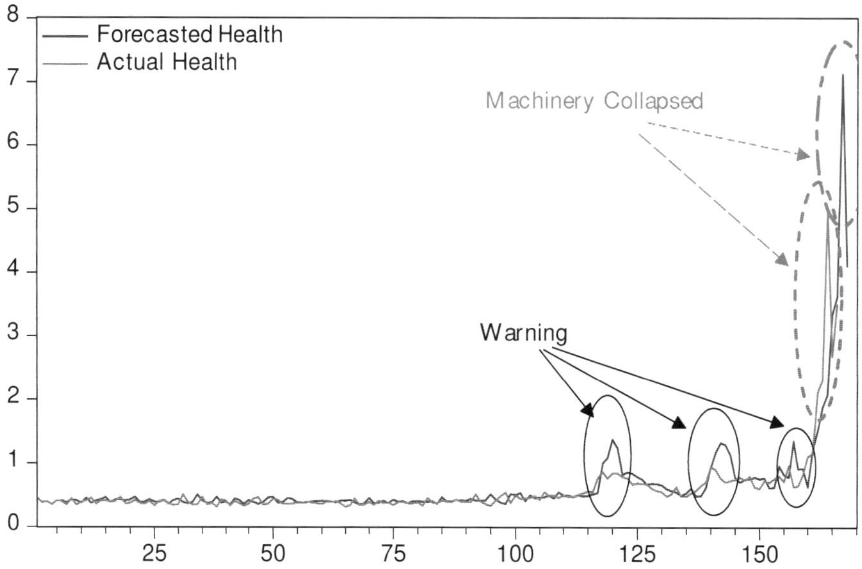

Figure 6. Actual and forecasted health indicator signifying volatility in health parameter

The forecasted performance indicator with its estimated statistics are illustrated in 'Table 3' which authenticates the predicted model:

Table3. Forecasting performance statistics

Forecasting Performance Indicator	**Estimated Statistics**
RMSE	0.279202
MAE	0.108040
Theil Inequality Coefficient	0.183851
Bias Proportion	0.004359
Variance Proportion	0.098221

Conclusion:

In this communication two state of time-varying Markov-switching model is utilized to machinery health variable as it permits the shift likelihood to fluctuate thoroughly with information predictors which reflects the future health state of machinery. This investigation signifies that machinery stayed in healthy state for a period of 116 hours before becoming unhealthy. The real time performance of the machinery is also forecasted by employing the estimated model. The forecasted performance statistics signify the capability of the constructed model. Finding of this study will serve as a basis for making suitable decision for machinery future operation.

References:

[1] H. M. Society, *Phm 2013*. 2013.

[2] J. Lee, F. Wu, W. Zhao, M. Ghaffari, L. Liao, and D. Siegel, "Prognostics and health management design for rotary machinery systems - Reviews, methodology and applications," *Mech. Syst. Signal Process.*, vol. 42, no. 1–2, pp. 314–334, 2014.

[3] A. Heng, S. Zhang, A. C. C. Tan, and J. Mathew, "Rotating machinery prognostics: State of the art, challenges and opportunities," *Mech. Syst. Signal Process.*, vol. 23, no. 3, pp. 724–739, 2009.

[4] M. S. Kan, A. C. C. Tan, and J. Mathew, "A review on prognostic techniques for non-stationary and non-linear rotating systems," *Mech. Syst. Signal Process.*, vol. 62, pp. 1–20, 2015.

[5] X.-S. Si, W. Wang, C.-H. Hu, and D.-H. Zhou, "Remaining useful life estimation – A review on the statistical data driven approaches," *European Journal of Operational Research*, vol. 213, no. 1. pp. 1–14, 2011.

[6] J. K. Kimotho, T. Meyer, and W. Sextro, "PEM fuel cell prognostics using particle filter with model parameter adaptation," *2014 Int. Conf. Progn. Heal. Manag.*, pp. 1–6, 2014.

[7] M. Yu, D. Wang, A. Ukil, V. Vaiyapuri, N. Sivakumar, C. Jayampathi, A. K. Gupta, and V. Nguyen, "Model-based failure prediction for electric machines using particle filter," *2014 13th Int. Conf. Control Autom. Robot. Vis.*, vol. 2014, no. December, pp. 1811–1816, 2014.

[8] E. Walker, S. Rayman, and R. E. White, "Comparison of a particle filter and other state estimation methods for prognostics of lithium-ion batteries," *J. Power Sources*, vol. 287, pp. 1–12, 2015.

[9] K. Medjaher, D. Tobon-mejia, and N. Zerhouni, "Remaining useful life estimation of critical components with application to bearings.," *Ieee Trans. Reliab.*, pp. 292–302, 2012.

[10] M. Dong and D. He, "Hidden semi-Markov model-based methodology for multi-sensor equipment health diagnosis and prognosis," *Eur. J. Oper. Res.*, vol. 178, no. 3, pp. 858–878, 2007.

[11] X. S. Si, W. Wang, C. H. Hu, M. Y. Chen, and D. H. Zhou, "A Wiener-process-based degradation model with a recursive filter algorithm for remaining useful life estimation," *Mech. Syst. Signal Process.*, vol. 35, no. 1–2, pp. 219–237, 2013.

[12] X. S. Si, C. H. Hu, M. Y. Chen, and W. Wang, "An adaptive and nonlinear drift-based Wiener process for remaining useful life estimation," in *2011 Prognostics and System Health Management Conference, PHM-Shenzhen 2011*, 2011.

[13] J. G. De Gooijer and R. J. Hyndman, "25 years of time series forecasting," *Int. J. Forecast.*, vol. 22, no. 3, pp. 443–473, 2006.

[14] I. El-Thalji and E. Jantunen, "Dynamic modelling of wear evolution in rolling bearings," *Tribol. Int.*, vol. 84, pp. 90–99, 2015.

[15] I. El-Thalji and E. Jantunen, "A summary of fault modelling and predictive health monitoring of rolling element bearings," *Mech. Syst. Signal Process.*, vol. 60, pp. 252–272, 2015.

[16] J. D. Hamilton, "A new approach to the economic analysis of nonstationary time series," *Econometrica*, vol. 57, no. 2. pp. 357–384, 1989.

[17] F. X. Diebold, J.-H. Lee, and G. C. Weinbach, "Regime switching with time-varying transition probabilities," *Nonstationary Time Ser. Anal. Cointegration*, pp. 283–302, 1994.

[18] M. Bazzi, F. Blasques, S. J. Koopman, and A. Lucas, "Time-Varying Transition Probabilities for Markov Regime Switching Models," *Journal of Time Series Analysis*, 2016.

Biographies:

1. Murtaza Hussain is a PhD student in Mechanical Engineering at National University of Sciences and Technology, Pakistan. His research area is related to machinery health dynamics. His topics of interest includes, model and data driven machinery prognostics, machinery health monitoring in non stationary conditions and application of statistical tools in machinery prognostics. He has a wide hand on experience on condition monitoring tools including vibration analysis, endoscopy, oil analysis and infrared thermography.

2. Senior Assistant Professor Dr. Asif Mansoor has PhD in Applied Mathematics from University of Karachi and is presently in Post graduate teaching faculty at Bahria University Karachi Campus. He has 12 international and 06 national research publications to his credential besides he has supervised 08 of students for research. He is regularly invited by eminent national universities and others to guide doctoral students on research methodologies, simulation and modeling and current trends in mathematics. His main research interest includes Probabilistic Method in Data Analysis, Stochastic Processes and CFD.

3. Qaisar Ali received Masters Degree in Thermo Fluids from NUST in 2010. He has over 07 years of research experience and 15 years Reliability Maintenance Management experience of wide range of industrial equipment/ machinery. He is certified Project Management Professional from Project Management Institute (PMI), USA. His 05 research papers have been published/ presented in International Journal/ Conferences. His areas of research interest are Heat transfer and fluid mechanics analysis across Porous Materials, Renewable Energy.

AN ANALYSIS OF THE DIFFERENT ASPECTS OF INTELLIGENCE IN MACHINES

Gabriele Veltre, Vukica Jovanovic, Anthony W. Dean, and Charles Daniels
Old Dominion University
Frank Batten College of Engineering and Technology
102 Kaufman Hall
Norfolk, VA 23529
Telephone: (937) 256-2285
Gabriele Veltre gvelt001@odu.edu

Abstract: The purpose of this research paper is to technically and philosophically analyze the behaviors of the current intelligent machines and try to think about an eventual artificial evolution. Will Artificial Intelligence (AI) bring us more advantages or more disadvantages? The paper analyzes the different types of intelligent technologies and sees how they are used in security controls, medical researches, military operations, house holding, and maintenance. It will take a look also at the cybernetic and robotic aspects of the topic, and see how they affect the military conflicts taking as points of reference the cyberterrorism and the design of new war-robots. The research will explain how everything started, how everything is quickly developing, and how the technological progress is affecting the human being. How will we react to this new intelligent machines in the future? Will they ever be able to overcome the human intelligence with an eventual artificial consciousness, more advanced learning skills and faster times of execution? Is it really coming up a new era? Are we really creating our successors or it is just science fiction?

Key words: Intelligent machines, artificial intelligence

Introduction: Artificial Intelligence (AI) is a topic that deals with various types of areas starting from medicine, aerospace, manufacturing, logistics, transportation, and manufacturing, passing by the everyday life comfort and facilities ending up to science fiction and philosophy. AI touches many different types of articulated subjects such as, Cybernetics, Neurophysiology, Mechatronics, and Robotics. Cybernetics is a term that derives from the Greek "kybernetes", which literally means "governor"[1]. The word was created by a group of American mathematicians, mainly guided by Norbert Wiener in the 1947 [2, 3]. Based on the Greek bases, they wanted to give to the word the meaning of "the science of the leadership" [4]. Essentially, they wanted to unify different emerging scientific movements of the time like digital electronic computing, neural networks, psychology, and social sciences. Lately, this subject has kept absorbing many other movements that differ between each other making the cybernetics the best unification between humanistic and scientific subjects. The cybernetics is an important aspect for the development of artificial intelligence, since it also describes the differences between a real brain and an artificial one (positronic brain vs human brain). The paper will also include, as the cybernetics, both humanistic and scientific aspects, which, as always, get in conflict between each other due to the arising ethical and moral issues.

Different Components of Cybernetics in Intelligent Machines: Neurophysiology is the physiology of the nervous system [5]. Neurophysiology is incredibly useful for the study of the emotions and the understanding of the feelings that the human being has, in addition it is extremely essential to discover the real derivations of consciousness in humans and in intelligent machines [6].

Mechatronics is the branch of technology concerned with the integration of mechanics, electronics, and embedded computer control [7]. The term of mechatronics started existing in the late 50s, but just about in the 2010s started getting an important role in the industries and the technological progress. Mechatronics is an emerging subject in the engineering field, it is the union of mechanics and electronics, and it is an important matter in the AI technology development, since the robots are designed and built basing all the procedures on the principles of mechanical, electrical, computer, and industrial engineering [8]. The current mechatronic systems are defined as smart, since they all present sensors, actuators and computer control systems that put them on the highest level of technology also due to their incredibly fast and accurate performances.

Robotics is the technology or science of design, construction, operation, and use of robots and similar automatic devices [9]. Robotics is an advanced branch of mechatronics, the step from to the other is very short, in fact they both present similarities, but robotics is defined as a different subject, since it differs from mechatronics because it presents the design of information technology, interpretation of electric signals, interaction between user and machine, and most importantly artificial intelligence.

But what is really artificial intelligence? During everyday life, we face many issues that are not solvable with basic or conventional mathematical methods, therefore, the human being tended, as always, to adapt and to indirectly develop its mind's abilities, but this time through technological advancements making a machine to learn from and to solve problems. This is the exact moment of when the artefact became intelligent creating Artificial Intelligence. To better understand the AI concept, we could divide it in two main groups such as soft and hard AI. Soft AI gives solutions that are too complex for the hard AI, but it can only provide "heuristic solutions" not guaranteeing optimal or fully correct solutions, however it can give very good starting points and useful results. Hard AI on the other hand is the part that allows the human being to completely solve problems through things like professional systems, formal logic, or Bayesian networks, nevertheless it can't solve very complex problems like soft AI does, but it can provide exact solutions to many problems [10-12].

Applications of AI: AI robots are used in various areas like medicine, manufacturing, agriculture, civil, commercial, logistics and transportation, however the AI progress is mainly concentrated in healthcare, military, logistics, and commercial fields [11]. Therefore, it is easy to find it in the medical, military, aerospace and everyday comfort development areas AI [13]. They are ideal for the technological and scientific developments, but at the same time they could also present negative aspects, that may lead to the infraction of some important ethical or moral principles. Most of the times, as already said, the engineering development focuses mostly in the medical and military improvement matters, and as always the aim of the engineers is the one of creating a machine that has to improve the human work, or even try to make it faster and easier. In the last decades, most of the times, a great part of the intelligent machines totally replaced the human intervention in entire processes. This led to the developments in the capabilities of the intelligent machines.

In numerous problems in the medical sciences it is essentially required the use of intelligent computational systems comparable to the human mind. In medical applications AI is also known as Bioinformatics, which is the point where medicine and technology meet and benefit the medical research developments [14, 15]. In fact, nowadays, the medical studies are done mostly with the help of machines, and thanks to them have been done several discoveries that saved and improved millions of human lives, favoring also the medical progress to grow every day faster.

A clear example of that is the use of AI in the computational prediction of toxicity in chemicals, an important aspect that is important for the control of the many emerging industries that produce thousands of new industrial substances every year [16]. The Office of Toxic Substances (OTS) of the United States Environmental Protection Agency (EPA) counted more hat 70,000 industrial chemicals in the 90s, and it has estimated that about 1000 compounds are added every single year, and most of them don't have a reliable toxicological information available. This is why EPA has started a project called ToxCast, which has the goal of finding reliable computational methods of toxicity prediction in chemicals to avoid poisoning circumstances in animals and humans or to decrease the environmental pollution [17]. AI has been a great help in this case, since many machine have been created to develop and to improve the quality of the researches basing all the experiments on computational kinds of Bayesian algorithms that really helped to simplify the predictions.

In the last decade, AI has been fundamental also for the prediction of cancer in various individuals [18, 19]. In fact, to monitor and to control the DNA micro-technologies are almost essential for the best efficiency in the medical analysis. Today, DNA microarrays are increasingly used to analyze the right organization and physiological functions of organs, since they provide highly precise diagnostics techniques allowing the discovery of genetic diseases or prevent cancers formations.

The logistic field is one of the areas that is achieving the best technological improvements building robots, that can travel miles and miles in all kinds of conditions without remote-controlled human interventions, or flying drones that can scan and control entire territories in seconds. The drones' development is one of the areas that is most subjected to the studies and inventions. The most famous and effective progresses in AI have been achieved by NASA, which uses robots for space station repairs, planetary explorations such as the ones on Mars or on the Moon.

"It is predicted that robotic technologies will become dominant in coming decades influencing every aspect at work or at home and in the process transforming lives and work practices, raise efficiency and safety levels and provide enhanced levels of service. It will create jobs and there would be lose interaction with people", said Y.H. Gharpure in his article called Metamorphic Robots[20]. In the same article he also said that the current robots market is 22 billion € and it is supposed to increase up to 56-62 billion by 2020, and the enormous progress will be that big that a whole robotic fast food restaurant will be a reality in 2030 bringing to an anguishing jobs halving in the U.S. by 2050.

AI Robots Description: Mainly the AI robots interact with the main user through programming, or tele-operation, but a self-sufficient robot should be autonomous, and require minimum supervision necessities. The AI Robots are composed by three main characteristics, which are mechanical construction, electrical aspects, and programming. They are all designed principally considering power sources, actuators, electric motors, piston and wiring muscles, electroactive polymers, elastic nano tubes, touch sensors, and vision sensors. In the AI Robots there are three main important skills that the robot should have in order to be an efficient machine, those are Manipulation, Locomotion, and Working abilities. (1) The manipulating skills are achieved by mechanical and vacuum grippers, or totally humanoid hands, in the robots the manipulation aspect primarily involves all the actions that affect picking up, modifying or destroying objects. (2) The locomotion are the second fundamental skills that have to characterize the AI Robots, and they are usually achieved by the use of one, two, four or even six wheels, actual mechanical legs, or tank tracks usually made out of metals or plastics. (3) The working skills are the most difficult to apply on robots, since they are skills that have to incorporate hopping, balancing, passive resistance, climbing, skating, swimming, or snaking. Moreover, robots are actually allocated in two categories, Industrial and Service Robotics based on what their purposes are. The Industrial Robots are the ones that are used in manufacturing fields, and their main purposes are to substitute the traditional manufacturing processes (handmade or artisanal ones) and do the same kind of works more precisely and efficiently. The Service Robots are also called the domestic ones, the ones that have the aim to make every daily difficulties easier and faster to do [20].

Consciousness and Philosophical Aspects: The Oxford English Dictionary, (OED), literal definition of consciousness is "Internal knowledge or conviction; the state or fact of being mentally conscious or aware of something". Instead, if we look for a scientific definition and we want to try to find where it comes from, why it makes an individual feel like an individual, and what it gives it the awareness of being a living creature, it would be hard to define because it is still today one of the greatest mysteries of science, and maybe it won't ever be defined scientifically, since it is something that can't be very explained. The two writers introduced in the previous paragraph are the most revolutionary ones, since their ideas are still the most brilliant ever seen. Spike Jonze analyzed a current situation, which sees the first real interaction between humans and machine, while on the other hand Asimov created from nothing a concept that gains substance every day more.

Spike Jonze in his movie "Her" wanted to highlight this new smartphone obsession creating a near probable future world, where everybody will be almost dependent on a machine. In an interview made by Nichole Holofcener, he said that he "was trying to write about was the way we long to connect with each other" admitting also that he is a sort of addicted to his phone like everybody else saying, "I definitely check my phone for texts a lot like, "Did anyone text me? Is anyone thinking about me? Does anyone love me?"". But not even he knows how to describe the actual relationship between the human and the machine, "I have a lot of contradictory feelings about it. It's so complex..." Jonze said. But why it is so complex? Is it really hard to talk about a conscious machine talking to you? Yes, it is. Because consciousness itself is something that can't be really recognized and studied, it is just something that the nature gave us. The application of this sense of existence to a machine would not be easy, it would be needed to know where it comes from, what it really does and how it does it. It will be required the creation of the Asimov's positronic brain. But why create a machine with feelings, it will not even be convenient, as Gunn J. But would the positronic robot really make our life better? The answer is no. A robot with a positronic brain might lead to an insurgence or to ideological conflicts between the machine and the human, this is why it wouldn't make sense to create a robot with feelings on purpose.

In fact, the real issue is that, what if we create a machine with consciousness without knowing that? Consciousness comes from perception of existence through the energy flux between physical space and mind space, like learning and reminding. Clear evidence of learning robots can be find easily today. Old Dominion University owns a NAO robot [21], a humanoid robot which is being taught from computer engineering students to know how to recognize their own body parts or individuals' faces and talk with them. An even better humanoid machines are the Honda Asimo Robot [22]. These last ones have great learning, reminding and moving skills, they are getting used to do not make the same mistakes and to improve their performances. One of the most ambitious AI project in the world is the COG robot, the robot that has implanted in its single body sight, sound, and touch. Scientists provided it with these three senses because they are trying to give it the perfect conditions to feel things and to study where consciousness really comes from. Humans are practically teaching to machines to learn from and to remind the mistakes that they do giving them a certain senses and a perception of what is wrong and what is right. This will lead them to have personal abilities, and an own developing brains with memories, that could get even better than the human ones. This doesn't mean that they will still have feelings, but they will still have a sort of awareness that would make them more intelligent and more powerful than us.

Conclusions: After a long analysis through the many aspects of artificial intelligence, we have seen that the opportunities that the human can give to a robot its own consciousness, its own thoughts, and its own ideas, are very low, and even taking in consideration the famous positronic brain that would be still hard to duplicate a humanlike brain capable of reasoning and have feelings. It would mean to put sensations into numbers. Science made huge steps, it has been able to solve very complex phenomena and to explain them mathematically. If we just think about the majestic theories of Einstein, Newton, or Hawking, we would truly realize that everything can be calculated or mathematically demonstrated, but it is also true that the step from the calculation to the real life application is really huge. Although, as the sleep psychologist Dr. Bob Stickgold said, "Consciousness is an entirely internal experience. It's why you say it's the core of our existence because it's absolutely private. If there's one thing about consciousness, it's that it's private. And frankly, it's so private that I as a scientist can't see in you. And I have no way to measure it, and where to look for it." Hence, consciousness would be hard to be applied on a machine, since it would require calculations, demonstrations and further applications.

Artificial Intelligence is an encouraging aspect of our era, it can give to humanity important tools for better futures and better lives, as long as we don't forget the moral and human values that have always been part of the human being. AI made us to discover cures for diseases that seemed impossible to be treated, it permitted better security systems that allow more peaceful times, it facilitated the calculation of complex engineering problems allowing engineers to build better structures, it brought the mankind on the moon and to study closely other planets, it made us to improve our performances and to fix big mistakes, it even makes our daily life easier than ever allowing us to interact with each other in very simple ways. But at the same time in many years AI might lead to a drastic jobs halving in the U.S., completely substituting the human work and maybe deleting the ability for us to further evolve. AI is a serious matter that if is not treated with the right attention, it might not lead to the creation of machines with sensations, but more probably, it will bring a worldwide crisis and a severe human collapse.

If we think about AI as a living machine that has feelings, we could probably confuse the great engineering evolution process with science fiction, but excluding a thinking machine with a growing brain, with better human, and a possible naturally spontaneous creation of consciousness in an artificial individual's skills would be still a big mistake.

NOTE: Artificial Intelligence already exists, it already presents self-developing brains, advanced processing skills, and very low physical limitations, but it is essentially impossible and mainly inconvenient the replica of consciousness behaviors in a machine by the human being, we would only be able to give the machine an intelligence that would lead to consciousness by itself.

"Consciousness and emotion, those things are hard to work on directly because they seem to emerge from intelligence somehow." – Prof. Patrick Winston Director, Artificial Intelligence Lab

References:

[1] F. Heylighen and C. Joslyn, "Cybernetics and second order cybernetics," *Encyclopedia of physical science & technology,* vol. 4, pp. 155-170, 2001.

[2] N. Wiener, *Cybernetics*: JSTOR, 1948.

[3] N. Wiener, *Cybernetics or Control and Communication in the Animal and the Machine* vol. 25: MIT press, 1961.

[4] B. Scott, "Second-order cybernetics: an historical introduction," *Kybernetes,* vol. 33, pp. 1365-1378, 2004.
[5] J. C. Eccles, "The neurophysiological basis of mind: the principles of neurophysiology," 1953.
[6] J. L. Krichmar and G. M. Edelman, "Brain-based devices for the study of nervous systems and the development of intelligent machines," *Artificial Life,* vol. 11, pp. 63-77, 2005.
[7] J. Wikander and M. Törngren, "Mechatronics as an engineering science," in *Proc. of Mechatronics98 Int. conference. Published by Elsevier science ltd. ISBN 0-08-043339-1*, 1998.
[8] I. Senk, G. Ostojic, V. Jovanovic, L. Tarjan, and S. Stankovski, "Experiences in developing labs for a supervisory control and data acquisition course for undergraduate Mechatronics education," *Computer Applications in Engineering Education,* vol. 23, pp. 54-62, 2015.
[9] B. Siciliano and O. Khatib, *Springer handbook of robotics*: Springer, 2016.
[10] E. Charniak, *Introduction to artificial intelligence*: Pearson Education India, 1985.
[11] R. A. Brooks, "Achieving Artificial Intelligence through Building Robots," DTIC Document1986.
[12] M. Minsky, "Steps toward artificial intelligence," *Proceedings of the IRE,* vol. 49, pp. 8-30, 1961.
[13] N. J. Nilsson, "A mobile automaton: An application of artificial intelligence techniques," DTIC Document1969.
[14] A. Narayanan, E. Keedwell, and B. Olsson, "Artificial intelligence techniques for bioinformatics," *Applied bioinformatics,* vol. 1, pp. 191-222, 2002.
[15] Z. Ezziane, "Applications of artificial intelligence in bioinformatics: A review," *Expert Systems with Applications,* vol. 30, pp. 2-10, 2006.
[16] E. Benfenati and G. Gini, "Computational predictive programs (expert systems) in toxicology," *Toxicology,* vol. 119, pp. 213-225, 1997.
[17] D. J. Dix, K. A. Houck, M. T. Martin, A. M. Richard, R. W. Setzer, and R. J. Kavlock, "The ToxCast program for prioritizing toxicity testing of environmental chemicals," *Toxicological Sciences,* vol. 95, pp. 5-12, 2007.
[18] J. W. Catto, D. A. Linkens, M. F. Abbod, M. Chen, J. L. Burton, K. M. Feeley, *et al.*, "Artificial intelligence in predicting bladder cancer outcome," *Clinical Cancer Research,* vol. 9, pp. 4172-4177, 2003.
[19] E.-K. Ng, S. Fok, Y. Peh, F. Ng, and L. Sim, "Computerized detection of breast cancer with artificial intelligence and thermograms," *Journal of medical engineering & technology,* vol. 26, pp. 152-157, 2002.
[20] Y. H. Gharpure. (2014). *Metamorphic robots: Whither robots with artificial intelligence.*
[21] S. Shamsuddin, L. I. Ismail, H. Yussof, N. I. Zahari, S. Bahari, H. Hashim, *et al.*, "Humanoid robot NAO: Review of control and motion exploration," in *Control System, Computing and Engineering (ICCSCE), 2011 IEEE International Conference on*, 2011, pp. 511-516.

[22] M. Hirose and K. Ogawa, "Honda humanoid robots development," *Philosophical Transactions of the Royal Society of London A: Mathematical, Physical and Engineering Sciences,* vol. 365, pp. 11-19, 2007.

AN APPROACH TO PROCESSING CONDITION-BASED DATA FOR USE IN PROGNOSTIC ALGORITHMS

James P. Hofmeister, Ferenc Szidarovszky, and Douglas L. Goodman
Ridgetop Group, Inc.
3580 West Ina Road
Suite 200
Tucson, AZ 85741
Telephone: (520)-742-3300 x107
hoffy@ridgetopgroup.com

Abstract: Modern Prognostic Health Maintenance/Monitoring (PHM) and Integrated Vehicle Health Monitoring (IVHM) systems use algorithms to process Condition-based Data (CBD) to provide prognostic information and actionable imperatives to support Condition-based Maintenance for the system. Prognostic information comprises the following: estimates of remaining useful life (RUL); estimates of state-of-health (SoH); estimates of prognostic horizon (PH) – also called time-to-failure (TTF) or end-of-life (EOL) estimates. Algorithms, such as Kalman Filtering, are used to condition CBD for further processing to accurately project a future time when data reaches an amplitude level indicative of failure. This paper discusses techniques and methods to first transform CBD into a Degradation Progression Signature (DPS) and then into a Functional Failure Signature (FFS): the latter is particularly amenable to processing by prognostic algorithms to produce estimates, such as RUL, that rapidly converge to alpha (α) accuracy bounds of ten percent or better with a convergence efficiency (χ) of at least 50% of the prognostic distance (PD). [1] [1]

Key words: Prognostics; health management; prognostic distance; alpha accuracy; condition-based data; degradation-progression signature; functional-failure signature; remaining useful life.

Introduction: An important objective of PHM/IVHM systems (Prognostic Health Maintenance or Monitoring/Integrated Vehicle Health Monitoring), such as the example in Figure 1, is to monitor selected nodes to collect Condition-based Data (CBD) to produce prognostic information to create actionable imperatives and information useful to maintain health of the system, especially for systems that support condition-based maintenance (CBM). The example system uses a node-based design: monitored nodes are defined; each node definition contains information and pointers; node information is node-sensor specific and, for example, defines the sampling rate, data format and type, and noise margin; pointers identify the modules such as drivers, feature extraction, prediction, and system health and services; and the names of files used for checkpoint/restart, data output, and data input. The heart of the PHM/IVHM system comprises routines that poll and use the node definitions to control and manage the flow of the data and system services. Not shown or discussed is a framework to perform health, maintenance, and fault management. [1] [2]

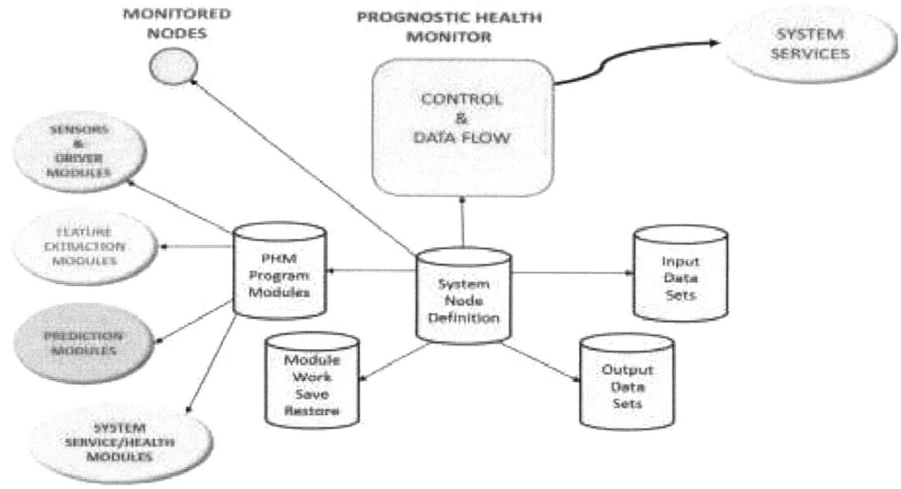

Figure 1: Framework for an Example PHM System.

Referring to Figure 2, sensor CBD is conditioned, fused, and transformed as necessary to extract feature data (FD), to create a CBD-signature. Such signatures are correlated to degradation damage and become data points in a Degradation Progression Signature (DPS), which is further conditioned, fused, and transformed to become data points in a Functional Failure Signature (FFS) that is input to prediction algorithms to produce prognostic information for health management. [2] [3]

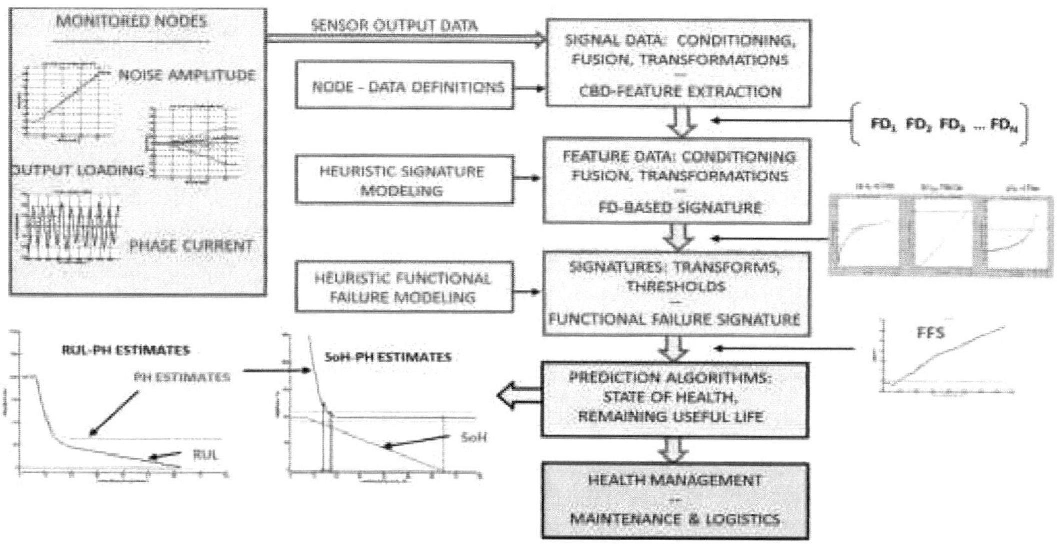

Figure 2: Functional Diagram of an Example PHM System.

Sensor Conditioning of CBD: Conditioning of CBD is almost always required because raw sensor data is noisy, distorted, and otherwise varied in ways not related to degradation as illustrated by examples of analog data in Figure 3. Rather than continuous reading and processing of analog signals, data sampling typically consists of “bursts” of digital sampling wherein analog data is sampled and converted to digital data. For example, a smart sensor might use a stimulus pulse to elicit a damped-ringing response in the output of a DC voltage power supply; capture and convert the analog response (see Figure 4) to digital data; use digital signal processing (DSP) to extract a frequency; average the frequency of a burst of samples; and then and produce scalar value of that frequency. That type of sampling is a form of low-pass filtering; and averaging is a form of data smoothing.

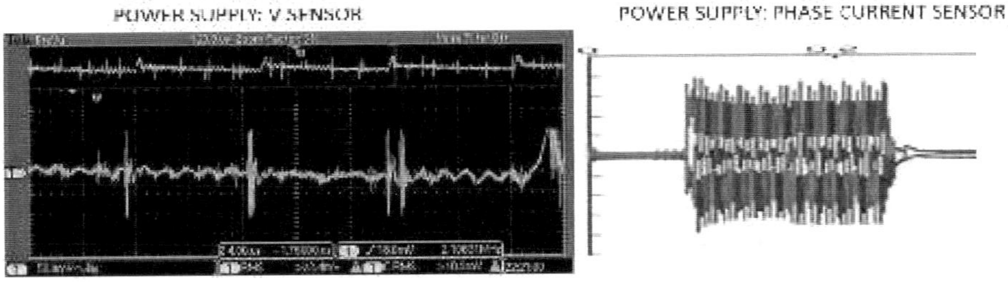

Figure 3: Examples of CBD - Voltage and Current

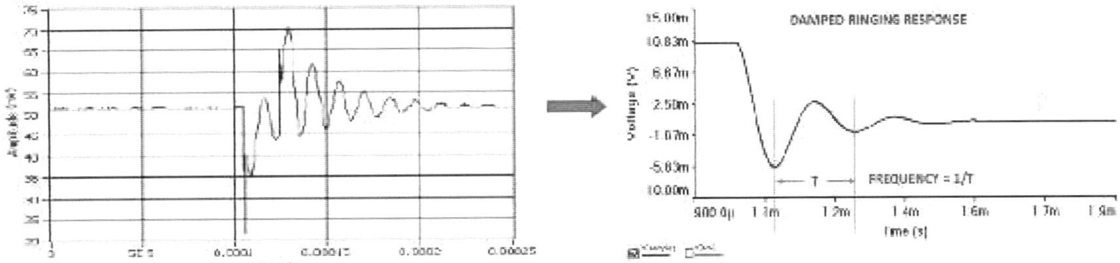

Figure 4: Example of a Damped-ringing Response to a Pulse Excitation

It can be shown that the resonant frequency of a damped-ringing response is highly correlated to the capacitance in the output circuit of, for example, a switch-mode power supply (SMPS). Physics-of-failure (PoF) analysis shows the resonant frequency can be modeled as shown by Equation (1) through Equation (4). [4]

$V_U = V_D + A_R\{e^{-t/\tau}\}\{\cos(\omega + \theta)\} + v_A$	Signal model	(1)
$\omega_0 = \sqrt{A+1}\left(\frac{1}{\sqrt{LC}}\right)$	Where A is the gain of the feedback loop	(2)
$\omega \approx \omega_C/\sqrt{(C/(C-dC))}$	Where dC is the loss in capacitance	(3)
$f = (\omega/2\pi)$	Where f is the extracted FD	(4)

The sampling design might call for a burst of 100 successive samples in a period of 3 ms from which an FD value (frequency) is extracted. Burst sampling might occur once every hour. A collection of FD values (feature data points) forms an FD-based signature: the absolute amplitude of the signature changes as degradation progresses (see Figure 5). Although the FD-based frequency response can be used as input to a prediction algorithm to produce prognostic information, use of the FD-based phase currents is somewhat problematic. Additional signal conditioning should be performed for both FD-based signatures.

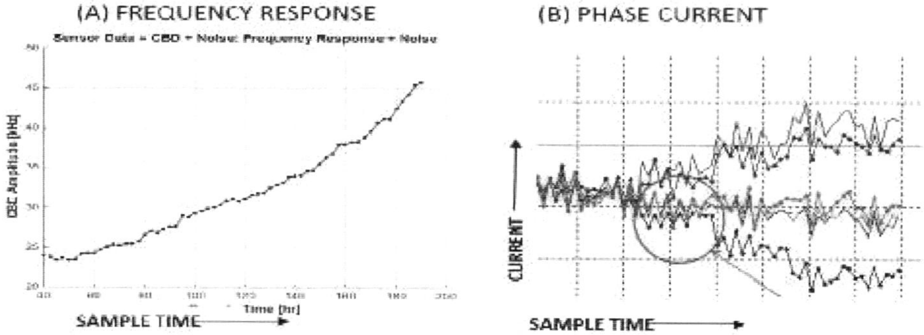

Figure 5: Examples of FD-based Signatures Extracted from CBD

Conditioning of FD-based Signatures: We will use the data shown in Figure 5 (A), Frequency Response, as our illustration example. Close examination of that data reveals noise issues/concerns as shown in Figure 6: noise is defined as any variation in a signal not caused by degradation. [2]

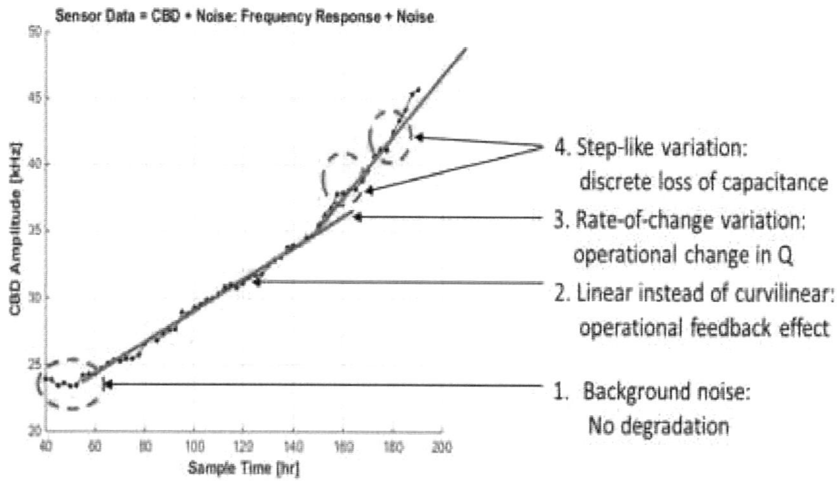

Figure 6: Noise in a FD-based Signature (Frequency Response)

The amplitude (y-axis) and time (x-axis) values of sensor-based FD have differing units of measure such as mA, kHz, V, min, day, and so on, which complicates the processing of such data to produce prognostic information. The first thing we want to do is to account for noise and normalize the data, so after examining experimental and simulated data, we apply data-smoothing (averaging method), we choose a noise margin (NM) of 1.5 kHz and a nominal FD frequency value (FD_0) of 24 kHz.

$$F \quad (c \qquad) = [C \quad (s \quad he \ ) - N \ ] \qquad (5)$$

The result of applying Equation (5) is shown in Figure 7 (A). We then transform FD-based signature data into DPS data by using a $(d \ /F)$ transform: d is a change in a parameter (P) because of degradation. For example, Equation (3) is a power function where n = 0.5 and P = C (the capacitance). The transform is Equation (6).

$$D \quad = d \quad / P = d \quad / C_C = [1 - (F \quad _C / F \quad)^2] \qquad (6)$$

The result of applying Equation (6) is shown in Figure 7 (B). After examining the original, experimental, and simulation data and performing Failure Mode Effects Analysis (FMEA), you conclude functional failure, operating out of specifications, occurs when the DPS is at or above 0.6 and Equation (7) is used to transform DPS data into Functional Failure Signature (FFS) data, which is plotted in Figure 7 (C).

$$F \quad = 100 * (D \quad / F \quad) \qquad (7)$$

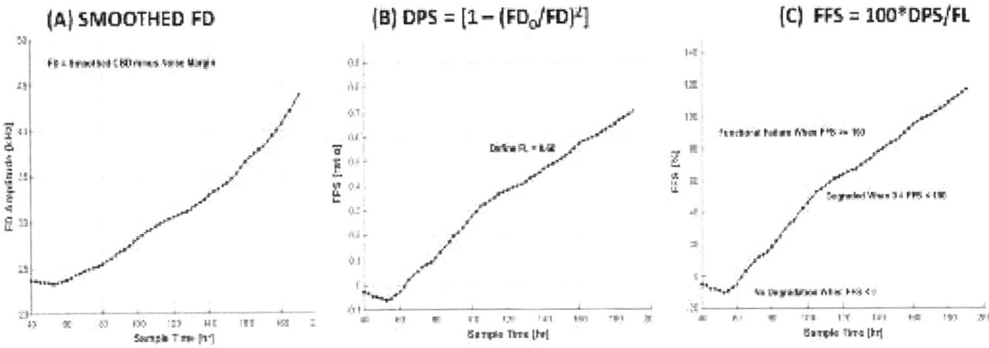

Figure 7: Example of Conditioning and Transforming an FD-signature

Amenability of an FFS: We transform data into an FSS because the characteristics of that signature is very amenable to processing by prediction algorithms:

- A transform of ideal linear, power, or exponential data is an ideal straight line
- Non-ideality in the data is "preserved" in the FFS curve – compare Figure 6 to the FFS curve in Figure 8
- All FSS curves have the same y-axis values
  - When a data point is < 0, state-of-health (SoH) is defined as 100%
  - When a data point is >= 100, SoH is defined as 0%
  - Else 0% > SoH < 100%
- An FFS transfer curve is more linear compare to its original CBD, which is amenable to prediction algorithms that use Kalman-like filtering and/or random-walk methods to estimate when data is likely to reach a value of 100%.

DPS Transforms: CBD curves related to degradation, a change in a parameter d are generally the result of power or exponential functions that cause those curves to be convex or concave. To transform decreasing curves to increasing curves, use the following equation:

$$C \quad = D \quad 0 - D \qquad \text{Transform decreasing data} \qquad (8)$$

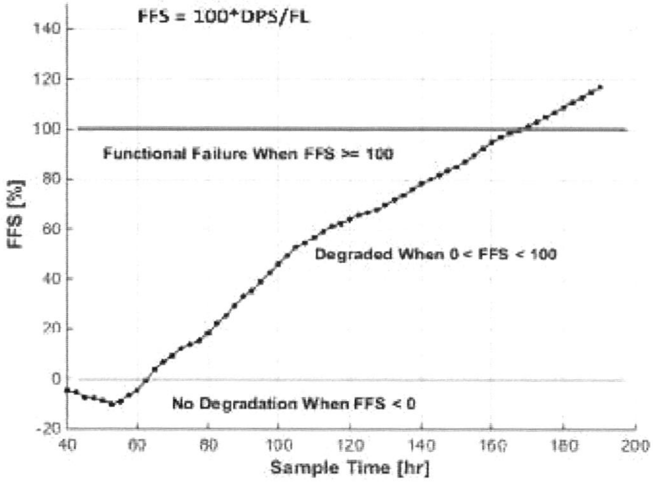

Figure 8: FFS Plot and Degradation Zones

Power functions: Directly-proportional functions can be convex, linear (n=1), or concave. Inversely-proportional curves are concave. The FD and DPS models are the following:

$F \quad = F_{c}[(P + d\ )/P]^{n}$	Directly proportional	(9)
$D \quad = (F\ /F_{c} - 1)^{(1/n)}$		(10)

$F \quad = F_{c}[P/(P - d\ )]^{n}$	Inversely proportional	(11)
$D \quad = 1 - (F_{c}/F\ )^{(1/n)}$		(12)

Note the DPS linearity in Figure 9, Figure 10, and Figure 11). When there is no degradation ($d \quad = 0$), FD equals FD_0: a constant.

Exponential functions: Exponential curves are either concave or convex. The FD and DPS models are the following:

$F \quad = F_{c}[\exp(d\ /P)]$	Concave, increasing	(13)
$D \quad = \ln(F\ /F_{c})$		(14)

$F \quad = F_{c}[\exp(-d\ /P)]$	Concave, decreasing	(15)
$D \quad =$ use Equation (8) then apply Equation (17)		

$F \quad = F_{c}[1 - \exp(-d\ /P)]$	Convex, increasing	(16)
$D \quad = \ln[F_{c}/(F_{c} - \mathrm{FD})]$	For FD <= FD_0	(17)

Note the DPS linearity in Figure 12 and Figure 13. When there is no degradation ($d \quad = 0$), FD equals FD_0: a constant.

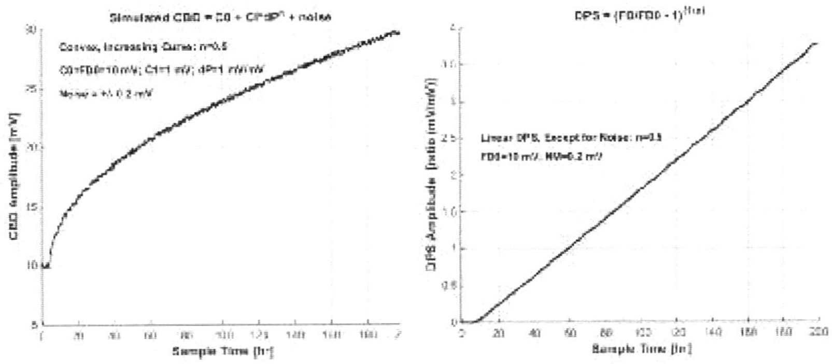

Figure 9: Power Function Example for n < 1, Directly Proportional

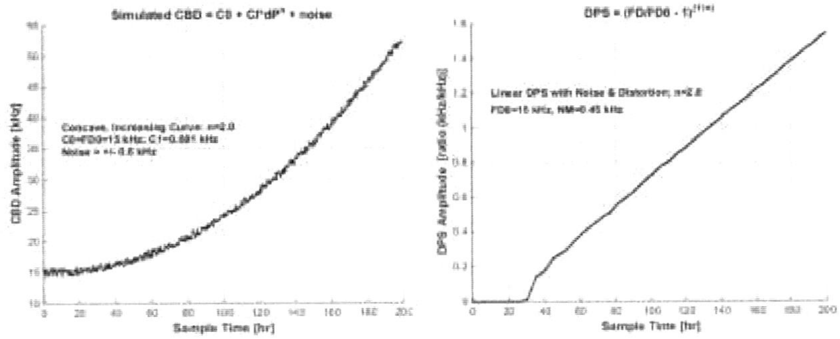

Figure 10: Power Function Example for n > 1, Directly Proportional

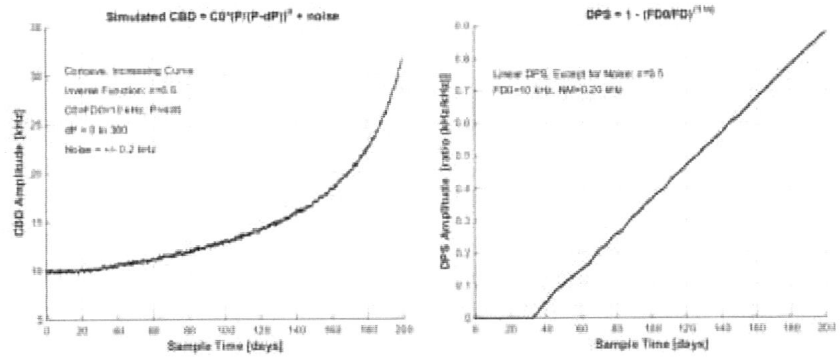

Figure 11: Power Function Example, Inversely Proportional

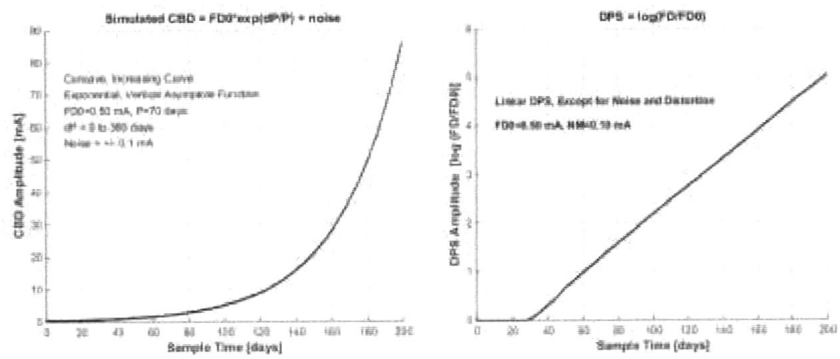

Figure 12: Exponential Function Example, Concave Curve

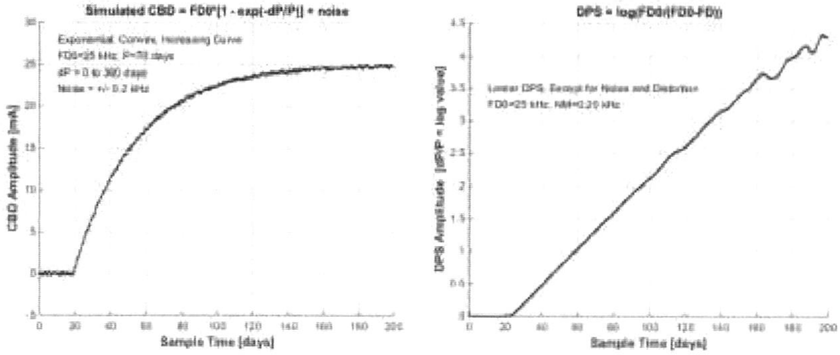

Figure 13: Exponential Function Example, Convex Curve

Data Fusion and Comparison: It is often the case that two or more FD-based signatures need to be fused to condition and/or transformed to create a single FD-based signature. For example, the six FD-based current signatures plotted in Figure 5 (B) can be fused and transformed into three different DPS as seen in Figure 14. This state is occurs when one of the six power-switching diodes in an H-bridge power supply driving a rotor degrades.

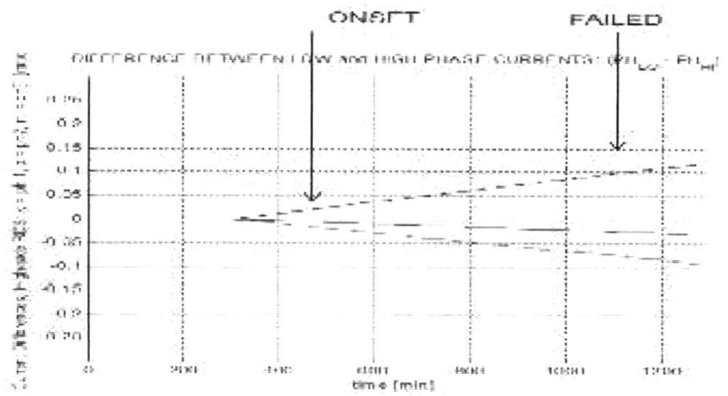

Figure 14: Three FD-based Signatures

Domain Transform: It is sometimes necessary to use domain transforms. For example, the voltage and current data in Figure 15 can be transformed into resistance by fusing the two: $R = V/I$ – but the result, shown in Figure 16 (A), is temperature-dependent. Another transform results in temperature-independent DPS data points, plotted in Figure 16 (B), that is further conditioned (smoothed) and transformed into an FFS as plotted in Figure 16 (C).

Prognostic Information: An FFS is input to a prediction algorithm to produce prognostic information: State-of-Health (SoH), Remaining Useful Life (RUL), and Prognostic Horizon (PH): PH is an estimate equal to current relative time of the sampled CBD plus RUL (see Table 1). Figure 17 (A) shows the plots for PH, RUL, and SoH produced from the input FSS shown in Figure 16 (B): the Convergence Efficiency (χ) for a relative accuracy of 10% is better than 98% (refer to Table 1 and Figure 17 (A).

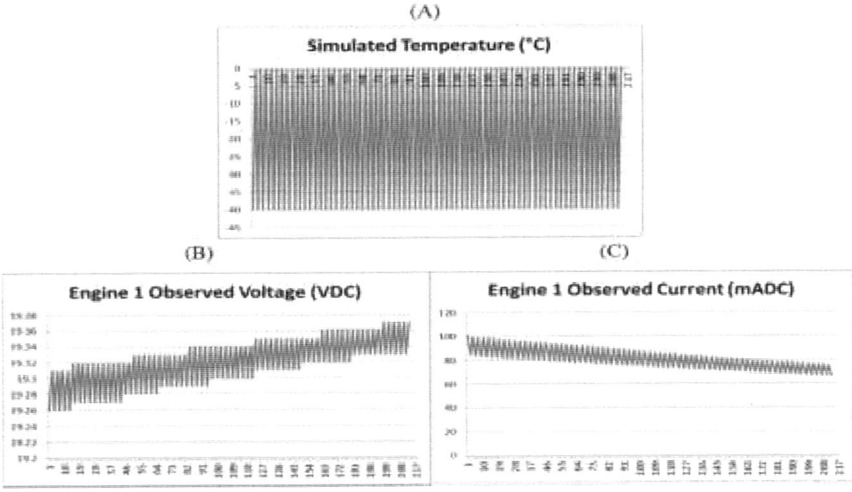

Figure 15: Example of CBD – Temperature, Voltage, and Current.

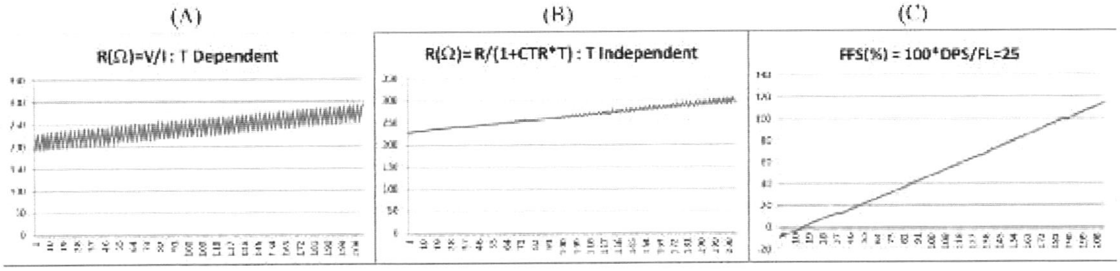

Figure 16: Example of Fusing, Transforming, and Smoothing to Produce an FFS.

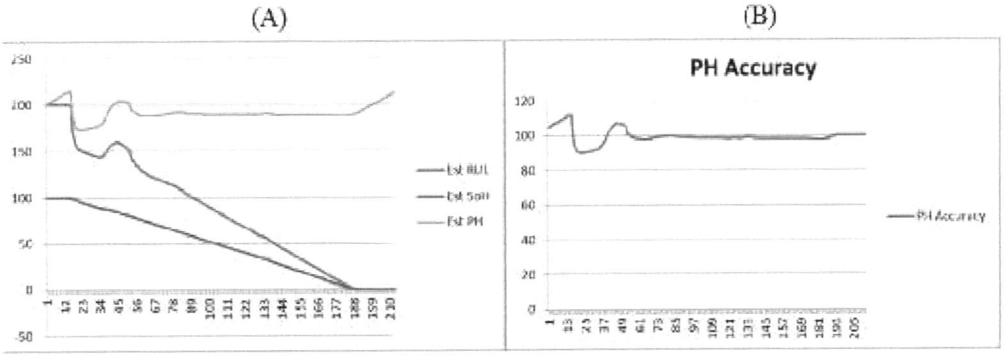

Figure 17: Prognostic Information Plots (A) and PH Accuracy Plot (B)

Table 1: Terminology and Definitions

Term	PHM System Definition
EOL	End-of-Life, time of failure
Prognostic Horizon (PH)	T_S + RUL (sample time + RUL)
Error Margin (α)	Percent
Relative Accuracy (α–λ)	(α–λ) = (PH /EOL)
Prognostic Distance (PD)	PD = EOL – time when degradation first detected
Convergence Efficiency (χ)	χ = (RULα / PD) when RUL within α accuracy

Conclusion: In this paper we used an example PHM system based on node definitions and described sensor conditioning of CBD from FD can be extracted. We then described conditioning of FD-based signatures and showed how FD-based signatures are transformed into DPS data and how DPS data is transformed into FFS data. We then described why FFS data is useful as input to prediction algorithms. We asserted that degradation can be modeled as power or exponential functions and we listed the FD and DPS models for those functions. We presented examples of data fusion, comparison, and transforms. We concluded the paper by showing and describing the prognostic information that we obtained using an FFS from example data we used in the paper.

References:

[1] Prognostics Health Management for Electronics. (Accessed 2015, Nov.) Auburn University, Auburn, Alabama. [Online]. Available: http://cave.auburn.edu/rsrch-thrusts/prognostic-health-management-for-electronics.html.

[2] James Hofmeister, Douglas Goodman, and Robert Wagoner, 2016, "Advanced Anomaly Detection Method for Condition Monitoring of Complex Equipment and Systems," *2016 Machine Failure Prevention Technology,* Dayton, OH, 24-26 May 2016.

[3] James Hofmeister, Robert Wagoner, and Douglas Goodman, 2013, "Prognostic Health Management (PHM) of Electrical Systems using Conditioned-based Data for Anomaly and Prognostic Reasoning," *Chemical Engineering Transactions*, Vol. 33, pp. 992-996, 2013.

[4] Judkins J.B, Hofmeister J., Vohnout S., 2008, "Non-Invasive Prognostic Sensor for Voltage Regulated Switch-Mode Power Supplies," IEEE Aerospace Conference, Big Sky, MT, 3-10 March 2008, Track 11-0804, pp. 1-8.

[5] A. Saxena, J. Celaya, B. Saha, S. Saha, and K. Goebel, 2009, "On Applying the Prognostic performance Metrics," *Annual Conference of the Prognostics and Health Management Society (PHM09),* CA, 2009.

Bibliography:

[1] Engineering Statistics Handbook, 1. Exploratory Data Analysis, 1.3 EDA Techniques. (Accessed 2015, Dec.). [Online]. Available: http://www.itl.nist.gov/div898/handbook.htm.

[2] Engineering Statistics Handbook, 8. Assessing Product Reliability, (Accessed 2015, Dec.). [Online]. Available: http://www.itl.nist.gov/div898/handbook.htm.

ANALYSIS AND BEARING-DAMPER REDESIGN FOR A $100 MILLION HIGH PRESSURE COMPRESSOR FAILURE

Edgar J. Gunter
RODYN Vibration Analysis, Inc.
Professor Emeritus, Mechanical and Aerospace Engineering
Former Director, Rotor Dynamics Laboratory, University of Virginia
1501 Gordon Avenue
Charlottesville, VA 22903

Brian K. Weaver
Research Associate, Mechanical and Aerospace Engineering,
Rotating Machinery and Controls Laboratory, University of Virginia
122 Engineer's Way
P.O. Box 400746
Charlottesville, VA 22904
Telephone: (434) 243-4934
bkw3q@virginia.edu

Abstract: The Kaybob compressor failure of 1971 was an excellent historic example of rotordynamic instability and the design factors that affect this phenomenon. In the case of Kaybob, the use of poorly designed bearings produced unstable whirling in both the low and high pressure compressors. This required over five months of vibration troubleshooting and redesign along with over 100 million modern U.S. dollars in total costs and lost revenue. The problems began with the use of inadequate five-pad tilting pad journal bearings which produced large ratios of bearing stiffness to shaft stiffness that contributed heavily to the instability. Subsequent designs included attempts at using bearing asymmetry and squeeze film dampers to improve the stability of the machine, however these solutions produced little improvement. The ultimate solution, a redesign of the compressor to a shorter, stiffer machine, resulted in the significant costs incurred as a result of these failed bearing designs. Another reason for the costly nature of this project was a lack of accurate bearing and rotordynamic analysis tools that could have identified the instability during the design process and aided in the redesign of the fluid film bearings supporting the machine. Since the 1970s, a wide range of bearing and rotordynamic analysis tools have been developed that accurately predict the performance of fluid film bearings and rotordynamic systems via the use of practical lubrication theory, finite element analysis, and simple relationships derived as a result of the physical insights gained from using these methods.

In this paper, the history of the Kaybob compressor failure is discussed in detail including a discussion of the ineffective bearing designs that were considered. Modern bearing and rotordynamic analysis tools are then employed to study both designs that were considered along with new designs for the bearings that could have ultimately restored stability to the machine. These designs include four-pad, load-between-pad bearings and squeeze film dampers with a central groove. Simple relationships based on the physics of the system are also used to show how the bearings could be tuned to produce optimum bearing stiffness and damping of the rotor vibration, producing insights which can inform the designer as they perform more comprehensive analyses of these systems.

Key words: Compressor failure; stability failure; compressor design; rotordynamics; fluid film bearings; squeeze film dampers

Introduction: The rotordynamic stability of high-speed compressors and turbines has been a critical element of their design for decades. With the ever-increasing demand for greater output through an increased number of stages, higher speeds, and higher pressures, these machines are continuously being pushed to the design limits of the previous generation of machines. As they are being pushed to these new extremes of operation the designs must also adapt to accommodate the consequences of these extremes including increased rotor flexibility and an increased likelihood of subsynchronous whirl. This whirl can be excited by a number of sources in turbomachinery including seals and even the bearings supporting the rotor. The stability of that whirl can then be reduced by destabilizing cross-coupled stiffness forces from the bearings, interstage seals, balance pistons, and Alford-type aerodynamic cross-coupled forces around blades.

The increasing use of tilting pad journal bearings in the 1960s and 70s brought about new opportunities for compressor design as these bearings eliminated the self-exciting oil whirl commonly found in fixed geometry bearings. They also essentially eliminated the destabilizing cross-coupled stiffness forces produced by these bearings. However, despite the fact that a number of important papers were being published on these topics by Lund [1] and others, it wasn't until 1974 that Lund published his landmark paper on damped eigenvalue solutions for stability analysis [2]. It was because of this lack of accurate design and analysis tools that a significant number of high-speed machines were not designed appropriately for the conditions in which they were intended to operate.

The Kaybob compressor failure of 1971 was an historic case of rotordynamic instability from self-excited, subsynchronous whirl. The many months of troubleshooting that followed this failure served as a valuable lesson to the team involved but even to this day this classic example serves as an important motivator for understanding the underlying principles that govern the design of these machines, many of which that were developed after the incident and troubleshooting took place. With the advent of advanced bearing and rotordynamic analysis tools, designers now have the ability to fine tune their designs to avoid these failures, however it is critically important that the underlying principles and physical insights gained from the analysis of these machines are properly considered as even modern bearing technologies can contribute to stability problems when not implemented correctly.

In this study, modern bearing and rotordynamic analysis tools are applied to the Kaybob compressor instability case to highlight the effects of the design choices made with this machine and how they contributed to its ultimate failure. By combining useful design principles with these analysis tools it also shown not only why the costly solution to this failure worked at the time, but also how a number of alternative and much simpler design changes to the support structure could have restored stability to the machine including the use of properly designed tilting pad journal bearings and squeeze film dampers. It is the goal of this study to provide practical information and design approaches that designers can then use to avoid costly failures like the one experienced in Kaybob.

Stage Case History - Kaybob Plant, Alberta: The case of the Kaybob compressor instability began in November of 1971, a detailed summary of which is provided by Kenneth Smith of Cooper-Bessemer [3]. Three natural gas re-injection trains were being commissioned at the Kaybob South Beaverhill Lake plant in Fox Creek, Alberta operated by Chevron-Standard.

These duplicate trains (Figure 1) consist of a gas turbine drive, a speed-increasing gearbox, a 9-stage low pressure compressor, and a 5-stage high pressure compressor. Both compressors have back-to-back impeller arrangements with centrally located balance pistons to control thrust and are supported by tilting-pad thrust and radial bearings. The tilting pad journal bearings were designed as five-pad, load-on-pad bearings, a design that has been commonly employed in compressors over the years. Floating oil film ring seals were employed in both compressor casings. The trains were designed for a rated speed of 10,200 rpm with a maximum continuous operating speed of 11,400 rpm.

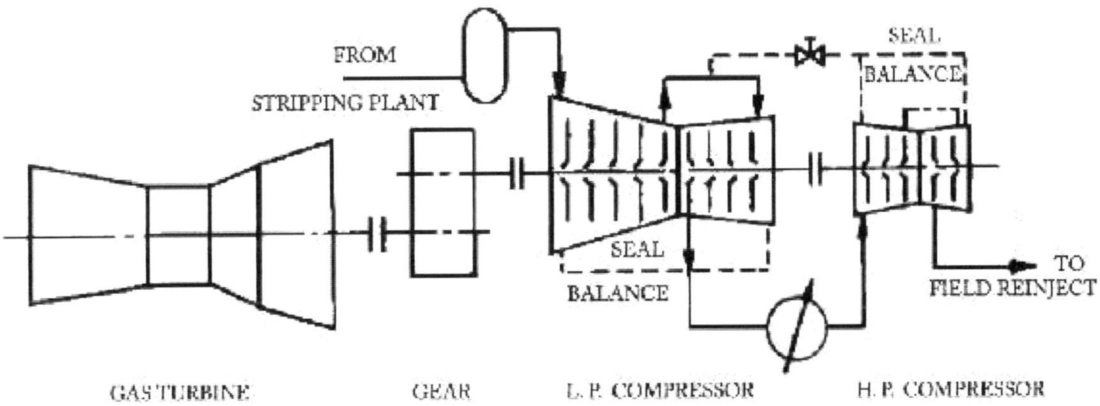

Figure 1: Kaybob Compressor Train Schematic. From Smith [3].

As site testing began, a significant vibration problem became apparent in the low pressure compressor. A first train was operated at speeds up to 10,000 rpm with inlet pressures of 700-750 psi and discharge pressures of 1,300-1,500 psi. Under these conditions a significant vibration was present at a frequency of approximately 4,250 cpm, producing 3 to 4 mil peak-to-peak amplitudes in one of the bearings which had a 6 mil clearance.

A second train was then tested to rule out possible unexplained sources of resonance from the first train. This machine was run up to its maximum speed of 11,440 rpm, followed by increases to the inlet and discharge pressures to 1,120 psi and 3,300 psi.

At this point the machine became highly unstable, resulting in a violent 5,100 cpm whirl of 9+ mils peak-to-peak amplitude. Figure 2 shows the 20-second time lapse of the instability onset in the thru-drive end bearing along with the orbit shape. Smith [3] notes that "the rotor assembly and internal labyrinths were severely damaged as a result of this violent, high-energy motion" and that the "vibration could be heard and felt in the control room some 50 feet away from the unit."

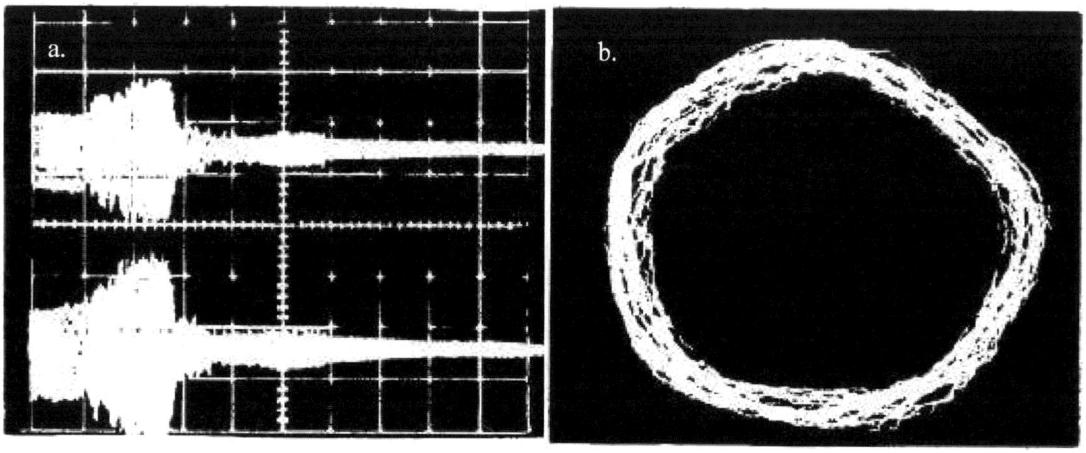

Figure 2: (a) Time-lapse and (b) orbit of the instability onset. From Smith [3].

The five months that followed included a thorough investigation into the causes of this instability as well as tests of various potential solutions to the problem including investigations into both seals and journal bearings. The seal investigation involved the consideration of four different seal configurations to determine whether seal lockup or effective damping had any significant effects on the instability. It was concluded, though, that the seals were in fact floating and their configuration overall had little effect on the problem.

A number of design modifications were considered for the tilting pad journal bearings as well. These modifications included increasing the specific load of the bearing, introducing asymmetry to the bearing design, different pad-load configurations, offset pads, variable preloads, the introduction of squeeze film dampers, and combinations thereof. Two of the configurations tested are shown in Figure 3. Overall it was found that all of these bearing design-based solutions produced marginal improvements in rotordynamic stability. Additional analysis and discussion of these and other bearing designs is presented in the following sections.

Note that in Fig. 3a, the side pads, B, are reduced in width in an attempt to incorporate some bearing asymmetry. This modification had a negligible influence on stability. In Fig. 3b, the A pads are reduced in arc length and a squeeze film damper was added without a central groove. This design results in a damper with excessive damping and stiffness.

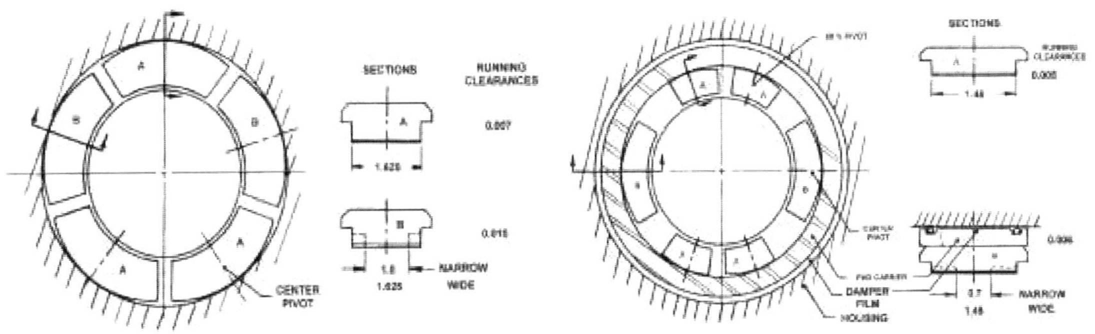

Figure 3: (a) Asymmetric Tilt Pad Bearing and (b) Squeeze Film Damper Design. From Smith [3].

The ultimate solution to this vibration problem was a redesign of the rotor to a shorter, stiffer configuration with increased diameters beneath the impellers. This stiffer shaft would thereby accommodate the stiff bearing designs that were used and that will be demonstrated in the following sections. Figure 4 shows the initial and final shaft designs. Overall, these significant changes resulted in a redesign of most of the primary machine components, a process which would cost the plant over 100 million modern U.S. dollars.

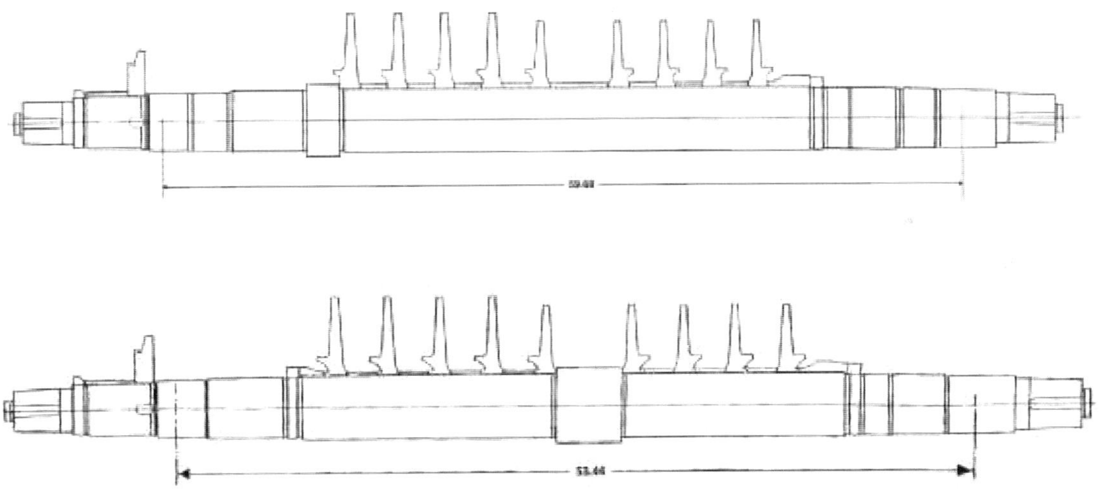

Figure 4: Rotor redesign. From Smith [3].

In the original design of the Kaybob rotor, the bearing span was 59.688 inches. A modal analysis of the rotor indicated that the shaft first critical speed modal stiffness was approximately 115,000 lb/in. This fundamental shaft modal stiffness is extremely important for the proper tuning of the bearings to the shaft. From previous research studies of the optimum bearing stiffness, is has been determined that the optimum bearing stiffness is approximately one half the rotor shaft stiffness. When the bearing stiffness greatly exceeds this fundamental design value, the first mode damping is drastically reduced causing the rotor to be very susceptible to self excited whirl effects.

The fundamental stiffness of a uniform beam is given by the following equation:

$$K_{shaft} = 48\ EI/L^3\ \text{lb/in} \quad (1)$$

In the final redesign as shown in the lower rotor of Figure 4, the bearing span has been reduced to 53.438 inches. In addition to the reduced bearing span, the center section diameter has also been increased. This then leads to an improved shaft modal stiffness of 200,000 lb/in.

Rotordynamic Characteristics: In revisiting the design and troubleshooting of the Kaybob compressor, the first step was to establish the overall rotordynamic characteristics of the system. A finite element model of the compressor was first developed in the Dyrobes software suite used for this analysis and is presented to scale in Figure 5. The rotor spans 78 inches in total length with a bearing span of 60 inches. The model consists of 25 elements and 26 nodes with 9 rigid disks representing the mass and inertial properties of the 9 compressor stages, the details of which are provided in Table 1. The total rotor weight is 400 lbs. The bearings are represented as linear stiffness and damping coefficients at nodes 5 and 23.

After building the shaft model, the analysis began with an assessment of the rotor critical speeds in the operating range of the compressor (Figure 6). Two modes were found in the operating range: a first bending mode at approximately 4,600 rpm (Figure 7) which proved to be the unstable mode for the machine, and a well-damped conical mode at approximately 8,500 rpm. Overlaying the critical speed map is the vertical stiffness K_{yy} of the original 5-pad load-on-pad tilting pad bearings supporting the rotor. Its position on the curve for the first bending mode provides an early indication of a relatively stiff support structure for the compressor.

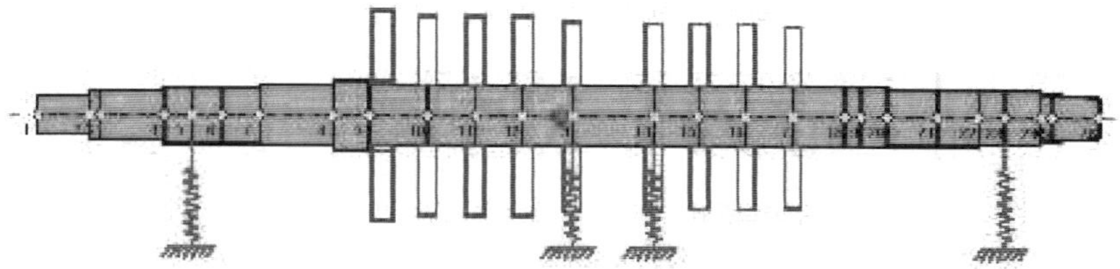

Figure 5: Kaybob 9 Stage Compressor Model with Aerodynamic Cross Coupling at Stations 13 and 14.

In order to evaluate the stability characteristics of this compressor, a nominal amount of cross coupled stiffness of 25,000 lb/in was assumed acting at stations 13 and 14. This value was estimated based on experience and the description of the stability characteristics of the machine. These cross coupling effects are known as the Alford effect [4] and can occur in all gas turbines and high pressure compressors. These effects are often not observed on a test stand when the unit is operated under low pressure ratios. When operated in the field under high pressure conditions, the Alford effects become apparent resulting in self excited whirling which can then lead to damage of seals and labyrinths. Due to a lack of information on the design of the seals these were not modeled separately, with their contribution to the destabilization of the compressor being included in the cross coupled stiffness applied.

Table 1: Kaybob compressor model disk properties.

Disk #	Station	Mass (lbm)	Transverse Inertia ($lbm \cdot in^2$)	Polar Inertia ($lbm \cdot in^2$)	Length (in)	Inner Diameter (in)	Outer Diameter (in)
1	9	15	760	1400	1.5	4.36	15.0
2	10	15	760	1400	1.5	4.36	14.3
3	11	15	760	1400	1.4	4.36	14.3
4	12	15	760	1400	1.4	4.36	14.3
5	13	5	760	1400	1.3	4.36	13.6
6	14	5	400	750	1.3	4.36	13.3
7	15	5	400	750	1.3	4.36	13.1
8	16	5	400	750	1.2	4.36	13.1
9	17	5	400	750	1.2	4.36	13.0

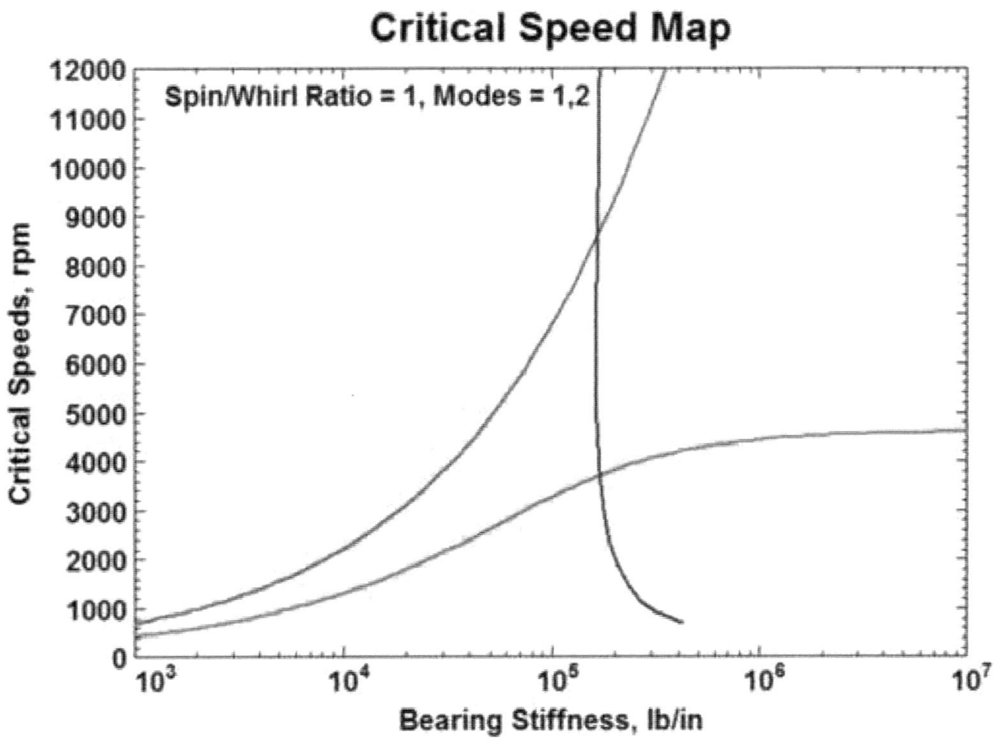

Figure 6: Compressor Critical Speed Map.

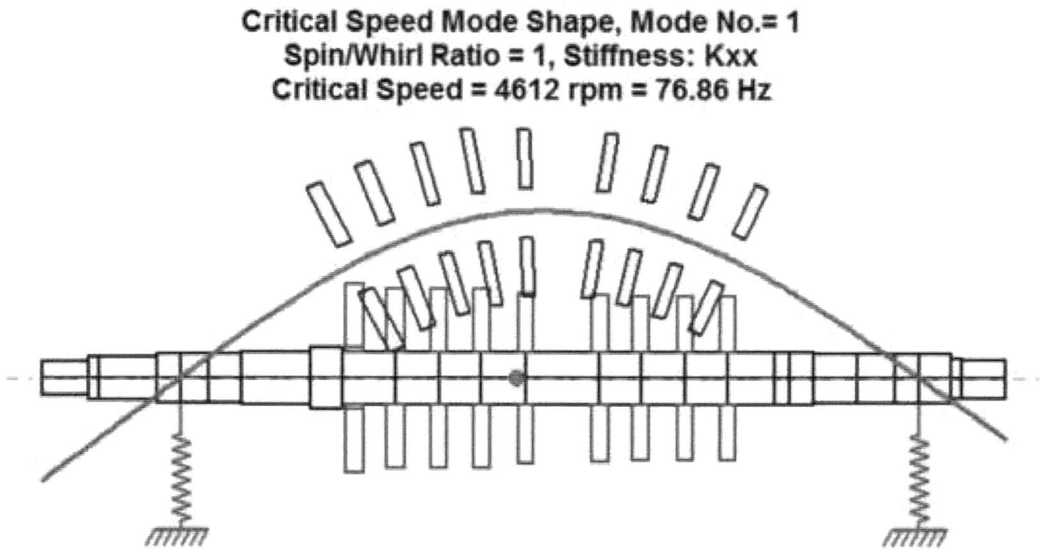

Figure 7: First Bending Mode of the Compressor.

The shaft modal stresses and stiffness were also assessed for the first bending mode assuming rigid bearings with a stiffness of 10^8 lb/in (Figure 8). Peak stresses of 1,709 psi were predicted along with a shaft modal stiffness of 115,850 lb/in. This magnitude of stiffness will prove to be very important to the stability of the machine as it is compared to the stiffness of the bearings in the following sections. It was presented by Barrett, Gunter, and Allaire [5] that the optimum ratio of twice the bearing stiffness to the shaft modal stiffness for maximizing modal stability is one.

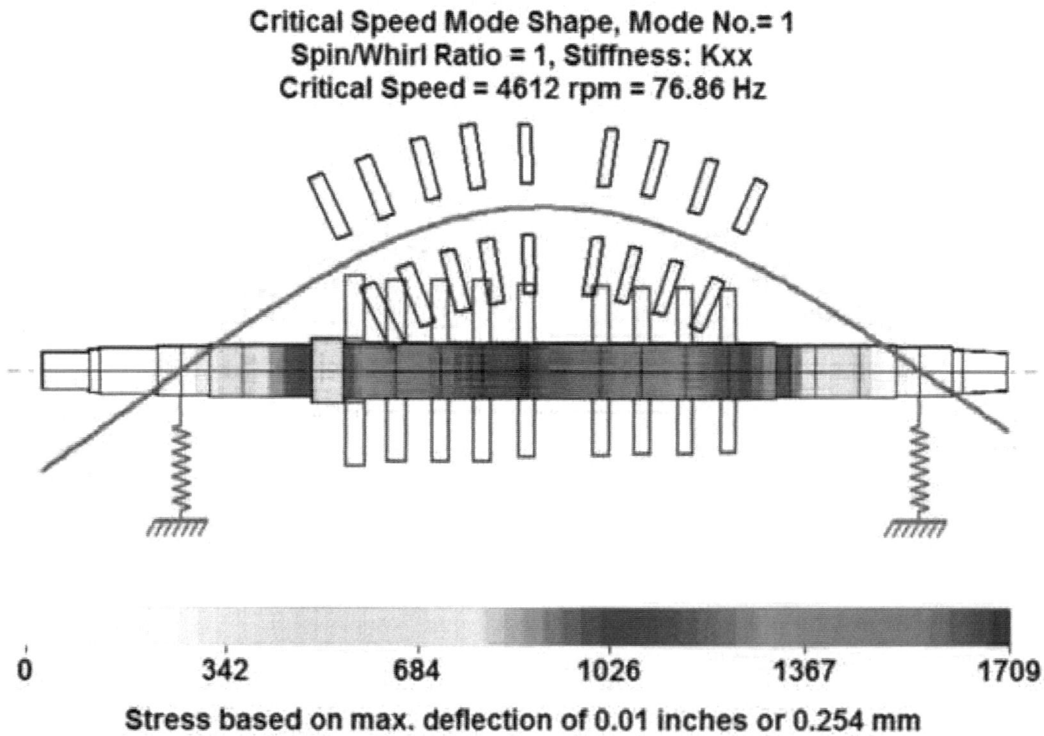

Figure 8: Shaft Modal Stresses of the First Bending Mode.

To further support this important relationship in rotordynamics, a plot was created showing the amplification factors of the first bending mode under various sets of bearing stiffness and damping (Figure 9). It is shown that the amplification factors are reduced as the bearing stiffness approaches this ideal ratio. The plot also highlights the importance of designing the bearings in such a manner that underdamped and overdamped conditions can also be avoided through proper design. The following sections on various bearing designs for the compressor will further show how various design variables affect these bearing properties and the stability of the compressor.

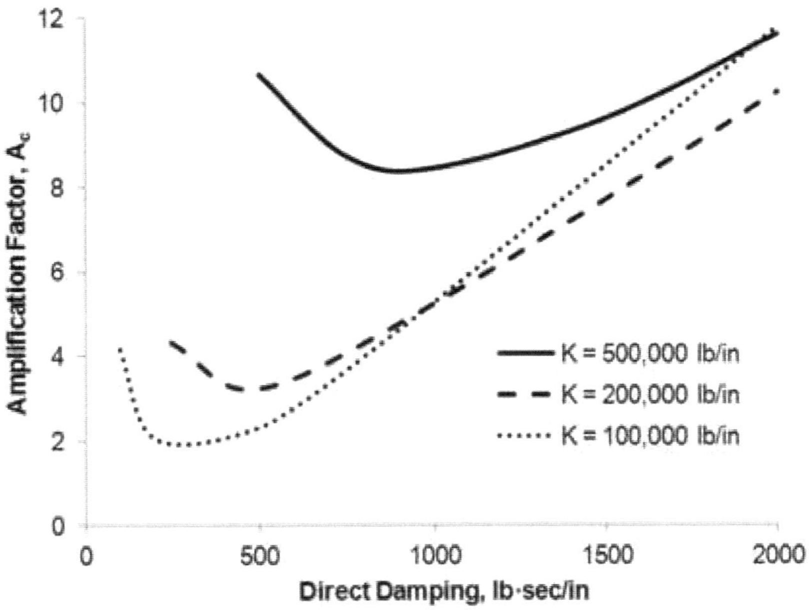

Figure 9: Amplification Factors of the First Bending Mode.

Figure 9 shows the importance of tuning the bearing stiffness to the rotor shaft stiffness. For example with the original rotor design, it is seen that with a vertical bearing stiffness of 100,000 lb/in, the optimum damping is about 250 lb-sec/in with an amplification factor of two. As the bearing stiffness increases, the required optimum damping increases to 500 lb-sec/in and the minimum amplification factor is over three. For the case of very high stiffness bearing of 500,000 lb/in, it is seen that the optimum bearing damping is over 800 lb-sec/in and the amplification factor is now over eight. Therefore it is very apparent that it is critical to select the bearing stiffness in relationship to the predicted first modal shaft stiffness. This concept will often lead to the requirement of an extended width bearing in order to have a reduced vertical bearing stiffness in order to properly tune the bearings to the fundamental shaft stiffness.

Original Bearing Design: The analysis of the Kaybob compressor continued with the development of various bearing models to better understand how the design of the bearings would impact the stability of the machine. The original design of the compressor bearings was a 5-pad load-on-pad configuration with an L/D of approximately 0.4, depicted in Figure 10. This model utilizes a two-dimensional Reynolds equation solver that for the current analysis neglects thermal effects and deformation for simplicity. All bearing analyses were performed at the maximum continuous operating speed of 11,400 rpm.

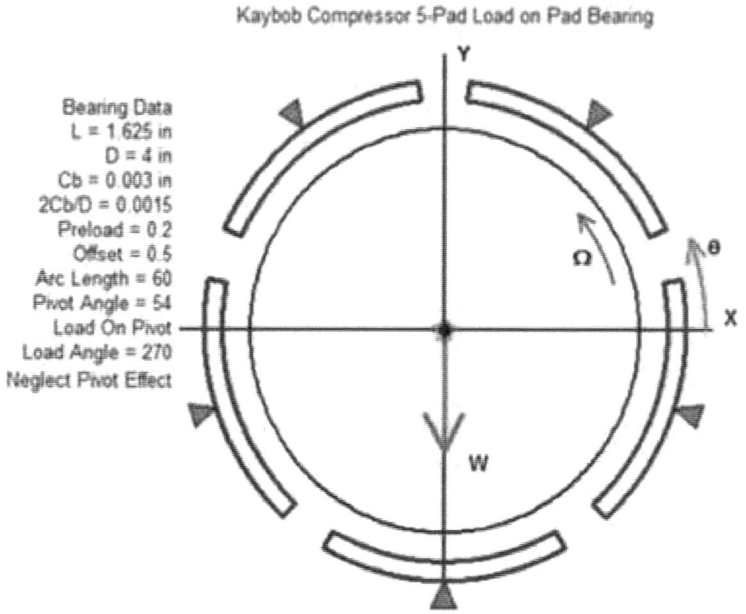

Figure 10: Original Compressor Bearing Design.

The results of this initial bearing analysis are shown in Figure 11. It was found that for this configuration that the tight clearance, preload, and load-on-pad configuration together produced a relatively stiff bearing, resulting in a vertical bearing stiffness of 169,500 lb/in. Recall from the previous section that the ideal ratio of twice the bearing stiffness to the shaft modal stiffness is unity. When comparing twice the vertical stiffness of the original compressor bearing to the calculated shaft modal stiffness, this results in a ratio of over 2.9. These results alone are indicative of an overly stiff support structure for the compressor that could contribute to an unstable whirl motion.

The next step in analyzing the original rotor-bearing system was to perform a stability analysis of the compressor with the original bearings. The results of the damped eigenvalue analysis are presented in Figure 12. It is shown that the poor choices made in the bearing design process resulted in an unstable forward bending mode in the operating range. The combination of high stiffness and damping in the bearings resulted in less shaft motion at the bearings and therefore a greater distribution of the system energy into the shaft, resulting in the unstable mode. A critical speed analysis using the bearing vertical stiffness reveals the relative energy distribution in the shaft and bearings (Figure 13).

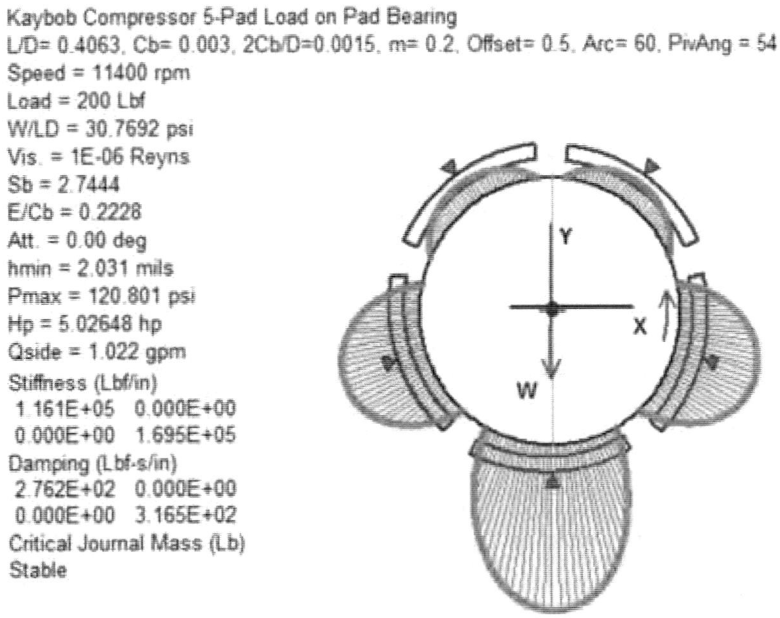

Figure 11: Original Compressor Bearing Analysis Results.

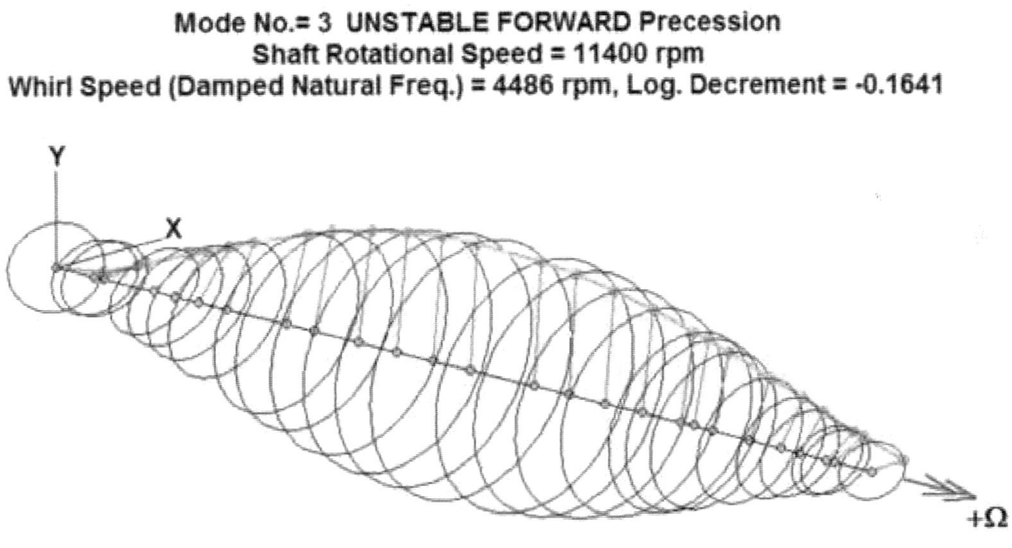

Figure 12: Stability Results for the Original Bearing Design.

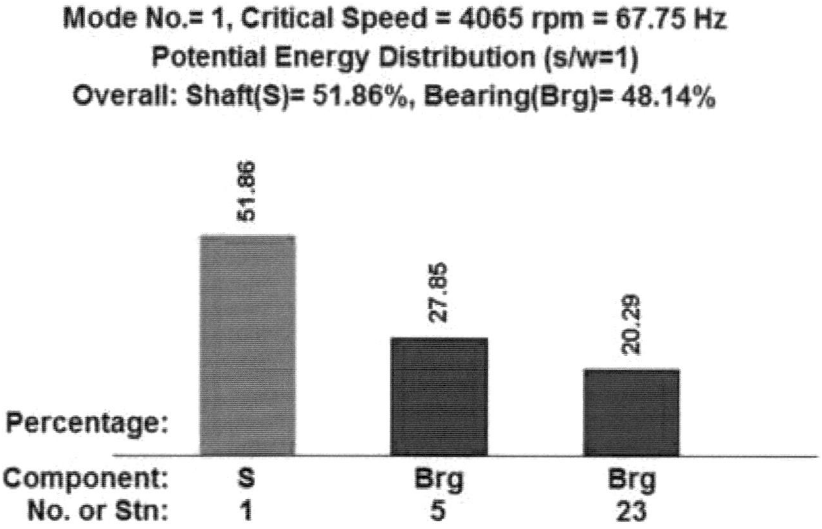

Figure 13: First Mode Shaft and Bearings Energy Distribution with the Original Bearing Design.

A Lesson in Bearing Asymmetry: A second bearing design considered when trouble-shooting the Kaybob compressor was an asymmetric bearing. Smith [3] cited two papers by Gunter [6,7] as evidence that introducing bearing asymmetry could lead to an improvement in rotor stability. However, this evidence was misinterpreted as asymmetry has little effect in cases of high bearing stiffness and large quantities of aerodynamic cross-coupling [8]. This is demonstrated in the below analysis of an asymmetric bearing design similar to that considered by the team at Cooper-Bessemer.

The model below (Figure 14) shows how the bearing was clocked to a load-between-pad position. The bearing clearance was also increased in the upper-left and upper-right pads from a 3.5 mil radial clearance to 5 mils. These two pads were also narrowed from an axial length of 1.625" to a length of 1", though this modification is not captured by the model. This proves to have little effect, however, as shown in the results (Figure 15). The pads with the increased clearance show no participation in controlling the shaft motion, resulting in predicted vertical stiffness values that are still quite large when compared to the shaft stiffness. A significant drop in damping is also produced as a result of this design.

A stability analysis of the bearing-rotor system (Figure 16) in fact shows a decrease in the log decrement of the first bending mode when compared to the original bearing design. This is due to the fact that not only is the bearing stiffness still too large, but now the system is also likely underdamped with this bearing design.

The team at Cooper-Bessemer tried multiple bearing designs that utilized bearing asymmetry, but with such high levels of bearing stiffness and aerodynamic cross-coupling they were unable to achieve any significant increases in modal stability.

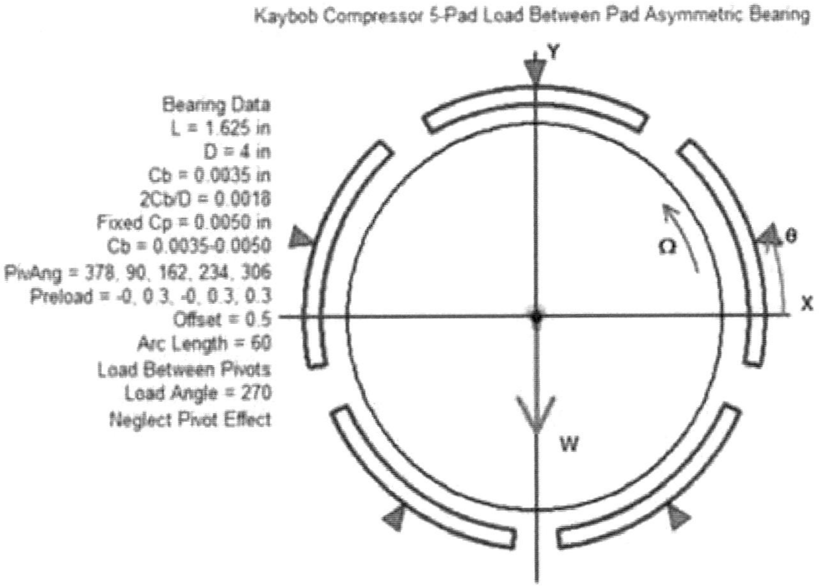

Figure 14: Asymmetric Bearing Model.

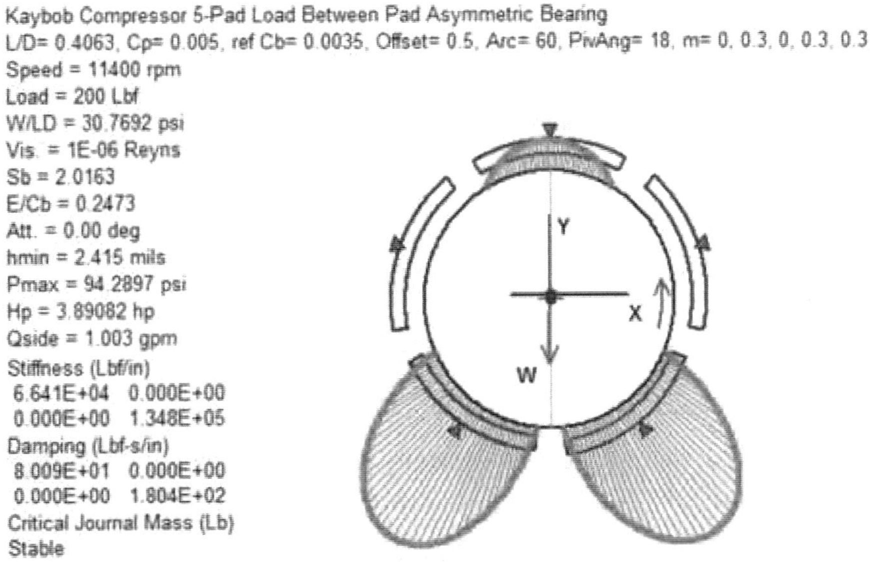

Figure 15: Results for the Asymmetric Bearing.

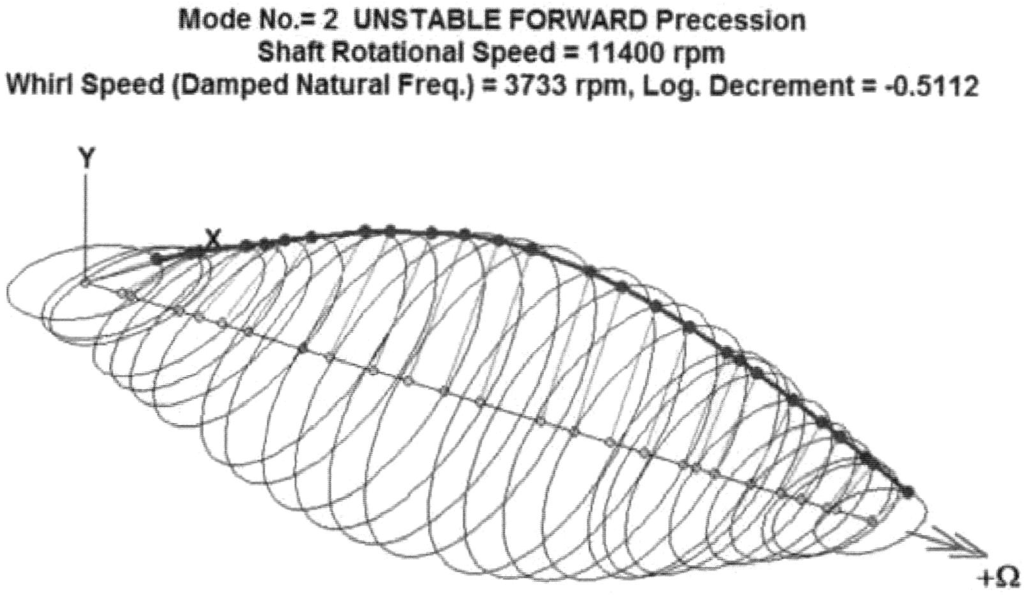

Figure 16: Stability Results for the Asymmetric Bearing.

Figure 16 represents the damped eigenvalue analysis of the nine stage Kaybob compressor with the aerodynamic cross coupling included at stations 13 and 14. It has also been shown that in addition to turbines and compressors developing aerodynamic cross coupling, a center span balance piston can also create aerodynamic cross coupling effects. Therefore small values of aerodynamic cross coupling of 25,000 lb/in were placed at the center stations 13 and 14 at the balance piston location.

It is seen that at the operating speed of 11,400 rpm, there now occurs a self excited whirl component at 3,733 cpm. The log decrement shows a value of -0.51 which represents a highly unstable system. Non-contact probes are often placed at the bearings to observe the rotor motion. The motion observed at the bearing locations is often quite small as compared to the large orbital motion occurring at the rotor center due to the added whirl component. This large motion occurring at the center causes extensive rubbing and damage to seals.

4-Pad Bearing Design: A design not considered by the team at Cooper-Bessemer was a 4-pad tilting pad bearing. These bearings, particularly when oriented in a load-between-pad position, are known to be beneficial from a rotordynamic standpoint due to their ability to more evenly distribute stiffness forces in the vertical and horizontal directions. To see how this design might impact the Kaybob compressor, an initial 4-pad load-between-pad bearing design was analyzed with other design parameters that were kept constant from the original 5-pad bearing including the axial length, bearing clearance, and preload as shown in Figure 17.

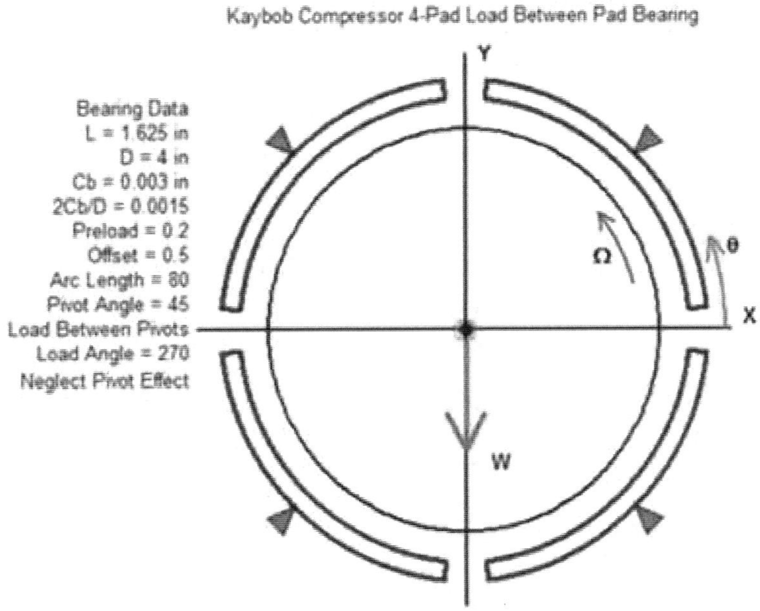

Figure 17: Initial Four Pad Bearing Design.

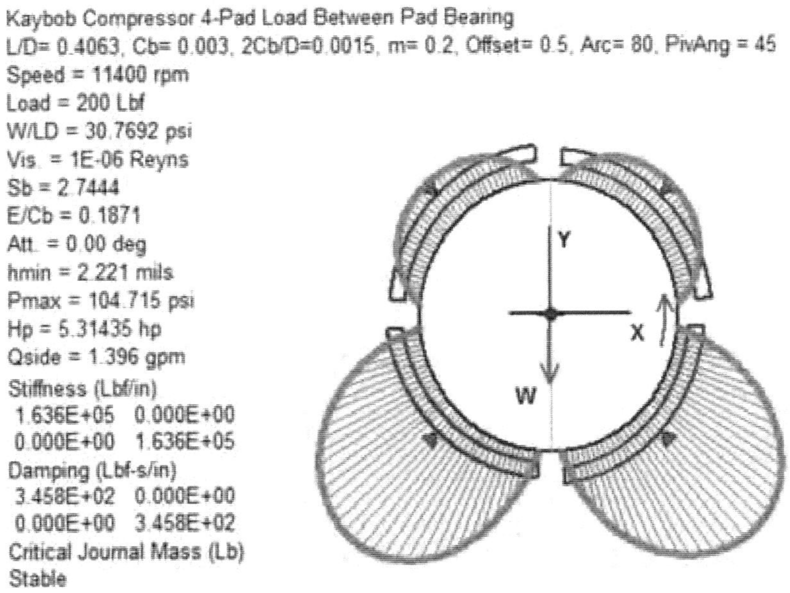

Figure 18: Results of the Four Pad Bearing Analysis.

The results of the bearing analysis (Figure 18) show that the 4-pad design does produce equal values of vertical and horizontal stiffness. When compared to the original 5-pad bearing, the vertical stiffness is slightly lower along with an increase in direct damping. However, in general this design is still very stiff for this machine.

A stability analysis (Figure 19) reveals a reduction in the log decrement of the first bending mode when compared to the original design, likely due to increases in damping and horizontal stiffness preventing the bearings from absorbing the shaft vibrational energy. These results indicate that more drastic changes to the bearing design are necessary in order to restore stability to the system.

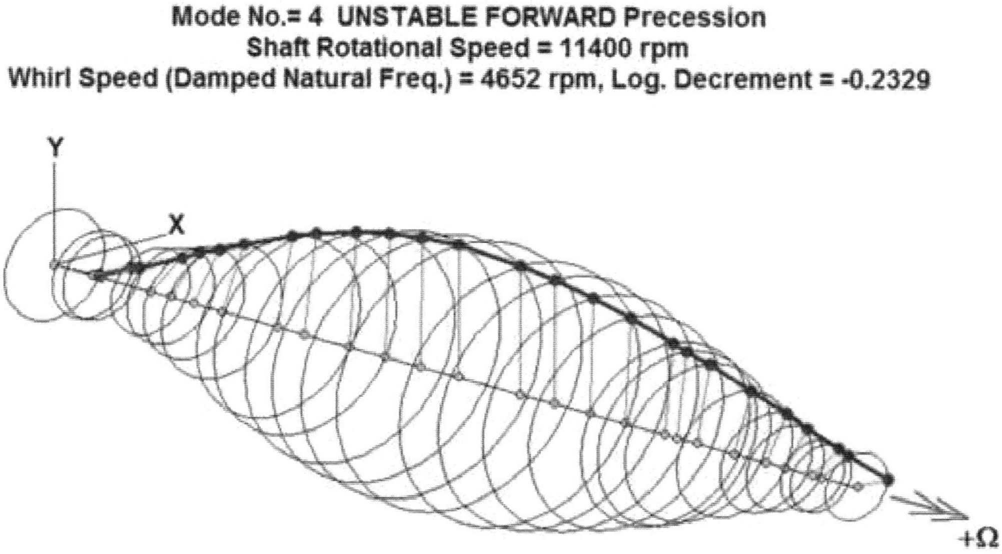

Figure 19: Stability Results for a Four Pad Bearing Design.

The design for the 4-pad bearing is superior to the asymmetric 5-pad bearing in that the log decrement has improved. However the dynamic analysis of the compressor with the 4-pad design as shown above is still unstable. The system has been improved over the asymmetric 5-pad design with a reduction of the log decrement from -0.51 for the 5-pad case as compared to -0.233 with the 4-pad design. In order to improve the rotor dynamics with the 4-pad bearings, it will be necessary to further reduce the vertical bearing stiffness in order to bring the values more in line with the fundamental shaft stiffness value. This requires a longer pad with a reduced bearing preload.

The paradox of this type of design is that the noncontact probes monitoring the bearing motion indicate higher amplitudes of motion, although the shaft center motion has been reduced. This phenomenon led one chief engineer of a compressor company to remove the redesigned 4-pad bearings and replace them with the original 5-pad design because of the increased motion observed at the bearings. Total rotor failure occurred shortly after the 5-pad bearings were installed!

Improved 4-Pad Bearing: In an effort to determine what bearing design choices were necessary in order to restore stability to Kaybob compressor, a number of modifications were made to the previously analyzed 4-pad bearing. Because it has been revealed that all of the designs considered thus far have produced bearings that are overly stiff, a number of design parameters must be considered to reduce the stiffness of the bearing while maintaining damping so that the bearings can better absorb the energy of the first bending mode. Design parameters including the bearing clearance, pad preload, load orientation, and lubricant properties have all been shown by many authors [9-11] to affect the bearing stiffness. Therefore, for this design, it was decided to increase the bearing clearance from 3 mils radial to 6 mils and reduce the pad preload from 0.2 to 0.1 while maintaining the load-between-pad orientation to reduce the stiffness of the bearing. In order to maintain adequate damping the bearing axial length was also increased from 1.625 inches to 4 inches, resulting in an increase in the L/D ratio from 0.4 to 1.

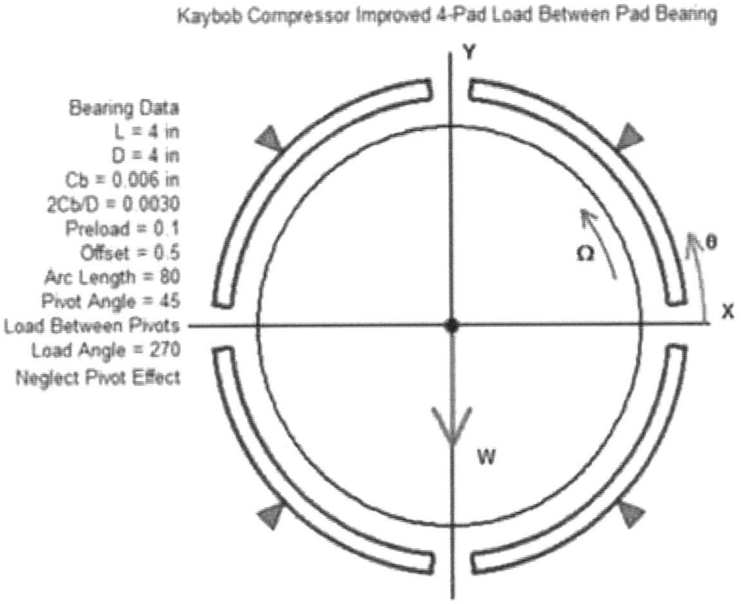

Figure 20: Improved Four Pad Bearing Design.

The results from the bearing analysis show a significant change in bearing performance when compared to the previous designs (Figure 21). The bearing stiffness has now been reduced to 48,960 lb/in. When comparing twice the bearing stiffness to the shaft modal stiffness this now results in a ratio of 0.85, much closer to the desired value of unity. The increase in bearing axial length also results in reasonable levels of damping. A stability analysis of this system shows a drastic increase in the log decrement of the first bending mode (Figure 22), resulting in a fully stable mode for the compressor. An energy distribution plot (Figure 23) shows how the improved bearing design does a much better job of allowing the bearings to absorb the vibrational energy, resulting in an increase of 32% of the energy being transmitted to the bearings when compared to the original 5-pad design.

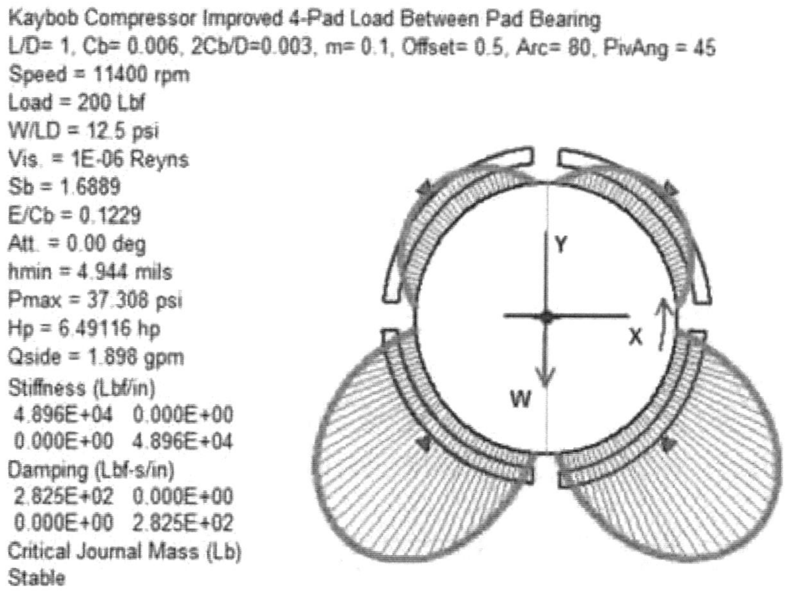

Figure 21: Analysis of the Improved Four Pad Bearing.

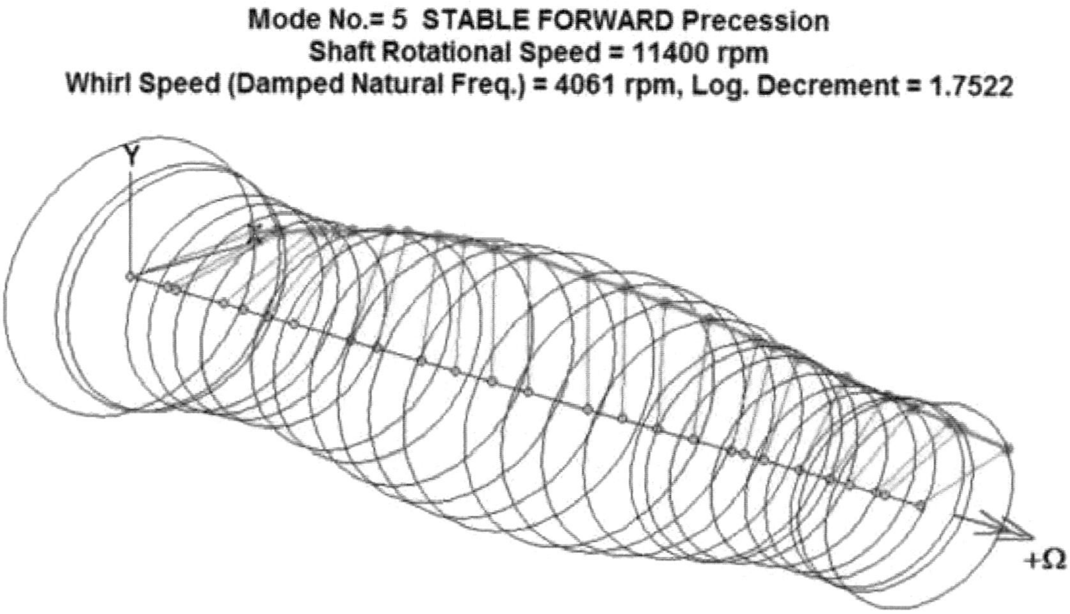

Figure 22: Stability Analysis of the Improved Four Pad Bearing.

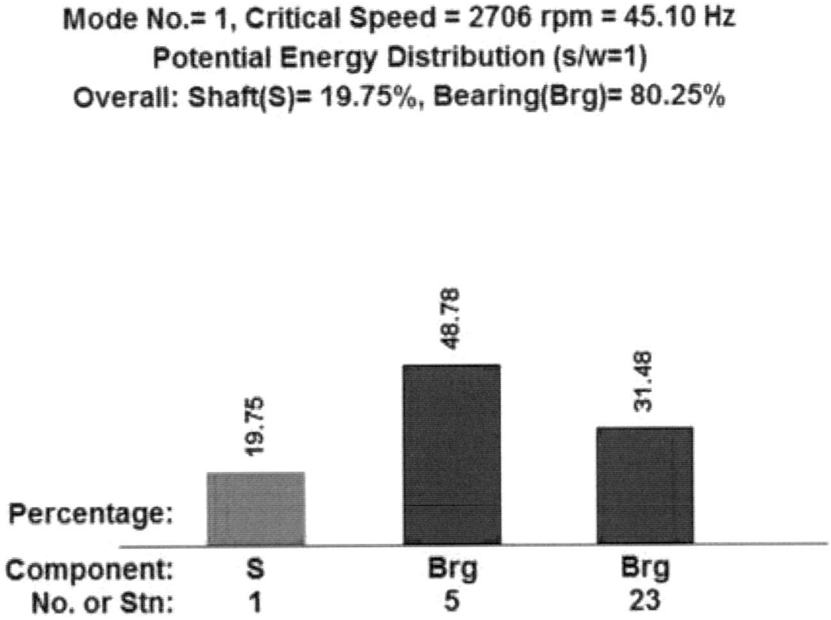

Figure 23: Energy Distribution in the Shaft and Bearings for the Improved Four Pad Bearing Design.

Squeeze Film Dampers: Another solution considered by the team at Cooper-Bessemer was a squeeze film damper. Smith [3] describes the overall design as a long damper surrounding a number of different bearing designs that were considered including a unique design that involved six tilting pads with asymmetries in the pad clearance, axial length, and arc length. However, none of these designs produced an ultimately stable solution for the machine and the use of a long squeeze film damper was likely a key factor in this.

To further make this point, the below analysis was performed on two damper designs surrounding the original 5-pad bearing (Figure 24): a long damper design and a damper with a central groove of 0.15 inches in axial length. The non-grooved damper is treated as a long bearing, but because an added groove to the damper results in a different pressure distribution within the damper the grooved damper is modeled as a short bearing.

With all other inputs equal it can be seen that the groove has a significant effect on the predicted stiffness and damping of the damper. The longer damper design - similar to the one used by Cooper-Bessemer - produces very high levels of stiffness and damping while the grooved design produces much more reasonable values.

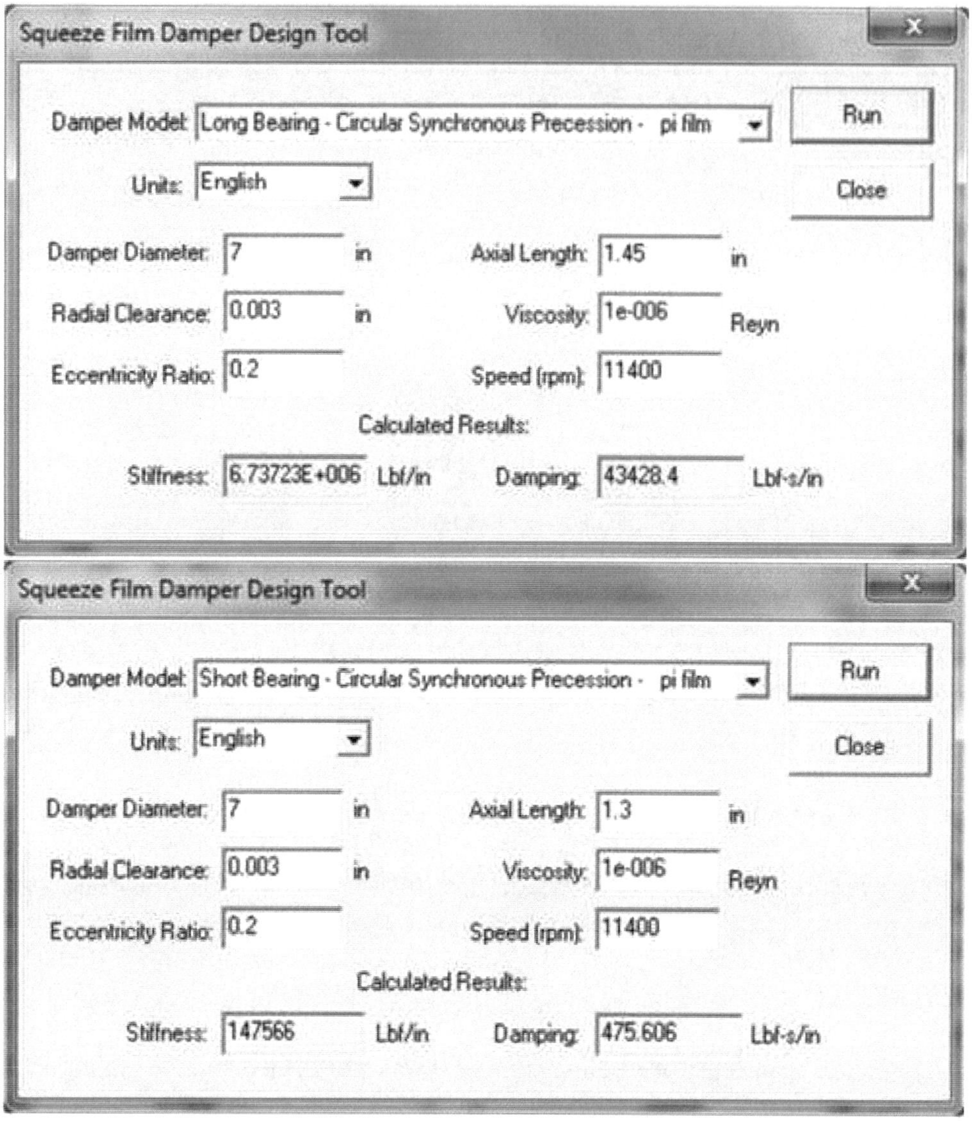

Figure 24: Non-Grooved and Grooved Squeeze Film Damper Analysis.

Stability analyses of these two support scenarios also reveal the significant effect of the central groove (Figures 25 and 26). It is shown that the non-grooved damper produces only a very slight increase in the log decrement of the unstable mode when compared to the original bearing design. The grooved damper, however, provides a substantial increase in the log decrement, resulting in a marginally stable system. Had the team at Cooper-Bessemer considered adding a central groove to their squeeze film damper design they may have actually avoided the costly rotor redesign discussed in the next section.

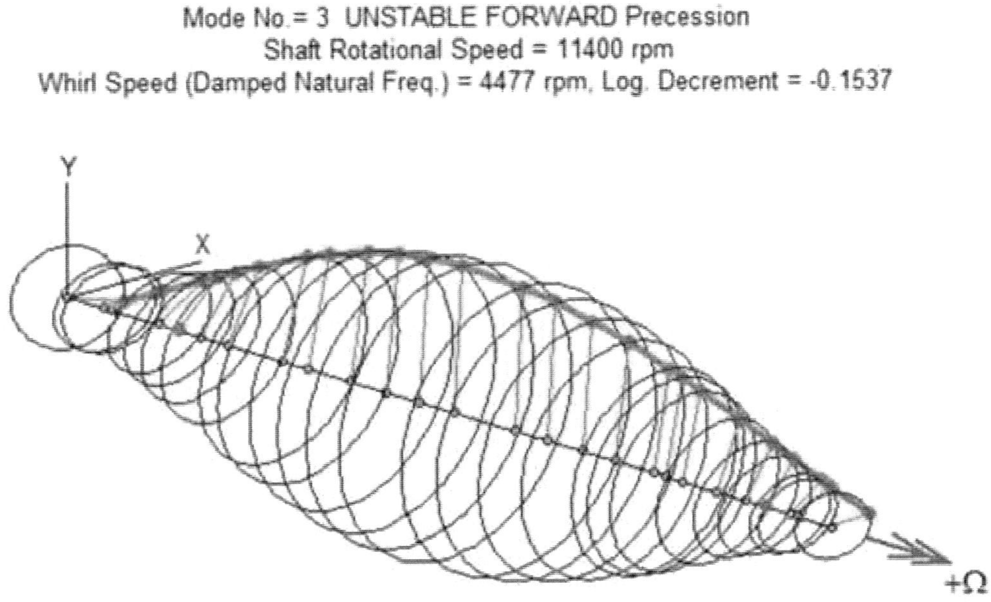

Figure 25: Stability Analysis of the Non-Grooved Squeeze Film Damper Supported System.

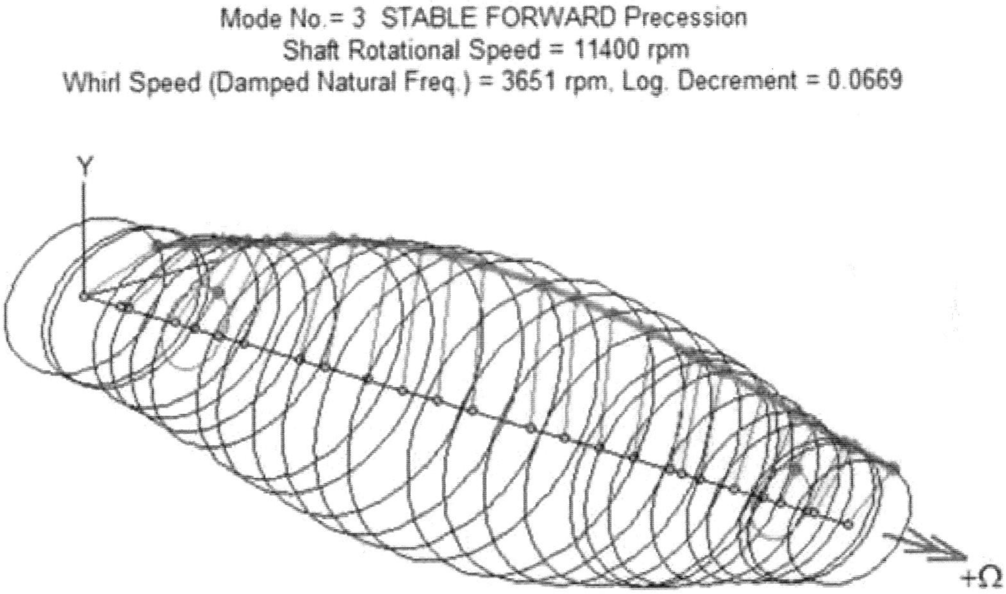

Figure 26: Stability Analysis of the Grooved Squeeze Film Damper Supported System.

Rotor Redesign: To produce a stable operating environment the team at Cooper-Bessemer ultimately had to redesign the rotor. This involved multiple design iterations that resulted in a final shaft design with a bearing span reduced from about 60 inches to approximately 53 inches, modified shrink fits, an integral center seal section, and an increased shaft diameter under the impellers (Figure 4). This design, along with either a squeeze film damper or a 5-pad tilting pad bearing with no preload, was enough to restore the full operational capabilities of the compressor.

To show how these shaft design modifications made such a difference in the performance of the machine, a shaft model was developed to analyze with the original bearing design and the original design with a long squeeze film damper (Figure 27). A critical speed analysis assuming rigid bearings with a stiffness of 10^8 lb/in predicted shaft stresses at the first bending mode of 2,371 psi (Figure 28), an increase of 39% when compared to the original rotor. The shaft modal stiffness was calculated to be 201,780 lb/in, a 74% increase when compared to the original rotor. This significant increase in shaft modal stiffness made the new rotor design much more suited to the high-stiffness bearings utilized.

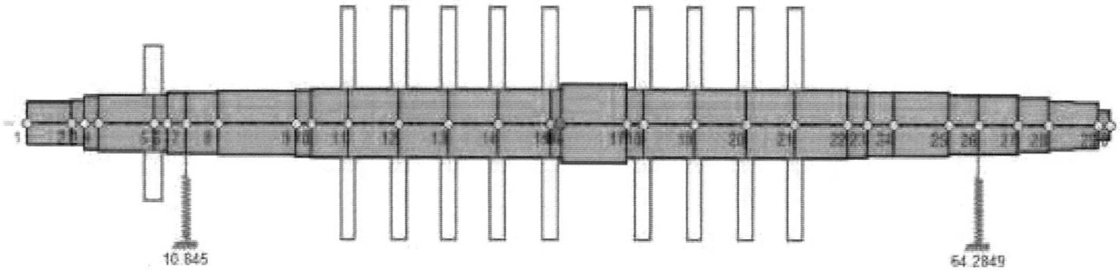

Figure 27: Reduced Bearing Span Shaft Model.

A stability analysis of the original bearing design with the new shaft design (Figure 29) demonstrates how the new rotor accommodates the stiffer bearing design, resulting in a stable first bending mode. Note that the ratio of twice the bearing stiffness to the shaft modal stiffness in this case is 1.7, a significant improvement from the original design.

Since the long squeeze film damper was also considered along with the new rotor design, this case was analyzed for stability as well (Figure 30). Similar to before, however, the addition of the damper without a central groove proves to have little effect on the stability of the system.

Conclusions: The Kaybob compressor failure of 1971 was an historic case of rotordynamic instability and has taught us many things about the design of these machines. Tilting pad journal bearings have become a new normal in compressor stabilization due to their innate capacity for avoiding self-excited oil whirl, however great care must be taken in the design of these bearings for both ensuring a stable machine and meeting API specifications. Many have been tempted to try and avoid critical speeds and reduce motion at the bearings by increasing bearing stiffness, however this approach can and has often created a stability disaster as demonstrated by the Kaybob case. By designing for stability, however, these modes can also be eliminated from the operating range to the extent that they are not even detected because they are so well-damped.

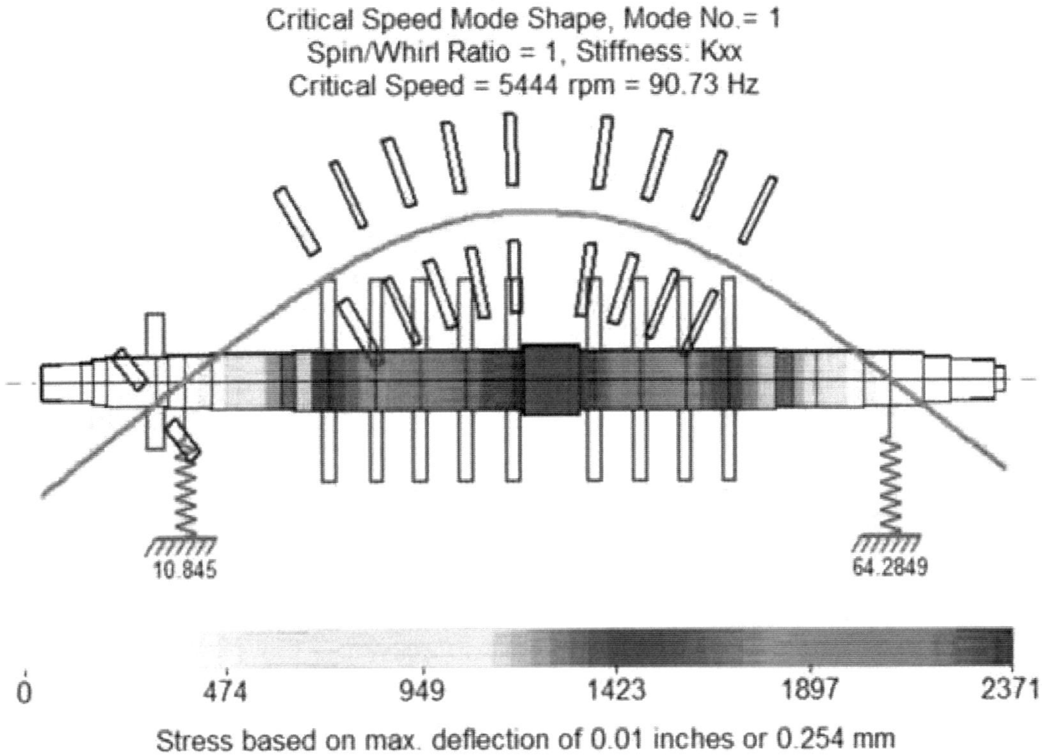

Figure 28: Shaft Modal Stresses of the First Bending Mode.

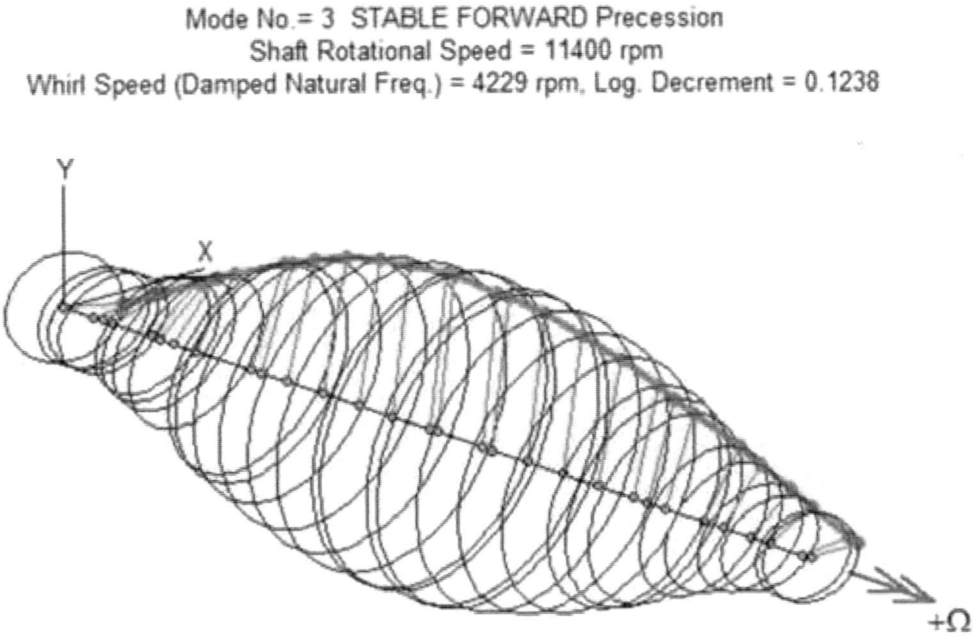

Figure 29: Stability Analysis of the New Rotor Design with the Original Five Pad LOP Bearings.

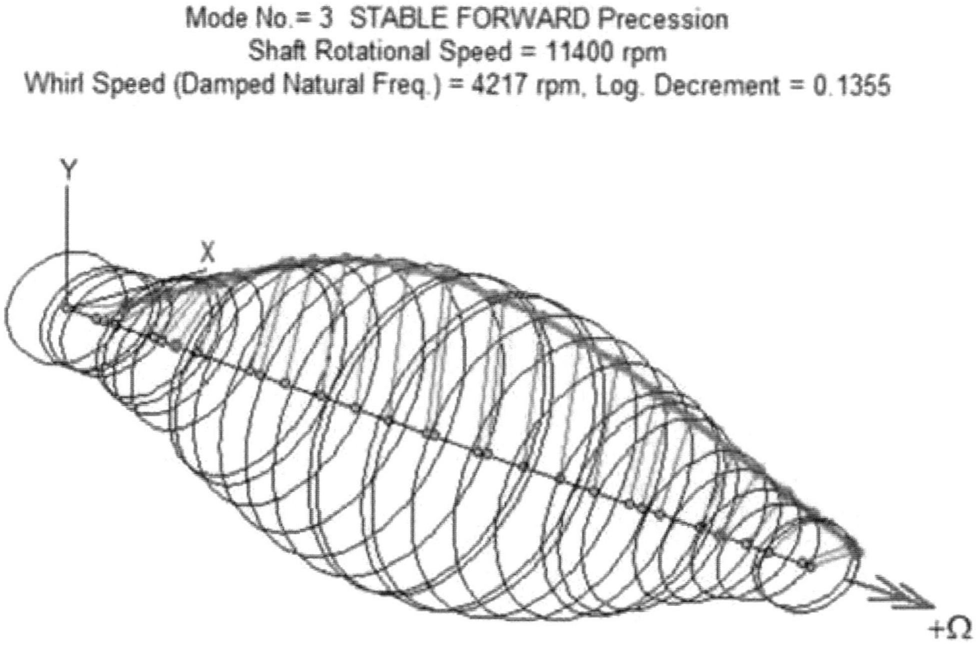

Figure 30: Stability Analysis of the New Rotor Design with the Original Bearing and a Long Squeeze Film Damper.

In particular the Kaybob case has revealed the importance of designing bearings such that the bearing stiffness is tuned in to be in proper proportion to the shaft modal stiffness. By using this simple relationship and designing for proper damping as well, the bearings can properly absorb shaft vibrational energy and produce highly stable modes. While bearing asymmetry has been shown in past cases to produce some increases in stability - particularly in the presence of internal shaft friction or moderate levels of aero-dynamic cross-coupling - the bearing stiffness must still be considered as a primary design variable as asymmetry will not benefit a system with a large bearing to shaft modal stiffness ratio. The presented design cases for 5- and 4-pad bearings demonstrate the importance of these concepts.

Squeeze-film dampers have also been a significant source of compressor stabilization in decades past and will continue to do so when properly designed. It was shown in the context of the Kaybob failure that important features such as central grooves can make a large difference in the stiffness and damping characteristics of the damper, which can ultimately have as large an effect on compressor stability as the design of the bearing itself. While the redesign of the compressor rotor also produced a stable machine, this solution was far more costly than the simple use of a properly designed bearing or squeeze-film damper. However, the team at Cooper-Bessemer cannot be faulted for these errors as they did not have the luxury of high-speed analysis tools that are available to designers today. Overall, the failure of the Kaybob compressor and their work to investigate the instability has helped create the knowledge and methods necessary to avoid these kinds of failures in the future.

Acknowledgments: This work was supported by the Rotating Machinery and Controls Laboratory and industrial consortium at the University of Virginia and Rodyn Vibration Analysis, Inc.

References:

1. Lund, J. W., 1964, "*Spring and Damping Coefficients for the Tilting Pad Journal Bearing*," ASLE Transactions, Vol. 7, No. 4, pp. 342-352.
2. Lund, J. W., 1974, "*Stability and Damped Critical Speeds of a Flexible Rotor in Fluid Film Bearings*," Journal of Engineering for Industry, Vol. 96, No. 2, pp. 509-517.
3. Smith, K. J., 1975, "*An Operation History of Fractional Frequency Whirl*," Proceedings of the Fourth Turbomachinery Symposium, Texas A&M Univ., pp. 115-125.
4. Alford, J., 1965, " Protecting *Turbomachinery From Self–Excited Rotor Whirl*," Journal of Engineering for Power, pp. 333–334.
5. Barrett, L. E., Gunter, E. J., and Allaire, P. E., 1978, "*Optimum Bearing and Support Damping for Unbalance Response and Stability in Rotating Machinery*," Journal of Engineering for Power, Vol. 100, No. 1, pp. 89-94.
6. Gunter, E. J., 1966, "*Dynamic Stability of Rotor-Bearing Systems*," NASA SP-113.
7. Gunter, E. J. and Trumpler, P. R., 1969, "*The Influence of Internal Friction on the Stability of High Speed Rotors with Anisotropic Supports*," Journal of Engineering for Industry, Vol. 91, No. 4, pp. 1105-1113.
8. Gunter, E. J., 2012, "*The Influence of Bearing Asymmetry on Rotor Stability*," Proceedings of the Vibration Institute Training Conference, June 19-22, Williamsburg, Virginia, pp. 1-28.
9. Nicholas, J. C. and Kirk, R. G., 1979, "*Selection and Design of Tilting Pad and Fixed Lobe Journal Bearings for Optimum Turborotor Dynamics*," Proceedings of the Eighth Turbomachinery Symposium, Texas A&M Univ., College Station, Texas, pp. 43-57.
10. Nicholas, J. C. and Kirk, R. G., 1982, "*Four Pad Tilting Pad Bearing Design and Application for Multistage Axial Compressors*," Journal of Lubrication Technology, Vol. 104, No. 4, pp. 523- 539.
11. Weaver, B. K., Zhang, Y., Clarens, A. F., and Untaroiu, A., 2015, "*Nonlinear Analysis of Rub Impact in a Three-Disk Rotor and Correction via Bearing and Lubricant Adjustment*," Journal of Engineering for Gas Turbines and Power, Vol. 137, No. 9, pp. 092504-1-092504-8.

Bibliography:

1. Allaire, P. E, Nicholas, J. C., and Gunter, E. J., 1977, "*Systems of the Finite Elements for Journal* Bearings" , ASME Journal of Lubrication Technology, pp. 187 – 197.
2. Chen, W. J. and Gunter, E. J., 2007, "*Introduction to Dynamics of Rotor-Bearing Systems*," Trafford Publishing, ISBN: 978-1-4120-5190-3, pp. 1-469.
3. Gunter, E. J., Barrett, L. E. and P. E. Allaire, 1977, "*Design of Nonlinear Squeeze–Film Dampers for Aircraft Engines*", Journal of lubrication technology, pp. 57 – 66.

CALLIBRATION CYBERSECURITY RISKS

James McGlone
Kenexis
3366 Riverside Drive, Suite 200
Upper Arlington, Ohio 43221 USA
Telephone: (614) 643-2473
james.mcglone@kenexis.com

Abstract: Hardly a day goes by anymore without hearing about a cyber security event in the news. In this paper/session, we will analyze the risk associated with smart devices, calibration, and maintenance that could compromise the cyber security status of operations and possibly our entire companies.

As a nuclear power plant operator and electronic technician, we never considered the security risk of having an instrument calibrated by a third party. Today however, things have changed significantly and it is common to find microprocessors in virtually everything including the instruments we are using in our plants. The microprocessor has made significant impacts in performance and communications and consequently might be weakening your organization's security.

Analysis presented will discuss the real risk associated with different types of instrumentation and devices, network topologies, and technologies used in smart devices and systems. Additionally, we will analyze the possible effects on safety systems and how to compensate for them and how to protect your systems.

Key words: Calibration Risks, Calibration Cybersecurity Risks, Malware Insertion

Introduction: It is common to find microprocessors in virtually everything including the instruments we are using in our facilities. The microprocessor has made significant impacts in performance and communications and might be weakening your organization's security. The simple act of sending the instrument to a third party for calibration might be compromising your facility's cybersecurity posture.

The smart instruments include a microprocessor and memory where firmware that runs code to operate the instrument resides. Many of these devices were designed and built without much concern for cybersecurity. If an instrument designed without following strict cybersecurity rules for both the firmware, memory management and communications is calibrated by connecting to a system where malware is present, then the malware could potentially migrate to the instrument. Reconnecting the instrument after calibration might introduce malware into your facility, thereby placing your entire operation at risk.

Smart Instruments: Many instruments are simple devices and lack the ability to store and pass a payload like malware. Malware, unwanted and often malicious code, will require a certain amount of memory to store itself in and then a communications method to get from the instrument into the network. Looking for potential risk can be as easy as looking at the specifications or in the manual for directions like the image below in Figure1.

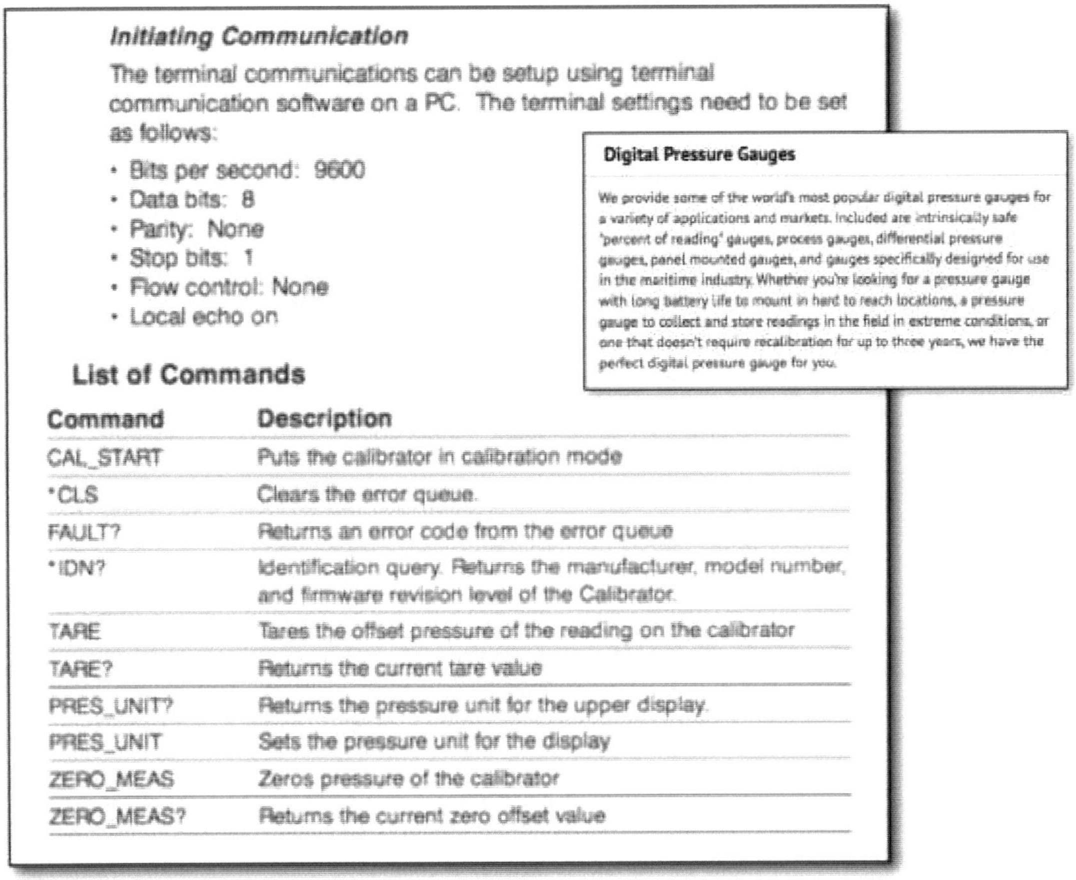

Initiating Communication

The terminal communications can be setup using terminal communication software on a PC. The terminal settings need to be set as follows:

- Bits per second: 9600
- Data bits: 8
- Parity: None
- Stop bits: 1
- Flow control: None
- Local echo on

Digital Pressure Gauges

We provide some of the world's most popular digital pressure gauges for a variety of applications and markets. Included are intrinsically safe 'percent of reading' gauges, process gauges, differential pressure gauges, panel mounted gauges, and gauges specifically designed for use in the maritime industry. Whether you're looking for a pressure gauge with long battery life to mount in hard to reach locations, a pressure gauge to collect and store readings in the field in extreme conditions, or one that doesn't require recalibration for up to three years, we have the perfect digital pressure gauge for you.

List of Commands

Command	Description
CAL_START	Puts the calibrator in calibration mode
*CLS	Clears the error queue.
FAULT?	Returns an error code from the error queue
*IDN?	Identification query. Returns the manufacturer, model number, and firmware revision level of the Calibrator.
TARE	Tares the offset pressure of the reading on the calibrator
TARE?	Returns the current tare value
PRES_UNIT?	Returns the pressure unit for the upper display.
PRES_UNIT	Sets the pressure unit for the display
ZERO_MEAS	Zeros pressure of the calibrator
ZERO_MEAS?	Returns the current zero offset value

Figure 1

Words like digital, smart, wireless, and IoT (Internet of Things) in the description of a device, are clues that their might be a method to accidently or deliberately load a malware payload onto the instrument. If the instrument has a digital communications port or capability (wireless), it is a potential threat.

Background: To our knowledge, malware has not migrated using instruments as of the writing of this paper. However, it is entirely possible for this to occur and it requires an awareness to prevent the malware insertion from occurring.

Malware has been inserted into many facilities to date and has caused significant damage including complete loss of control. Malware easily migrates across digital connections including wired and wireless communications. It would be easy enough for a nation state or sponsored entity to create malware to take advantage of instrumentation to gain access to secure systems. Unfortunately, every threat vector must be considered in today's environment.

We already know malware can be carried in using a USB memory device. We also know from www.KrebsOnSecurity.com distributed denial of service (DDoS) attack, that the malware called "Mirai" that caused to DDoS attack resided in IoT devices like routers, IP cameras, and digital video recorders (DVRs). This type of malware scans connected devices looking for suitable devices to install itself on and spread further. This type of code typically uses a brute force username and password method to access the device.

We discovered with "Mirai" that it took advantage of devices that had weak or published user names and passwords. Most of the infected devices had no way to change the username or password built into their firmware and even if they did, most people do not change them because they are simple devices. An IP camera's username and password seems irrelevant in a world where most people do not have passwords on their cellphones. Unfortunately, this is the same problem we see in many industrial facilities. Many operation personnel and management insist that the same password and username be used for everyone, or they insist that none be used, and in many cases the user name and password is the default so that they can find it in an emergency. Unfortunately, this is exactly the weakness that Mirai utilized and it will make it easier for the next malware to take advantage of your control system.

If unwanted code known as malware finds its way into your instrumentation, some malware types can cause harm by itself like erasing the boot sector on a hard disk drive, but malware commonly communicates with a command and control (CIC) server somewhere in the world plugged into the Internet. The command and control server may just collect intelligence or it might facilitate the insertion of a larger payload (more powerful malware) or allow for an attacker to take remote control. The Ukraine power grid attack was a remote-control man-in-the-middle attack which removed control and visualization from operators followed by destroying the disk drives afterwards. The Saudi Aramco attack called Shamoon (or Disttrack), uploaded files from 35,000 computers before erasing the files and then destroying the disk drives. Both attacks had devastating results. And if you are assuming instrumentation isn't a target because of it small processor and memory foot print, the US Government Aurora Generator Test which destroyed a power generator in 2007, was only 21 lines of code.

Best Practices: Protecting your systems is one more ongoing engineering project. It requires common sense and good procedures designed to prevent the insertion. We will not stop using the microprocessor based devices, so we need to get smarter about how we use them and protect them.

Practical steps include identifying what instruments you have onsite and which ones are digital followed by which ones are calibrated externally. We might need to define the calibration instruments we use in calibration too, especially if they are sent out or connected externally to calibrate or update.

Once each instrument is identified and classified, the threat from each could be assessed also. An instrument inside of a critical process that is carefully isolated and sent out for calibration, should warrant further investigation.

Investigation should include understanding the vendor's methods for insuring the smart instrument is secure and will remain secure. This should include verifying methods used during design and programming and any third-party certification granted. If the instrument or calibration tool is updated through a website, then what effort has the vendor made to insure the website and update does not become corrupted.

The IEEE Standard for Intelligent Electronic Devices Cyber Security Capabilities should be followed for the design of the embedded electronic devices that are utilized in critical infrastructure. If you consider your process critical, even if the government does not, then you might want to consider using devices that meet this criterion also.

Likewise, the ISA100 Wireless Compliant registration is a very good list of vendor products that are taking wireless communications seriously.

Unfortunately, none of these guarantee that your instruments and connected systems are safe from malware. Good practice would dictate we use caution in addition to using instruments from trusted and verified sources. Caution should include verifying the testing facility and the procedures used both internally and externally.

Conclusion: Awareness of cybersecurity is growing. As more and more vendors bring products out that have better security, we will not be able to relax our own security efforts because the threat will also evolve. Thinking logically about devices, especially devices that have a communications port other than a 4-20ma or 0-10v signal, will need to be treated as potentially a threat.

A practical reality is that a sophisticated instrument with massive memory and computing power that doesn't connect to a communication network is not really a risk the systems. But a simple instrument with a microprocessor connected to the system might bring the entire process down.

COMPARISON BETWEEN TWO VERY EFFICIENT SIGNAL PROCESSING APPROACHES FOR VIBRATION-BASED CONDITION MONITORING OF ROLLING ELEMENT BEARINGS: THE MED-SK AND CS-BASED APPROACHES

Dany Abboud[(1)] and Mohammed Elbadaoui[(1)(2)]
[(1)]SafranTech
Rue des Jeunes Bois
Châteaufort 78772 Magny –les- Hameaux, France
Telephone: (+33) 6 95222573
dany.abboud@safrangroup.com

[(2)]Univ Lyon
UJM-St-Etienne, LASPI
EA3059, F-42023, Saint-Etienne, France

Abstract: This paper compares the most efficient signal processing approaches used for vibration-based condition monitoring of rolling element bearings. The first is based on pre-processing the vibration signal through the maximum entropy deconvolution method (MED) followed by the spectral kurtosis (SK), before analyzing the spectrum of the signal envelope. The MED aims at maximizing the signal impulsivity by deconvolving the system transfer function through an optimization approach that maximizes the kurtosis of the output. Then, the spectral kurtosis (SK) is applied to conceive the optimal filter to be applied before computing the envelope spectrum. The second approach is based on a cyclostationary modeling of the bearing signal. It applies the spectral coherence to the signal with a special attention on setting the estimation parameters. The spectral coherence is a bi-variable map of the cyclic frequency, α, and the spectral frequency, f. The former variable describes the cyclic content of the modulations, while the former describes the properties of the carrier. The improved envelope spectrum is then computed by simply projecting its squared-magnitude with respect to the f-variable. These methods are evaluated according to their potentiality to detect the fault in its earliest stage. The comparison is be made on real bearing vibration signals in run-to-failure tests.

Key words: Condition monitoring; spectral coherence; cyclostationarity; improved envelope spectrum; incipient fault detection; minimum entropy deconvolution; spectral kurtosis

1. Introduction: Rolling element bearings (REBs) are perhaps the most widely used elements in rotors. Because of their fragility, bearings are likely to be exposed to sudden failures causing system outage. Therefore, special care growths in last decades to develop appropriate diagnosis technique for incipient fault detection. Due to their non-invasive nature and their high reactivity to incipient faults, the development of vibration based techniques spiked the interest of the scientific committee. In this context, envelope analysis has been recognized for long as a powerful technique for REB diagnosis. Typically, it consists of a bandpass filtering step in a frequency band wherein the impulsive excitation is amplified followed by a demodulation that extracts the signal envelope. The spectrum of the envelope— known as the envelope spectrum— is expected to contain the desired diagnostic information, including the repetition rate of the fault and potential modulations [1]. It has been shown in Ref. [2] that it is preferable to use the squared envelope instead of the envelope inasmuch as the latter introduces extraneous components that appears as misleading peaks in the envelope spectrum. Since that time, the squared envelope spectrum (SES) has probably become the benchmark technique for bearing diagnostics and efforts were directed on how to choose the most suitable band for demodulation. Historically, this band was estimated through a hammer tap testing by finding the bearing housing resonances. Today, this problem is almost solved by the use of the spectral kurtosis (SK) [3]. Being based on a statistical approach, the SK provides a way to determine which frequency bands are informative according to their impulsivity. It can be actually seen as the kurtosis of the signal bandpass filtered with a filterbank covering its frequency content. However, the SK has proven ineffective in high-speed applications. The reason is that the fault period is too small that the impulse responses triggered by the chocks overlap. This significantly reduces the signal impulsivity and jeopardize the effectiveness of the SK. Reference [1] has proposed to preprocess the signal before applying the SK by means of the "minimum entropy deconvolution" (MED). This technique is based on finding the optimal inverse filter which compensates the transfer function of the transmission path, with the aim of separating the excitation impulses. The combination of the MED and the SK has proven very effective in incipient fault detection in REBs and has become a leading procedure in this framework. Another interesting approach for REB diagnosis is based on the cyclostationary (CS) theory. In details, a new model of REB vibrations has been introduced in Refs. [4], providing an insightful understanding of the REB fault signature within the CS framework. Accordingly, it has been shown that the mechanical signature generated by a faulty REB is random in nature and has symptomatic properties that can be detected by means of second-order CS tools such as the spectral correlation, the spectral coherence, the cyclic modulation spectrum and others [5]. These techniques have also proven efficient and their use is in continuous growth. This paper comes in this context aiming at comparing these two efficient approaches on real REB vibration signals in run-to-failure tests. Section 2 recalls the structure of a vibration signal emitted by an incipient fault in a REB. Section 3 and 4 briefly review the MED-SK-based and CS-based approaches, respectively. Section 5 compares these methods on real vibration signals and discuss the results and the paper is sealed with a conclusion in section 6.

2. Vibration-based diagnosis of rolling element bearings: As a defect strikes a mating surface, an abrupt change in the contact stress occurs at the interface (called shock) which corresponds to a transfer of kinetic energy into elastic potential energy. These shocks occur on a pseudo-periodic basis due to the rotating kinematics of the bearing and exciting the system resonance. In facts, the generated vibration can be seen as a train of short-time impulses, periodic on average (with a fault period $T_d = 1/f_d$), locked to the period of the fault characteristic frequency. The inter-arrival times of the impacts are subjected to slight random fluctuations, δT, due to existing slips. Also, this pseudo-periodic train is likely to be modulated by slower modulation (with a period $T_m = 1/f_m$) due to the change of the load in the impact load. These modulations can either be caused by a possible shaft unbalance or misalignment, or a periodic change in the impulse response as the distance and orientation of the impacts moves towards and backwards the sensor. Figure 1 provides an academic illustration of a typical vibration signal generated by a faulty REB.

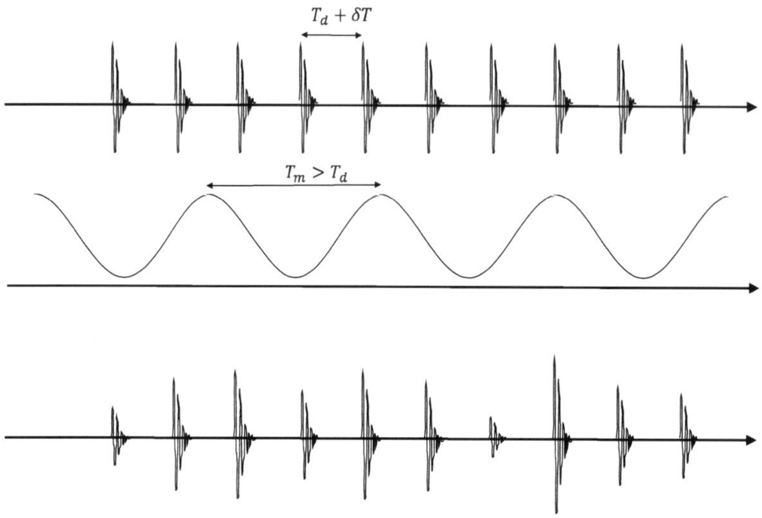

Figure 1: An illustration of the vibration signal emitted by an incipient fault in a REB.

Despite the presence of this periodicity, the power spectrum is continuous and, thus, unable to reveal the mechanical signature of the defect. This is a direct result of the randomness induced by the jitter between the impacts which, despite being in the order of few percent of the fault period, its presence change the whole statistical structure of the signal. The squared magnitude of the envelope, however, is indeed quasi-periodic. Therefore, its spectrum, called the squared envelope spectrum (SES) is discrete and do reveal the fault signature. In details, the fault signature appears as discrete harmonics located at the fault frequency f_d and its multiples, and modulated by the load change modulation frequency f_m and its multiples. This is clearly illustrated in Figure 2. These facts explain why the envelope analysis has been historically used for vibration-based fault detection in REBs.

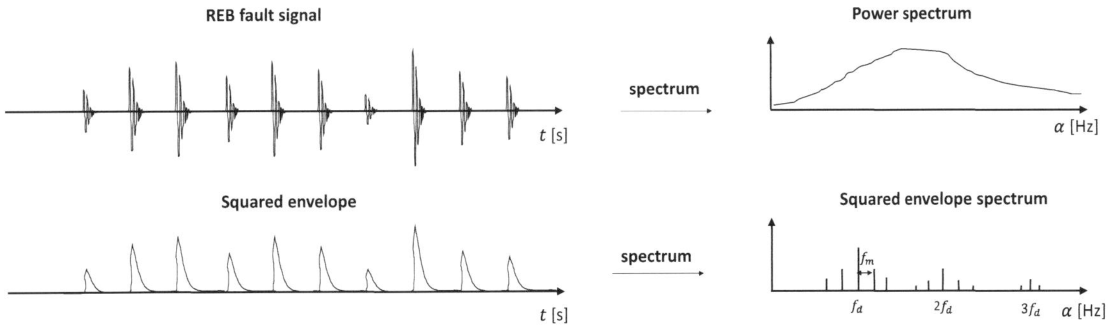

Figure 2: Spectral properties of REB signals.

In practice, the fault component is often masked by multiple contributions generated by multiple sources. These sources can be classified in two major groups. First, the deterministic contribution, being characterized by a discrete spectrum, can contribute in the envelope spectrum and mask the REB fault component. These contributions are typically generated by gears and are generally eliminated by numerous signal processing techniques such as the synchronous average, blind filters, Fourier calculation and others. The elimination of the deterministic component is out of the scope of the current paper. Even after removal of discrete frequency ''noise'', the bearing signal will often be masked in many frequency bands by other random noise. This effect leads to a reduction in the signal impulsiveness as the fault pulses are modified by the passage through a transmission path with a long impulse response. This is typically the case of high-speed application in which the fault period are too small that the impulse responses triggered by the chocks overlap. This jeopardizes the effectiveness of the classical envelope analysis and calls for different approaches to promote the fault signature when embedded in strong noise. The enhancement of the REB fault component and signature can be made through two very efficient approaches discussed in the following sections.

3. MED-SK-based approach: The first efficient approach is based on pre-processing the vibration signal by combining two efficient methods, namely the "minimum entropy deconvolution" (MED) followed by "the spectral kurtosis" (SK). The MED was first proposed in Ref. [6] to sharpen the reflections from different subterranean layers in seismic analysis. This technique is based on finding the optimal inverse filter which compensates the transfer function of the transmission path, assuming an impulsive excitation. This process is illustrated in Figure 3. The impulsive REB fault signal, $e[n]$, is likely to be corrupted by background noise, $b[n]$, before being convolved with the transfer function of the transmission path from the excitation source to the transducer. The obtained signal is denoted by $x[n]$. The MED aims at estimating the filter, $b[n]$, that inverts the system impulse response function, $g[n]$, such that

$$(h \otimes g)[n] = \delta[n - l_m] \tag{1}$$

where $\otimes$ denotes the numerical convolution, $\delta[n-l_m]$ is the Kronecker delta function. Note that l_m is allowed to make the inverse filter causal. The filter $g[n]$ is modeled as a FIR filter with L coefficients such that:

$$y[n]=\sum_{l=1}^{L} g[l]x[n-l]. \tag{2}$$

In vibration analysis, the only method used for filter coefficient estimation is the objective function method provided in Ref. [7], where the objective function to be maximized is the kurtosis of the output signal $y[n]$:

$$O_k(g)=\frac{\sum_{n=0}^{N-1} y[n]^4}{\left(\sum_{n=0}^{N-1} y[n]^2\right)^2}. \tag{3}$$

The optimal filter is that whose coefficients maximize $O_k(g)$ or, said differently, set to zero its first derivative, i.e.

$$\frac{\partial O_k(g)}{\partial g}=0. \tag{4}$$

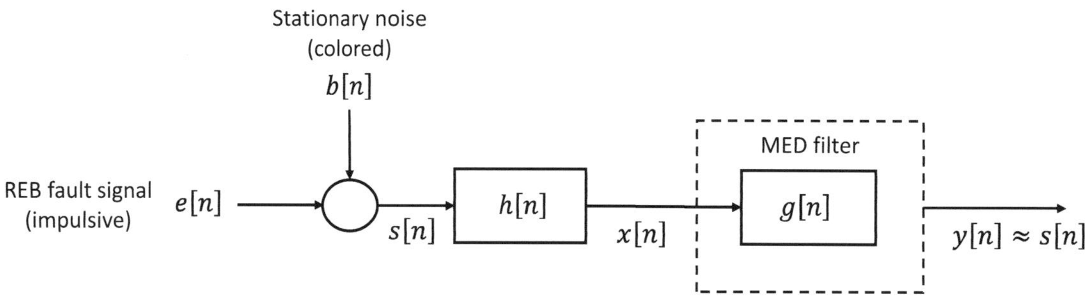

Figure 3: Principle of the MED technique.

The MED has for object to enhance the signal impulsivity, yet there is still a need to enhance the REB signal with respect to residual background noise before analyzing the SES. In other words, there is a need to define a filter which maximizes the signal-to-noise ratio and, consequently, enhances the fault signature in the SES. As previously indicated, the SK is a very good candidate to accomplish this task. Let's denote by $Y(n;f)$ the short-time Fourier transform of $y[n]$, the SK is defined as:

$$K_y(f) = \frac{\left\langle |Y(n;f)|^4 \right\rangle}{\left\langle |Y(n;f)|^2 \right\rangle^2} - 2. \tag{5}$$

with $\langle \bullet \rangle$ stands for the average operator over the time index n. Note that the subtraction of 2 is used to enforce a zero value for a complex Gaussian process. Since the SK returns high values in the frequency bands related to an impulsive REB fault, it can be used to conceive a filter which maximizes the signal impulsivity. Since the aim is to demodulate the signal to apply the SES, the matched filter is perhaps the most appropriate as maximizes the SNR of the filtered signal without regard to its shape. Let's denote by $g'[n]$ the optimized matched filter based on the SK, the filtered signal can be simply expressed as:

$$z[n] = \sum_{l=1}^{L} g'[l] y[n-l] \tag{6}$$

Eventually, the SES is applied on $z[n]$ to reveal the hidden cyclic content:

$$SES_z(\alpha) = DTFT_{n\to\alpha}\left\{ z[n]^2 \right\}. \tag{7}$$

For a faulty REB, $SES_z(\alpha)$ is expected to exhibit harmonics at the fault frequency f_d with sidebands spaced by the load change modulation frequency f_m:

$$SES_z(\alpha) = \sum_{i,j} SES_z^{i,j} \delta(\alpha - i f_d + j f_m). \tag{8}$$

The SK-based filtering operation followed by the SES is illustrated in Figure 4.

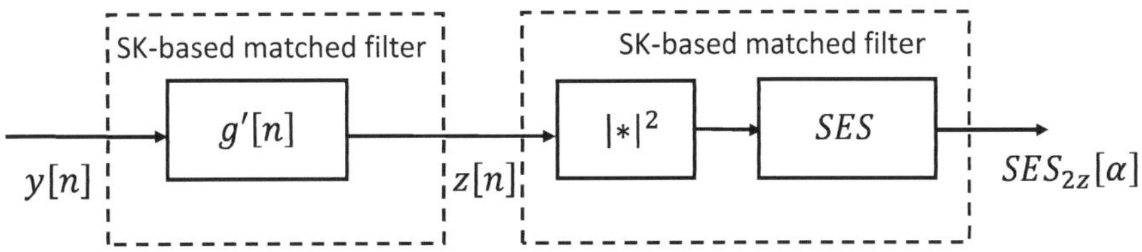

Figure 4: SK-based filtering and envelope analysis (continuation of Figure 3).

4. CS-based approach: As asserted in precursory works [3,4] vibrations produced by faulty bearings are random CS (i.e. second-order CS). This is principally due to the presence of slight random fluctuations in inter-arrival times of the impacts caused by the rolling element slippage. The properties of such a signal are fully preserved in autocorrelation function, defined for a stochastic signal $x[n]$ as:

$$R_{2x}[m,n] = E\{x[n+m/2]x[n-m/2]^*\}. \tag{9}$$

where m and n respectively denotes the time and time-lag indexes. The same information can be more prominently represented through the spectral correlation, defined as being the double Fourier transforms of the autocorrelation function;

$$S_{2x}(f,\alpha) = \lim_{N\to\infty} \frac{\Delta_t}{N} \sum_{n=0}^{N-1} \sum_{m=0}^{N-n} R_{2x}[m,n] e^{j2\pi\Delta_t(mf+n\alpha)}. \tag{10}$$

where Δ_t stands for the sampling period, N is the signal length, f is the spectral frequency variable in Hz, α is the cyclic frequency variable in Hz. Alternatively, the spectral correlation can be expressed as:

$$S_{2x}(f,\alpha) = \lim_{N\to\infty} \frac{1}{N\Delta_t} E\{X_N[f-\alpha/2]X_N[f+\alpha/2]^*\}. \tag{11}$$

where $X_N[f] = \Delta_t \sum_{n=0}^{N-1} x[n] e^{j2\pi fn\Delta_t}$ is the signal discrete time Fourier transform over N samples. This expression provides another interpretation of the SC, which actually justifies its name: it is a measure of the correlation between the frequency components of the signal at $f-\alpha/2$ and $f+\alpha/2$. A related operator is the spectral coherence (SCoh) which has the property of measuring the degree of correlation between two spectral components independently of the signal power spectrum. This quantity is defined as being the coherence (instead of the correlation) between the frequency components located at f and spaced by α :

$$\gamma_{2x}(f,\alpha) = \lim_{N\to\infty} \frac{E\{X_N[f-\alpha/2]X_N[f+\alpha/2]^*\}}{\left(E\{|X_N[f-\alpha/2]|^2\}E\{|X_N[f+\alpha/2]|^2\}\right)^{1/2}}. \tag{12}$$

Just like the regular correlation coefficient, the squared magnitude of the SCoh, $|\gamma_{2x}(f,\alpha)|^2$, is normalized between 0 and 1: the closer it is to the unity, the stronger the CS component at cyclic frequency α. Interestingly, when a CS signal of cyclic frequency α is embedded in stationary noise, its $|\gamma_{2x}(f,\alpha)|^2$ can be seen as an indication of the signal-to-noise ratio. Often, vector indicators are preferred in industry as their readability is generally simple and does not require advanced skills in signal processing. Accordingly, one may sacrifice the optimality of the SCoh by integrating over the spectral frequency variable preferably in a high SNR frequency band (or simply over the whole frequencies). Precisely, one may devise a suboptimal quantity, henceforth called the "improved envelope spectrum" (IES)—as:

$$IES_{2x}(\alpha) = \frac{1}{F_2 - F_1} \int_{F_1}^{F_2} |\gamma_{2x}(f, \alpha)|^2 df, \tag{13}$$

where F_1 and F_2 are respectively the lower and upper bounds of the integration frequency domain. For a faulty REB, $IES_{2x}(\alpha)$ is expected to exhibit harmonics at the fault frequency f_d with sidebands spaced by the load change modulation frequency f_m:

$$IES_{2x}(\alpha) = \sum_{i,j} IES_{2x}^{i,j} \delta(\alpha - if_d + jf_m). \tag{14}$$

This approach is illustrated in Figure 5.

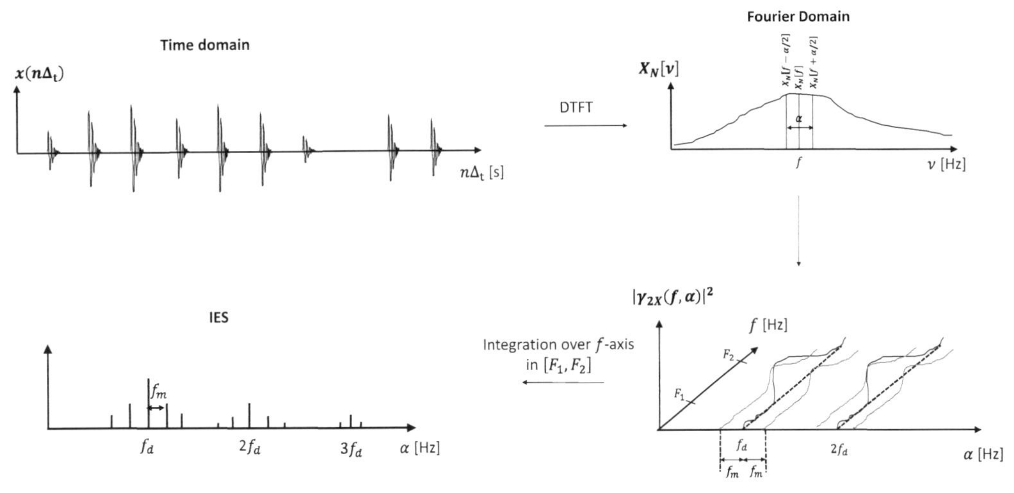

Figure 5: Illustration of the cyclostationary-based approach for REB fault detection.

5. Application: The approaches discussed above are tested on real vibration signal emitted from a benchmark including faulty REB. The data are generated by the NSF I/UCR Center for Intelligent Maintenance Systems (IMS –www.imscenter.net) with support from Rexnord Corp. in Milwaukee, W [8]. The test rig consists of four bearings installed on a shaft. The rotation speed was kept constant at 2000 RPM by an AC motor coupled to the shaft via rub belts. A radial load of 6000 lbs is applied onto the shaft and bearing by a spring mechanism. All bearings are force lubricated. Rexnord ZA-2115 double row bearings were installed on the shaft as shown in Figure 6. PCB 353B33 High Sensitivity Quartz ICP accelerometers were installed on the bearing housing. All failures occurred after exceeding designed life time of the bearing which is more than 100 million revolutions. More details can be found in Ref. [9].

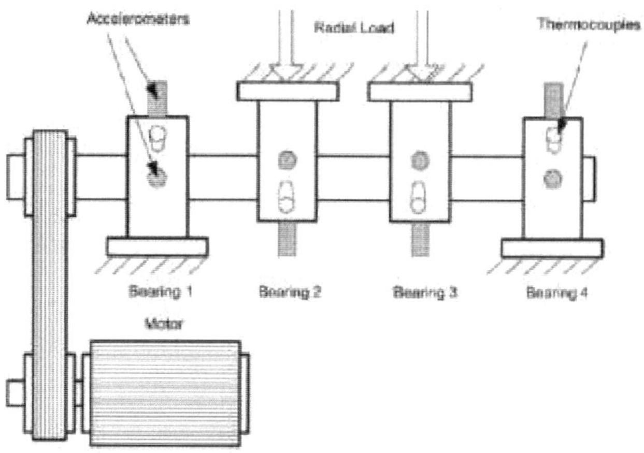

Figure 6: Bearing Test-rig.

Each of the REB is a double row bearing consisting 16 rollers, having a pitch diameter of 2.815 in., roller diameter of 0.331 in., and a tapered contact angle of 15.17 degrees. The data sampling rate is 20 kHz and the data length is 20480 points. For every record of 1s duration, the RMS value and the kurtosis are computed for a general evaluation of the test. The SK-MED based approach is evaluated by evaluated by taking the maximum of the SES around the theoretical fault frequencies (outer race fault frequency BPFO, inner race fault frequency BPFI, ball fault frequency BSF) as indicators. The CS approaches is evaluated through the same indicators applied on the IES instead of the SES. These values are computed in the vicinity of the theoretical frequency as the actual fault frequency values are likely to deviate from the theoretical one.

During the test, an outer race fault in bearing 1 occurs at the end. The indicators are calculated on the vibration signals captured from Acc1 (located at the bearing 1 housing). The indicators are computed with a threshold equal to the mean of the first 200 records (assuming a healthy bearing state) plus six of their standard deviation. The obtained results are displayed in Figure 7. The RMS was able to detect a defect at the record number 542, whereas the kurtosis detects at the record 702 but subjects significant fluctuations after the record that may jeopardize the detectability (notice that the value of the kurtosis has decreased under the threshold at the record 778). It is worth noting that the RMS was efficient in this detection due to the near position of the accelerometer to the defect source, however, the RMS can't specify the fault type. The MED-SK-SES was able to detect the fault at the same record as the RMS, yet only the BPFO value has exceeded the threshold indicating the fault type (i.e. outer race fault). Interestingly, the IES was able to detect the fault earlier at the record 533.

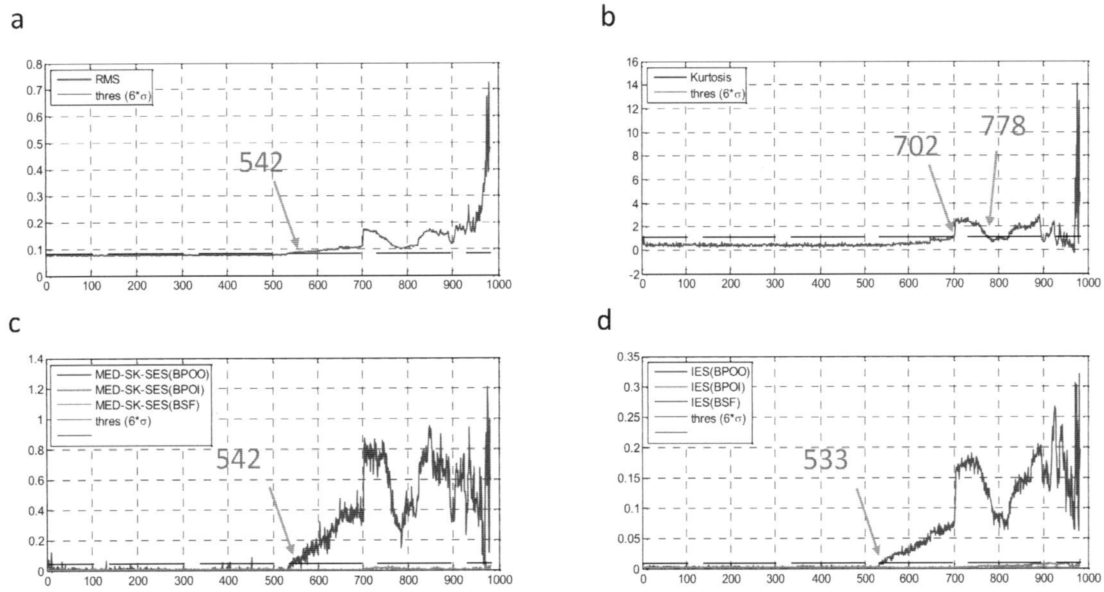

Figure 7: Indicators applied on Acc1 signals. (a) RMS, (b) kurtosis, (c) MED-SK-SES at the fault frequencies, (e) its zoom, (d) IES at the fault frequencies, (f) its zoom.

In the following, these indicators are applied on Acc3 signal (mounted on bearing 3) with the aim of evaluating the fault detectability from a remoted accelerometer. The indicators are exposed in Figure 8. The RMS was able to detect a global energetic change at the record 904 whereas the kurtosis and the MED-SK-SES have totally failed to detect any sign of failure. However, the IES was able to detect the outer race fault at the record 838.

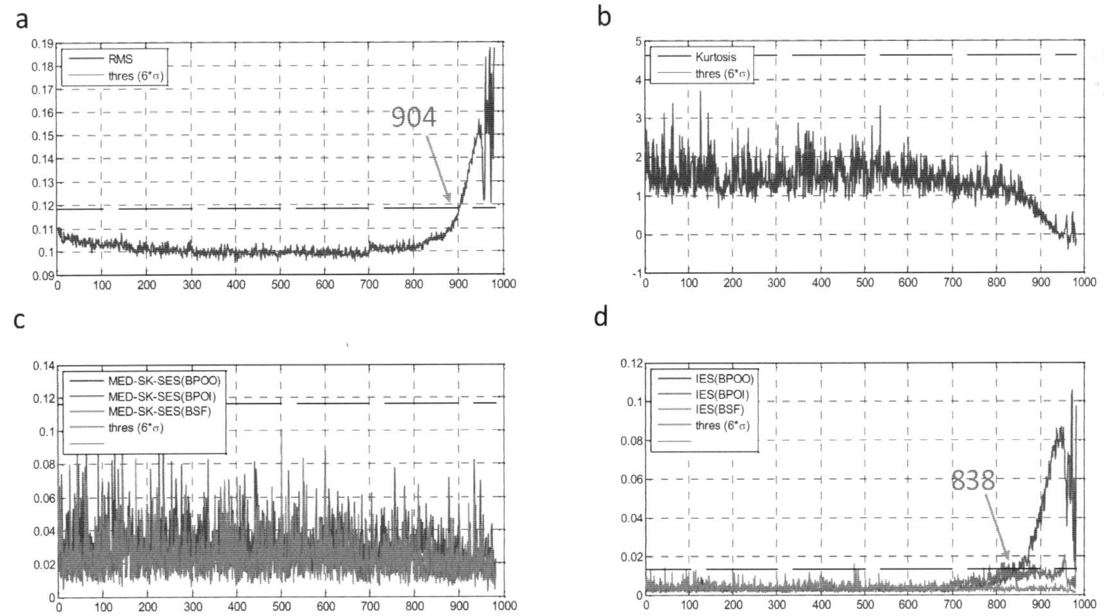

Figure 8: Indicators applied on Acc1 signals. (a) RMS, (b) kurtosis, (c) MED-SK-SES at the fault frequencies, (e) its zoom, (d) IES at the fault frequencies, (f) its zoom.

Conclusion: This paper has addressed the issue of incipient fault detection in REBs. Two leading techniques are tested on a run-to-failure test bench and compared in terms of precocity of detection and fault detectability of a remoted source. The first approach preprocess the signal by an optimized filter based on the minimum entropy deconvolution method, followed by the spectral kurtosis which design another filter that increases the signal-to-noise ratio before computing the squared envelope spectrum. The second approach is based on the cyclostationary theory. It consists of calculating the improved envelope spectrum through a projection over the spectral frequency of the squared-magnitude of the spectral coherence. Condition monitoring indicators are then designed based on these techniques. The results has shown a clear superiority of the cyclostationary approach which has proven very effective in detecting and identifying the bearing fault in its earliest stage, even when vibrations are captured from a remoted accelerometer.

References:

[1] R.B. Randall, J. Antoni, Rolling element bearing diagnostics—A tutorial, Mech. Syst. Sig. Process., Volume 25, Issue 2, p. 485-520.

[2] D. Ho, R.B. Randall, Optimisation of bearing diagnostic techniques using simulated and actual bearing fault signals, Mechanical Systems and Signal Processing 14 (5) (2000) 763–788.

[3] J. Antoni, R.B. Randall, The spectral kurtosis: application to the vibratory surveillance and diagnostics of rotating machines, Mechanical Systems and Signal Processing (2006) 308–331.

[4] R.B. Randall, J. Antoni, S. Chobsaard, The relationship between spectral correlation and envelope analysis for cyclostationary machine signals: application to ball bearing diagnostics. Mechanical Systems and Signal Processing. 15 (5) (2001) 945–962.

[5] J. Antoni, Cyclic spectral analysis of rolling-element bearing signals: Facts and fictions, Journal of Sound and Vibration 304 (2007) 497–529

[6] R.A. Wiggins, Minimum Entropy Deconvolution, Geoexploration, vol. 16, Elsevier Scientific Publishing, Amsterdam, 1978, pp. 21–35.

[7] H. Endo, R.B. Randall, Application of a minimum entropy deconvolution filter to enhance autoregressive model based gear tooth fault detection technique, Mechanical Systems and Signal Processing 21 (2) (2007) 906–919

[8] J. Lee, H. Qiu, G. Yu, J. Lin, and Rexnord Technical Services (2007). IMS, University of Cincinnati. "Bearing Data Set", NASA Ames Prognostics Data Repository (http://ti.arc.nasa.gov/project/prognostic-data-repository), NASA Ames Research Center, Moffett Field, CA

[9] Hai Qiu, Jay Lee, Jing Lin. "Wavelet Filter-based Weak Signature Detection Method and its Application on Roller Bearing Prognostics." Journal of Sound and Vibration 289 (2006) 1066-1090

CONDITION MONITORING BASED ON FMECA: A CASE STUDY OF SENSORS SPECIFICATIONS FOR MARITIME

Shan Guan,
Materials Technology Development
DNV GL (USA)
Dublin, OH 43017
Telephone: (614) 397-1062
shan.guan@dnvgl.com

Knut Erik Knutsen, and Øystein Åsheim Alnes,
Group Technology & Research
DNV GL
Oslo, Norway

Abstract:

For Maritime, one technical challenge associated with ship machinery condition monitoring is to select the best suitable sensors technology as ship owners always desires an economically viable, maintenance-free while technically reliable monitoring system. In this case study, condition monitoring of a tunnel thruster based on Failure Mode, Effects and Criticality Analysis (FMECA) was chosen to demonstrate the basic approach to overcome this challenge. Based on potential failure modes, four types of condition monitoring technologies including Vibration Monitoring, Acoustic Emission Monitoring, Wear Debris and Water in Oil Monitoring, and Thermal Monitoring are recommended. Out of these technologies, the sensor specification for a reliable vibration monitoring is discussed in detail as an example.

Key words: Condition Monitoring, Sensor, Vibration, Failure Modes.

Introduction:

In Maritime, the motivation for introducing condition monitoring (CM) is to increase the overall safety level in order to reduce the risk of loss of life and property, also minimizing the costs associated with maintenance. This can be achieved through increased reliability of the component or system being monitored and reduced consequences of failures. All these goals require a deeper understanding of condition monitoring technologies including the immediate needs to define data flow, sharing and hardware specifications of sensors. The technology for condition monitoring is directly related to failure modes, which can be varied greatly for different industries. Technical challenges associated with ship machinery condition monitoring include selecting the suitable technology for sensor data flow, sharing and hardware of monitoring systems.

The process and data flow in a ship machinery condition monitoring system is sketched in Figure 1, where in this paper sensor specifications was rendered through a case study carried out between DNV GL and Rolls-Royce Marine (RRM) involving condition monitoring of tunnel thrusters. [1]

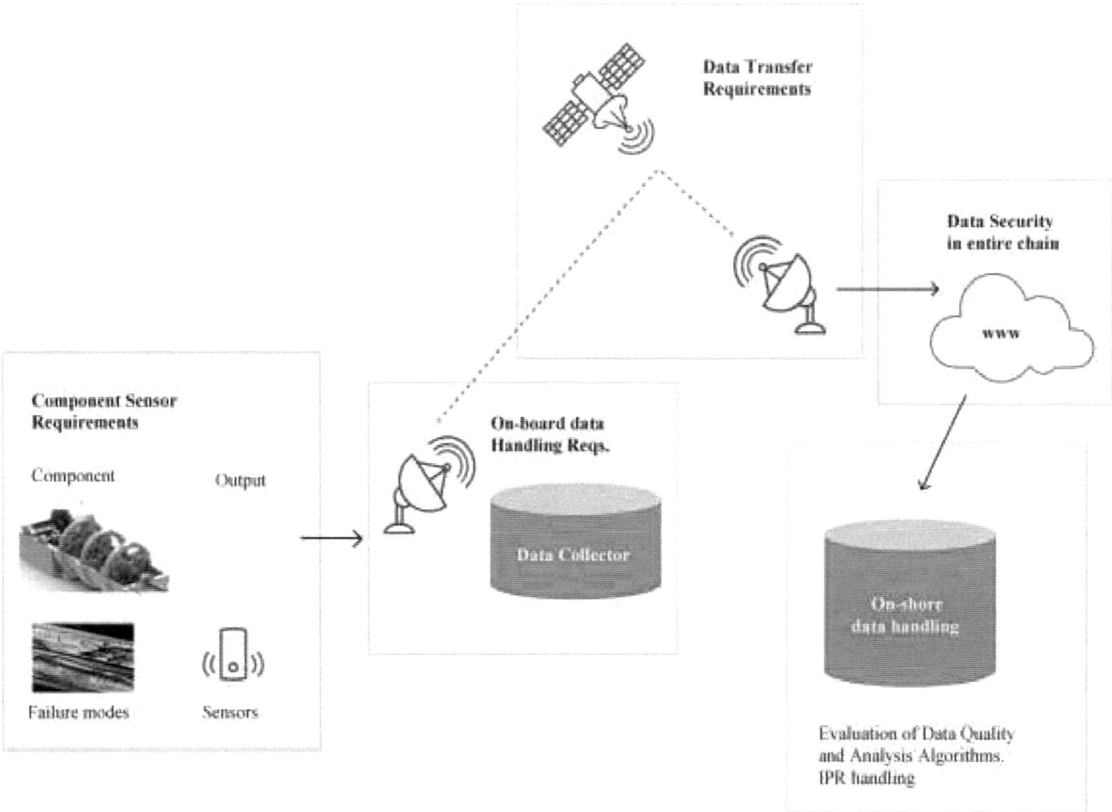

Figure 1: An overview of data flow of ship machinery condition monitoring

Case Study and Failure Modes

A case study involving condition monitoring of tunnel thrusters was performed to demonstrate the basic sensors specifications of CM system based on FMECA concept. Rolls-Royce TT2200 DPN FP thrusters are used as the system for condition monitoring. To carry out an FMECA of tunnel thrusters, many considerations must be given beyond the components themselves, i.e. a systematically analysis need to be performed based on different types (class) of vessels they were installed on and their functionalities. For example, if in a vessel, two identical bow thrusters were installed, then loss of one thruster is acceptable for a short period of operation. In this case, failure of a single component resulting in a system failure can be identified as Medium to High. However, if there is only one bow thruster installed on a vessel, then any failure of a single components resulting in the loss of thruster (the system) is unacceptable and must be prevented from happening in the maintenance plan. In addition, an immediate reaction plan to these incidents must be in place in the condition monitoring plan (maintenance plan). In this situation, any failure mode resulting in the system failure will be identified with a criticality of Very High. Figure 2 illustrated the definition of criticality of this case study and the recommended actions. It should be understood that during this case study, the criticality of each component/subsystem in the tunnel thruster determined was approximately, but conservatively, and only one thruster was considered (for example, loss of this thruster will affect the dynamic position of the vessel). The criticalities as graded in Table 1 were in five categories of Very Low, Low, Medium, High, and Very High. Equipment with Very Low or Low criticality can be subject to corrective maintenance. Equipment with Very High criticality must be considered for re-design/redundancy to decrease criticality. The remaining equipment of Medium to High criticality is to be evaluated further. [2]

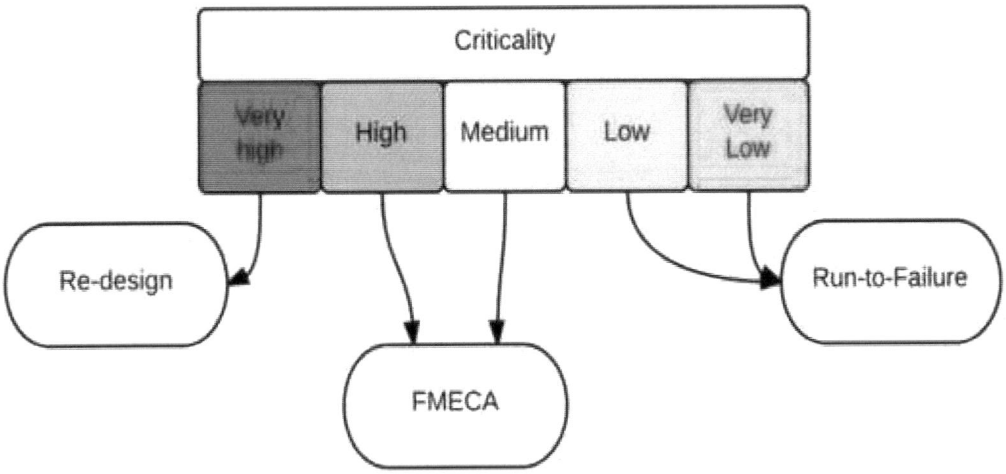

Figure 2: The Criticality definition and the associated actions of the case study

Table 1: Important Failure Modes, Failure Criticality and Monitoring Technology of Tunnel Thruster Case Study.

Component	**Failure Modes**	**Failure Effects**	**Failure Criticality**	**Sensors**
Electric Motor	Overheating	Short motor life to motor failure	High to Very High	Temperature
Frequency Converter	Temperature induced	Components/ system failure, Won't start	High to Very High	Thermography

Shaft	Shaft failure	Components/ System failure	High to Very High	Vibration, Acoustic Emission, UT
	Sheared Shaft, Shaft failure	Seized	High to Very High	Vibration, Acoustic Emission, UT
Tooth Coupling	Teeth wear away	High vibration to system failure	Medium to Very High	Vibration, Torque, Particule analysais/ Wear Debris
	Tooth fatigue failure	High vibration to system failure	High	Vibration, Torque
Rolling Bearing	Rolling contact fatigue	Seized to system failure	High to Very High	Vibration, Temperature, Oil analysis (off site)/ Wear Debris
	Plastic deformation	Noisy/Excessive vibration, Seized Motor, Loss of torque	Medium to Very High	Vibration
Bevel Gear	Plastic deformation	Noisy, Vibration, System Failure	Medium to Very High	Oil analysis (off site)/ Wear Debris, Vibration
	Tooth flank contact fatigue	Vibration, System Failure	Medium to Very High	Oil analysis (off site)/ Wear Debris, Vibration
Propeller Blade	Fatigue failure	Loss of Torque, Vibration, System Failure	Medium to Very High	Torque, Vibration, Ultrasonic
Lubrication System	Pressure drop	Components/System failure in long term	Medium to Very High	Oil Pressure
	Overheating	Components/System failure in long term	Medium to Very High	Temperature

SENSOR SPECIFICATIONS

It is essential to select suitable sensors for ship machinery condition monitoring, not only need to consider the physical elements they measure, but also need to evaluate the installation, cost, maintenance and calibration. Although sensors differ greatly in their physical principles, their selection can be guided by the following procedures:

1. Determine the variables to be measured,
2. Determine the technical specification of sensors for each measurement,
3. Determine the availability and affordability of sensors,
4. Determine the installation, maintenance plan and calibration procedure of sensors.

Based on FMECA and testing results, four types of monitoring technologies are identified as Vibration Monitoring, Acoustic Emission Monitoring, Wear Debris and Water in Oil Monitoring, and Thermal Monitoring. In this paper, the sensor specification of vibration monitoring will be given as an example.

Machine vibration can be measured by three representing motion parameters: acceleration, velocity and displacement. The corresponding vibration sensors are accelerometers, velocity vibration sensors, and proximity vibration sensors, respectively. Selection of a suitable vibration sensor depends on the frequencies of interest and signal levels that are involved. For example, DNV GL Guidance for Condition Monitoring, Classification Note 10-2, summarized the relationships between sources of vibration under fault condition and the corresponding frequencies that significant vibration velocity peaked. These types of table are the basics for determining the technology specification of vibration sensor, and should be prepared in the CM planning phase such as through an FMECA process. Each type of these sensors is mature and can be easily customized to meet ship owner's requirement for condition monitoring. Table 2 is a brief comparison of vibration sensors for the measuring parameters, sensing mechanism, advantages/disadvantages etc. To select the correct types of vibration sensors for condition monitoring, the following criteria can be considered:

1. The source of vibration,
2. The vibration level and frequency range,
3. The temperature range,
4. The environment factors (e.g. corrosive condition or combustible atmosphere),
5. The interference with electromagnetic fields (electromagnetic compatibility) or acoustic fields,
6. Electrostatic discharge (ESD) conditions,
7. Sensor size, cost, packaging and weight considerations.

Accelerometer type of vibration sensors measures the degree of vibration through measuring acceleration. It can measure mechanical vibration levels through a single axis, or determine the type of vibration or the direction of acceleration through tri-axial. Accelerometers designed for measuring vibration are based on the piezoelectric effect. Piezoelectric accelerometer is the mostly used vibration sensors that rely on piezoelectric effect by capturing acceleration data at a faster rate. In a piezoelectric accelerometer, a force applied to a crystal will create a high-impedance charge resulting in a voltage across the crystal. These sensors require an external amplifier (note that most of the modern sensors are designed with integrated amplifier) to amplify the piezoelectric charge and to provide an impedance buffer. They can respond to vibrations with frequencies down to 4 Hz. However, due to the temperature reaction of the material, these devices can lead to thermal noise. Nowadays, many types of commercially available MEMS accelerometers are the better choices for vibration monitoring, because the integrated (built-in) digital signal processor (DSP) and band-pass filter can reduce or even eliminate the noise. [3, 4] Velocity sensors are used for low to medium frequency measurements (0.5 to 1000 Hz) which are useful for vibration monitoring of rotating machinery. Traditional velocity sensors use an electromagnetic coil and magnet system to generate a velocity signal. Today, piezoelectric velocity sensors are becoming popular due to improved capabilities and their rugged nature. However, most of these sensors are bulky and heavy making them unsuitable for vibration measurement of delicate parts. [5] A commonly used displacement vibration sensor is the proximity vibration sensors such as an Eddy-Current probe, which determines the vibration through measuring the change of eddy currents. An Eddy-Current sensor is a noncontact vibration measurement tool with high frequency response (up to 80 kHz) to accommodate high-speed motion. In addition, the ability to determine relatively small motions (i.e. very weak vibration) and non-contact motion make it a suitable tool for engine rolling bearing vibration, shaft motion and internal clearances monitoring. The major drawback of Eddy-Current probe is that they are sensitive to electrical and mechanical noise. In addition, the installation and maintenance of this type of sensors is complex. [6]

Table 2. A Comparison of Three Types of Sensors for Vibration Monitoring.

	Accelerometers	**Velocity Vibration Sensors**	**Displacement Vibration Sensors**
Measuring Parameters	Acceleration	Velocity	Displacement
Sensing Mechanism	Piezoelectric Sensors	Electromagnetic transducer	Capacitance sensors or Eddy-current probe
Major Advantages	Good response at high frequencies; Able to stand high Temperature; Small size	Good response in middle range frequencies; Able to stand high temperature; Low Cost; No external power needed.	Non-contact, No wear; Able to measure both static and dynamic displacements; Good response at low frequencies

Major Disadvantages	Sensitive to high frequency noise; Higher cost	Low resonant frequency and phase shift; Large footprint; Cross noise	Bounded by high frequencies; Sensitive to Electrical and mechanical noise
Wireless Capability	Commercially available	Not available	Possible, not yet commercially available.
MEMS-based devices available	Commercially available	Not available	Commercially available

In the tunnel thruster CM case study, accelerometer type vibration sensors, SPM SLD144S were applied for vibration monitoring of "stern tunnel thruster TT2200 DPN CP" electric motor housing (Non-drive end horizontal direction, and top of horizontal direction, as well as "stern tunnel thruster TT2200 DPN FP" electric motor housing (Non-drive end horizontal direction, and top horizontal direction). The installation locations of these sensors are illustrated in Figure 3. The same type of sensors was also applied for vibration monitoring of starboard propulsion line PTI (power take-in) motor housing and reduction gear housing, and port propulsion line reduction gear housing. The SLD144S sensor is a piezoelectric accelerometer of compression type with built-in preamplifier, designed for vibration monitoring of industrial machinery. It was designed with side entry which is especially suitable for use where installation space is limited, and can be used in harsh industrial environments where moisture/water poses constant challenge, for example, the marine environment where tunnel thrusters were in. The electrical signal is isolated from the transducer housing. The transducer is mounted against a smooth, flat surface on the machine. In the following table, major technical specifications of this sensor are listed. The SLD144S was a type of vibration sensor based on the patented shock pulse method (SPM), which was applied to detect the short duration shock waves caused by rolling element bearing defects. The major difference between conventional accelerometers and SPM sensors compared is that SPM sensors are operated at 32 kHz (resonant frequency). In this way, even weak shock pulses can generate large output signals are generated from. The signals of SPM sensors include both the magnitude of the shock peaks (dBm) and of signal level between the peaks (dBc). In addition, by performing FFT (fast Fourier transform) analysis on the signals. Some degree of the causes of fault can be revealed. [7] The major limitation is that by using SPM sensors, the user needs to use SPM's data analysis protocol. Based on the manufacturer technical data sheet, a list of general requirement of vibration sensors used for the ship machinery condition monitoring is rendered. However, it is worth to point out that the parameters specified in Table 3 have not been verified and can only be used as general reference based on references. [8-14]

Table 3. Technical Specification for Accelerometers Type Vibration Sensors for Tunnel Thruster CM

Tech Specification	**SLD144S Vibration Sensor**	**General Vibration Sensors for Ship Machinery CM**
Nominal sensitivity, main axis	100 mV/g	10-500 mV/g
Transverse sensitivity	Max. 10%	Max. 10%
Typical base strain sensitivity	0.01 m/s2/ strain	0.01-0.05 m/s2/ strain
Linear frequency range	2 Hz - 10 kHz (±1dB) (–3 dB at 0.7 Hz)	2 Hz - 10 kHz (±1dB) (–3 dB at 0.7 Hz)
Max. peak acceleration	600 m/s2 = 60 g	100 g
Settling time	3 sec	1-5 sec
Bias point	11 to 13 V (typical 12 V)	11 to 13 V (typical 12 V)
Temperature range	40° C to +125° C (–40° F to 260° F)	0° C to +175° C
Power requirements	24 V /2 - 5 mA	5-24 V (1-10 mA)
Casing	Stainless acid proof steel	Anti-Corrosion/Marine Environment Resist materials preferred
Isolation	Case isolated, > 1 Mohm	Case isolated, > 1 Mohm

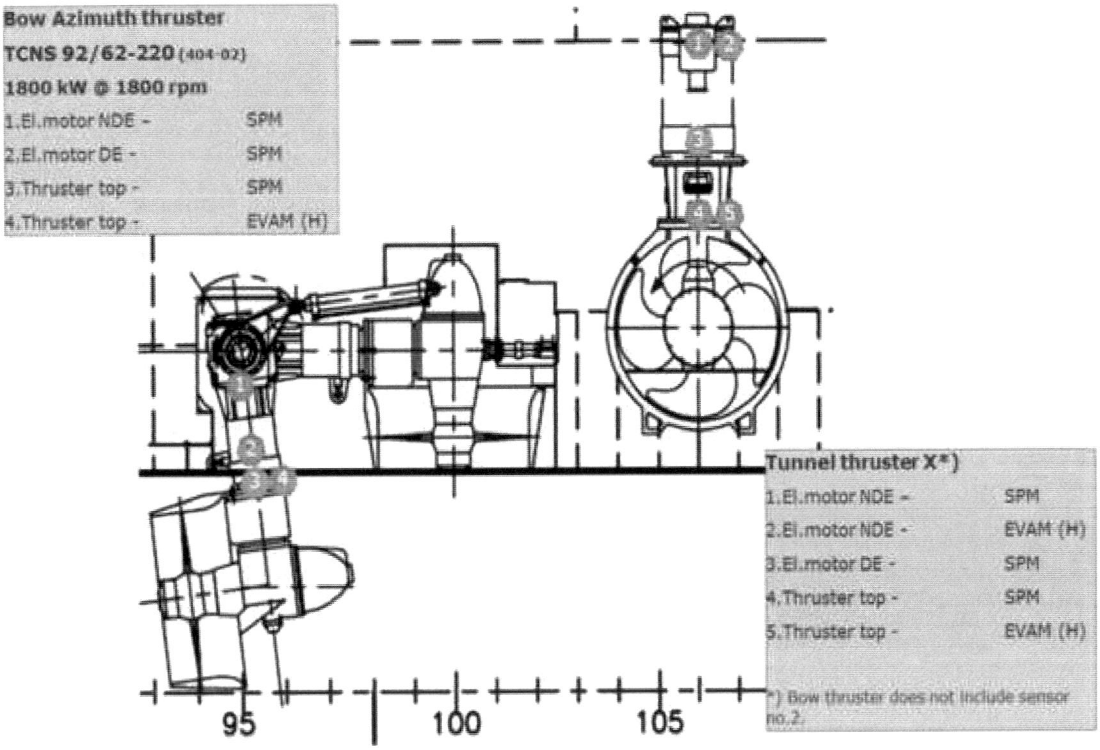

Figure 3. The Sensors Installation Locations for the Tunnel Thruster CM Case Study.

CONCLUSIONS

A case study on CM of tunnel thrusters was carried out as the effects to define the general sensor specification in CM for Maritime applications. Based on the testing results, four types of CM technologies are identified as Vibration Monitoring, Acoustic Emission Monitoring, Wear Debris and Water in Oil Monitoring, and Thermal Monitoring. The sensor specifications of vibration monitoring were given as an example.

References:

[1] K. E. Knutsen, "COMPASS – Condition and operation monitoring for performance and ship safety, Final Report," DNV GL Report. 2016.

[2] S. Guan, "Sensors Requirements for Condition Monitoring and Data-driven Classification: A Case Study," DNV GL Report No.: GGRUS915SGUA (91515017), May 8, 2015.

[3] M. Serridge, and T. R. Licht. "Piezoelectric accelerometer and vibration preamplifier handbook". Brüel & Kjaer, 1987.

[4] Z.F. LU, M. Qin, H. Chen, L. Lin, and Y. Zhang. "Application and Investigation of Piezoelectric Accelerometer on Vibration Measurement System", Instrument Technique and Sensor, 7, 2007, p: 3.

[5] C.W. De Silva, ed. "Vibration monitoring, testing, and instrumentation". CRC Press, 2007.

[6] G. Y. Tian, Z. X. Zhao, and R. W. Baines, "The research of inhomogeneity in eddy current sensors," Sensors and Actuators A: Physical, 69 (2), 1998, p. 148-151.

[7] L. E. Morando, Technology Overview: Shock Pulse Method, Defense Technical Information Center, 1996.

[8] SPM Instrument AB, Technical Specification Sheet, Vibration Transducer SLD144S/ SLD144B /SLD144F.

[9] Connection Technology Center Inc., Technical Specification Sheet: CTC Accelerometers.

[10] Colibrys Inc., Technical Specification Sheet: MS7000 and VS9000 accelerometers.

[11] MISTRAS Inc., Technical Specification Sheet: ACCELEROMETERS & VIBRATION SENSORS, 100 mV/g: Model 6061-IS, Model 6051C, Model 6051, and Model 1020S; High Sensitivity: Model 8002HS; High Shock: Model 1018.

[12] Kistler Inc., Technical Specification Sheet: Piezoelectric vibration sensor, ±50 g, 100 mV/g, Model 8141A.

[13] Meggitt Sensing Systems, Technical Specification Sheet: 100 mV/g, 1 Hz - 12 kHz, Model 785A.

[14] Measurement Specialties Inc., Technical Specification Sheet: Piezo Sensor - MiniSense 100, MEMS Based Vibration Sensor.

Bibliography:

[1] Shan Guan is a Principle Consultant and Senior Project Manager at DNVGL, where his research focus was on managing the risks of pipelines and oil & gas production systems, as well as applications of sensors on Maritime, Oil and Gas and Energy industry. He obtained a Ph.D. in Mechanical Engineering in 2004, and a Masters of Electrical Engineering in 2003, both from the University of Minnesota. He obtained a M.S. and B.S. in Materials Science from BeiHang University, in 2000 and 1997, respectively. He has published over 50 papers and has been awarded more than 10 patents.

DETECTION OF LUBRICATION STARVATION IN BALL BEARINGS AND PRELOAD EFFECTS.

Jorge A. Mijares, Bryan P. Rasmussen and Alexander G. Parlos

Texas A&M University
Department of Mechanical Engineering
3123 TAMU
College Station, TX 77843-3123
Telephone: (979) 255-2459
jorge.mijares@tamu.edu

Abstract: Detection of lubricant starvation in ball bearings is essential to maximize their lifetime. The lack of lubrication introduces impulses in the vibration signal related to fundamental train and ball spin frequencies. The phenomenon is attributed to the lack of a damping to dissipate the impacts of the balls when entering to the load zone together with the gap increase between bearing components, previously filled by lubricant. This work presents an experimental vibration analysis to study the effect of preload as a mean to decrease the extra clearance in dry conditions and eliminate the source of impulses that characterize lubrication starvation. Furthermore comparison between RMS, Crest Factor and Kurtosis reveals that an increase in preload indeed reduces the signal impulsivity and kurtosis proves to be an effective method to detect lubricant starvation. Finally, the application of fast-kurtogram algorithm is used as a pre-processing method to improve the performance of kurtosis in detecting of lubrication starvation.

Key words: Ball bearing; crest factor; fast-kurtogram; kurtosis; lubrication starvation, RMS.

Introduction: Bearings is the most common failure mode for electric motors and pumps [1], therefore condition monitoring of ball bearings is fundamental for good preventive maintenance program for these type of systems. Bearing lifetime has increased with the advanced manufacturing technologies, but eventually all bearings fail due to fatigue which cause spalling on the surfaces in contact [2]. Current technology can detect the initial stages of spalling months in advance, but most bearings have premature failures due to improper lubrication (contaminants, starvation, and wrong selection) [3] and 90% never reach their expected lifetime [4]. The early detection and diagnosis of improper lubrication could maximize bearing lifetime and avoid downtime problems. The benefits of lubrication are not only to decrease friction between components but also to dissipate high temperatures, avoid corrosion and prevent contamination with foreign debris. Additionally, the lubricant could be modeled as a spring-damper system for vibration analysis, this was studied by Wijnant et al. [5].

The process of condition monitoring is classified by some authors in three stages [6], the first stage is to detect the presence of a fault in the system (fault detection or screening). Based on the impulsive nature of bearing faults, some of the most common screening indicators are Crest Factor (CF), high frequency RMS, and Kurtosis [7]. Pachaud [8] shows that kurtosis is a more sensitive indicator than crest factor but also is highly sensitive to noise, which requires a pre-treating process. Once a fault has been found, the next stage is to determine the cause of the fault (fault diagnosis), in this stage is important to isolate the signal of interest and relate it to a known characteristic fault signature. Diagnosis of ball bearing faults is mainly based on the characteristic signatures for bearings presented by Tandon and Choudhury [9]. These signatures are used to identify mainly localized cracks or spalls in the raceways, rollers, or defects in the cage. The finally stage, called prognosis, is to predict the time of failure of the specific damaged component or system.

Efforts have been made to address the problem of lubrication starvation. Singh and Kazzas [10] found an increase of RMS value of wavelet components related to dry bearing conditions proving that it is possible to isolate the frequencies associated with lubrication starvation. Using a more complex analysis, Boškoski et al. [11], found an increment in ball spin (BSF) and fundamental cage frequencies (FTF) by using spectral coherence analysis and fast-kurtogram to identify the frequency band affected by impacts, and envelope analysis to the selected band. Boškoski proposed that fault signature on FTF and BSF is caused due the lack of movement restriction of the cage, created by the absence of lubricant filling the gaps between components.

Besides the use of lubricant, one method to reduce clearance of bearing components is to apply a certain preload. Based on the experimental results of Tallian and Gustafsson, that proves ball pass frequency (BPF) amplitude is proportional to clearance, Akturk et al. [12] studied the effect of different preload on ball bearings and results showed the BPF could be reduced by selecting an adequate preload. The disadvantage of increasing preload is the friction increase and consequently the temperature rises, which is an accelerating factor for bearing degradation [13], [14].

In this work a preload was applied to bearings in dry conditions in order to reduce the impulsivity of the signal based on the hypothesis that extra clearance and gap between the cage and bearing balls are the cause of the characteristic signature of lubrication starvation. Besides the preload test, experiments with minimum levels of lubrication (5%) were done to determine a detection methodology useful not only for dry conditions, but for minimum level of lubrications. Kurtosis, CF, and RMS indicators performance were compared for each experiment to assess advantages and disadvantages of lubrication starvation indicators. The algorithm of fast-kurtogram was used to identify the band frequency most affected by the impacts. The band frequency was filtered to improve the performance of kurtosis indicator.

Experimental Setup: The test bench is shown in
Figure 1. The bearings used for the experiments are the standard series 6204-2RS-10. Bearing #2 is used for all experiments since it shows lower frequency perturbations from the variable frequency drive (VFD) and motor than bearing #1. The signal is sampled at 60

kHz, analog filtered at 20 kHz, and acquired for 4 seconds to achieve a frequency resolution (d) of 0.25. The rotational speed was set at 1680 rpm. A triaxial accelerometer is used but only the Y-Axis is acquired. To measure the shaft-speed a torque meter with a speed analog output is used.

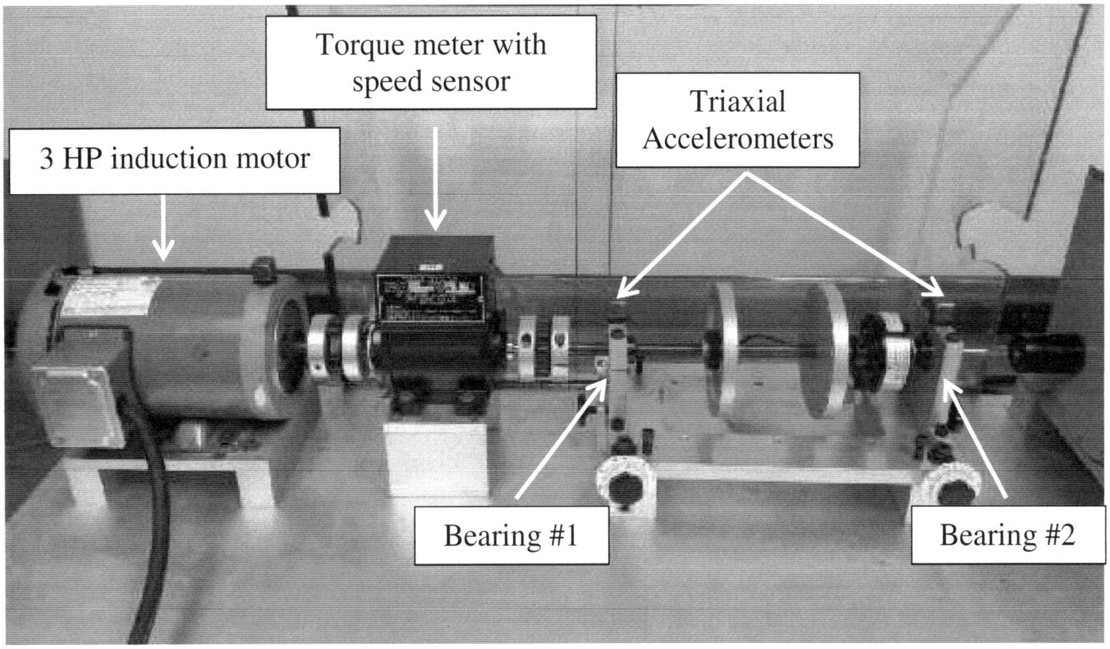

Figure 1: Test Bench Used for Experimentation.

A total of six experiments were conducted and each experiment consisted of four tests on a new bearing. The first test was realized under dry conditions, a careful and systematic procedure was done during the preparation of the bearings; mineral spirit was used to remove the lubricant and completely clean the bearings, special attention was made to avoid any foreign particles from entering the bearing and affecting the experiments. The bearing was temperature-mounted on the shaft. To mount the rotor on the bearing housings all screws were tightened to 15 lb-ft with the objective to control the preload on all tests.

Once the test with dry conditions was completed, the test was repeated but with a preloaded bearing, the preload was increased by reducing the housing diameter using a shim of .001". For the final two tests, the shim was removed from the bearing and two levels of lubricant were tested (5% level and 100%) to evaluate the efficiency of the methods on detecting dry, low, and fully lubricated conditions. Grease LGMT/2 was added to the bearing using a syringe, distributing even quantities around all rollers. The total amount of grease (1.2 g) was determined previously by weighting five bearings before and after removing the lubricant.

The machine was warmed up for two minutes before starting to acquire data, and ten continuous acquisitions were taken for each test. In all analysis, a 95% confidence interval was used to obtain the uncertainty of the ten measurements. All acquired signals were processed with a digital filter to remove a VFD perturbation at 15 kHz and band-passed between 1 kHz to 20 kHz, in order to keep only high frequency content and remove any residuals of the analogue filter.

Indicators Performance Comparison: RMS is one of the most common vibration indicators, it corresponds to the square root of the average of square values in a given signal.

Figure 2 compares RMS values at the different conditions for all experiments, for each experiment the value on dry conditions is higher than at lubricated conditions, but the range is different for each bearing, thus, it is not possible to establish a limit to detect lubrication starvation for all bearings without compromising the indicator with false alarms or missed faults. Besides, the indicator increases under preload even if the impulsivity of the signal decreases as shown further by the other indicators. The problem associated with preload is the change of resonance frequencies due a change in stiffness, and the possible excitation of the bearing housing due the increase in dry friction, which is random excitation generated from surface roughness. Hence, even if the impulsivity of the signal decreases the RMS cannot detect the change when a preload is applied.

Figure 3 shows a cascade comparison for experiment #2, where it is observed a switch of frequency due increase of stiffness from 4 to 6 kHz (frequency range associated with the rotor-assembly resonance frequency), and an increase in activity at 11 kHz (vertical resonance of house bearing).

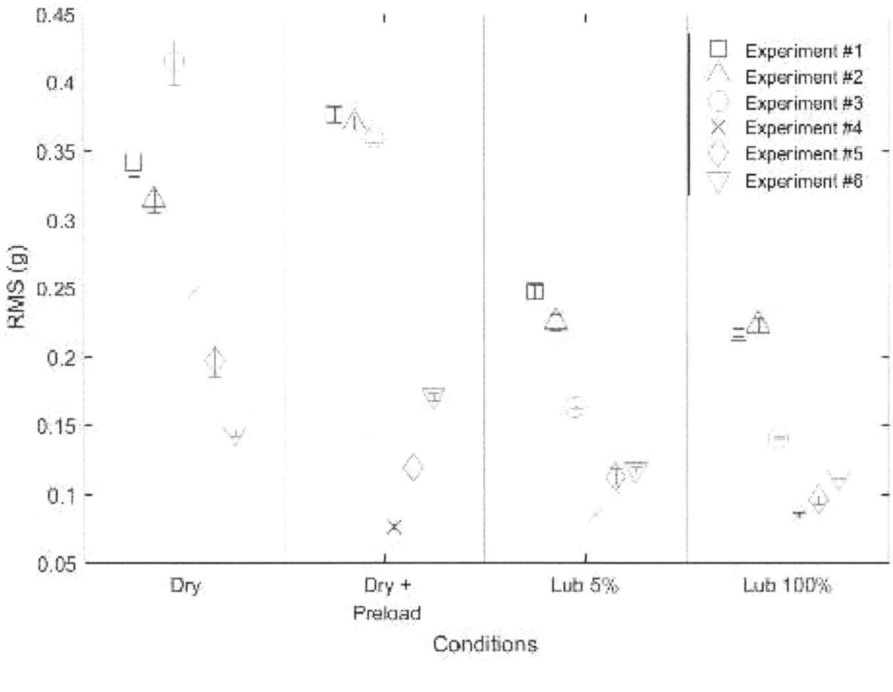

Figure 2: RMS Performance Comparison for Experimental Conditions.

Crest Factor is the ratio between the absolute-maximum peak value and the RMS value of a given waveform, the value for a sinusoidal signal is 1.414. When applied to detection of bearing starvation, it is a better indicator than RMS since is more related to impulsivity. The problem is the similarity between lubricated and dry conditions, as shown in

Figure 4, if a severity level is established based on dry condition results, some of the full lubricated results may trigger a false alarm. An advantage of CF is the high sensitivity for low lubrication levels.

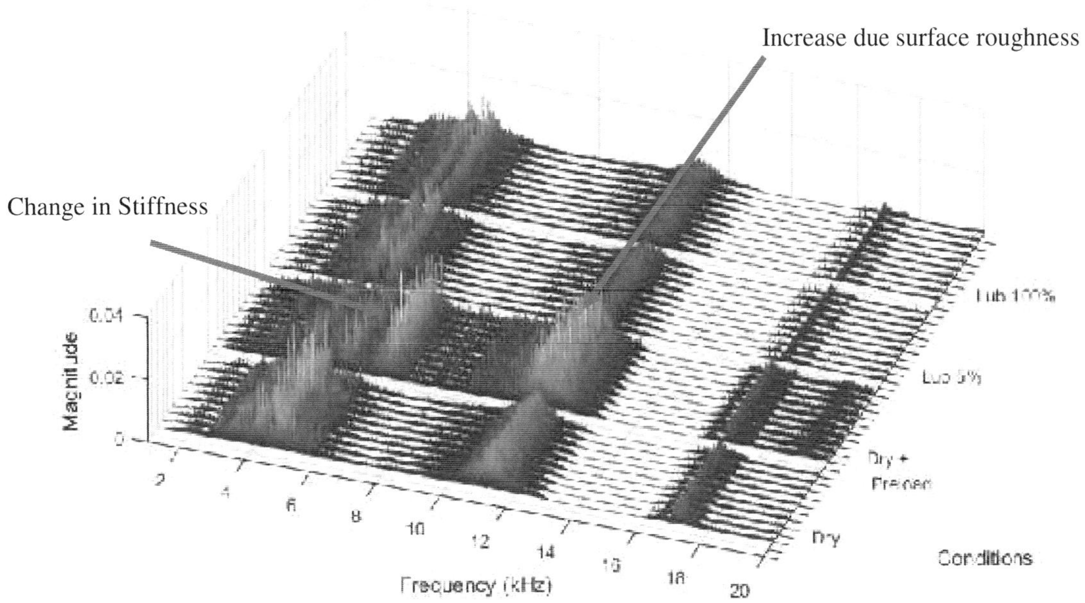

Figure 3: Cascade Comparison and Preload Effects on Experiment #2.

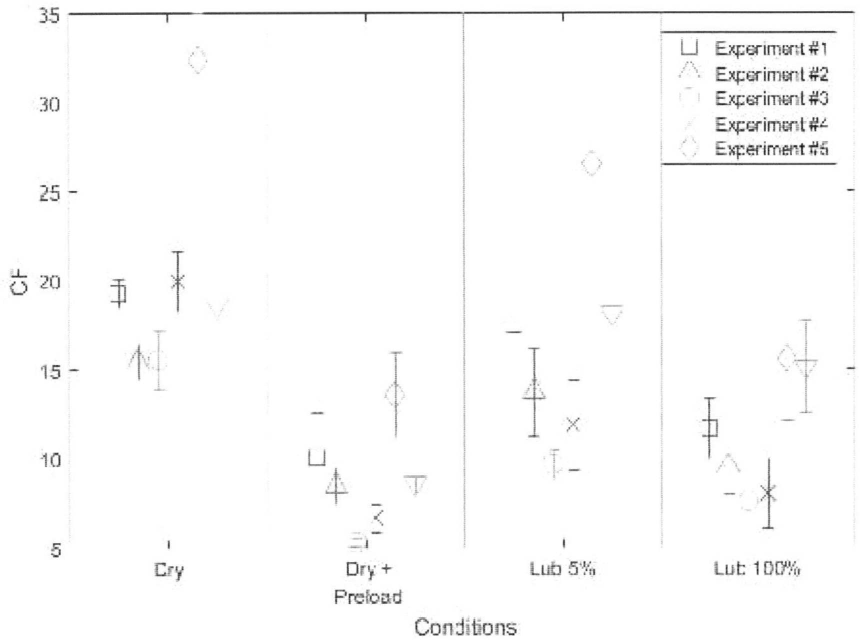

Figure 4: Crest Factor Performance Comparison for Experimental Conditions.

Kurtosis is a statistical analysis called "fourth moment" that represents the deviation of signal's amplitude from a Gaussian distribution and is useful to detect impulsiveness in a signal, e.g. faults in ball bearings. From a statistic point of view, it is a measure of the sharpness of the tail in a frequency-distribution curve. Kurtosis results, presented in Figure 5, show a better performance than RMS and CF to detect lubrication starvation, the consistency between experiments allow proposal of severity levels: 0-3 good, 3-7 satisfactory, 7-10 unsatisfactory, 10- unacceptable.

The main disadvantage is the variability between acquisitions of the same experiment, therefore more than one acquisition is recommended for an effective evaluation, which could be a problem when a large number of machines must be assessed. In both CF and Kurtosis indicators it is possible to observe a decrease in impulsivity of the signal when a preload is applied, as explained by Sawalhi and Randall [15], the impulsivity could be attributed to each time a ball enters the load zone and a small impact is produced, generally these excitation are small and quickly damped by the lubricant which also acts as an isolation mechanism between the balls and the bearing structure, but with the absence of lubricant and the increase in clearance, the impact amplitude increases and the wave is propagated to the bearing housing, hence the increase in FTF and BSF. When a sufficient preload is applied, the clearance between the balls, train, and the races is reduced as an effect of load deformation and in the case of light loads the deformation due preload dominates over the deformation due to dynamic or static load, therefore the ball never enters or leaves the "load zone". Both factors reduce the source of impacts that characterize the lack of lubrication.

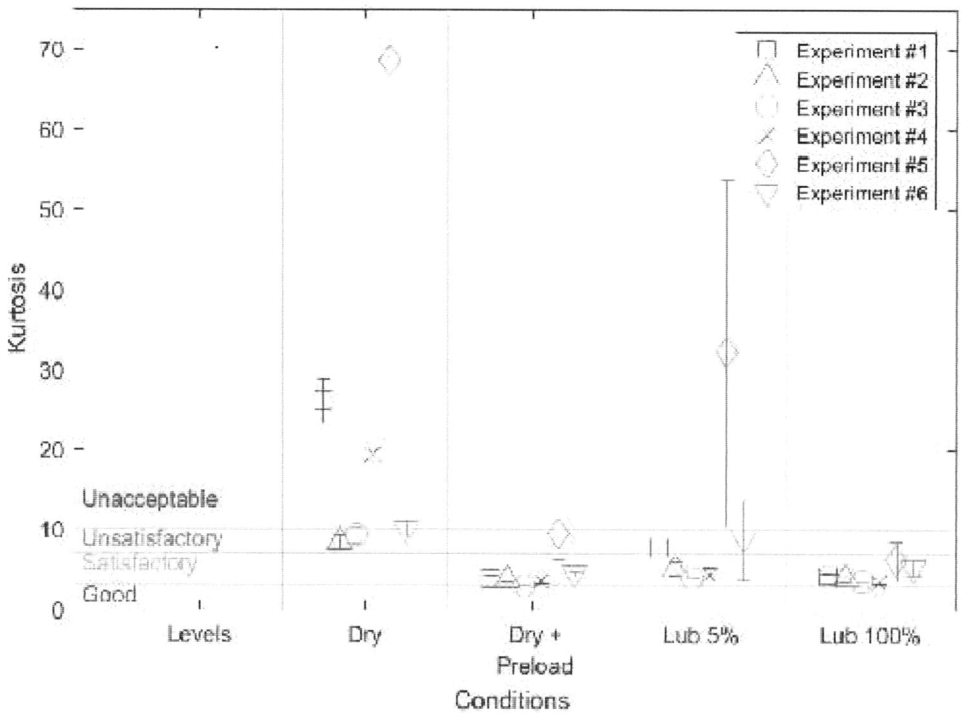

Figure 5: Kurtosis Performance Comparison for Experimental Conditions with Severity Levels.

Fast-Kurtogram: The use of kurtosis is maximized when the resonance frequency excited by bearing impulses is isolated from other excitations. Introduced by Dywer in 1983 [16], Spectral Kurtosis (SK) is a time-frequency plot used to obtain a kurtosis value of a specific frequency band in the time direction and is useful to detect frequency bands with more content of impulsive signals. The main issue with SK is the selection of an optimum value for the bandpass filter bandwidth and the center frequency. A map can be used where the kurtosis value is calculated for each frequency in Time-Frequency analysis and a range of bandwidths. This map is called a "Kurtogram". The disadvantage of this method is computation time; is expensive for online systems. One solution for this problem is called "Fast-Kurtogram" (FK) and was presented by Antoni in 2007 [17]. The technique is based on a series of digital filters instead of time-frequency analysis and the decomposition of the signals is based in a dyadic decomposition similar to the discrete wavelet packet transform. The FK creates a map where one of the axes is a progressive split of the frequency range in bands and the other axis is the center frequency of the band. A typical recommended to do it with a 1/3-binary tree where the division is a sequence of 1/2, 1/3, 1/4, 1/6, 1/8, and so on.

The results of FK exhibit a predominant central frequency (f_c) at 17.8 kHz with a bandwidth (B) of 1.8 kHz (level 4), the region is associated with a resonance frequency of the bearing housing.
Figure 6 shows the results for dry conditions of experiment #1, these parameters could be used to filter all signals and apply an envelope analysis to find the origin of the impulses and how the amplitude reduces. This work presents the advantages of using FK analysis to improve kurtosis as a fault detection method for lubrication starvation.
Figure 7 compares the results of kurtosis before and after using FK. The mayor improvement is the sensitivity to detect low levels of lubricant where the kurtosis of the entire high frequency spectrum does not differs from full to low lubrication. Moreover, a decrease of impulses on the signal is corroborated on these results as well.

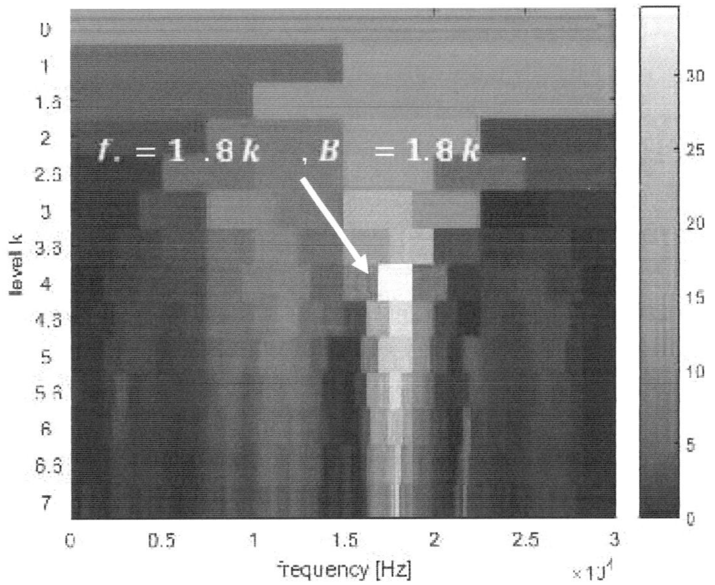

Figure 6: Fast Kurtogram for Experiment #1 on Dry Conditions.

The levels proposed for the kurtosis of high frequency spectrum were also used in the kurtogram results. Experiments 1,2,5, and 6, detected low lubrication as an unacceptable fault. For condition monitoring it is important to minimize false alarms without compromising the sensitivity of the method to detect a problem which, as in this case, is low levels of lubrication.

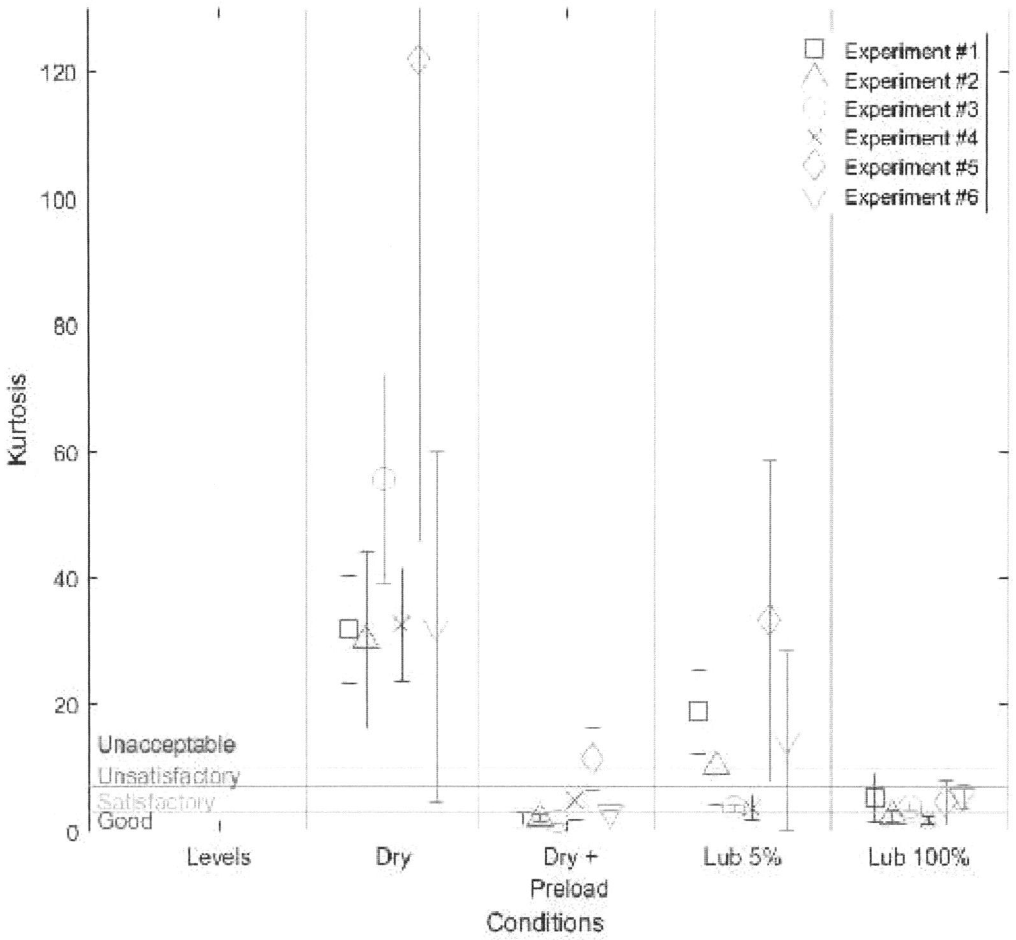

Figure 7: Kurtosis of Filtered Frequency Band Localized by Kurtogram.

Conclusions: Three main indicators, RMS, CF, and Kurtosis were evaluated to detect lubrication starvation in a ball bearing under light load. While RMS detects the presence of a fault in the system, it is not recommended to be used alone to isolate faults related to bearings. It must be used with an additional indicator, sensitive to impulsivity, such as kurtosis or CF. Kurtosis proves to be a more consistent indicator than CF to discriminate between lubricated and dry conditions.

Levels of severity are proposed for kurtosis, the levels are based on experimental results, but a formal methodology to select levels is needed. Instead of using fixed levels based only on kurtosis values a machine learning technique could be applied to include RMS and CF to improve the fault detection for lubrication starvation.

A Fast Kurtogram was used to improve the fault detection based on kurtosis. An increase in sensitivity to low lubrication levels and a better discrimination between dry and lubricated conditions are some of the benefits. In future work, determining if FTF value decreases with preload is important as it supports the idea that the characteristic signal of lubrication starvation is generated due to the increase of clearance gap between bearing components. This could be done by applying an envelope analysis to the predominant frequency detected by the fast-kurtogram as explained previously by Boškoski.

Acknowledgments: The authors would like to thank Jeromy Antoni for the script used to generate the fast-kurtogram.

References:

[1] H. Penrose, "Large Electric Motor Reliability: What Did The Studies Really Say?," 2012. [Online]. Available: http://www.maintenancetechnology.com/2012/03/large-electric-motor-reliability-what-did-the-studies-really-say/.

[2] T. A. Harris, "Rolling bearing types and applications," in *Rolling Bearing Analysis*, 4th ed., 2001, pp. 1–44.

[3] C. Radu, "The most common causes of bearing failure and the importance of bearing lubrication," 2010. [Online]. Available: http://www.rkbbearings.com/admin/gest_docs_res/attach1/1271884609.pdf.

[4] B. P. Graney and K. Starry, "Rolling Element Bearing Analysis," *Mater. Eval.*, vol. 70, no. 1, pp. 78–85, 2012.

[5] Y. H. Wijnant, J. A. Wensing, and G. C. Nijen, "The influence of lubrication on the dynamic behaviour of ball bearings," *J. Sound Vib.*, vol. 222, no. 4, pp. 579–596, May 1999.

[6] R. B. Randall, *Vibration-based condition monitoring: industrial, aerospace and automotive applications*. John Wiley & Sons, 2011.

[7] T. Williams, X. Ribadeneira, S. Billingon, and T. Kurfess, "Rolling element bearing diagnostics in run-to-failure lifetime testing," *Mech. Syst. Signal Process.*, vol. 15, pp. 979–993, 2001.

[8] C. Pachaud, R. Salvetat, and C. Fray, "Crest factor and kurtosis contributions to identify defects inducing periodical impulsive forces," *Mech. Syst. Signal Process.*, vol. 11, pp. 903–916, 1997.

[9] N. Tandon and A. Choudhury, "A review of vibration and acoustic measurement methods for the detection of defects in rolling element bearings," *Tribol. Int.*, vol. 32, no. 8, pp. 469–480, 1999.

[10] G. K. Singh and S. A. S. Al Kazzaz, "Isolation and identification of dry bearing faults in induction machine using wavelet transform," *Tribol. Int.*, vol. 42, no. 6, pp. 849–861, 2009.

[11] P. Boškoski, J. Petrovčič, B. Musizza, and Đ. Juričić, "Detection of lubrication starved bearings in electrical motors by means of vibration analysis," *Tribol. Int.*, vol. 43, no. 9, pp. 1683–1692, Sep. 2010.

[12] N. Akturk, M. Uneeb, and R. Gohar, "The Effects of Number of Balls and Preload on Vibrations Associated With Ball Bearings," *J. Tribol.*, vol. 119, no. 4, p. 747, 1997.

[13] M. N. Kotzalas and T. A. Harris, "Fatigue Failure Progression in Ball Bearings," *J. Tribol.*, vol. 123, no. 2, p. 238, Apr. 2001.

[14] S. A. Spiewak and T. Nickel, "Vibration based preload estimation in machine tool spindles," *Int. J. Mach. Tools Manuf.*, vol. 41, no. 4, pp. 567–588, 2001.

[15] N. Sawalhi and R. B. Randall, "Simulating gear and bearing interactions in the presence of faults," *Mech. Syst. Signal Process.*, vol. 22, no. 8, pp. 1952–1966, Nov. 2008.

[16] R. Dwyer, "Detection of non-Gaussian signals by frequency domain Kurtosis estimation," in ICASSP '83. IEEE International Conference on Acoustics, Speech, *and Signal Processing*, vol. 8, pp. 607–610.

[17] J. Antoni, "Fast computation of the kurtogram for the detection of transient faults," *Mech. Syst. Signal Process.*, vol. 21, no. 1, pp. 108–124, 2007.

FAILURE ANALYSIS OF A REACTOR POOL COOLING PUMP USING MODAL AND VIBRATION ANALYSIS

Thomas J. Hazelwood, Larry D. Phillips, and Blake W. Van Hoy

Oak Ridge National Laboratory

1 Bethel Valley Road

Oak Ridge, TN 37831

Telephone: (865) 576-0340

vanhoybw@ornl.gov

Abstract: Two 75 HP pumps redundantly supply cooling water to the reactor pool of the High Flux Isotope Reactor (HFIR) at the Oak Ridge National Laboratory (ORNL). Due to a recent history of premature bearing failures, one of these pumps has undergone maintenance to deal with possible issues of misalignment and base looseness. Vibration analysis and modal analysis including steady state spectrum, operational deflection shape, run up and down order tracking, and modal impact have been utilized to verify the effectiveness of the maintenance and identify possible remaining failure modes. The studies conclude that the pump is, after the maintenance, in an overall good conditional state as per ISO 10816, but a few failure modes remain. These modes consist of some shaft unbalance, considerable shaft misalignment intensified by piping movement, possible motor ground fault, hydrodynamic issues such as cavitation with modal interaction, and base looseness. These failure modes and their supporting data have been used to make suggestions for future maintenance, to verify the effectiveness of the previous maintenance, and to provide a base on which to check future data. This report will cover the testing setup, methodology, analysis results, and maintenance suggestions.

Keywords: Condition monitoring; diagnostics; failure prevention; fault analysis; signal analysis

Introduction: Two 75 HP pumps redundantly supply cooling water to the reactor pool of the High Flux Isotope Reactor (HFIR) at the Oak Ridge National Laboratory (ORNL). One of these pumps, referred to as PU-9A, has a history of premature bearing failures; therefore, it has undergone maintenance. The bearing failures were thought to be produced by shaft misalignment and base looseness. These issues were addressed during the maintenance. The purpose of the testing summarized in this report is to evaluate the effectiveness of the maintenance and to diagnose any remaining failure modes. The data acquisition equipment and sensor locations will be denoted. The vibration analysis and modal analysis included as part of this report are steady state spectrum, steady state operational deflection shape, run up waterfall, run down waterfall, stationary spectrum, and modal impact tests. The results of these tests will be summarized and used to provide suggestions for future pump maintenance.

Testing Setup: A laptop with commercial data acquisition and analysis software was used to acquire the data, post process the data, and display the results. The results include spectrum charts, waterfall graphs, operational deflection shape videos, modal shape videos, and time data audio clips. Four commercial, 8 input/2 output channel data acquisition hardware units were used in conjunction with the software (See 'Figure 1').

Figure 1: Data Acquisition Hardware

Twenty-five simultaneous channels of accelerometers were epoxied to the pump, motor, base, and piping. The outboard motor and outboard pump both received horizontal, vertical, and axial accelerometers (See 'Figure 2' and 'Figure 3').

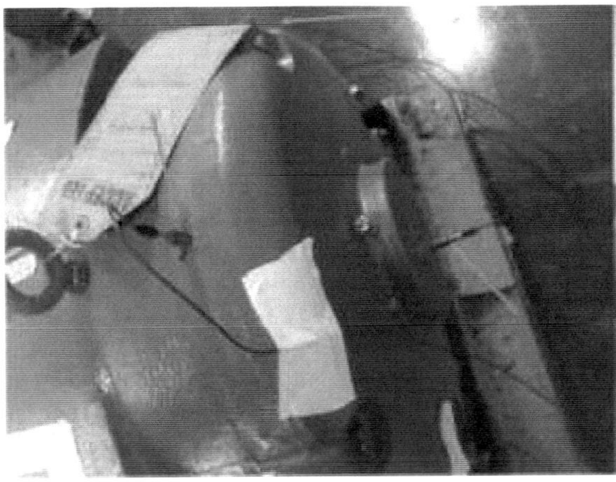

Figure 2: Outboard Motor Accelerometers

Figure 3: Outboard Pump Accelerometers

The inboard motor and inboard pump both received horizontal and vertical accelerometers (See 'Figure 4' and 'Figure 5').

Figure 4: Inboard Motor Accelerometers

Figure 5: Inboard Pump Accelerometers

The base received six vertical accelerometers on and between the mounting bolts (See 'Figure 6' and 'Figure 7').

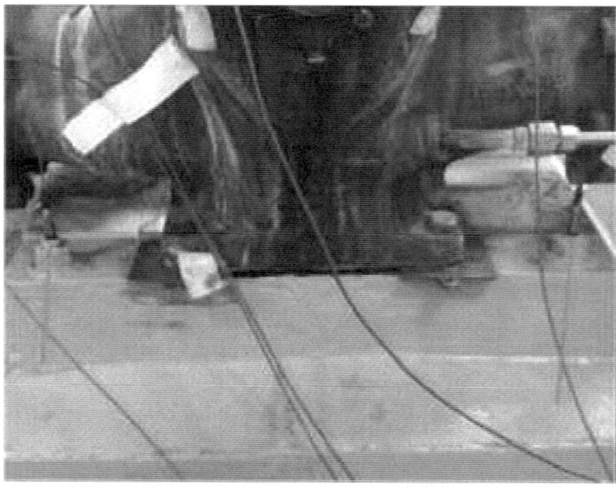

Figure 6: Base Bolt Accelerometers

Figure 7: Base Between Bolts Accelerometers

The suction pipe flange, suction mid-pipe, and discharge elbow each received a triaxial accelerometer (See 'Figure 8', 'Figure 9', and 'Figure 10').

Figure 8: Suction Flange Accelerometer

Figure 9: Suction Mid-Pipe Accelerometer

Figure 10: Discharge Elbow Accelerometer

Steady State Spectrum: The steady state spectrum data set consists of several minutes of time waveform data. The data was sectioned into blocks and each block was transformed into the frequency domain using the Fast Fourier Transform (FFT). Each FFT was then averaged into a single set of spectrum data. This process was repeated for each accelerometer. Significant and representative data will be displayed and discussed within this report. It is important to note that the run speed of the motor is 59.3 Hz. One and two times run speed (1x and 2x) peaks with harmonics show up in the motor's spectrum (See 'Figure 11" and 'Figure 12').

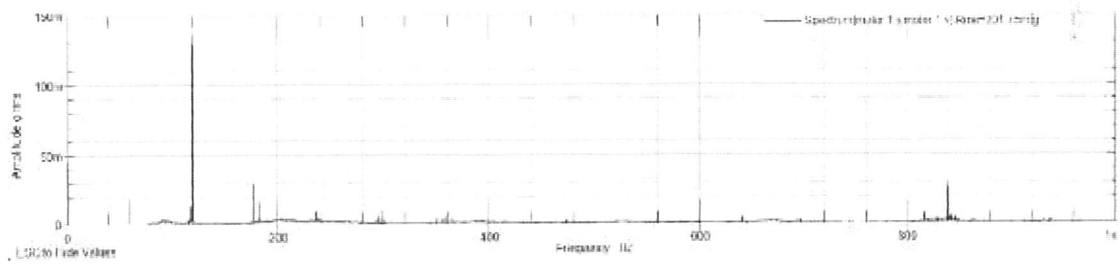

Figure 11: Steady State Outboard Motor Spectrum (Horizontal Radial)

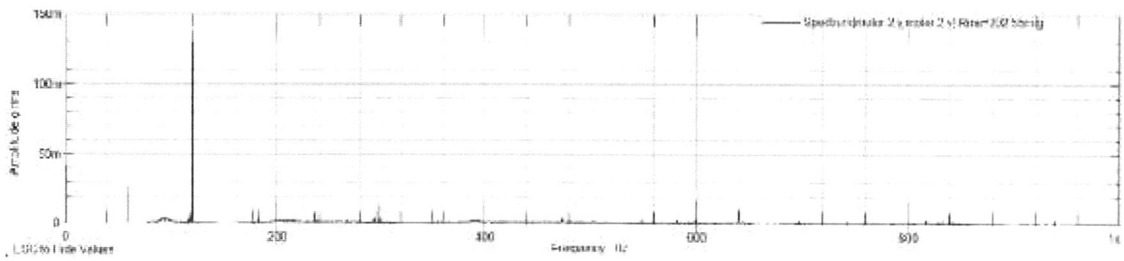

Figure 12: Steady State Inboard Motor Spectrum (Horizontal Radial)

One and two times run speed peaks with harmonics also show up in the pump's spectrum with the addition of broadband noise (See 'Figure 13" and 'Figure 14'). Also, note the twin peak in 'Figure 13' around 720 Hz, as this will be discussed in the Modal Impact section.

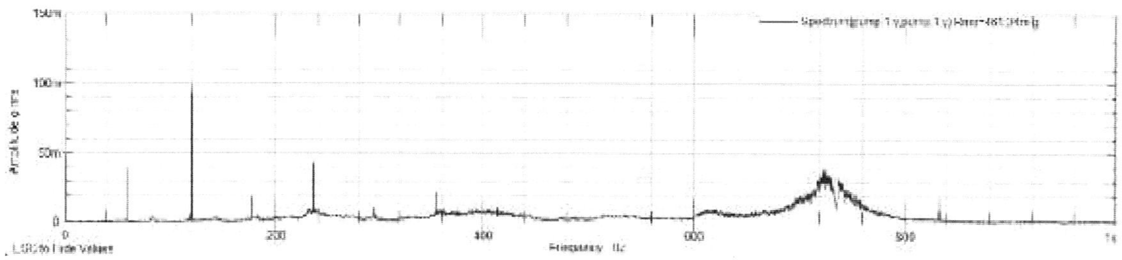

Figure 13: Steady State Inboard Pump Spectrum (Horizontal Radial)

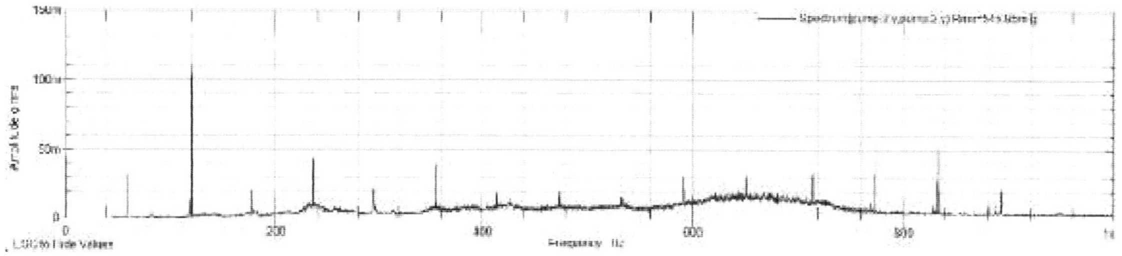

Figure 14: Steady State Outboard Pump Spectrum (Horizontal Radial)

The 1x vibration is indicative of some shaft unbalance, but it is much lower than the 2x vibration which is indicative of shaft misalignment. Care needs to be taken analyzing the

2x vibration in separating the two times run speed (118.6Hz) from two times the line frequency (120Hz). The bearing frequencies can be seen but are not significantly higher that the noise floor. This should be expected since they have recently been replaced. The broadband noise of the pump is indicative of hydrodynamic effects such as cavitation. Audio files of the pump and motor along with impulses within the time domain back up the impression of cavitation.

Steady State Operational Deflection Shape (ODS): Magnitude and phase data as a function of time from each accelerometer is combined with a simple representation of the system geometry to illustrate the operational deflection shape. The first ODS of interest, seen in 'Figure 15', utilizes the spectrum data from 1x the run speed. All deflections are scaled to the largest motion, which is 1 mil peak-peak of the discharge elbow. The dotted lines represent the non-deformed geometry. The shaft and the base have a relatively small deflection at this frequency.

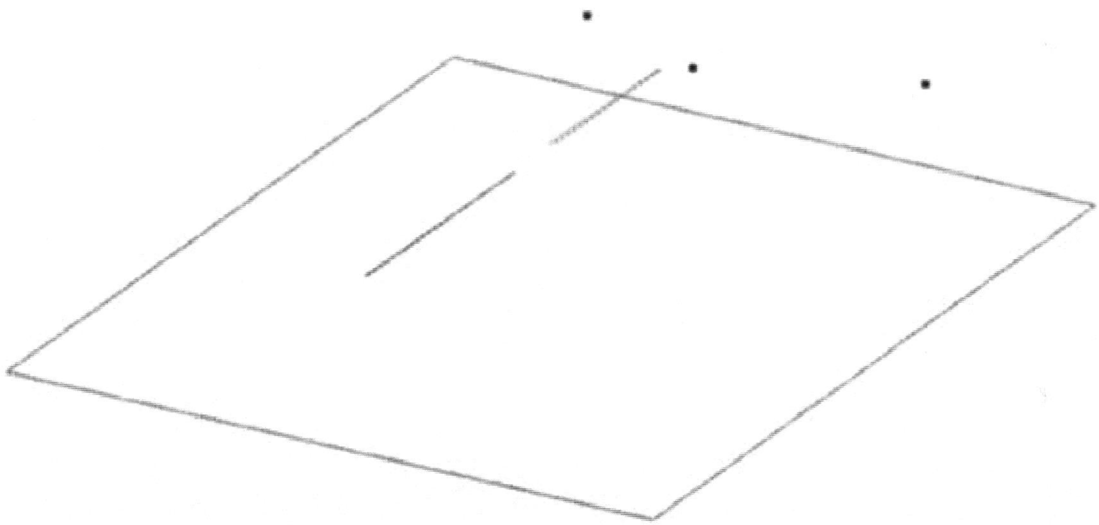

Figure 15: 1x Run Speed ODS

The Second ODS of interest, seen in 'Figure 16', utilizes the spectrum data from 2x the run speed. All deflections are scaled to the largest motion, which is 1 mil peak-peak of the suction mid pipe. There is high deflection of the shaft that also transfers to the base. A 2x run speed vibration of the shaft is often due to misalignment of the shaft, which the movement of the pipes can amplify.

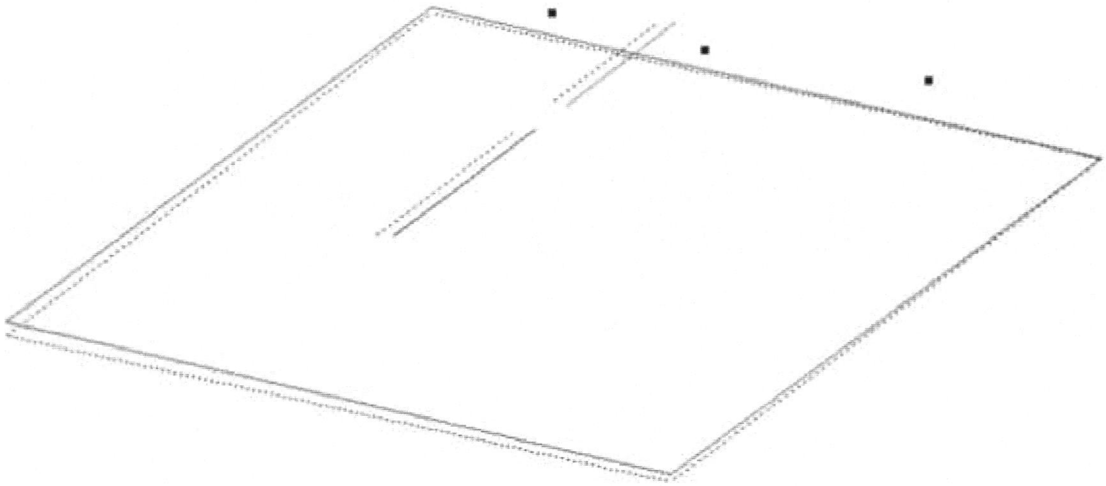

Figure 16: 2x Run Speed ODS

Run Up Waterfall: The run up waterfall data set is similar to the steady state spectrum except that the data is taken during a transient event of the system; therefore, the data is not averaged. Instead, the data is plotted repeatedly stacking up with time. This data tends to be noisier since an average is not calculated. The transient in this case is the pump going from rest to operational speed. The run up of this pump occurred quickly and the data was not significantly different than the steady state spectrum data, aside from the additional noise.

Run Down Waterfall: The run down waterfall data set is taken in the exact same manner as the run up waterfall. The transient in this case is the pump going from operational speed to rest. 'Figure 17' is a representative measurement from the motor, plotted in reverse to aid in viewing. The lines in the front occurred last and the peaks occurred first. A 2x run speed peak is clear and some additional noise is present. The energy dissipates quickly, but some energy remains. This could be cross talk from the already running redundant pump.

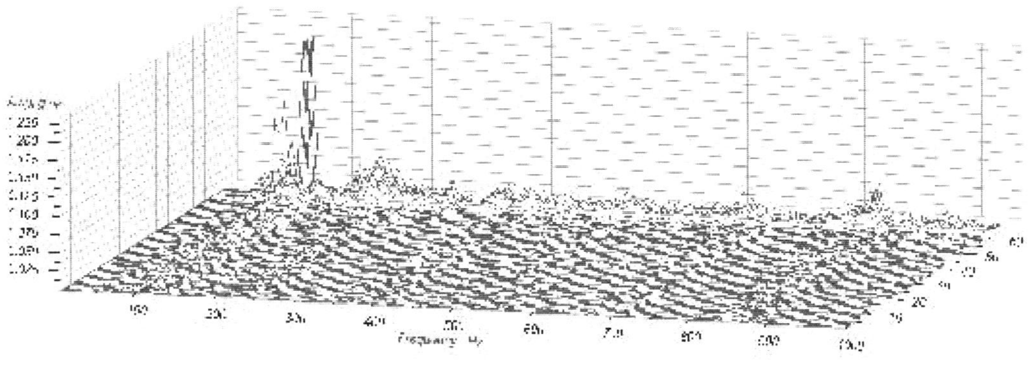

Figure 17: Motor Run Down Waterfall

"Figure 18' is a representative measurement from the pump, also plotted in reverse. A 2x run speed peak is clear. Broadband noise is prevalent and a general peak around 700Hz is present. This peak will be discussed in the modal impact section.

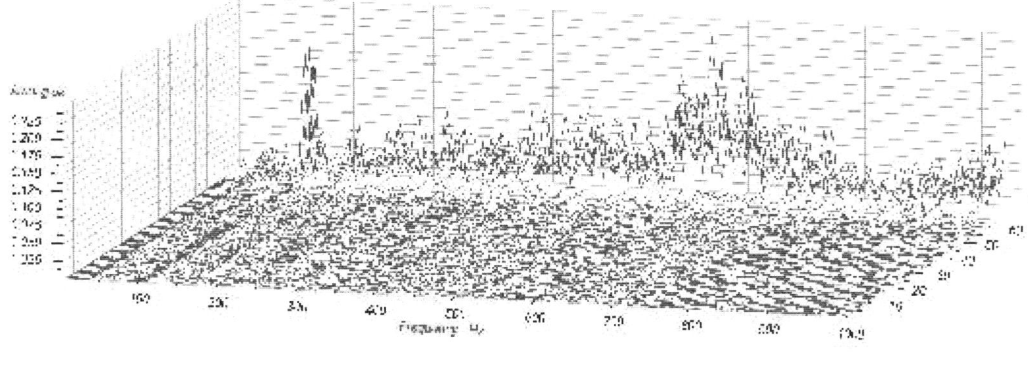

Figure 18: Pump Run Down Waterfall

Stationary Spectrum: The stationary spectrum data set was taken in a similar manner as the steady state spectrum except that the pump has been turned off and is at rest. The neighboring pump, PU-9B, is about ten feet away and has been turned on to continue the supply of cooling water. This data measures cross talk from PU-9B to the accelerometers on PU-9A. 'Figure 19' shows that the magnitude of the crosstalk present in the pump is about six percent of the pump during normal operation.

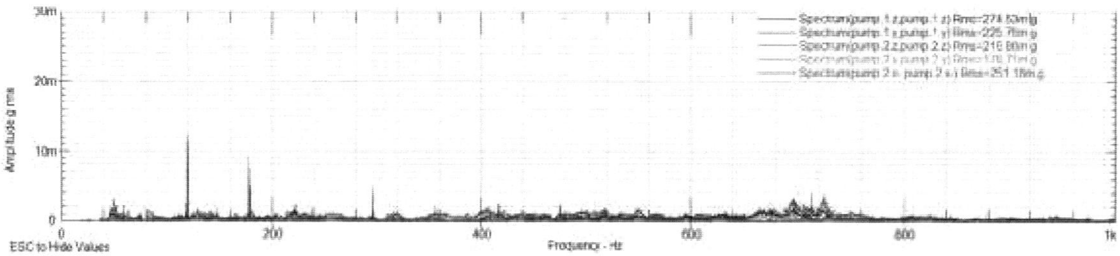

Figure 19: Stationary Pump Spectrum

'Figure 20' shows that the magnitude of the crosstalk present in the piping is about twenty percent of piping during normal operation.

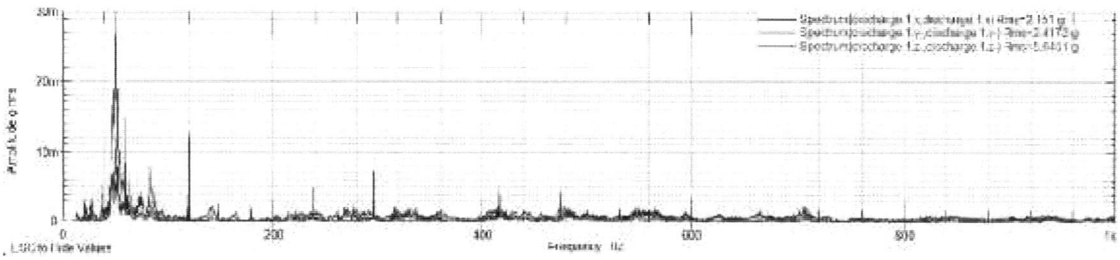

Figure 20: Stationary Discharge Spectrum

These results are expected, given the proximity of the two systems and that they share some common piping.

Modal Impact: The modal impact data set is taken while the pump is at rest. Modal impact analysis was preformed utilizing an additional channel of the data acquisition and a commercial modal impact hammer. Four averages where taken at each impact location and all twenty-six channels were acquired simultaneously. Using the hammer channel as the reference, a Frequency Response Function (FRF) was calculated for each channel. 'Figure 21' is a representative data set for the motor. The first modal resonance of the motor occurs at around 700Hz, but the mode does not appear to be excited during normal operation as seen in the steady state spectrum section of this report.

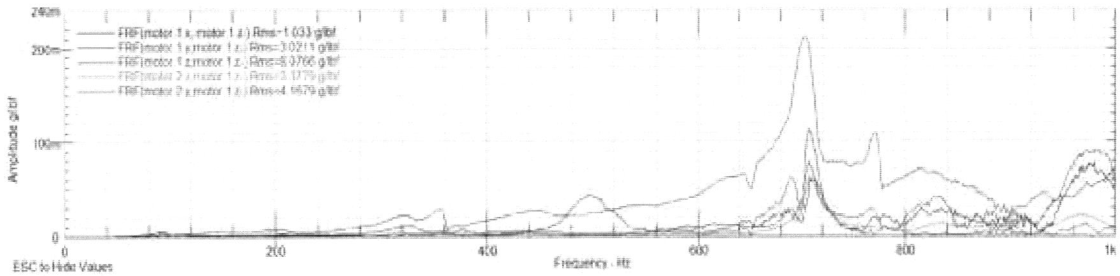

Figure 21: Motor FRF

'Figure 22' is a representative data set from the pump. The first modal resonance of the pump occurs at around 740Hz. Some modal interaction with this mode occurs during normal operation as seen in the steady state section of this report, but it is barely above noise level.

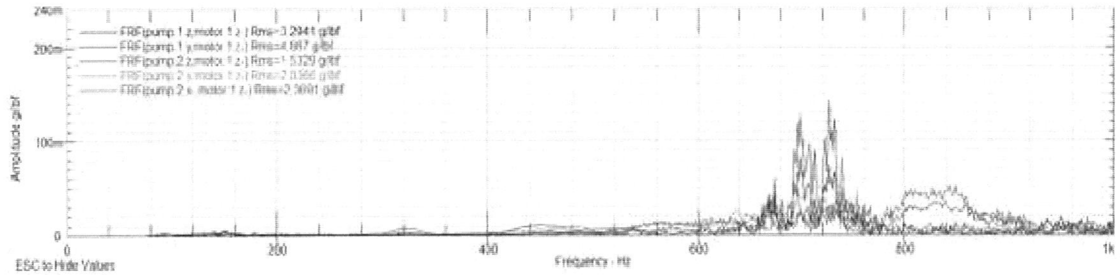

Figure 22: Pump FRF

Analysis Results Summary and Maintenance Suggestions: Overall, the pump is in a good conditional state as per ISO 10816 vibration standard. The 2x run speed vibrations seen in the steady state spectrum data and ODS data are indicative of misalignment in the shaft during operation. The motor's electrical grounding could be checked to rule out electrical fault causing a 120Hz peak and validating that it is in fact misalignment alone. The new bearing frequencies are present but not excessive, therefore, they have not worn out and were not damaged during installation. The high energy broadband noise in the pump spectrum is indicative of hydrodynamic issues such as cavitation. This could possibly be due to inadequate suction head pressure. It is suspected that this is due to piping layout. A piping engineer could verify this and suggest changes. If a piping teardown, redesign, and rebuild takes place, it is suggested that a new concrete and steel base be considered. The only modal interaction observed from the impact test is around 740 Hz at the pump and it is barely above noise level. This interaction could get worse if the cavitation were to get more severe. It is out of range of most other mechanical issues. Some cross talk from PU-9B can be measured on PU-9A when 9B is running, but it is less than twenty percent of the energy of operational 9A.

Conclusion: Due to a recent history of premature bearing failures, one of the two redundant cooling water supply pumps of HFIR at ORNL has undergone maintenance including base looseness and shaft misalignment work. The testing summarized within this report corroborate that the maintenance was effective and that the pump is in a good conditional state. However, some failure modes do persist. The testing setup section lists the equipment used and the locations of the accelerometers on the system. The steady state spectrum section identifies a 2x run speed vibration indicative of shaft misalignment, and broadband noise in the pump is indicative of hydrodynamics issues. The steady state ODS section corroborates the shaft misalignment and identifies base and pipe movement. The run up waterfall section did not add additional information. The run down waterfall section corroborates the 2x run speed vibration in the system and broadband noise in the pump. The stationary spectrum section describes some cross talk from the PU-9B while PU-9B is powered down. The modal impact section, along with the steady state spectrum section, highlight some modal integration during normal operation of the pump at around 740Hz. The analysis results summary and maintenance suggestions section shows that the analysis results do not immediately warrant system maintenance, but indicate areas of future work. The motor's electrical grounding should be checked. The piping layout should be checked by a piping engineer and, if piping work is done, a new concrete and steel base is recommended.

Acknowledgments: This research used resources of the High Flux Isotope Reactor, which is a DOE Office of Science User Facility.

FAULT DETECTION AND ESTIMATION IN A CLASS OF DISTRIBUTED PARAMETER SYSTEMS

Hasan Ferdowsi
Texas A&M University – Texarkana
Electrical Engineering Department
7101 University Ave,
Texarkana, TX 75503
Telephone: (903) 223-3181
Hasan.Ferdowsi@mst.edu or Hasan.Ferdowsi@mst.edu

Abstract: In this paper, the problem of fault detection and isolation in a class of distributed parameter systems (DPS) will be investigated. The behavior of distributed parameter systems is best described by partial differential equation (PDE) models. However, due to complex nature of DPS, a PDE model is traditionally transformed into a finite set of ordinary differential equations (ODE) prior to the design of control or fault detection schemes by using significant approximations thus reducing the accuracy and reliability of the overall system. By contrast, in this paper, the PDE representation of the system is directly utilized to design the fault diagnosis scheme for DPS. Faults that can occur anywhere in the domain of the DPS (referred to as state faults) are considered, rather than only actuator and sensor faults. State faults are significantly more complicated to deal with in the case of DPS since they can be initiated anywhere within a continuous range of space, while in practice sensors are only available at limited locations which in many cases only include the input and/or output sides of the DPS. This problem is tackled by using an observer structure, which includes input, and output filters directly based on the PDE model of the system. A fault is detected by comparing the detection residual, which is the difference between measured and estimated outputs, with a predefined detection threshold. Once the fault is detected, an online approximator is activated to learn the fault function. An update law is introduced for updating the unknown parameters of the online approximator. The stability of the observer along with the online approximator will be discussed analytically in the paper. It is shown that one sensor is satisfactory for fault detection and approximation if the fault function has only one unknown parameter or can be expressed as linear in the unknown parameters. However, additional sensors are required for fault approximation or isolation in the general case. For example, a leakage fault in a pipeline has magnitude and location as unknown parameters and these parameters cannot be approximated by using one sensor. An algorithm is designed to approximate the location of a state fault with unknown magnitude by using multiple sensors. The distributed parameter systems considered in this paper are modelled by parabolic partial differential equations with Neumann or Dirichlet boundary conditions. Heat transfer systems and fluid pipelines are examples of systems in this class of DPS. The scheme is verified in simulations on the aforementioned systems.

Key words: Diagnostics; prognostics; distributed parameter systems

Introduction: Fault diagnosis has become an attractive research topic in the past couple of decades due to increased complexity of industrial systems and safety for such systems has become more important than ever. Among the different methods of fault diagnosis, model-based methods have become both popular and more suitable when a mathematical model of the system under consideration is either available or can be obtained [1] because they do not need extensive amount of offline data and can operate online without requiring additional sensors. Therefore, model-based fault diagnosis schemes have been developed for lumped parameter systems based on their ordinary differential equation (ODE) representation by using sliding mode observers [2, 3], online approximators [4], geometric approach [5], and adaptive observers [6, 7].

A large number of industrial systems which involve heat transfer, fluid dynamics, and others are classified as distributed parameter systems (DPS). Application examples of such DPS include hydraulic systems, chemical processes, flexible robots, and aerospace systems. Due to the wide range of such systems and their important and sensitive role in the industry, reliable fault detection and diagnosis schemes are required to guarantee their safe operation.

The variables in DPS are defined over a continuous range of space [8], which makes them different from lumped parameter systems where each variable only evolves in time. The most comprehensive and accurate mathematical representation for these systems is given in terms of partial differential equations (PDEs). Limited work has been done on DPS when compared to the systems with ODE models, because dealing with PDEs is much more complicated due to boundary conditions and infinite number of states [8].
In order to simplify the problem of dealing with PDE model, an infinite set of ordinary differential equations (ODEs) is employed and then the Galerkin's method is applied to obtain an approximate finite dimensional ODE model [9, 10] which is subsequently utilized for fault detection and diagnosis. A learning systems approach is introduced in [11] for fault detection of such systems whereas fault detection and isolation of such DPS with actuator faults are discussed in [12], while a geometric approach is proposed for fault detection and isolation of dissipative parabolic PDEs [13].

Although this method has opened new doors to the problem of control and fault detection of DPS, it has a number of challenges. First of all, these methods [11-13] can possibly render inaccurate results since they utilize an approximate system dynamics by neglecting a significant portion of the system dynamics. In addition, there is no guarantee that the system output can be generated as a function of only the states of the finite dimensional ODE system. Further, when a fault happens in the system, the PDE dynamics will change, which can make the approximated ODE model even more inaccurate.

In this paper, a class of DPS modeled by linear parabolic PDE is investigated. Faults that are considered here are state faults which can occur anywhere in the domain of the DPS and affect the PDE dynamics of the system rather than faults which only affect input or outputs through boundary conditions. State faults are significantly more complicated to deal with in the case of DPS since they can be initiated anywhere within a continuous range of space, while in practice sensors are only available at limited locations which in many cases only include the input and/or output sides of the DPS. This problem is tackled by using an observer structure which includes input, output, and fault filters. The combination of input and output filters provides an estimate of the system behavior under healthy conditions while the fault filters are added upon detection to help in fault approximation based on the assumption that the fault can be represented as a linear combination of known basis functions. Therefore, in contrast with existing ODE-based schemes [11-13], the detection observer is designed directly based on the PDE model and the aforementioned filters. It is shown that the proposed observer can estimate both measured and unmeasured system states in the healthy operating conditions with a bounded error. Detection residual is generated by comparing the measured and estimated system outputs. Since the residual is bounded in healthy conditions, a fault can be detected by comparing it with a predefined detection threshold. It should be noted that in earlier works of the author fault diagnosis and fault accommodation schemes have been developed for DPS directly based on the PDE model of the system [14,15]. However, the schemes presented in [14,15] require the distributed state to be measureable at all points, which is a very strong assumption for DPS. In contrast, the scheme presented in this paper only requires the boundary values of the system state, providing a more practical method for fault diagnosis in DPS.

Upon detecting a fault, a number of fault filters, which reflect the effect of fault basis functions on the system behavior are activated and incorporated in the PDE observer with adaptive weights. The particular fault that has occurred will be approximated as a weighted summation of fault basis functions, where the adaptive weights are tuned by an adaptive update law, which seeks to minimize the error between observer output and measured output. The stability of both the proposed observer and fault approximator is investigated analytically.

The effectiveness and stability of the proposed scheme is verified by using a heat system and a fluid pipeline. The dynamics of both systems are described by parabolic PDEs. The proposed PDE observer is utilized to estimate the system states and provide an estimation of the system output which is used to generate detection residual. Fault detection and estimation are successfully performed in simulations and the results are presented in the last section of this paper.

The paper is organized as follows: First the system and its model are described mathematically. Then the PDE observer and online estimation of fault dynamics are presented, and finally the verification of proposed scheme in simulations is provided.

System Description: Consider the class of linear system described by the following normalized PDE

$$\frac{\partial \bar{x}(\bar{z},t)}{\partial t} = \bar{a}\frac{\partial^2 \bar{x}(\bar{z},t)}{\partial \bar{z}^2} + \bar{b}\frac{\partial \bar{x}(\bar{z},t)}{\partial \bar{z}} + \bar{c}\,\bar{x}(\bar{z},t), \tag{1}$$

for $(\bar{z},t) \in (0,L) \times \mathbb{R}^{+}$, with the following boundary conditions

$$\begin{cases} \dfrac{\partial \bar{x}(0,t)}{\partial \bar{z}} = \bar{q}\,\bar{x}(0,t), \\ \dfrac{\partial \bar{x}(L,t)}{\partial \bar{z}} = \bar{u}(t) \qquad or \qquad \bar{x}(L,t) = \bar{u}(t) \end{cases} \tag{2}$$

where $\bar{x} \in (0, L) \times \mathbb{R}$ is the state of the system, $\bar{a}, \bar{b}, \bar{c}$, and $\bar{q}$ are scalar constants, and $\bar{u} \in \mathbb{R}$ is the control input applied as either Dirichlet or Neumann boundary control at $\bar{z} = L$. Suppose that the output of the system is at the opposite end of the actuation, i.e. the output $\bar{y} \in \mathbb{R}$ which is the only available measurement given by

$$\bar{y}(t) = \bar{x}(0,t). \tag{3}$$

Now the PDE in (1) can be normalized and $\bar{b}$ can be eliminated from the equation by using the following transformation [16]

$$\bar{x}(\bar{z},t) \to x(z,t)e^{-\frac{bz}{2a}}.$$

After applying this transformation, the system will be described by

$$\frac{\partial x(z,t)}{\partial t} = a\frac{\partial^2 x(z,t)}{\partial z^2} + cx(z,t), \tag{4}$$

$$\begin{cases} \dfrac{\partial x(0,t)}{\partial z} = qx(0,t), \\ \dfrac{\partial x(1,t)}{\partial z} = u(t) \qquad or \qquad x(1,t) = u(t), \end{cases} \tag{5}$$

$$y(t) = x(0,t), \tag{6}$$

for $(z, t) \in (0,1) \times \mathbb{R}^+$, where $x \in (0,1) \times \mathbb{R}$ is the system state, and a, c, and q are scalar constants such that $c < a\pi^2/4$. Next consider a fault in the domain of the DPS which is modeled by $h(z, t)$. Then the system representation in (4) and (5) can be rewritten in the presence of a fault as

$$\frac{\partial x(z,t)}{\partial t} = a\frac{\partial^2 x(z,t)}{\partial z^2} + cx(z,t) + h(z,t), \tag{7}$$

subject to the boundary conditions

$$\begin{cases} \dfrac{\partial x(0,t)}{\partial z} = qx(0,t), \\ \dfrac{\partial x(1,t)}{\partial z} = u(t) \qquad or \qquad x(1,t) = u(t). \end{cases} \tag{8}$$

Moreover, the fault function can be represented by

$$h(z,t) = \Omega(t - t_0)\bar{h}(z,t),$$

where $\Omega(t - t_0)$ is the time profile of the fault defined by

$$(\tau) = \begin{cases} 0 & , \ if\, \tau < \\ 1 - e^{-\kappa\tau} & , \ if\, \tau \geq \end{cases},$$

with $\kappa \in \mathbb{R}^+$ is an unknown constant that is determined by the growth rate of the fault. Although, this time profile is basically used to model incipient faults, it can also address abrupt faults for large values of κ. The following standard assumption is needed in order to proceed.

Assumption 1: The fault function can be expressed as linear in the unknown parameters (LIP) [17], i.e. $h = w^T o(y,z)$ where $w \in \mathbb{R}^p$ is the unknown parameters vector and $o(\cdot)$ is a known nonlinear vector function which is bounded by $\|o\| \leq o_m$.

Fault Detection Observer: In order to detect a fault, an observer is utilized to estimate the system output in healthy conditions. Then the estimated and measured outputs will be compared to generate fault detection residual. For the purpose of observer development based on the PDE model of the system in healthy operating conditions, we first investigate the healthy system dynamics given in (4) and (5). Since (4) and (5) define a linear PDE system, its solution is the summation of system response due to initial conditions plus the system response due to external inputs. If we consider both the boundary terms $q\ (0,t)$ and $u(t)$ as external inputs, then the solution $x(z,t)$ can be obtained as

$$x(z,t) = x_0(z,t) + x_1(z,t) + x_2(z,t), \tag{9}$$

where $x_C(z,t)$ is the response of the system due to initial conditions with zero external inputs, i.e. the solution of

$$\frac{\partial x_0(z,t)}{\partial t} = a\frac{\partial^2 x_0(z,t)}{\partial z^2} + cx_0(z,t), \tag{10}$$

$$\begin{cases} \dfrac{\partial x_0(0,t)}{\partial z} = 0, \\ \dfrac{\partial x_0(1,t)}{\partial z} = 0 \quad or \quad x_0(1,t) = 0, \end{cases} \tag{11}$$

$$x_0(z,0) = x(z,0), \tag{12}$$

with $x_1(z,t)$ is the response of the system due to $q\ (0,t) = q\ (t)$, i.e. the solution of

$$\frac{\partial x_1(z,t)}{\partial t} = a\frac{\partial^2 x_1(z,t)}{\partial z^2} + cx_1(z,t), \tag{13}$$

$$\begin{cases} \dfrac{\partial x_1(0,t)}{\partial z} = qy(t), \\ \dfrac{\partial x_1(1,t)}{\partial z} = 0 \quad or \quad x_1(1,t) = 0, \end{cases} \tag{14}$$

$$x_1(z,0) = 0, \tag{15}$$

and $x_2(z,t)$ is the response of the system due to $u(t)$, i.e. the solution of

$$\frac{\partial x_2(z,t)}{\partial t} = a\frac{\partial^2 x_2(z,t)}{\partial z^2} + cx_2(z,t), \tag{16}$$

$$\begin{cases} \dfrac{\partial x_2(0,t)}{\partial z} = 0, \\ \dfrac{\partial x_2(1,t)}{\partial z} = u(t) \quad or \quad x_2(1,t) = u(t), \end{cases} \tag{17}$$

$$x_2(z,0)=0. \tag{18}$$

The system defined by (13-15) is called an output filter [16] since it is derived by the output of the actual system, while the system defined by (16-18) is called an input filter [16] since it is derived by the input of the actual system.

The fault detection observer is now designed based on this discussion. Since the system dynamics are known and the external inputs that are deriving the systems (13-15) and (16-18) are both available, $x_1(z,t)$ and $x_2(z,t)$ can be obtained for $0 \le z \le 1$ when$t > 0$. If the initial condition $x(z,0)$ was available for all values of $0 \le z \le 1$, then $x_c(z,t)$ could also be obtained, giving us an exact estimation of system state using equation (9). However, the initial conditions are generally not known, thus $x_c(z,t)$ cannot be used in the observer development. But if the system defined by (10) and (11) is asymptotically stable, then its solution $x_c(z,t)$ will converge to zero, so the state and output estimates provided by the following observer

$$\hat{x}(z,t)=x_1(z,t)+x_2(z,t), \tag{19}$$

$$\hat{y}(t)=\hat{x}(0,t), \tag{20}$$

where $\hat{x} \in (0,1) \times \mathbb{R}$ and $\hat{y} \in \mathbb{R}$ are the observer state and output respectively, will converge to the true value of the system state and output asymptotically.

The detection residual $e \in \mathbb{R}$ is defined as the difference between actual and estimated system outputs, i.e.

$$e(t)=y(t)-\hat{y}(t). \tag{21}$$

Fault detection is performed by comparing this detection residual with a predefined detection threshold which is derived. A fault is detected when the absolute value of the detection residual $|e|$ exceeds the detection threshold $\rho \in \mathbb{R}^{+}$. The next theorem discusses the boundedness of residual under healthy conditions and selection of detection threshold.

Theorem 1 (Fault Detection Observer Performance): Let the observer in (19) and (20) be used to monitor the system in (4)-(6). If the threshold for the detection residual is selected as

$$\rho=|e(0)|\left(e^{-\frac{a\pi^2-4c}{4}t}+\eta\right), \tag{22}$$

then a fault will not be detected as long as the system is working under healthy operating conditions.

Proof: First define the state residual $\tilde{x} \in (0,1) \times \mathbb{R}$ as the difference between the actual and estimated state$\tilde{x}=x-\hat{x}$. When the system is working in the healthy condition, the state residual is given by $\tilde{x}(z,t)=x_0(z,t)$. Therefore, the state residual dynamics in the case of Neumann boundary are described by

$$\frac{\partial \tilde{x}(z,t)}{\partial t}=a\frac{\partial^2 \tilde{x}(z,t)}{\partial z^2}+c\tilde{x}(z,t),$$

$$\begin{cases} \dfrac{\partial \tilde{x}(0,t)}{\partial z} = 0 \\ \dfrac{\partial \tilde{x}(1,t)}{\partial z} = 0 \end{cases},$$

with initial condition $\tilde{x}(z,0) = x(z,0)$. To show the stability of residual dynamics during the healthy conditions, select the following positive definite and radially unbounded Lyapunov function candidate

$$V(t) = \frac{1}{2}\|\tilde{x}\|^2 = \frac{1}{2}\int_0^1 \tilde{x}^2(z,t)\,dz.$$

The derivative of this Lyapunov function candidate is given by

$$\dot{V}(t) = \int_0^1 \frac{\partial \tilde{x}(z,t)}{\partial t}\tilde{x}(z,t)\,dz = \int_0^1 \left(a\frac{\partial^2 \tilde{x}(z,t)}{\partial z^2} + c\tilde{x}(z,t) \right)\tilde{x}(z,t)\,dz.$$

By applying integration by parts and then Poincare inequality [18], it can be shown that

$$\dot{V}(t) = -a\left\|\frac{\partial \tilde{x}(z,t)}{\partial z}\right\|^2 + c\|\tilde{x}\|^2 \le -\frac{a\pi^2 - 4c}{4}\|\tilde{x}\|^2.$$

Since $c < a\pi^2/4$, it can be concluded that the state residual is globally exponentially stable. Therefore, the residual always satisfies

$$|e(t)| \le |e(0)|\, e^{-\frac{a\pi^2 - 4c}{4}t}.$$

To avoid false alarms, a constant threshold can be selected as $\rho = (\eta + 1)|e(0)|$, or a time varying threshold can be selected as

$$\rho = |e(0)|\left(e^{-\frac{a\pi^2 - 4c}{4}t} + \eta \right),$$

where η is a small positive constant which is used in the threshold to ensure that $|e(t)|$ remains below the threshold at all times during healthy operation.

Remark 1: Theorem 1 defines a detection threshold in the absence of modeling uncertainty or noise. In the presence of uncertainties and noise, their upper bounds must be included in the derivation of detection threshold in order to prevent false alarms. Therefore, dealing with modeling uncertainties is part of the future work.

The next step in fault diagnosis is to determine the behavior of fault or approximating its dynamics, which allows further analysis of fault as well as estimation of remaining useful life of the system. To this end, an online approximator is added to the observer, which is discussed in the next section.

Online Fault Estimation: As mentioned above, the proposed detection observer is able to estimate the distributed system states and output with an asymptotically decreasing error in healthy operating conditions. When a fault occurs, the residual is no longer bounded and the fault is detected when the residual exceeds detection threshold. Upon detecting a fault, adaptive filters are added to the detection observer in order to estimate the fault parameters.

In the presence of fault, the system dynamics are described by the following PDE

$$\frac{\partial x(z,t)}{\partial t} = a\frac{\partial^2 x(z,t)}{\partial z^2} + cx(z,t) + \sum_{i=1}^{p} w_i\sigma_i(y,z), \tag{23}$$

for $(z,t) \in (0,1) \times \mathbb{R}^+$, where $x \in (0,1) \times \mathbb{R}$ is the system state, $\sigma = [\sigma_1 \dots \sigma_p]^T$ is the vector of fault basis functions, and $w = [w_1 \dots w_p]^T$ is the vector of unknown fault parameters. Similar to the design of input and output filters in the previous section, we now consider the terms $w_i^T \sigma_i(y,z)$ for $i = 1, \dots, p$ as external outputs to the system. Now we construct p fault filters to determine the response of system to each fault basis function. The ith fault filter is designed as

$$\frac{\partial \phi_i(z,t)}{\partial t} = a\frac{\partial^2 \phi_i(z,t)}{\partial z^2} + c\phi_i(z,t) + \sigma_i(y,z), \tag{24}$$

$$\begin{cases} \dfrac{\partial \phi_i(0,t)}{\partial z} = 0, \\ \dfrac{\partial \phi_i(1,t)}{\partial z} = 0 \quad or \quad \phi_i(1,t) = 0, \end{cases} \tag{25}$$

$$\phi_i(z,0) = 0 \tag{26}$$

for $(z,t) \in (0,1) \times \mathbb{R}^+$ and $i = 1 \dots p$, where $\phi_i \in (0,1) \times \mathbb{R}$ is the ith fault filter state. With this definition the system state in the presence of fault can be represented as

$$x(z,t) = x_0(z,t) + x_1(z,t) + x_2(z,t) + \sum_{i=1}^{p} w_i\phi_i(z,t). \tag{27}$$

The fault basis functions are known, so the solution of fault filters can be obtained and used in the observer design. However, the fault parameter w_i is not known, thus its estimated value $\hat{w}_i$ must be used in the diagnosis observer as follows

$$\hat{x}(z,t) = x_1(z,t) + x_2(z,t) + \sum_{i=1}^{p} \hat{w}_i(t)\phi_i(z,t), \tag{28}$$

$$\hat{y}(t) = \hat{x}(0,t), \tag{29}$$

for $(z,t) \in (0,1) \times \mathbb{R}^+$ and $i = 1 \dots p$. The adaptive update law for tuning the adaptive parameters to show boundedness of the residual and parameter estimation errors is provided in the next theorem.

Theorem 2 (Fault Diagnosis Observer Performance): Let the observer defined by (28) and (29) be utilized to monitor the DPS upon detecting a fault. If parameter update law is selected as

$$\dot{\hat{w}}(t) = \alpha\Phi(t)e(t) - \gamma\hat{w}(t), \tag{30}$$

where $\Phi(t) = [\phi_1(0,t), \dots, \phi_p(0,t)]^T$, $\hat{w} = [\hat{w}_1 \dots \hat{w}_p]^T$, γ is a positive constant, and α is the learning rate chosen such that

$$\alpha < \frac{2a\pi^2 - 8c}{\pi^2 + 4}, \tag{31}$$

then the residual e and the parameter estimation error $\tilde{w} = w - \hat{w}$ are uniformly ultimately bounded.

Proof: To prove the boundedness of estimation errors, a Lyapunov function candidate is chosen as

$$V(t) = \frac{1}{2}\|x_0\|^2 + \frac{1}{2}\sum_{i=1}^{p}\|\phi_i\|^2 + \frac{1}{2}\tilde{w}^T\tilde{w}.$$

Derivative of this Lyapunov function with respect to time is given by

$$\begin{aligned}
\dot{V}(t) &= \int_0^1 \frac{\partial x_0(z,t)}{\partial t} x_0(z,t)dz - \tilde{w}^T(t)\dot{\hat{w}}(t) + \sum_{i=1}^{p}\int_0^1 \frac{\partial \phi_i(z,t)}{\partial t}\phi_i(z,t)dz \\
&= \int_0^1 \left(a\frac{\partial^2 x_0(z,t)}{\partial z^2} + cx_0(z,t)\right)x_0(z,t)dz - \tilde{w}^T(t)\dot{\hat{w}}(t) \\
&\quad + \sum_{i=1}^{p}\int_0^1 \left(a\frac{\partial^2 \phi_i(z,t)}{\partial z^2} + c\phi_i(z,t) + \sigma_i(y,z)\right)\phi_i(z,t)dz.
\end{aligned}$$

By applying integration by parts and Poincare inequality, and using the adaptive update law given in (30), it can be shown that

$$\begin{aligned}
\dot{V}(t) \le &-\frac{a\pi^2 - 4c}{4}\|x_0\|^2 - \frac{a\pi^2 - 4c}{4}\sum_{i=1}^{p}\left(\|\phi_i\| - \frac{2\sigma_{i_{\max}}}{a\pi^2 - 4c}\right)^2 \\
&-\alpha\tilde{w}^T(t)\Phi(t)e(t) + \gamma\tilde{w}^T(t)\hat{w}(t) + \frac{\sigma_{\max}^2}{a\pi^2 - 4c}
\end{aligned}$$

where $\sigma_m^2 = \sum_{i=1}^{p}\sigma_{im}^2$. Since $e(t) = x(0,t) - \hat{x}(0,t) = x_c(0,t) + \tilde{w}^T(t)\Phi(t)$, the above inequality can be rewritten as

$$\begin{aligned}
\dot{V}(t) \le &-\frac{a\pi^2 - 4c}{4}\|x_0\|^2 - \frac{a\pi^2 - 4c}{4}\sum_{i=1}^{p}\left(\|\phi_i\| - \frac{2\sigma_{i_{\max}}}{a\pi^2 - 4c}\right)^2 \\
&-\alpha\left(e(t) + x_0(0,t)\right)e(t) + \frac{\gamma}{\alpha}\tilde{w}^T(t)\left(w - \tilde{w}(t)\right) + \frac{\sigma_{\max}^2}{a\pi^2 - 4c} \\
\le &-\left(\frac{a\pi^2 - 4c}{4} - \frac{\pi^2 + 4}{8}\alpha\right)\|x_0\|^2 - \frac{\gamma}{2\alpha}\|\tilde{w}\|^2 \\
&-\frac{a\pi^2 - 4c}{4}\sum_{i=1}^{p}\left(\|\phi_i\| - \frac{2\sigma_{i_{\max}}}{a\pi^2 - 4c}\right)^2 + \frac{\gamma}{2\alpha}w_{\max}^2 + \frac{\sigma_{\max}^2}{a\pi^2 - 4c}
\end{aligned}$$

With α being selected such that condition (31) holds, the derivative of Lyapunov function will be less than zero if the following conditions are satisfied:

$$\|x_0\| > \sqrt{\frac{8D}{2a\pi^2 - 8c - \alpha\pi^2 - 4\alpha}} \quad or$$

$$\|\tilde{w}\| > \sqrt{\frac{2\alpha D}{\gamma}} \quad or$$

$$\|\phi_i\| > \sqrt{\frac{4D}{\alpha\pi^2 - 4c} + \frac{2\sigma_{i_{\max}}}{a\pi^2 - 4c}} \quad (i = 1, ..., p),$$

where $D = \frac{\gamma}{2\alpha} w_m^2 + \frac{\sigma_m^2}{\alpha\pi^2 - 4c}$. Therefore, x_c, $\tilde{w}$, and ϕ_i for $i = 1, \ldots, p$ are uniformly ultimately bounded, based on which the boundedness of $e(t) = x_c(0,t) + \tilde{w}^T(t)\Phi(t)$ is immediately followed.

Simulation Results: To verify the proposed fault detection and prognosis scheme, it has been applied in simulations on two different distributed parameter systems that are modeled by parabolic partial differential equations. First a heat system and then a fluid pipeline will be investigated.

Heat system: The normalized dynamics of a heat transfer system is given by

$$\frac{\partial x(z,t)}{\partial t} = \frac{\partial^2 x(z,t)}{\partial z^2} + 0.2x(z,t), \tag{33}$$

$$\begin{cases} \dfrac{\partial x(0,t)}{\partial z} = -0.4x(0,t), \\ x(1,t) = u(t), \end{cases} \tag{34}$$

$$y(t) = x(0,t), \tag{35}$$

for $z \in (0,1)$ and $t > 0$. The distributed system state $x(z,t)$ is the temperature. The input $u(t)$ in the form of a heater is applied at the position $z = 1$ while the only available measurement $y(t)$ is taken at the opposite side. Simulation is accelerated and runs for a total time of 10 seconds while a heat leakage fault is initiated at time $t = 3s$, which is modeled by

$$h(y,z,t) = -5y(t)e^{-200(z-0.5)^2}\left(1 - e^{-0.2(t-3)}\right) \quad for \ t \geq 3.$$

The following observer is used to monitor the system state and output

$$\hat{x}(z,t) = x_1(z,t) + x_2(z,t) + \sum_{i=1}^{p} \hat{w}_i(t)\phi_i(z,t),$$

$$\hat{y}(t) = \hat{x}(0,t),$$

with the filters defined in (13),(14), (16),(17), and (24),(25). The vector of fault parameters $\hat{w}_i$ is initialized at zero and the vector of fault basis functions is selected in the form of RBF that can span the entire space $z \in (0,1)$ as follows

$$\sigma_i(y,z) = y(t)e^{-2p^2(z-i/p)^2} \quad for \ i = 1, ..., p-1,$$

where p is selected to be 10. The estimated fault parameters are not updated before the detection of a fault, while they will be tuned with the update law in (30) after the detection.

Figure 1 shows the solution of the distributed parameter system defined by (33), (34) with a constant input. It can be observed that the system state reaches a steady state after about 2 seconds, but when the fault is initiated at time $t = 3s$, the system state behavior changes. As expected, the leakage fault results in lower temperature at the output side.

The filter-based observer is utilized to estimate the system state and output online. The evolution of estimated state with respect to time and position is shown in Figure 2. The state estimation is initially inaccurate due to different initial conditions, but the estimation converges to the actual values in a short time. In order to perform fault detection, residual is generated by comparing the estimated and actual states at position $z = 0$.

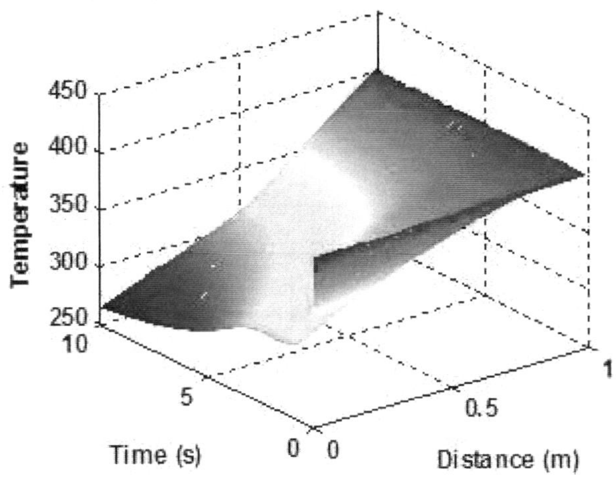

Figure 1: Actual system response

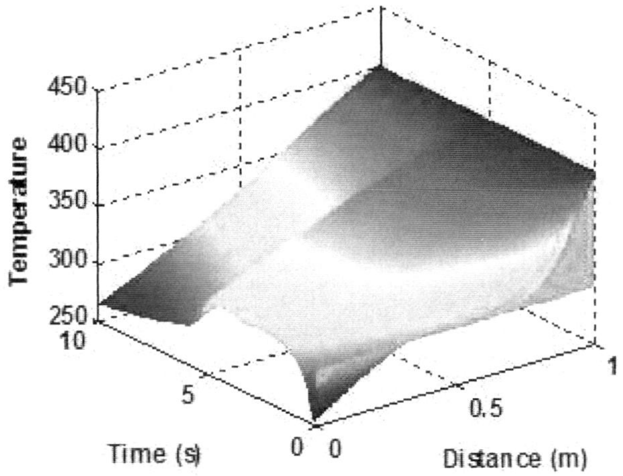

Figure 2: Estimated distributed temperature.

The residual is plotted in Figure 3 along with the detection threshold which is obtained from (22) by using $\eta = 0.2$. The residual is initially large due to incorrect initial conditions, but the stable observer derives the residual to zero under healthy operating conditions. The time-varying detection threshold prevents a false alarm in the beginning while increasing the possibility of detection when the observer error reaches its steady state value. The residual starts to increase upon initiation of fault at $t = 3s$ and the fault is declared active when the residual exceeds the threshold which happens at $t = 4.6s$. The adaptive update law for tuning the estimated fault parameters is activated immediately after detection. Subsequently, the learning algorithm captures the effect of fault on the system which causes the residual to decrease again, even in the presence of fault. The actual and estimated fault magnitudes are shown in Figure 4. The estimated fault remains at zero until the fault is detected, since the fault parameter vector is initiated at zero and the update law is only activated upon detection.

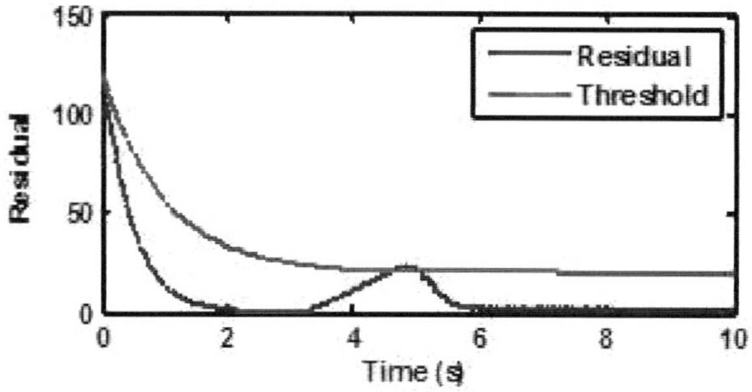

Figure 3: Detection residual and threshold

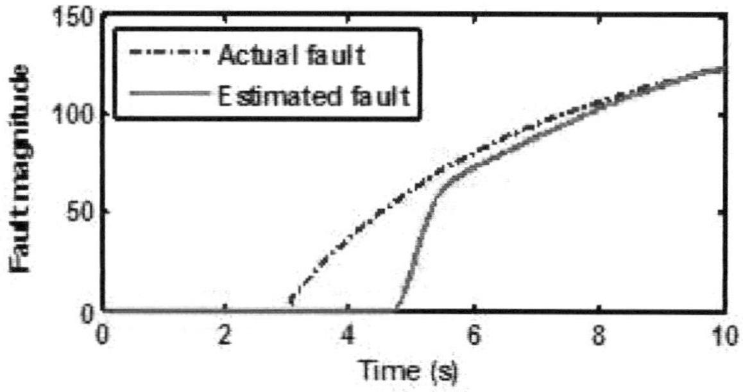

Figure 4: Actual and estimated fault magnitude

Fluid pipeline: Fluid flow and fluid pressure dynamics in a pipe are modeled by hyperbolic partial differential equations. However, assuming that the flow variation through the pipe is slow, the behavior of the system can be represented by a parabolic PDE in the following form

$$\frac{\partial p^2}{\partial t} = \frac{D^3(ZRT)^2\pi}{16fQ}\frac{\partial^2 p^2}{\partial z^2} + \frac{2g\pi D^3 sin\theta}{16fQ}\frac{\partial p^2}{\partial z}$$

where p is the distributed variable, pressure, and z is the distance from the start point of the pipe. A horizontal gas duct with the length of 100km is considered here. Table 1 provides the names of other parameters in the equation along with the values used in the simulations.

Table 1: Fluid pipeline system parameters

Parameter	Description	Value
D	Pipe diameter	0.6 m
f	Friction factor	0.003
R	Gas constant	392 m^2/s^2K
Z	Compressibility factor	0.872
T	Temprature	278 K

The pressure applied by pump at the start of line and flow demand at the output determine the boundary conditions for the PDE equation. A pump fixes the inlet pressure at 50 bars and Figure 5 shows the output flow demand. Assume the only available measurement is the output pressure. A leakage fault is seeded 50km away from the start of pipe, 50 hours into the simulation. The fault is created by $h_{l_l}(z) = Me^{-\frac{(z-l)^2}{5\times 1^{7}}}$, where $l = 50k$ and M is the magnitude of leakage fault.

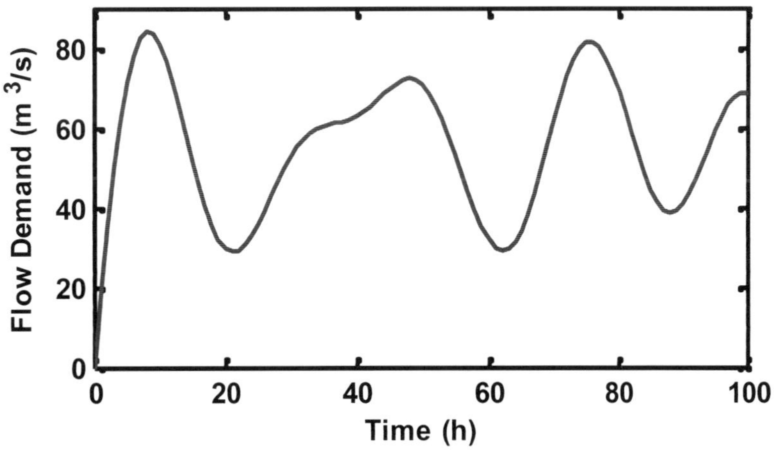

Figure 5: Flow demand at the pipe outlet

The filter-based fault detection observer is designed based on equations (13)-(20). The observer runs in parallel with the actual system and provides estimate of pressure at all points in the pipe even though the only available measurement is the pressure at the end of the pipe. Figure 6 shows the actual fluid pressure inside the pipe and Figure 7 shows the estimated fluid pressure provided by the observer with inactive fault estimator. It can be observed from the figures that the observer provides an accurate estimation of pressure before the occurrence of fault ($t < 50$), while the estimated pressure cannot follow the changes in the actual pipe in the presence of fault ($t > 50$). This difference in the estimation of state is used for fault detection. The residual which is calculated as the difference between actual and estimated output pressure is plotted in Figure 8 along with the detection threshold. The residual exceeds the threshold about 26 hours after the occurrence of fault which is the time of detection.

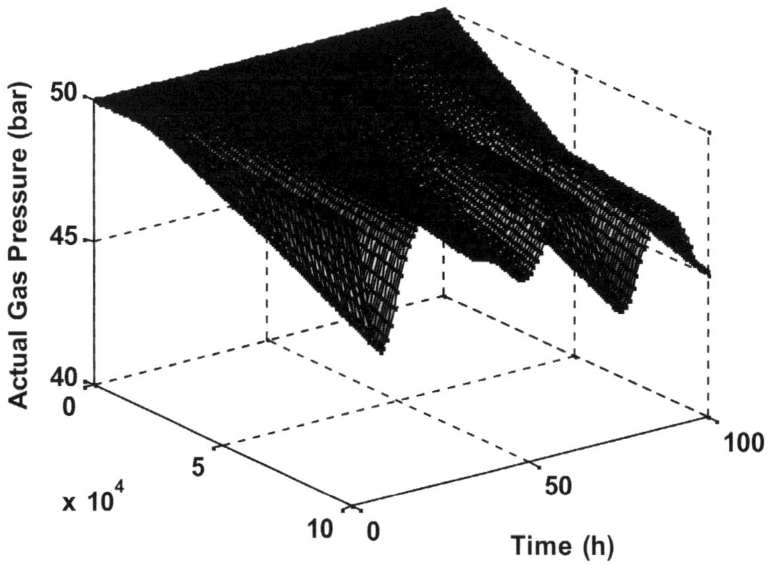

Figure 6: Actual fluid pressure along the pipe

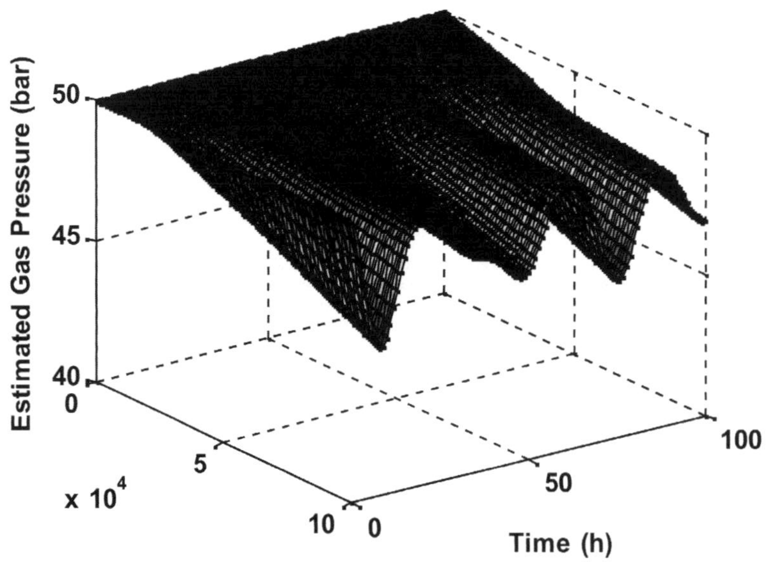

Figure 7: Estimated fluid pressure along the pipe

In this example, the magnitude of fault is assumed to be known and the estimation is performed on leakage location. This might be a strong assumption, but the proposed method cannot accurately estimate more than one fault parameter in a distributed system by only using one measurement. Figure 9 shows the actual and estimated fault location with the aforementioned assumption. The estimation is very accurate, nevertheless, future work on this topic includes investigation of fault estimation by using multiple sensors.

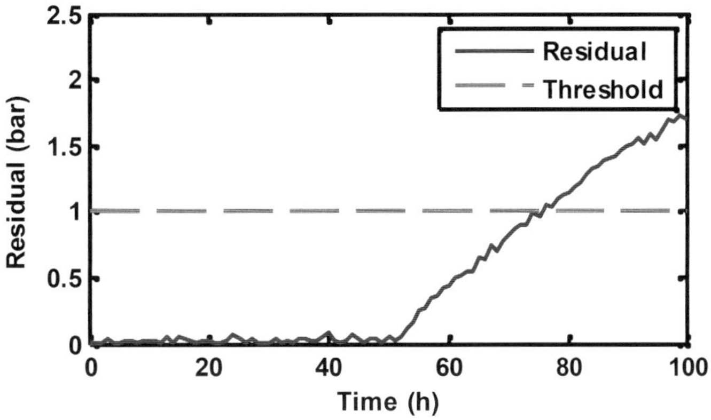

Figure 8: Detection residual and threshold

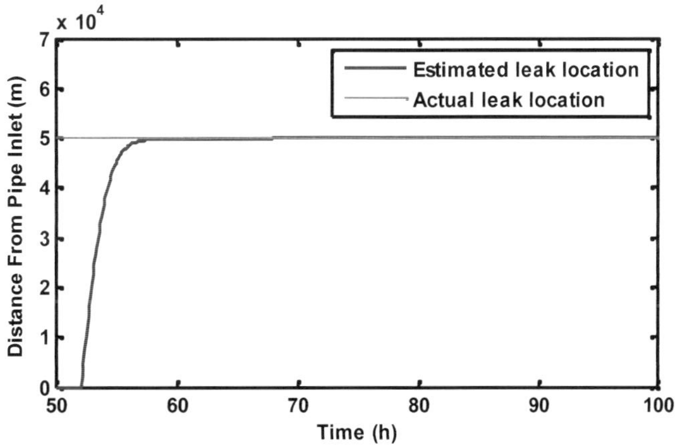

Figure 9: Actual and estimated leak location

Conclusions: Since the PDE model is directly used to construct the detection observer, the proposed scheme is more accurate in estimating the system states and thus highly reliable in performing fault detection than the existing fault diagnostic methods for distributed parameter systems. As seen in the simulation examples, it can also provide useful information about the unavailable system states. It was shown that if the stability conditions are satisfied with proper selection of design parameters, the observer will track the actual system states in healthy conditions and the adaptive update law will learn the unknown parameters of fault with a bounded error. Accurate and fast fault detection enhances the reliability, decreases the maintenance costs, and increases the system availability.

References:

[1] Hammouri, H., Kinnaert, M., and El Yaagoubi, E. H., 1999, "Observer-based approach to fault detection and isolation for nonlinear systems," Automatic Control, IEEE Transactions on, 44(10), pp. 1879-1884.

[2] Edwards, C., Spurgeon, S. K., and Patton, R. J., 2000, "Sliding mode observers for fault detection and isolation," Automatica, 36(4), pp. 541-553.

[3] Zhang, X., and Pisu, P., 2014, "Sliding Mode Observer and State-Machine-Based Fault Diagnosis With Application in a Vehicle Chassis Steering System," Journal of Dynamic Systems, Measurement, and Control, 136(4), pp. 041026-041026.

[4] Demetriou, M. A., and Polycarpou, M. M., 1998, "Incipient fault diagnosis of dynamical systems using online approximators," Automatic Control, IEEE Transactions on, 43(11), pp. 1612-1617.

[5] De Persis, C., and Isidori, A., 2001, "A geometric approach to nonlinear fault detection and isolation," Automatic Control, IEEE Transactions on, 46(6), pp. 853-865.

[6] Ferdowsi, H., and Jagannathan, S., 2013, "A unified model-based fault diagnosis scheme for non-linear discrete-time systems with additive and multiplicative faults," Transactions of the Institute of Measurement and Control, 35(6), pp. 742-752.

[7] Caccavale, F., Pierri, F., and Villani, L., 2008, "Adaptive Observer for Fault Diagnosis in Nonlinear Discrete-Time Systems," Journal of Dynamic Systems, Measurement, and Control, 130(2), pp. 021005-021005.

[8] Omatu, S., and Seinfeld, J. H., 1989, Distributed parameter systems: theory and applications, Clarendon Press.

[9] Christofides, P. D., 2001, Nonlinear and Robust Control of PDE Systems: Methods and Applications to transport reaction processes, Springer.

[10] Friedman, A., 2008, Partial Differential Equations, Dover Publications, Incorporated.

[11] Armaou, A., and Demetriou, M. A., 2008, "Robust detection and accommodation of incipient component and actuator faults in nonlinear distributed processes," AIChE Journal, 54(10), pp. 2651-2662.

[12] Ghantasala, S., and El-Farra, N. H., 2009, "Robust actuator fault isolation and management in constrained uncertain parabolic PDE systems," Automatica, 45(10), pp. 2368-2373.

[13] Baniamerian, A., and Khorasani, K., "Fault detection and isolation of dissipative parabolic PDEs: Finite-dimensional geometric approach," Proc. American Control Conference (ACC), 2012, pp. 5894-5899.

[14] H. Ferdowsi and S. Jagannathan, "Fault diagnosis of a class of distributed parameter systems modeled by parabolic partial differential equations," 2014 American Control Conference, Portland, OR, 2014, pp. 5434-5439.

[15] J. Cai, H. Ferdowsi and S. Jagannathan, "Model-based actuator fault accommodation for distributed parameter systems represented by coupled linear PDEs," 2015 IEEE Conference on Control Applications (CCA), Sydney, NSW, 2015, pp. 978-983.

[16] Smyshlyaev, A., and Krstic, M., 2010, Adaptive control of parabolic PDEs, Princeton University Press.

[17] Jagannathan, S., 2006, Neural Network Control of Nonlinear Discrete –time Systems, CRC publications, NY.

[18] Hardy, G. H., Littlewood, J. E., and Polya, G., 1952, Inequalities, Cambridge university press.

FAULT DETECTION IN BEARINGS USING AUTOCORRELATION

Rushit N. Shah, Michael H. Azarian and Michael G. Pecht
Center for Advanced Life Cycle Engineering (CALCE)
University of Maryland
College Park, MD 20742.
Telephone: (301) 405-7555
mazarian@calce.umd.edu

Abstract: Autocorrelation is a special case of cross-correlation wherein a signal is correlated with a time-lagged version of itself – the resulting signal comprises only the periodic information from the original signal whilst eliminating noise. This property of autocorrelation can be particularly useful in analyzing bearing faults since vibration data from a bearing, with local faults/defects, consists of cyclostationary acceleration signals usually contaminated with noise from sensors and other environmental factors. This study introduces a method which provides early failure warning in rolling element bearings by applying an autocorrelation operation to vibration data. The Sequential Probability Ratio Test (SPRT) is used to detect anomalies indicative of incipient failure. The results from the autocorrelation analysis are compared with results from a simple moving-RMS analysis of the acceleration data. The developed method is shown to provide an earlier warning of failure than the RMS-based method. This method can detect early stages of degradation in bearings – which in turn allows earlier scheduling of maintenance and the avoidance of system failures.

Keywords: Autocorrelation; bearings; diagnostics; health management; prognostics; SPRT;

Introduction: Bearings are load-bearing elements in machines, the smooth functioning of which proves to be critical in the failure-free operation of the system of which they are a part. Rolling element bearings, especially, are some of the most widely used components in mechanical systems and the failures of such systems are most frequently attributed to the failure of these bearings [1]. Owing to their criticality, early detection of faults in bearings is desired. Consequently, fault detection in bearings has been studied extensively. Vibration-based fault detection is the most common and reliable approach to this problem [2]. The emphasis on the vibration-based diagnostic techniques stems from the fact that a force impulse is generated every time a rolling element strikes a fault on a supporting bearing surface (inner race or outer race of the bearing). The series of impulses generated by such interactions between the fault and the rolling elements generally generates an amplitude modulated cyclostationary signal which can then be analyzed for fault signatures. The frequency spectrum of such a signal consists harmonics of resonance frequencies excited by the impulses and spaced at the bearing fault frequency. However, raw vibration signals are usually riddled with noise and disturbances produced by the system itself, as well as other external sources. In the domain of vibration-based fault detection, envelope analysis [3] is an established method of obtaining the bearing fault frequencies from the vibration signal. The general procedure for envelope analysis involves initial pre-processing using a high-pass filter to remove low-frequency noise and then inspecting the envelope spectrum of the resulting signal for bearing fault frequencies. Several improvements have been suggested over the original envelope analysis method. For instance, fast Fourier transform (FFT) is the most widely used technique to analyze spectra. Employing Hilbert transform for demodulation also allows effective extraction of the relevant section of the signal spectrum [4]. However, most methods in the domain of vibration-based fault detection focus on the raw vibration signal in the frequency domain. The frequency domain approach is suitable since extraction of fault frequencies provides detailed information about the nature and the location of the fault because fault frequencies are intrinsically related to the geometry of the bearing.

However, our study focuses on the early detection of a bearing fault (rather than gleaning information about location, nature or the extent of the fault) by examining the vibration signal time domain rather than the frequency domain. Analyses in the time domain involve extraction of various features from the temporal vibration signal such as mean, standard deviation, skewness, kurtosis, etc. Changes in these features over time are then used as indicators of health of the bearing. Advanced approaches in the time domain analysis involve fitting temporal signals to parametric time series models, followed by extraction of features from the parametric model. Popular methods of this type are the autoregressive (AR) model and autoregressive moving average (ARMA) model. Poyhonen et al. [5] fit vibration signals to an AR model and used the model coefficients as features for fault detection. The technique proposed by Garga et al. [6] uses AR modeling followed by dimensionality reduction. However, the AR and ARMA models are complex to model. Baydar et al. [7] studied the use of Principal Component Analysis (PCA) for analyzing time domain vibration signals from gears. For analyzing non-stationary signals, time-frequency domain analysis may be applied to such vibration signals. Time-frequency domain methods represent vibration signals in both time and frequency domain. The resulting distributions represent the energy of the signal and can help understand how the energy of the signal is distributed across different frequency bands over time. Popular time-frequency domain techniques include short-time Fourier transform [8] [9] and Wigner-Ville distribution [10] [11] [12] [13]. In this paper, a fast, new and computationally inexpensive method based on the autocorrelation function and SPRT is introduced. The rest of this paper is organized as follows. The next section provides theoretical background of autocorrelation and SPRT. Then, the developed bearing fault detection methodology is delineated - a new parameter called normalized autocorrelation function range (R_{range}) is then introduced and the methodology explained. Experimental results are provided to demonstrate the effectiveness of the developed method in the context of bearing fault detection. Finally, conclusions about the developed method are discussed.

The next section discusses the autocorrelation function, normalized autocorrelation function and their properties. A theoretical background of SPRT is then provided. Finally, the method is tested using an open-source dataset and the results are compared with an equivalent fault-detection method based on the RMS of the vibration of the signal.

Autocorrelation Function: Given a signal $f(t)$, the continuous autocorrelation function $R_f\ (\tau)$ is defined as the continuous cross-correlation integral of $f(t)$ with itself, at lag τ :

$$R_f\ (\tau) = \int_{-\infty}^{\infty} f(t) f(t-\tau) d \tag{1}$$

Alternatively, the autocorrelation function given above can be modified by including a normalizing factor to obtain the normalized autocorrelation function $R_f\ (\tau)$:

$$R_f\ (\tau) = \frac{\int_{-\infty}^{\infty} f(t) f(t-\tau) d}{\sigma^2} \tag{2}$$

where σ^2 is the variance of $f(t)$. The benefit of normalization is twofold. Firstly, it limits the range of autocorrelation function to $R_f\ \epsilon[-1,1]$. Secondly, it provides a scale-free measure of similarity between different moments of the signal $f(t)$.

A New Health Indicator for Bearing Fault Detection: When a bearing is operating in a defect-free state, the vibration signal collected from the bearing is composed primarily of noise from the system. However, the vibration signal from a faulty bearing is composed of different periodic components. This is due to the periodic nature of the impulses generated when the rolling-element impacts a defect on the outer or the inner race of the bearing. This effect has a direct bearing on the autocorrelation function. As these periodic components appear in the vibration signal, the magnitude of correlation between different moments of the vibration signal increases. In other words, the measure of similarity between different regions of the vibration signal increases. To capture this change in the magnitude of the autocorrelation function, a new parameter called *normalized autocorrelation function range* (R_r) is defined for a given signal $f(t)$ as:

$$R_r = \max\left(R_f\ (\tau)\right) - \min\left(R_f\ (\tau)\right) \tag{3}$$

Sequential Probability Ratio Test (SPRT): The sequential probability ratio test (SPRT) is used to track any change in the magnitude of R_r which is then used as an indicator of the health of the bearing. SPRT is a statistical hypothesis test which determines whether the test data falls into the probability density distribution of the base line training data [14] [15]. The SPRT detects changes in the test data by comparing a null and alternative hypotheses. The null hypothesis is the case in which the test data adheres to a Gaussian distribution with a mean of 0 and a variance of σ^2 extracted from the training data, which represents the normal test data. There exist four alternative hypotheses - the positive/negative mean test, the normal variance test, and the inverse variance test. For the positive/negative mean test, the alternative hypotheses are that the test data follows a Gaussian distribution with a mean of +M or -M and a variance of σ^2, where M is a pre-determined system disturbance level. The SPRT index is the logarithm of the ratio of the probability that the alternative hypothesis is true to the probability that the null hypothesis is true. Given the null and alternative hypotheses, the SPRT index is given as

$$S = \frac{M}{\sigma^2}\sum_{k=1}^{n}\left(x_k - \frac{M}{2}\right) \quad (4)$$

where x_k are successive observations of the test data. The SPRT index is continuously calculated and compared to the lower and upper threshold limits which are set by the user and determined by the desired level of sensitivity to detection. When the SPRT index is less than the lower threshold, it can be concluded that the test data is normal. When the SPRT index is greater than the upper boundary indicates that the test data is abnormal. The SPRT test can thus be used detect any deviation in the normalized autocorrelation function range from the base line data.

Methodology: Data from the bearing with outer race faults was analyzed for warning of an impending fault. The raw vibration signal in the time domain was used for the purpose of this analysis. The developed fault detection method is outlined in Fig. 2. The vibration signal is first divided into non-overlapping windows. The length of each window was fixed at 50000 samples. Thus, each window corresponds to a signal duration of 2.5 seconds. Selecting a window size of 50000 samples can be justified by the fact that the ball pass frequency for an outer race fault is ~236 Hz for this bearing. This implies that the interval between two consecutive impulses is ~1/236 = 0.0042 s. Since the sampling frequency is 20000 Hz, the interval between two consecutive impulses is ~86 samples. Hence, a window size of 50000 samples is sufficient to capture any fault related periodic components in the vibration signal. The normalized autocorrelation function was then computed for the first window. This autocorrelation function corresponding to the first window was then used to compute the normalized autocorrelation function range R_r . This value was recorded and saved. This constituted one iteration. This procedure was then repeated sequentially for every window, the R_r values for their corresponding windows were saved. After each iteration, the SPRT test was used to detect deviation in the R_r values. The first SPRT detection was used as the early warning of impending failure.

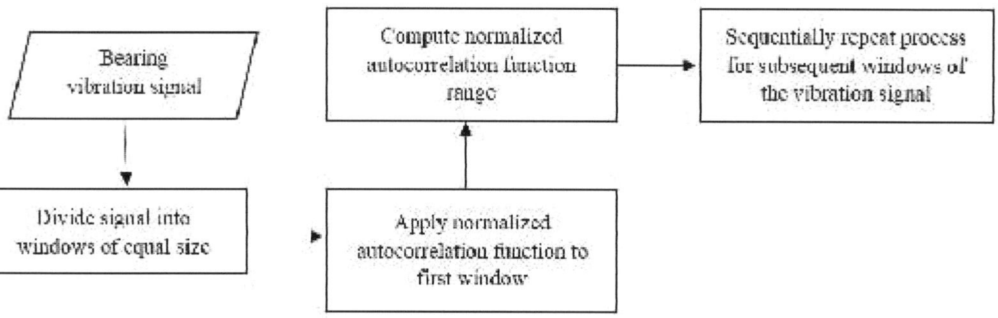

Figure 1 Flow chart of the developed fault detection method

Test Setup and Data Collection: Bearing run-to-failure tests under constant load conditions were performed on a specially designed test rig, as shown in Fig.1, with bearing data from a data repository [16]. The bearing test rig hosts four test Rexnord ZA-2115 double row bearings on one shaft. Table 1 lists the geometrical specifications of the bearing. The shaft is driven by an alternating current motor and is coupled by rub belts.

The rotation speed was kept constant at 2000 rev/min. A radial load of 6000 lb. was added to the shaft and bearing. A PCB 353B33 high-sensitivity quartz Integrated Circuits Piezoelectric (ICP) accelerometer was installed on each bearing housing. The sampling rate is 20 kHz. Vibration data were collected every 20 min. The full life data from five representative bearings (named bearings 1–5, respectively) whose faults include inner race, outer race, and ball defect are available.

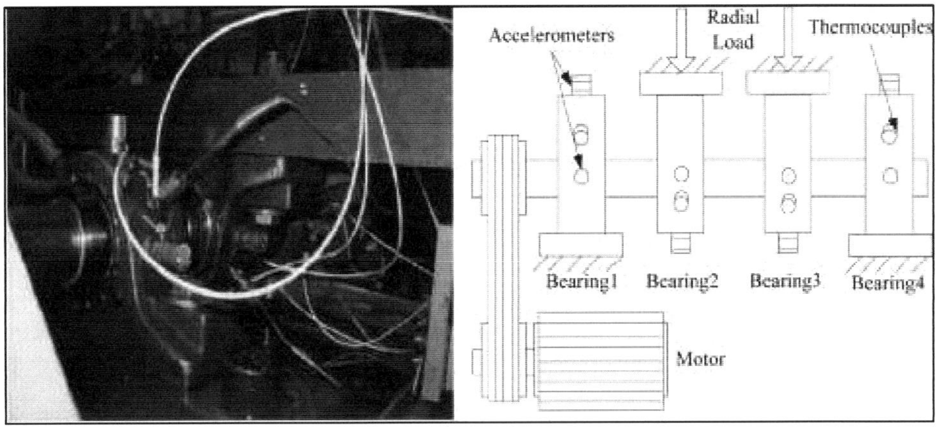

Figure 2 Test setup

Table 1 Geometrical specifications of the rolling element bearing

Number of rolling elements *n*	16
Diameter of rolling element *d*	8.4 mm
Pitch diameter of bearing *D*	71.5 mm
Contact angle γ	15.17

Results: The developed method was applied to the vibration signal corresponding to an outer race fault. Fig. 3 shows the normalized autocorrelation function range plotted as a function of the window number. Up to window number 200, the trend in R_r is uniform. The deviation in this trend is evident after window number 200. This deviation corresponds to the onset of the fault in the bearing. The process of damage propagation in a bearing involves onset of fault and damage aggravation followed by self-healing. Self-healing is the mechanism by which the spalls or cracks during damage propagation are smoothed out by continuous rolling contact which returns the bearing to a healthy or *near*-healthy state [17]. This mechanism continues until the bearing fails. In Fig. 3, the region between window numbers 300-350 corresponds to this self-healing mechanism. In order to automate the process of detection, SPRT was applied to this data. R_r values for the first 200 windows was used as the base line data since these windows correspond to the healthy section of the vibration signal from the bearing and the SPRT test is used to detect deviation from this base line data. It is expected that as the damage in the bearing accumulates, SPRT will indicate an alarm value of '1', indicative of a fault and return to '0' as the bearing self-heals. Hence, only the first SPRT detection is used as a fault indicator. For the developed method, the first detection was at window number 251. In order to compare this result with an equivalent method, the root mean squared (RMS) value of each window was used. Fig. 4 shows the RMS values for each window. For SPRT detection, the RMS values of first 200 windows were used as base line data. In this case, the first SPRT detection is observed at window number 308 i.e. once the bearing is already in the self-healing state [17].

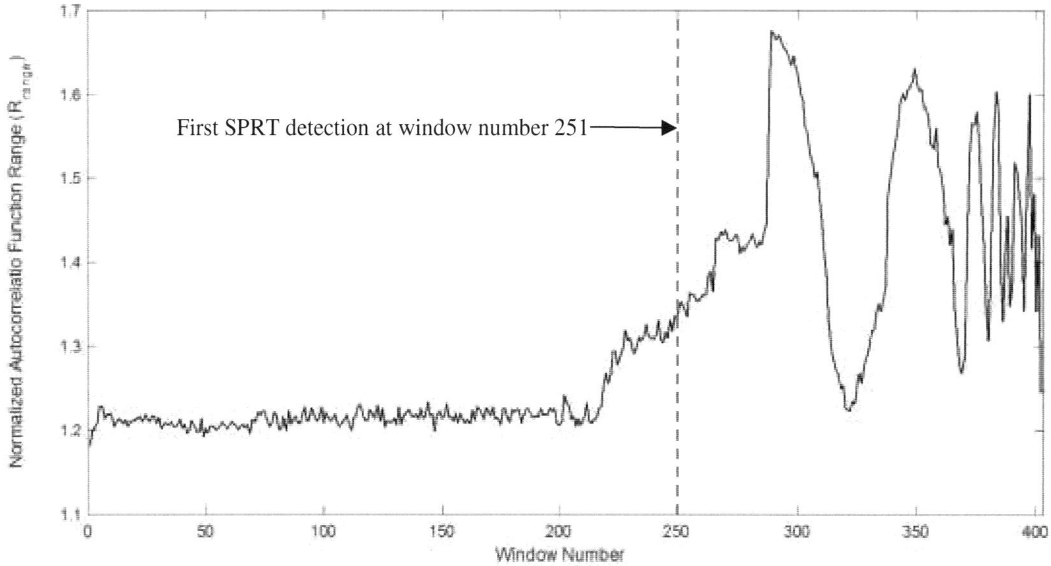

Figure 3 Normalized autocorrelation function range and first SPRT detection using the developed method.

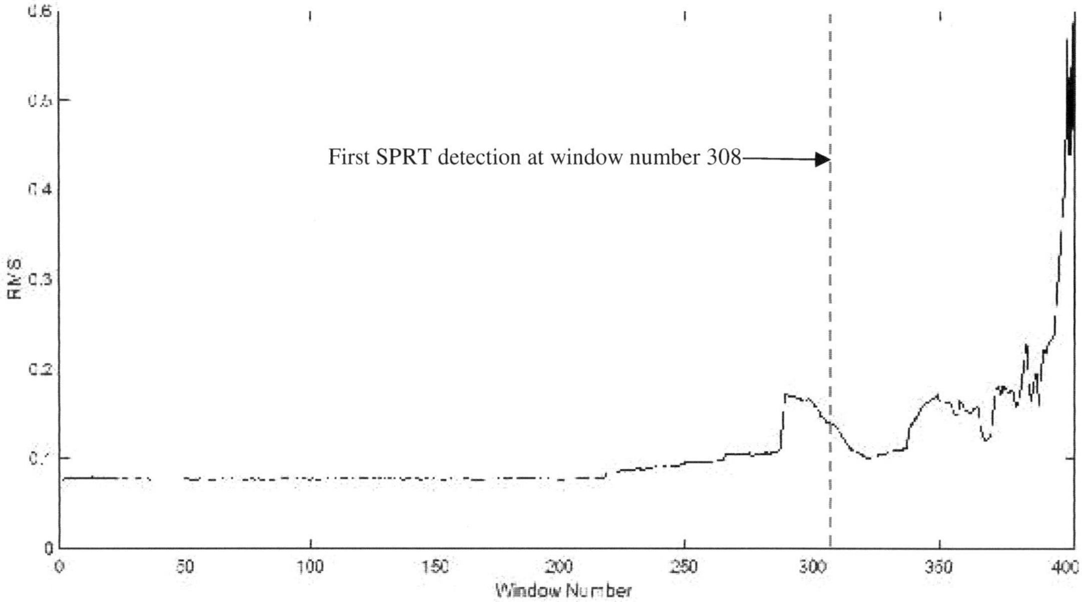

Figure 4 RMS-based fault detection method and first SPRT detection using this method

The results show that the developed method can detected the fault (at window number 251) much early compared to the RMS-based method which detected the identical fault only after window number 308. The difference between the first detection by both methods ~3 minutes.

Effect of Window Size: For the first study, a window size of 50000 samples was chosen. As discussed in the section on methodology, this window size was sufficient to capture any periodic impulse which may have occurred due to a fault in the bearing. To study the effect that selecting different window sizes has on the time of detection, the developed method was tested with seven different window sizes – 20000 to 80000 (at intervals of 10000 samples). For every window size, the percentage remaining useful life (% RUL) – time from first detection to bearing failure - was computed, for both, the developed method and the RMS-based method. The results are presented in Figure 5. As the window size was increased, the % RUL at the time of first detection decreased. Hence, a smaller window size is preferred. However, it must be ensured that the window size is large enough to capture the periodic impulses i.e. the size of the window must be larger than the time between two impulses. Also, with increasing window size the decrease in % RUL (at time of first detection) was more significant in the case of the RMS-based method than the developed method. This demonstrates the robustness of the developed method. In every case, the developed method based on autocorrelation had smaller detection times for the fault than the RMS-based method. In the case where the window size was 20000 samples, the developed method was able to detect the fault with ~40% of the life of the bearing remaining. In the case where window size was 80000 samples, the developed method detected the fault with ~35% of the life of the bearing still remaining. In the same case, the RMS-based method provided a failure warning with only ~15% of the bearing life remaining.

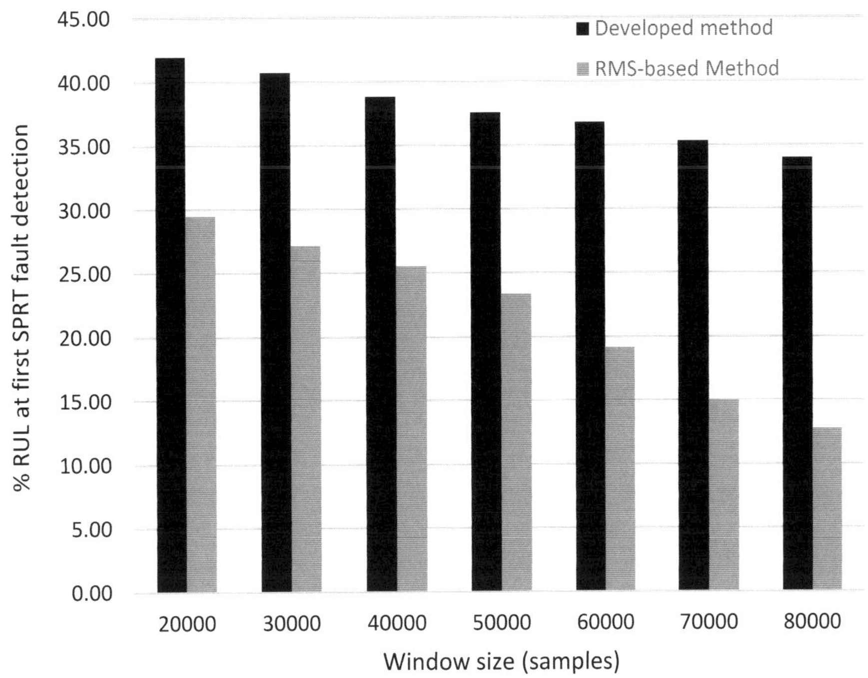

Figure 5 Effect of choosing different window sizes on the time of first SPRT fault detection

Conclusions: Bearings play a critical role in a mechanical system and early detection of bearing faults is desired to avoid system failure and to schedule early maintenance. This paper presents such a method for rolling element bearings. The developed method utilizes the normalized autocorrelation function to compute a new scale-free parameter called normalized autocorrelation function range (R_r) which is used to characterize the state of the system. Sequential probability ratio test (SPRT) is then used to detect a fault based on the deviation of R_r from base line data. The developed method has been compared with an equivalent RMS-based method using the same vibration data. The method, although rudimentary, provides a promising way of performing prognostics for rolling-element bearings. In the set of experiments performed, the developed method consistently detected the fault in the bearing with ~35-40% bearing life remaining. Since this method involves computation of only one parameter for determining the health of the system, it is computationally inexpensive, takes less than 5 seconds to execute (for the bearing vibration data used in this study) and can be easily integrated into larger system-level prognostics and health-management (PHM) setups.

References

[1] R. B. Randall and J. Antoni, "Rolling element bearing diagnostics - A tutorial," *Mechanical Systems and Signal Processing,* vol. 25, no. 2, pp. 485-520, 2011.

[2] R. Golfashan and K. Y. Sanliturk, "Rolling Bearing Fault Diagnosis Using Recursive Autocorrelation and Autoregressive Analyses," in *23rd International Congress on Sound and Vibration*, Athens, Greece, 2016.

[3] M. S. Darlow, R. H. Badgley and G. W. Hogg, "Application of high frequency resonance techniques for bearing diagnostics in helicopter gearboxes," Technical Report ,US Army Air Mobility Research and Development Laboratory, 1974.

[4] F. Bonnardot, R. B. Randall, J. Antoni and F. Guillet, "Enhanced unsupervised noise cancellation using angular resampling for planetary bearing fault diagnosis.," *International Journal of Acoustivs and Vibration,* vol. 9, no. 2, pp. 51-60, 2004.

[5] S. Poyhonen, P. Jover and H. Hyotyniemi, "Signal Processing of Vibrations for Condition Monitoring of an Induction Motor," in *2004*, New York, ICSSP: First International Symposium on Control, Communications and Signal Processing.

[6] A. K. Garga, B. T. Elverson and D. C. Lang, AR Modeling with Dimensino Reduction for Machinery Fault Classification, Haymarket, Critical Link: Diagnosis to Prognosis.

[7] N. Baydar, Q. Chen, A. Ball and U. Kruger, "Detection of Incipient Tooth Defect in Helical Gears Using Multivariate Statistics," *Mechanical Systems and Signal Processing,* vol. 15, pp. 303-321, 2001.

[8] W. J. Wang and P. D. McFadden, "Early Detection of Gear Failure by Vibration Analysis - I: Calculation of the Time-Frequency Distribution," *Mechanical Systems and Signal Processing,* pp. 193-203, 1993.

[9] F. A. Andrade, I. Esat and M. N. M. Badi, "Gearbox Fault Detection Using Statistical Methods, Time-frequency Methods (STFT and Wigner-Ville Distribution) and

Harmonic Wavelet - A Comparative Study," in *Proceedings of the COMADEM*, 99, 1999.

[10] Q. Meng and L. Qu, "Rotating Machinery Fault Diagnosis Using Wigner Distribution," *Mechanical Systems and Signal Processing,* vol. 5, pp. 155-166, 1991.

[11] M. C. Pan, H. Van Brussel, P. Sas and B. Verbeure, "Fault Diagnosisi of Joint Backlash," *Journal of Vibration and Acoustics,* vol. 120, no. 1, pp. 13-24, 1998.

[12] I. S. Koo and W. W. Kim, "The Development of Reactor Coolant Pump Vibration Monitoring and a Diagnostic System in the Nuclear Power Plant," *ISA Transactions,* vol. 39, pp. 309-316, 2000.

[13] N. Baydar and A. Ball, "A comparative Study of Acoustic and Vibration Signals in Detection of Gear Failures Using Wigner-Ville Distribution," *Mechanical Systems and Signal Processing,* vol. 15, pp. 1091-1107, 2001.

[14] W. Lu and K. Gross, "Early Detection of Signal and Process Anomalies in Enterprise Computing Systems," in *IEEE Int'l Conf. on Machine LEarning and Applications*, Las Vegas, NV, 2002.

[15] K. Humenik and K. Gross, "Sequential Probability Ratio Tests for Reactor Signal Validation and Sensor Surveillance Applications," *Nuclear Science and Engineering,* vol. 105, 1990.

[16] J. Lee, H. Qiu and G. Yu, *Rexnord Technical Services: Bearing Data Set,* IMS, Univ. Cincinnati: NASA Ames Prognostics Data Repository, NASA Ames, 2007.

[17] T. Williams, X. Ribadeneira, S. Billington and T. Kurfess, "Rolling element bearing diagnostics in run-to-failure lifetime testing," *Mechanical Systems and Signal Processing,* vol. 15, pp. 979-993, 2001.

FRAMEWORK FOR PHM IN THE SMART MANUFACTURING CONTEXT: INTEGRATION OF DIFFERENT APPROACHES

Luca Fumagalli and Marco Macchi
Politecnico di Milano
Department of Management, Economics and Industrial Engineering
P.zza Leonardo Da Vinci 32, 20133, Milano, Italy.
Telephone: (0039) 02-2399-2722
luca1.fumagalli@polimi.it
marco.macchi@polimi.it

Abstract: The technology has advanced at an exponentially high rate since the advent of Internet in the early 90s. The concepts like e-Maintenance, Internet of Things, Industry 4.0 are linked to this advancement in technology. All these have stimulated great potentials in industries and manufacturing. This will boost Prognostics and Health Management capabilities that will need to rely not only on consolidated algorithms and IT architectures, but also on new paradigms related with distributed computing, modularization of tools and development of new services. The paper will address such approach proposing a reference framework to highlight how predictive maintenance can be interpreted according to the new paradigm of Smart Manufacturing. The framework will be supported by an industrial case.

Key words: Condition based maintenance, Prognostics and Health Management, Smart Manufacturing, Reliability centered maintenance.

Introduction: Industrial asset management has always been a complex activity for the amount of information to be handled. Collection of data and information about maintenance events (i.e. maintenance interventions) is nowadays a key activity and it is generally carried out by Computerized Maintenance Management Systems (CMMS). Thanks to the availability of such data, other analyses can be carried out by tools for Reliability Centred Maintenance (RCM) analysis. Companies then adopt Condition Based Maintenance (CBM) programs and systems, with the purpose to monitor the industrial assets and prevent unexpected events. Such CBM systems are often used to monitor conditions as well as to generate alarms and warnings based on the conditions. Besides, a great number of techniques and methods have emerged during the last years, allowing to achieve an accurate information and knowledge about the systems' condition evolution and remaining useful life. These advances are recognized as outcomes of an innovative discipline, nowadays discussed under the term of Prognostics and Health Management (PHM); indeed, PHM is considered to lead to a "CBM enabled by PHM" [1]. Still, PHM is often not so widespread in industry; nevertheless the landscape is quite complex and requires the definition of a proper framework in order to provide new solutions for the integration of PHM in the industrial asset management practice. Some key issues are worth to be provided in order to build the proper background of the research.

The RCM approach considers maintenance engineering techniques, such as RBD (Reliability Block Diagram), ETA (Event Tree Analysis), FTA (Fault Tree Analysis), WA (Weibull Analysis), FMEA (Failure Mode and Effect Analysis), and is one of the most effective way to assess and optimise maintenance policies [2]. Besides, different works have studied the creation of a framework where RCM and CBM can be both addressed. Considering the integration with the CMMS, Gabbar [3] says that RCM is anyhow a time and effort consuming activity if not automated. It must be periodically performed, but the analysis is generally not aligned with real time data progressively stored in the CMMS. In other words, RCM results risk to be not always up-to-date. Lopez Campos et. al. [4] presented the possible interface between different systems, considering also the data model required to this end. Trapani et al. [5] also proposed how to integrate RCM and CBM considering the beneficial results of the analysis for risk assessment, while Colace et al. [6] propose how to exploit HAZOP to proper consider the risk in the analysis.
CBM is a relevant lever for maintenance transformation: in the recent decades, traditional maintenance models, considering run-to-failure or time-based maintenance, are evolving to CBM programs. In this evolution, PHM is considered one of the key factors to achieve system-level efficient maintenance and reduce life cycle costs [7]. Prognosis research field is in fact promising new capabilities to improve the system reliability, leveraging both on design and maintenance along the useful life [8][9]. Besides, PHM provides capabilities to achieve more proactivity in maintenance: in this regard, it is worth remarking that, as expectation for the future, the equipment data will be transformed by PHM solutions into valuable information to help not only maintenance managers, but also plant managers for optimising planning, saving cost and minimising equipment downtimes [10].
Thus, under the term of PHM, a body of knowledge has been nowadays created, leading to consider it as an engineering discipline [11]. This covers all methods and technologies to assess the reliability of a product in its actual life cycle conditions to determine the advent of failures, and mitigate system risks [12].
It is worth understanding the role of PHM in the modern maintenance systems. In particular, its contribution to more proactive approaches allow reaching operational excellence in manufacturing companies [1].
By pursuing the integration of different systems and approaches, the need of a comprehensive maintenance platform is crucial for the industrial business, in order to have an effective tool to support the daily activity and the periodic maintenance analysis. This aspect has been analysed by scientific and industrial literature since 2000, addressing the concept of e-Maintenance as a component of e-Manufacturing. In 2006, Muller et al. [13] stated that the term "e-Maintenance" was not yet consistently defined in maintenance theory and practice. Different engineers and scientists have considered e-Maintenance in different manners and recently it has been also linked to the concept of value creation [14]. Progressively, the concept of e-Maintenance has been exploited with different keywords, while the development of Internet of Things (IoT) has also created the background for a more pervasive technology exploitation in the shop floor for all the operations and not only maintenance activity. More recently, this trend has evolved in the Industry 4.0 concept (according to the European wave) or, with a more international perspective, in the Smart Manufacturing concept. For what concern the industrial asset management scope, Smart Maintenance is then the keyword that is going to be used in the present research.

In order to develop Smart Maintenance applications, it is necessary to have a methodology to express the processes to execute and the set of data required by the processes. Indeed, projects tent to deal with an important quantity of data and information: requirements, objects, relations, restrictions, functions, etc. In order to manage such information, a proper modelling approach is needed, to enable interoperable solutions (i.e. interoperable in existent ICT systems) since its conceptualization. IDEF0 is hereafter used for this modelling activity.

Background for the proposed framework: In order to support the deployment of PHM solutions, the IMS Center proposes the 5S methodology. This approach was devised by IMS Center in order to develop and research all aspects of future maintenance infrastructures. This systematic approach consists of five key elements, that is ground on the Watchdog Agent. The Watchdog Agent is an enabling technology that shall allow for a successful implementation of intelligent maintenance. The toolbox was developed by Center for Intelligent Maintenance Systems (Center for IMS). It is a collection of algorithms that can be used to assess and predict the performances of a process or equipment based on input from sensors, historical data and operating conditions [15].

The 5 key elements of the 5S methodology are:

- Streamline: this encompasses techniques for sorting, prioritizing, and classifying data into more feature-based health clusters. This may also include reducing large data sets (from both maintenance history and on-line data DAQ) to smaller dimensions, leading to correlation of the relevant data to feature maps for better data representation.
- Smart Process: using the right Watchdog Agent tool for the right application. This requires techniques for selecting appropriate prognostics tools based on application conditions, criticality of each condition for machine health, and system requirements.
- Synchronize: converting component data to component degradation information at the local level and further predicting trends of health using a visualized radar chart for decision-ready information. Maintenance data is transformed to health information and to an automated action.
- Standardize: consisting in the creation of a standardized information structure for equipment condition data and health information so that it is compatible with higher-level business systems and enables the information to be embedded in business ERP and asset management systems. The goal here is to keep the process as a standard approach for day-to-day practices.
- Sustain: utilizing the transformed data for information-level decision making. System information is then shared among all stages of product and business life.

The present work is based on the development of a Smart Maintenance Platform that focuses on the two last steps here mentioned in order to prepare a framework to include PHM into industrial solutions.
The complete IDEF0 diagram representing the smart maintenance platform is presented in Figure 1, which represents the new framework herein proposed. In Figure 2 the diagram is

divided in the modules that implement the proposed architecture. The focus of the present paper is on the Decision Support Tool (DST).

Figure 1: Proposed framework (designed with IDEF0 standard).

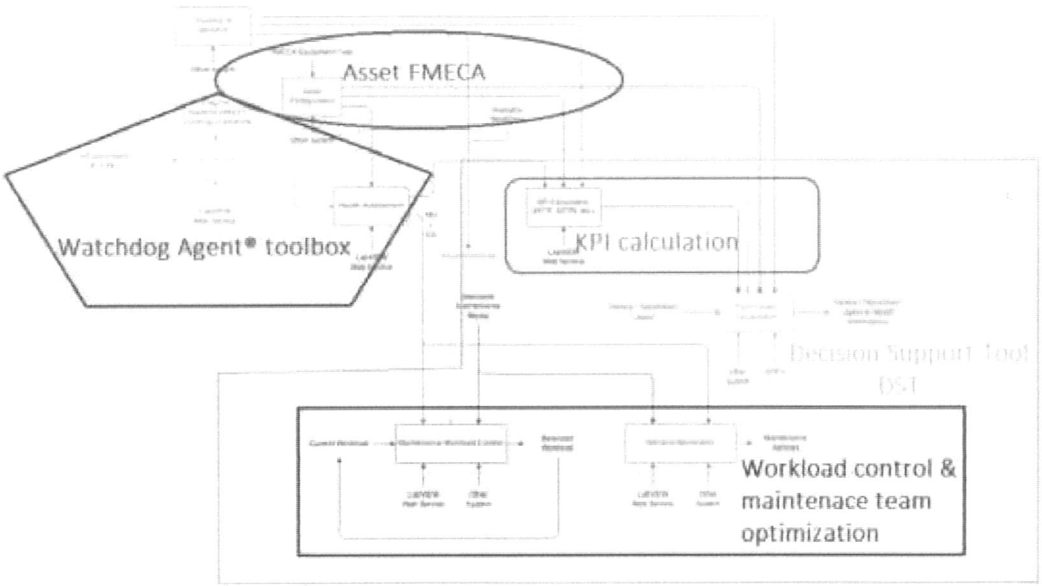

Figure 2: Key elements of the proposed framework.

Watchdog Agent tools have been considered for the proposed framework. They take care of State detection and Health assessment blocks. Decision Support Tool (DST), instead, must implement KPI calculation, Maintenance Workload Control and Advisory

Generation, considering also the appropriate Plant Cockpit Visualization activity in order to properly communicate with the operators and maintenance engineers.

Table 1: Summary of the literature analysis

Decision parameters & model	When it is applicable	Procedure	Source
Use of RPN of different failure modes	Simple model with detailed FMECA. It can be used also for those assets that are not involved in a PM policy	According to the values of RPN, one can decide if the intervention is needed (RPN high) or not (RPN low)	FMECA decomposition [5]
Cost model	All cost data are available: maintenance intervention cost, down- time cost (production losses, etc.), etc.	A cost analysis is done for each failure. By means of cost functions it is calculated if it is convenient to wait for the intervention or to implement the maintenance action	"An Options Approach for Decision Support of Systems with Prognostic Capabilities" [12]
RPN + ANP/AHP techniques	FMECA is detailed and interdependencies among failure modes must be considered. Moreover, an expertise knowledge is needed in order to assign weights. This model takes into account the domino effects of first and higher level	RPN is calculated in a detailed way by considering interdependencies among failure modes. A complex Multi-Criteria approach is used in order to have a more robust decision process	"ANP/RPN: A Multi Criteria Evaluation of the Risk Priority Number" [16]"
Model for production machines with buffers	Buffer levels are monitored and machines are prioritized	Machines are prioritized according to their production importance. Buffer levels are taken into account in order to implement Preventive and Opportunistic maintenance policies	"Simulation platform for anticipative plant- level maintenance decision support sys- tem" [17]
Model for production lines considering production rate and costs	Cost data are available both for production and maintenance activities	A Genetic algorithm is used in order to optimize maintenance scheduling. A cost function is set to highlight the most critical failure modes	"Maintenance scheduling in manufacturing systems based on predicted machine degradation" [18]
Prioritization by CRI index	Detailed information about the asset are available. Asset condition can be monitored	Two Fuzzy Inference Systems (FIS) are used. The former is able to calculate the Basic Condition index, the latter al- lows to compound the Basic Condition (BC) and Operating Condition (OC) values to calculate the CRI (Composite Risk Index). CRI is used to prioritize maintenance actions.	"Substations SF6 circuit breakers: Reliability evaluation based on equipment condition" [19]

The remainder focuses on the Maintenance Workload Control. Literature background has been considered looking for effective procedures for maintenance prioritization. This

assessment has allowed to identify the best suitable model for the specific problem under concern.
The decision-making process is not a simple activity to face: it must be chosen whether a maintenance intervention must be launched and what kind of actions needs to be implemented. Besides, it may happen that more than one intervention is needed, therefore prioritization is considered.
The research is focused also on whether to launch or not the work order according to specific conditions. Models have been investigated, in order to take into account as much cases as possible and have a wide-ranging application of the algorithms.
The Table 1 presents the summary of the literature analysis done to identify the model to support the decision-making process by exploiting different types of information data and parameters. The examined models are synthetized in order to identify in a faster way the most suitable one according to the different case characteristics.
The idea is to try to optimize the actions of maintenance team: the interventions can be performed contemporarily each other, if possible, according to the failure mode and/or the maintenance team competences and skills needed.
The vision can be thought more global and complete, taking into account the whole plant or even the company. Ideally, one can think to schedule the interventions of the maintenance team (also "external" and third party teams) according to the failure modes occurred or that are going to happen (monitored parameters). The objective is to try to associate more than one intervention to failure modes that need the competences and skills of the same team. This could be done in order to optimise the resources and reduce time for the following interventions.

Industrial case study: The equipment available for the industrial case study development is a vacuum and refrigerant charging machine used in the production line of a refrigerator manufacturer. The main function of this equipment is to fill with refrigerant fluid the circuit of refrigerators or freezers. It automatically performs discharge, leak tests and charging, together with all the diagnostic and self-diagnostic tests.
The identified model to apply for the case study implementation is described by Vianna et al. [19], and it is based on the use of two Fuzzy Inference Systems. Actually, the model defines three critical indexes (Basic Index, Operating Index and Composite Risk Index) to prioritize maintenance actions. On the other hand, the developed tool for this particular case study, exploits the benefits of Fuzzy Logic Systems by considering the already available information about criticality (RPN – Risk Priority Number) and maintenance plan.
The use of Fuzzy Logic, by the method proposed in the study of Vianna et al. [19], allows the application of knowledge of experts in drawing up rules. Instead, traditional Boolean logic is two-valued in the sense that a member either belongs to a set or does not. Values of one and zero represent the membership of a member to the set with one representing absolute membership and zero representing no membership.
Fuzzy logic allows for partial membership, or a degree of membership, which might be any value along the continuum of zero to one.
The system designed for the case study considers as Input Variables the RPN value of the failure and a quantity defined as TtP (Time to Preventive). This last index takes into account the left days between the failure date and the preventive maintenance intervention

date for that specific failure. In this way, the RPN and TtP indexes are the Input Variables of the Fuzzy System.
In particular, the RPN index can assume three linguistic terms (Low, Medium, High) according to the failure criticality and the ranges defined in Fuzzy Logic. On the other hand, the index TtP can adopt three linguistic terms (Close, Halfway, Faraway), that identify the temporal "distance" from the preventive maintenance action for that failure.

The Output Variable is called "Work order" and can assume three linguistic terms:

- Repair the failure (RF): it accounts for the case in which the failure needs to be repaired immediately;
- Repair the failure and implement preventive action now (BN): in this case, both the failure fixing and preventive intervention are implemented immediately;
- Wait preventive intervention (WP): this option considers that the identified failure will be fixed only when the preventive intervention will be executed.

The Input and Output Variables are represented in a matrix that is shown in Figure 3.

		RPN		
		Low	Medium	High
TtP	Close	WP	WP	BN
	Halfway	BN	WP	BN
	Faraway	WP	RF	RF

Figure 3: Matrix of Fuzzy System Rules.

The Rules are designed by considering the combination of RPN and TtP indexes. In particular, the AND (Minimum) antecedent connective is taken into account, that specifies to use the minimum degree of membership of the antecedents.
Figure 4 depicts a Test of the designed Fuzzy System, done through the Fuzzy System Designer available in LabVIEW. The platform user interface developed in LabVIEW environment, consists of several tabs, through which it can be possible to surf among the available information. Previous works developed the "Asset" tab, that allows to choose the asset and depicts the functional scheme of the equipment; "Threshold Values" tab that can be useful to change the threshold value of a specific failure and update its Risk Priority Number (RPN); and the "FMECA" tab that shows the Failure Modes, Components and related RPN of the selected asset. Previous researches with the use of such tools are reported in [20] and [21].
The tabs "Maintenance Work Orders" (Figure 5) and "Workload" are generated by referring to the optimization algorithms and Fuzzy Logic System developed. The idea is to embed all the available information about failure modes (indicated in the Figure 5 as Failure Code), failure criticality, and maintenance team skills in one screen.
LabVIEW environment is used in order to guarantee integration with Watchdog Agent tools.

The available information is presented in a readable fashion way and identifies for a specific failure mode the date and criticality data such as RPN and Time to Preventive.

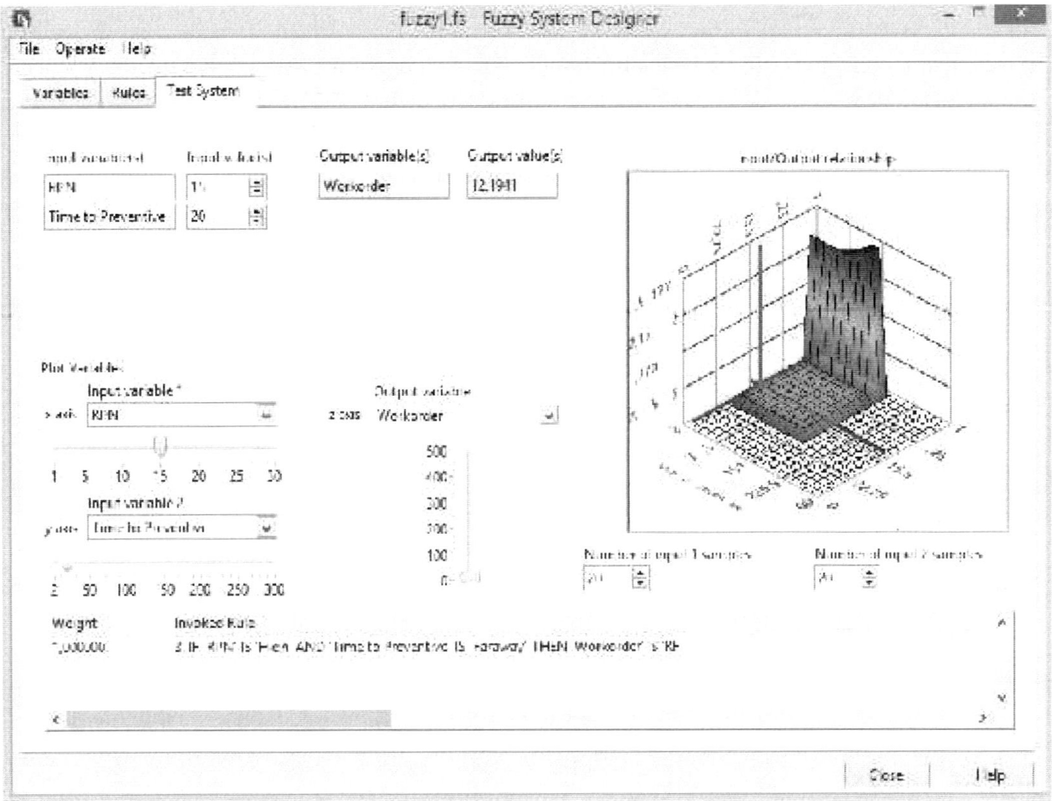

Figure 4: Fuzzy System Test through System Designer in LabVIEW.

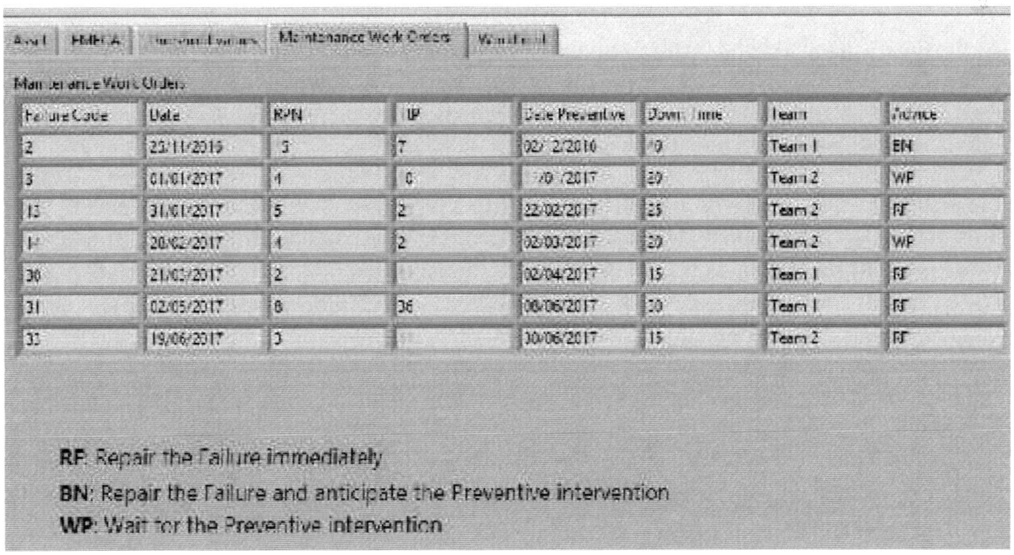

Figure 5: Maintenance Work Orders tab.

Moreover, the downtime for the failure is highlighted, together with the associated maintenance team and the advice resulting from the application of Fuzzy System. Finally, a brief legend reminds the meaning of advices' acronyms.
The last tab named "Workload" shows the maintenance plan for the equipment. In particular, it can be possible to surf among all the months and verify the already planned preventive interventions and the updating plan according to the output results of the decision support system. The availability of these information and the easiness of reading and understanding them are very useful, enabling to have a wide and clear picture of team workload and, eventually, of detection of problems or not standard behavior of the system.

Conclusions: This work presented the proposal of a Smart Maintenance framework to allow the integration of PHM, combining the features of different functions, into operational management activities. The analysis of Watchdog Agent Toolbox results a strategical step in order to understand the potentiality of right management of information and data captured by monitoring systems. The use of the described tools in order to assess equipment health state or implementing prognostic investigation, means having the availability of almost infinite solutions for any possible case. The development of a Decision Support Tool becomes essential to apply the knowledge about the system previously gained. Moreover, use of the solution by the operators could be further improved Human-Computer interaction is enhanced by mixed interface [22].

Then, the necessity of optimization algorithms for workload and maintenance team scheduling appears evident. It has been underlined how the algorithms become necessary in order to increase the efficiency of workload planning. The advices given by the system that takes into account both failure criticality and distance from preventive intervention allow the company to react in a proactive way managing failure events. The maintenance plan and maintenance team scheduling become "dynamic", by considering not only historical events and past analysis, but also updated data. To this end, future development can consider the proper integration of the proposed solution with smart sensors [23] [24], in order to complete the integration of the platform.

Acknowledgements: Part of the research work was performed within the context of Sustain-Owner ("Sustainable Design and Management of Industrial Assets through Total Value and Cost of Ownership"), a project sponsored by the EU Framework Programme Horizon 2020, MSCA-RISE-2014: Marie Skłodowska-Curie Research and Innovation Staff Exchange (RISE) (grant agreement number 645733 — Sustain-Owner — H2020-MSCA-RISE-2014). Part of the research work was performed within the context of Smart Manufacturing 2020, Italian funded project ID CTN01_00163_216744. The authors would like also to thank: Paolo d'Orso, Alice Giacomin and Alessandro Alfieri for the activities done in the scope of the presented research.

References:

[1] Guillén, A. J., Crespo, A., Macchi, M., & Gómez, J. (2016). On the role of Prognostics and Health Management in advanced maintenance systems. Production Planning & Control, 1-14.

[2] Tucci M., and G. Bettini. "Methods and tools for the reliability engineering: a plant maintenance perspective." Proceedings of the 2nd maintenance management MM2006, Sorrento, Italy, April (2006).

[3] Gabbar H. A., Yamashita H., Suzuki K., Shimada Y. (2003). Computer-aided RCM-based plant maintenance management system. Elsevier, Robotics and Computer Integrated Manufacturing, 19(5), pp. 449-458.

[4] Lopez Campos, M.A., Fumagalli, L., Gomez Fernandez J.F., Crespo Marquez A., Macchi M. (2010). UML model for integration between RCM and CBM in an e-Maintenance architecture. (pp. 133- 138). In: 1st IFAC Workshop, A-MEST'10, Advanced Maintenance Engineering, Services and Technology. Lisboa, Portugal.

[5] Trapani Natalia, Macchi Marco, Fumagalli Luca, Risk Driven Engineering of Prognostics and Health Management Systems in Manufacturing, in Proceedings of 15th IFAC/IEEE/IFIP/IFORS Symposium Information Control Problems in Manufacturing (INCOM 2015), Ottawa, Canada, 11-13 May 2015.

[6] Colace C., Fumagalli L., Pala S., Macchi M., Matarazzo N. R., & Rondi M. (2015). Implementation of a condition monitoring system on an electric arc furnace through a risk-based methodology. Proceedings of the Institution of Mechanical Engineers, Part O: Journal of Risk and Reliability, 1748006X15576441.

[7] Ly, C., Tom, K., Byington, C. S., Patrick, R., & Vachtsevanos, G. J. (2009, August). Fault diagnosis and failure prognosis for engineering systems: A global perspective. In 2009 IEEE International Conference on Automation Science and Engineering (pp. 108-115). IEEE.

[8] Sun, B., Zeng, S., Kang, R., & Pecht, M. G. (2012). Benefits and challenges of system prognostics. IEEE Transactions on reliability, 61(2), 323-335.

[9] Vachtsevanos, G., Lewis, F., Roemer, M., Hess, A., & Wu, B. (2007). Systems Approach to CBM/PHM. Intelligent Fault Diagnosis and Prognosis for Engineering Systems, 13-55.

[10] Lee, J., Holgado, M., Kao, H. A., & Macchi, M. (2014). New thinking paradigm for maintenance innovation design. IFAC Proceedings Volumes, 47(3), 7104-7109.

[11] Cheng, S., Azarian, M. H., & Pecht, M. G. (2010). Sensor systems for prognostics and health management. Sensors, 10(6), 5774-5797.

[12] Haddad, G., Sandborn, P. A., & Pecht, M. G. (2012). An options approach for decision support of systems with prognostic capabilities. IEEE Transactions on reliability, 61(4), 872-883.

[13] Muller, A., Crespo Marquez, A., & Iung, B. (2006). On the concept of e-maintenance: Review and current research. Reliability Engineering and System Safety (93), pp. 1165–1187.

[14] Macchi, M., Crespo Márquez, A., Holgado, M., Fumagalli, L., Barberá Martínez, L. 2014. Value driven engineering of E-maintenance platforms. Journal of Manufacturing Technology Management. Journal of Manufacturing Technology Management, Volume 25 (4) 2014, art. no.17110941, pp. 568-598.

[15] Djurdjanovic, D., Lee, J. and Ni, J. (2003). Watchdog agent- an infotronics-based prognostics approach for product performance degradation assessment and prediction. Advanced Engineering Informatics, 17, p.109-125.

[16] Zammori, F., & Gabbrielli, R. (2012). ANP/RPN: a multi criteria evaluation of the Risk Priority Number. Quality and Reliability Engineering International, 28(1), 85- 104.

[17] Chinnam, R. B., Ding, J., & May, G. S. (2001). Intelligent quality controllers for on-line parameter design. In In J. Wang.

[18] Yang, Z. M., Djurdjanovic, D., & Ni, J. (2008). Maintenance scheduling in manufacturing systems based on predicted machine degradation. Journal of intelligent manufacturing, 19(1), 87-98

[19] Vianna, E. A., Abaide, A. R., Canha, L. N., & Miranda, V. (2017). Substations SF6 circuit breakers: Reliability evaluation based on equipment condition. Electric Power Systems Research, 142, 36-46.

[20] Luca Fumagalli, Stefano Ierace, Emanuele Dovere, Marco Macchi, Sergio Cavalieri, Marco Garetti (2011), Agile diagnostic tool based on electrical signature analysis. In Proceedings of IFAC 18th World Congress, August 28th – September 2nd 2011, Milan, Italy.

[21] Fumagalli, L., Pala, S., Macchi, M. (2014). A web service-based toolbox for machine diagnostics based on statistical analysis. Proceedings of The 3rd International Workshop and Congress on eMaintenance, 17-18 June 2014, Lulea, Sweden.

[22] Espíndola D., Fumagalli L., Garetti M., Botelho S. and Pereira C. (2011) An Adaption of OSA-CBM Architecture for Human-Computer Interaction through Mixed Interface, INDIN 11 conference, Caparica, Lisbon, Portugal, 26-29 July 2011.

[23] Pala, S., Fumagalli, L., Garetti, M., Macchi, M. (2014). Smart Sensors for Condition Based Maintenance: a Test Case in the Manufacturing Industry. EUROPEAN CONFERENCE OF THE PROGNOSTICS AND HEALTH MANAGEMENT SOCIETY 2014, Nantes (France) July 2014.

[24] Fumagalli, L., Macchi, M. (2016). An E-maintenance platform based on SOA and smart sensors. 4th IMEKO TC10 Workshop on Technical Diagnostics 2016: New Perspectives in Measurements, Tools and Techniques for Systems Reliability, Maintainability and Safety 2016, Pages 96-101; Milan; Italy; 27 June 2016 through 28 June 2016.

MFPT 2017 Paper

Health and Usage Monitoring: Critical Components analysis

Ankit Patel and Dr. John M. Lacontora

Abstract:

The following paper is centered on the premise the Intelligent Transportation Systems (ITS), the self-driving automobile and the environment in which it operates will require the automobile to have an on-board Health and Usage Monitoring Systems (HUMS) to operate safely on roadways. HUMS are in their early stages of use to monitor vehicle components and critical structures to prevent routine and catastrophic failures. Vehicle health monitoring, prognostics, diagnostics and condition-based maintenance are the central tenants of a proposed HUMS for autonomous automobiles. Over the past decade, autonomous vehicles have been studied and developed for a real-world application. It is essential to understand the critical assets that need to be monitored in those systems to conduct safe operations in ITS. This paper identifies the components that possibly would be required to monitor and some methods that would be employed to monitor those components.

Keywords: Health and Usage Monitoring Systems, Prognostics, Autonomous Systems, Automobile

Introduction:

Autonomous vehicles are going to be an essential technology of the future. An autonomous car is a vehicle that has the capability to guide itself without direct human interaction. The interest in this technology has been growing in various sectors (academia, industry and government) since 2004. According to Amnon Shashua, chief technology officer of Mobileye, sensing the road, mapping the road, and negotiate your place on road are the vital features required in building an autonomous vehicle. Under which the first feature has matured through the driver assist technology. Whereas, the other two are immensely complicated features that requires time to mature in the right way. In order to achieve these, sensors can be considered as critical assets in this vehicular technology. There are many sensors in an autonomous car, which can be categorized in the following types: Radar technology, Ultrasound technology, Cameras, Navigational aid and LiDAR. This research paper will focus on LiDAR sensor to analyze and monitor its performance through health and usage monitoring.

1.0 <u>Identification of sensors:</u>

In figure 1, a list of possible sensory location is depicted on an autonomous vehicle. This would provide a understanding of possible locations of various sensors on the vehicle. As mentioned previously, there are many sensors on an autonomous vehicle and the figure below groups them under five major groups. First, vision which would comprise of the front camera. Second, Infrared/Thermal sensors followed by third, LiDAR sensor for a 360^0 view. Fourth, Radar sensors and its combination of long and short/midrange specification and finally, fifth, ultrasonic sensors that would work in conjunction with navigation/GPS for accurate position of the vehicle.

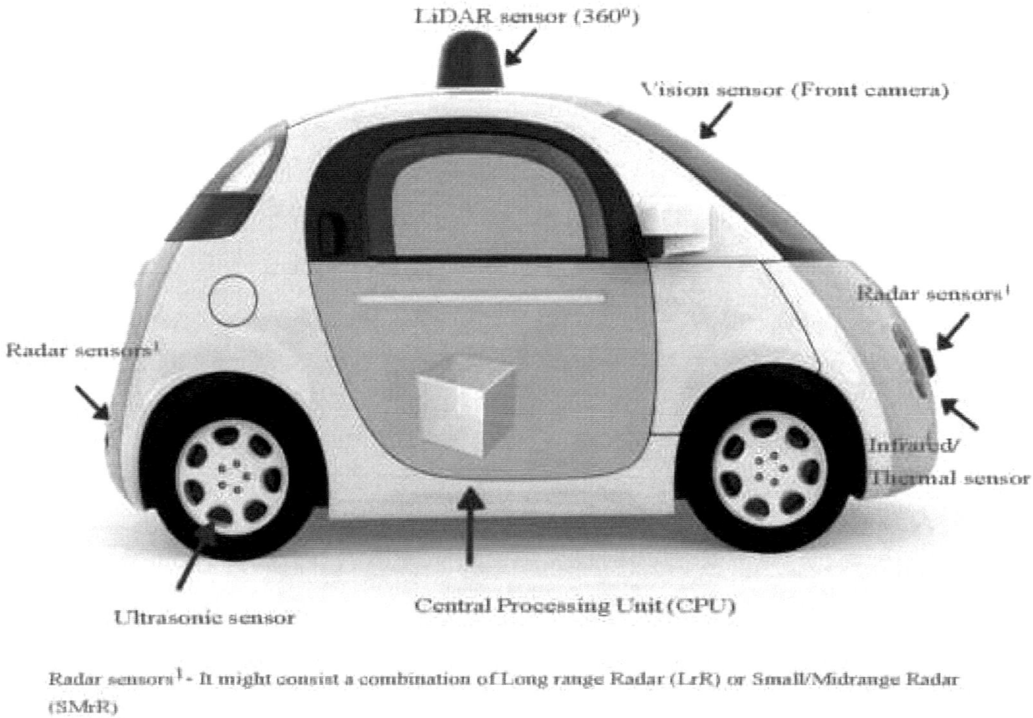

Figure 1 Sensor Identification in an autonomous vehicle

As mentioned above, there are multiple sensors coupled to a Central Processing Unit that enable an autonomous vehicle to operate efficiently. Let us identify and categorize.

1. Radar Technology:

It is primarily used for adaptive cruise control. The microwaves reflected from backside of the vehicles to the front side of the car behind it is used to adjust the speed. This type of technology does not use any satellite based information to set its control laws, it mainly uses on-board systems for processing the data. Co-operative adaptive cruise control

(CACC) on the other hand uses satellite and roadside infrastructure to make a decision about its speed and avoid the obstacle. They can be sub-categorized into Long range (78.6 MHz) and short/midrange (24.6 MHz/76.8 GHz) radars based on their applications.

2. Ultrasonic Sensor:

These sensors are basically used a position estimator for an autonomous vehicle. As mentioned before they work in conjunction with the Global Positioning System (GPS). These GPS's use real time geographical data received from several GPS satellites to calculate longitude, latitude, speed and course to help navigate a car. Eventually using this technology an auto-update feature can be created for the estimating the position of the vehicle in real-time.

3. Vision Sensor (Front cameras):

This technology is mainly used for lane-keeping and back up assistance. Image processing software can detect lane-stripes, signs, stop lights, road signs, and other objects.

4. Infrared/Thermal Sensor:

An autonomous car in the future will be traveling through the densely-populated city streets. This means they would be navigating through the stop signs and pedestrian zebra crossings. With the help of infrared/thermal sensor, there would be an added layer of detection of objects and in the case of thermal sensors for living beings.

5. LiDAR:

One of the most integral, expensive, and noticeable pieces of equipment found in an autonomous vehicle is the roof-mounted device called LiDAR, which stands for Light Detection and Ranging, is a remote-sensing technology that measures and maps the distance to targets, as well as other property characteristics of objects in its path. LiDAR essentially maps its surroundings by illuminating its targets with laser light and then analyzing that light to create a high-resolution digital image.

Google's autonomous vehicle research project uses a spinning range-finding unit, it has 64 lasers and receivers. The device creates detailed map of the car's surrounding as it moves. Software adds information from other sensors and compares the map with existing maps and notifies any differences [1].

Eventually all the data generated from these sensors would be complied in the Central Processing Unit (CPU). It would also be accountable for adjusting the vehicle dynamic functions such as steering, acceleration, braking and so on. Along with all of these various functionalities, it would also help AV in understanding the traffic laws with the help of certain control applications.

After understanding the sensor functionalities, it is vital to keep ourselves informed about how these sensors (or 5 major types in our case) would be applied or match to the new and upcoming or in some cases already existing technology (in regular cars) in comparison with the AV's. Knowing about these applications would help us evaluate the health of the sensor based on the all of the factors that could possibly work against it. Table 1, is one of many attempts towards the comparison of sensors and its applications.

Table 1 Sensors and its relative applications

Sensors /Application	Vision sensor (Front Camera)	Infrared/Thermal sensor	Long Range Radar	Short/Midrange Radar	LiDAR sensor
Adaptive Front Lighting (AFL)	X				X
Traffic Sign Recognition	X	X			X
Night Vision	X				X
Lane Departure Warning	X				X
Adaptive Cruise Control (ACC)	X		X	X	X
Emergency Brake Assist	X			X	X
Pedestrian detection	X	X		X	X
Blind Spot Detection	X			X	X
Rear Collision Warning	X			X	X
Park Assist	X			X	X
Camera Monitor System	X				X

2.0 **Reporting of sensor:**

As stated earlier, we are going to focus on a few LiDAR technology sensors as they are one of the most critical components in the functioning of an AV as seen in table 1. This section will focus on what they are and how they can be monitored to understand the sensors health and overall safety of the AV from LiDAR perspective.

2.1 LiDAR sensors: LiDAR technology is a combination of various functionalities such as optics, mechanics, analog front end and digital processing. The LiDAR emits a laser light which in return enables the stop watch. The light propagates and eventually diverges with an object, some of it scatters while few photons return back to the photo detector. The analog and digital chain then sense that a pulse is received and the disable the stop watch. Simply using the Distance = speed x time, formula the range of the object can be determined. In this technology, timing accuracy to carry out this operation is essential. According to Daniel Rosenband, in the Hot Chips 2016 conference, for every millimeter of precession that you give up that is equal to 60 picoseconds. So, if you want 1 centimeter of accuracy that is 60 picoseconds that you need of the timing control. In most common cases, Time Of Flight (TOF) is used to determine the range.

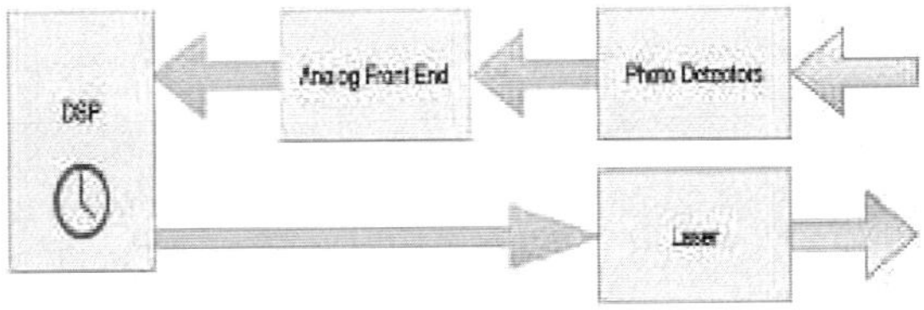

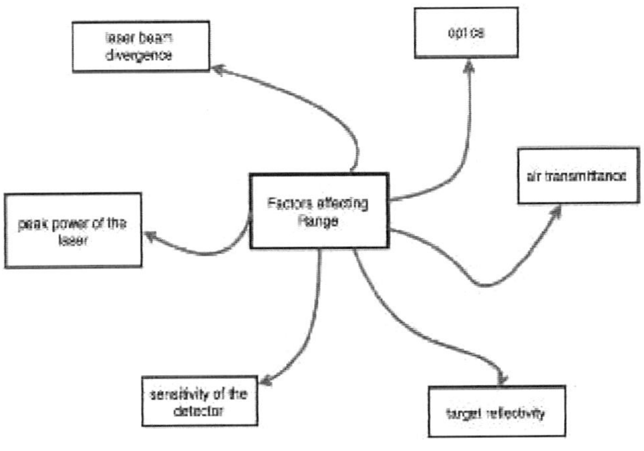

Figure 3 Mind Map for factors affecting while finding the distance between objects

Transmittance and reflectance parameters are usually imposed by the application. Design flexibility resides mainly in the selection of the laser source (power) and the receiver (sensitivity). The accuracy of TOF measurements depends on the pulse width of the laser and the speed and accuracy of the analog to digital converter (ADC) used. As seen in Figure 2, the factors affecting while finding the distance between two objects or its range gets its input from the light pulses detected by the photo detector and eventually implementing a range finding algorithm to determine the distance once the signal processing has been achieved through the combination of ADC and DSP. This range could be given by this formula as follows [2], it presents the range of the semiconductor pulsed laser based on its power in Watts and atmospheric conditions:

$$R \quad , R_S = \sqrt{\frac{L_P * R_A * T_A * T_U}{S_D * P * B}}$$

here, $L_P = L \quad P$

$R_A = R \quad O \quad A \quad (l \quad o \; m \quad)$

$T_A = T \quad o \;\; the \; at \quad he$

$T_U = T \quad o \;\; the \; o$

$S_D = D \quad S$

$B = B \quad d \quad in \; R$

The equation mentioned above can be used in a range identifying technique given by a flow chart below. The important things to make a note of are the use of Kalman filters and applying the on-board health and usage monitoring systems (HUMS) to understand the process efficiently.

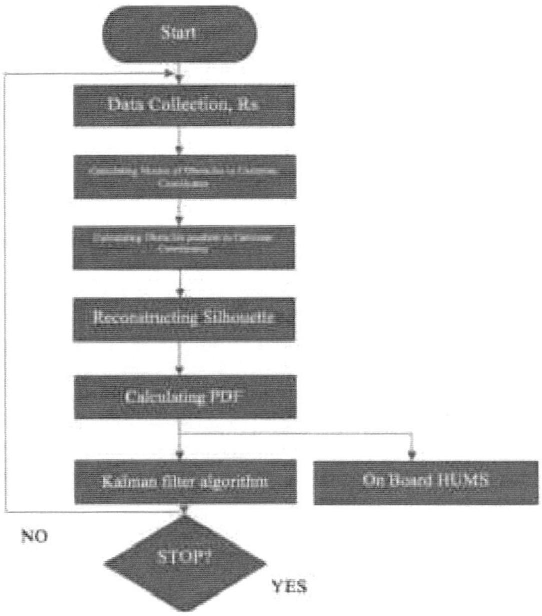

Figure 4 Flowchart for LiDAR sensor process flow and on-board HUMS application

Kalman filters algorithm is being used to transform the obstacle position from Cartesian co-ordinates to global coordinates for ease of use. Since, a Kalman filter can estimate a state of a linear system by using observation data online, this approach will not only transform the coordinates efficiently, but also enhance the accuracy of the obstacle position and motion detection. In this case, Kalman filter-based coordinates transformation algorithm consists of a system equation and an observation equation reported from the LiDAR sensor. On the other hand, the on-board HUMS system has the digital communication capability with possible options of USB, LAN, Ethernet ports [3] for data analysis and a possible sensor health management.

3.0 Sensor Fault Analysis:

The sensor groups mentioned above would be required to do certain tasks and comply with the performance standards set by their manufacturers. As we see, these sensors play an integral part in the functioning of an autonomous vehicle and monitoring them can be considered of utmost importance. Bad sensor readings can lead to catastrophic events on the roads. There are many factors/variables that would affect the sensor readings and eventually the health and operation of that specific sensor. Let us focus on LiDAR sensors and its performance criteria's. The performance of a LiDAR sensor can be judged by the following: (1) Range Specification (the maximum range at which it can see a target of a specified size). (2) Accuracy (the accuracy of its measurement of target location in range and angle). (3) Distinguishing ability (its ability of distinguish two or more target). (4) Form recognition (its ability of recognize form of the target). (5) Overall functionality (availability, reliability and maintainability).

The sensors can be evaluated on the basis of the fault modes it undergoes. There are various behavioral criteria's available for monitoring the sensor and its performance [5]. They are as follows: *Bias:* A constant offset from the nominal statistics of the sensor signal, *Drift:* A time varying offset from the nominal statistics of the sensor signal, *Scaling (or gain failure):* Magnitudes are scaled by a factor α(t) where the form of the waveform itself does not change, Noise: A random time series is observed, *Hard Fault:* The sensor output is stuck at a particular level expressed by $Y_f = C + noise$, where C is a constant, *Intermittent:* Deviations from normal readings appear and disappear several times from the sensor signal.

Based on these behavioral criteria's a decision matrix comparing the former with the specific sensors performance specifications. The purpose of this matrix would help HUMS to trigger its prognostics application to further determine sensors remaining useful life and the confidence interval of those predictions. The scoring in this matrix can be conducted through self-diagnostics capability of the system.

Table 2 Sensor performance decision matrix

Radar Sensor Performance Matrix	Range Specification *(Score 1 - 10) 10 being severe*	Accuracy *(Score 1 - 10) 10 being severe*	Distinguishing ability *(Score 1 - 10) 10 being severe*	Form Recognition *(Score 1 - 10) 10 being severe*	Overall Functionality *(Score 1 - 10) 10 being severe*
Bias					
Drift					
Scaling					
Hard Fault					
Intermittent					

Based on the scoring there would be signal generated on the severity of the fault and enable the prognostic algorithm in HUMS application for further evaluation. In order to imply these algorithm, it is of importance to understand how and what the sensor reports to the system which will be covered in the upcoming section.

4.0 Sensor Health:

Monitoring a system is certainly more than just data collection. The methodology for determining critical components, or sensors in this case, can be seen in figure 4 below [4].

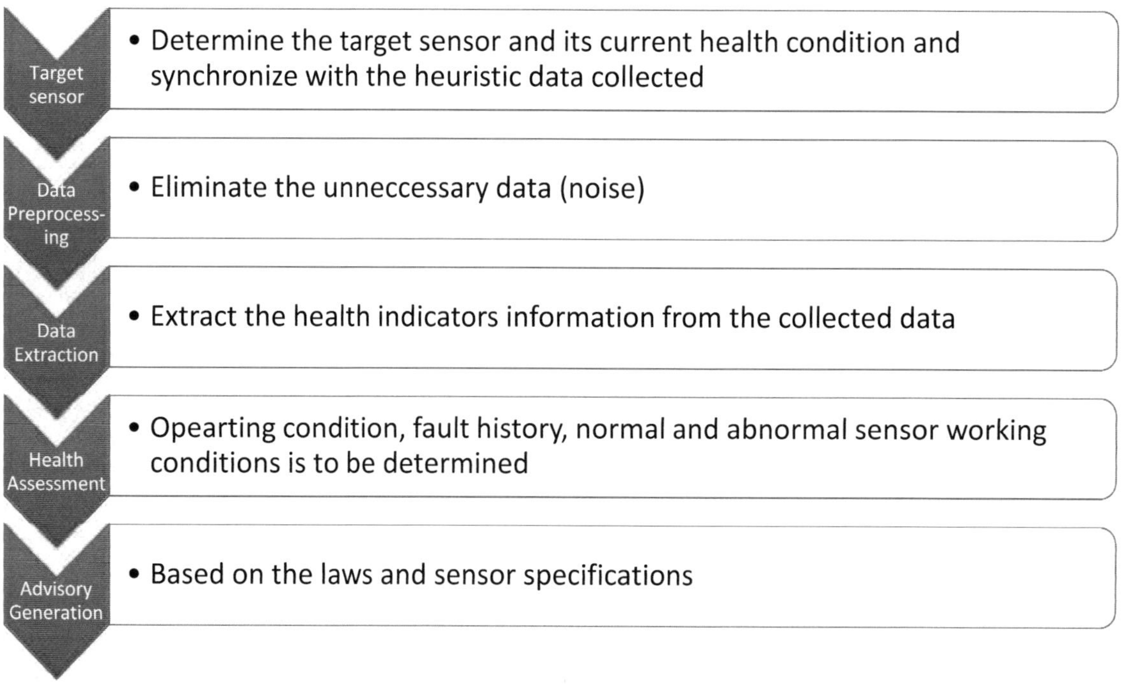

Figure 3 Health assessment for a critical sensor

As seen in the figure above, confidence intervals can be calculated. By determining the remaining useful life and trend of confidence intervals, the likelihood of failure in the near future can be estimated.

5.0 Conclusion:

As autonomous vehicles become more prevalent on our roadways the sensors that make them possible are at the keystone of their success or failure. The technology is expected to reach SAE level-5 by year 2021 and being promised by some automotive giants. Thus, the development of sensors and the on-board intelligence also needs to keep pace of AV implementation timelines. Understanding and maintain the sensor information and health is soon going to be a quintessential step in this process. As described in this paper, the process of determining the critical sensors, their functionalities and eventually their health will be of utmost importance in the upcoming years. This being said, it leaves us with a few questions:

1. How important is it for us to understand the health of all critical sensors in an autonomous vehicle while it is in operation?
2. How to tackle the understanding the current and future state of sensors with multiple variables for prognostics analysis?
3. How efficient are current algorithms in analyzing components and how reliable are the solutions?

6.0 Future Work:

To successfully create a functioning HUMS system for autonomous vehicle application, much more work is needed. As mentioned earlier, FMEA and FTA will need to be carried out for each autonomous vehicle model, and for the generic automated automobile as well, in order to pin-point the components and elements in a vehicle that require HUMS application. Following that, descriptive and precise mathematical models need to be developed that are capable of representing the founded failure mode and a mechanism will need to be developed to accurately interpret inputs from the system of sensors for autonomous vehicles in order to comply with SAE level-5. Even more mathematical models will need to be developed and tested for analyzing the data from HUMS and deriving practical insights to help predict the behavior of the components and to establish maintenance plans.

Reference:

1. "Technology and Costs." Google's Autonomous Vehicle. N.p., n.d. Web. 4 Aug. 2016.
2. McCormack, P. (2011). *LIDAR System Design for Automotive/ Industrial/Military Applications* (Literature Number: SNAA123). Texas Instruments.
3. Honeywell Aerospace. (2016). *ONBOARD VIBRATION MONITORING SYSTEM/HUMS* (N61-1549-000-000 I 04/16).
4. Advantech eAutomation. (n.d.). Smart Sensor and Prognostics for Self-Maintenance Machines.
5. D. C. Zimmerman, and T. L. Lyde, "Sensor failure detection and isolation in flexible structures using the eigensystem realization algorithm," in 33rd Structural Dynamics and Materials Conference, Dallas, TX, 1992, pp. 2156-2166.

HOW SENSOR MOUNTING SIGNIFICANTLY AFFECTS VIBRATION MEASUREMENTS

David A. Corelli
IMI Sensors
President, Vibration Institute
3425 Walden Avenue
Depew, NY 14043
dcorelli@pcb.com
dcorelli@vi-institute.org

Abstract: Sensor mounting can significantly affect the measured amplitudes of both overall vibration and spectral (FFT) data. This paper shows how the frequency response of a sensor system (accelerometer plus mount) is significantly different than the factory specified frequency response of the accelerometer itself. The sensor system frequency response, using various common mounting methods, such as stud, 2-rail (curved surface) magnet, and flat magnet, were measured under controlled laboratory conditions and then correlated with actual data collected on machinery in the field. The paper also shows the dramatic effect that mounting has on commonly used high frequency measurements such as Spike Energy™ and PeakVue® that are used for early warning of bearing and gear faults. Finally, it explains and shows generically, step-by-step how high frequency demodulation is calculated (e.g., Spike Energy spectrum and PeakVue spectrum).

Note: The words accelerometer and sensor are used interchangeably throughout this paper.

Key Words: Accelerometer; bearing analysis; demodulation; high- frequency; mounting; sensors; frequency response

Introduction: Making vibration measurements within the linear, specified, range of an accelerometer and its mount is generally perceived to be a relatively easy process. This is true as long as the vendor supplied sensitivity of the sensor is used and its frequency response is not altered by the mount. In this case, the data gathered is reasonably accurate, repeatable, and transportable. Transportability, a term coined by the author, implies that if the same measurement parameters are used, analysts should get the same values regardless of the sensor, mounting, or instrumentation used. This is the basic premise that Vibration Standards are based on. It turns out that this is often not the case because the mounting method changes the flat frequency response of the accelerometer or more accurately the sensor system (accelerometer plus mount), and thus, we are often making "high frequency" measurements, or measurements beyond the flat range of the sensor systems, without even knowing it.

High Frequency: For purposes of this paper, a high-frequency measurement is defined as any measurement whose frequency range is above either the specified range of the accelerometer or its mounted frequency response. Thus, if a sensor has a frequency range of 15 kHz but it is being used with a magnet that has a +3 dB gain at 5000 Hz, then any measurement above 5000 Hz will be considered a high-frequency measurement and will contain significant measurement errors. In the case of "real" high-frequency measurements such as demodulation, this is always the case. Since the measurement range is well above the linear range of the accelerometer, the processing techniques have significant differences, and the data collector parameters can vary widely, these measurements are neither accurate (as defined in this paper) nor transportable. At best, an analyst can only hope these measurements are repeatable, trendable, and understandable enough for reasonable interpretation.

Accuracy: Simply defined, accuracy is the closeness that a measurement comes to the actual physical quantity being measured. A more rigorous definition is "Accuracy or Uncertainty: Uncertainty is generally defined as the largest expected error between actual and ideal output signals."[1] For example, if a machine is physically vibrating at 1g and the sensor is measuring 0.97g, then there is an error or uncertainty of 3%.

Once the measurement range falls outside of the calibrated sensor system range, accuracy is generally gone. Additionally, some high frequency measurements such as demodulation don't actually have a real physical quantity that is being measured so who's to say what is and what is not accurate? This paper suggests that in high frequency data acquisition, the amplitudes measured will vary based on sensor, mount, data collector, and techniques used but the frequencies should be consistent. It is therefore left up to the analyst to determine what amplitudes are acceptable and which are not based on historical data. Additionally, even though it may be difficult to determine the actual severity of a fault from high frequency measurements, they will most likely provide an earlier warning of certain types of faults. The following statement expresses this very well. PeakVue® ". . . is a powerful complementary tool that can detect a range of faults and problem condition that techniques such as Vibration Analysis alone might miss under certain conditions."[2]

Sensor Frequency Response: Since the results obtained in high-frequency data acquisition are highly dependent on the frequency response of the sensor and mount, it is useful to examine this topic. A piezoelectric accelerometer can generally be modeled as a single degree of freedom system with a flat frequency response at lower frequencies, rising to a single resonance, and then dropping off as shown in Figure 1. This sensor has no internal filtering and a high gain (38 dB) at resonance which is excellent for providing very early warning of High Frequency Energy (HFE) caused by rolling element bearing faults, however, it can often be a poor design because it is subject to saturation (over driving the internal electronics) due to the high gain. Experience has shown this is particularly true for sensors with a natural frequency in the 25 kHz range as found in many industrial accelerometers.

To temper the saturation problem in accelerometers, many if not most, industrial sensors have a built in single or two-pole filter. The frequency response of an electrical 2-pole filter is shown in Figure 2. When this filter is combined with the response of the unfiltered sensor, the combined response in Figure 3 is obtained. Comparing the unfiltered and filtered response, it is seen that the gain at resonance is reduced from about 38.1 dB (gain of 80) to about 12.4 dB (gain of 4) a 20:1 reduction. Thus, the amplitudes of real HF measurements will differ widely between the two. That doesn't mean that one is right and one is wrong, it just means they will be different.

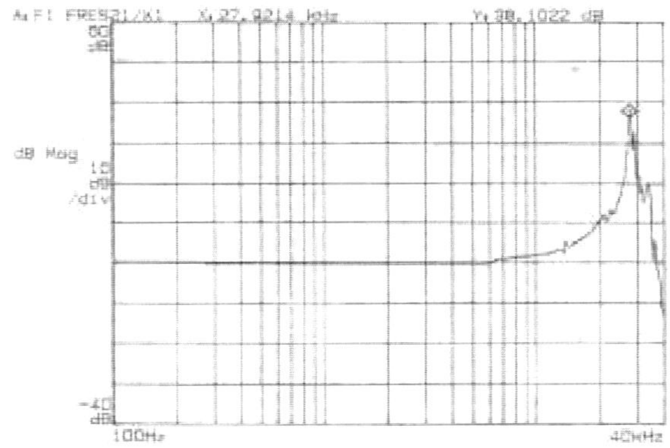

Figure 1. Unfiltered sensor response

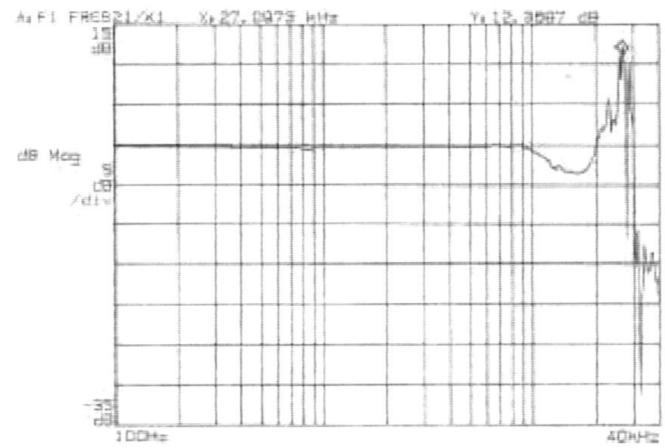

Figure 3. Frequency response of filtered sensor

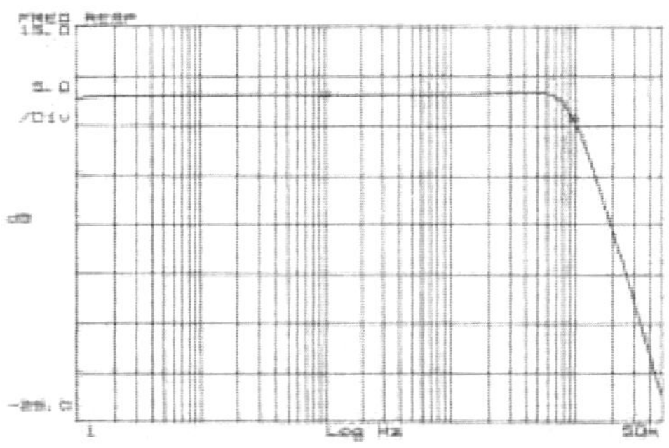

Figure 2. Two-pole electrical filter

Mounting Frequency Response: To complicate matters further, it is quite likely that the frequency response of the sensor mount, or mounted resonant frequency, will be lower than that of the sensor. A chain is only as strong as its weakest link and in general, the weak link is the mount. The responses of two accelerometers were examined with different mounts to see the effect of mounting method on frequency response, and the effect was dramatic.

IMI model 603C01 low cost and model 622B01 precision industrial accelerometers were used for these tests. The 603C01 has a 2-pole internal filter while the 622B01 has a single-pole filter. Frequency response tests were run using step sine analysis on an NIST traceable accelerometer calibration system. The frequency response was tested with the following mounts: stud, 35 lb. flat magnet, 35 lb. 2-rail or curved surface magnet, and 4" SS Probe (stinger). A typical test setup is shown in Figure 4. The lower sensor is a back-to-back calibration standard. Figure 5 shows the stainless steel probe mounted to a 603C01 accelerometer. The results of the tests are shown in Table 1.

Figure 4. Frequency response test setup

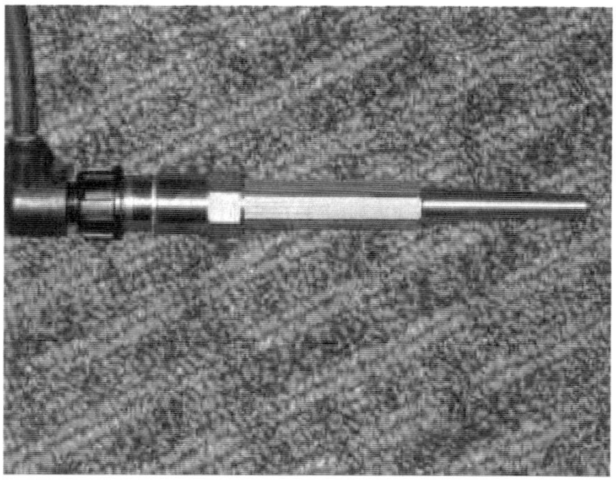

Figure 5. Stainless steel probe (stinger)

Sensor & Mount	3 dB Freq (Hz)	Resonant Freq (Hz)	Gain at Resonance	Total Mass (gm)
622B01 + Stud	15000	30000	35 to 40 dB	94
603C01 + Stud	12500	25000	15 dB	51
603C01 + Flat Magnet	6000	11245	21.5 dB	100
622B01 + Flat Magnet	5000	8000	20 dB	143
603C01 + 2-Rail Magnet	3554	6322	27.3 dB	128.1
622B01+ 2-Rail Magnet	3308	6000	22.9 dB	171.1
603C01 + 4" SS Probe	720	1300		

Table 1. Resonant frequency and 3 dB point for two sensors and 4-mounts

From the table it can be seen, for example, that a 622B01 stud mounted had a flat frequency range to 15k Hz, however, when mounted on a smooth flat surface with the 2-rail magnet, its range dropped all the way down to 3308 Hz, a dramatic reduction. Recall the definition of a HF measurement given in the introduction. Based on the results shown in Table 1, HF measurements can start as low as 720 Hz when using the SS Probe or as high as 15 kHz using a stud mounted 622B01 precision accelerometer. Note also that the 3 dB frequencies and resonant frequencies are different for the two sensors even though they are mounted using the identical magnets. The mass of the accelerometer plus the magnet is included, which shows that the useful range of the sensor decreases when the mass increases. This makes sense since the natural frequency is given by Equation 1, where m is the total mass of the sensor and magnet and k is the effective spring constant due to the magnetic pull.

$$f_n = \frac{1}{2\pi}\sqrt{\frac{k}{m}} \qquad \text{Equation 1}$$

This example also demonstrates that it is difficult at best to generalize on the frequency range of a sensor when used with a particular magnet. These results also do not take into account the condition of the magnet or surface that it's being used on, which can significantly affect the useful range.

Pump Mounting Example: This example demonstrates how the frequency response of the sensor mount can dramatically affect the HF measurement results. In this case, the sensor, an old IRD 970, has a flat frequency response well beyond 5 kHz as shown in Figure 6. The frequency response when used with the then popular 9" aluminum stinger, however, is only flat to about 550 Hz, Figure 7. By this paper's definition, anything above 550 Hz using the stinger in this example is a high-frequency measurement.

The natural frequency of the stinger is 42,750 CPM, Figure 7. It is clearly seen in the log amplitude plot that the measured data on the pump peaks at 42,750 CPM (the stinger resonance) and the data beyond about 90,000 CPM is significantly attenuated, Figure 8. At the higher frequencies, the stinger is acting as a mechanical isolator. This is obviously bad data, not accurate, and highly influenced by the stinger's frequency response. The analyst that originally collected the pump data had it plotted with a linear amplitude scale. He noticed that it lacked any HF data, which didn't make sense. The data was questioned and the frequency response plots of the sensor and stinger obtained from IRD. Upon closer examination, the problem became obvious.

While this is a quite dramatic example of how the sensor mount can adversely influence measurements, the same basic principle holds for all sensors and mounts. The morals of the story are to know your measurement system and to question your data.

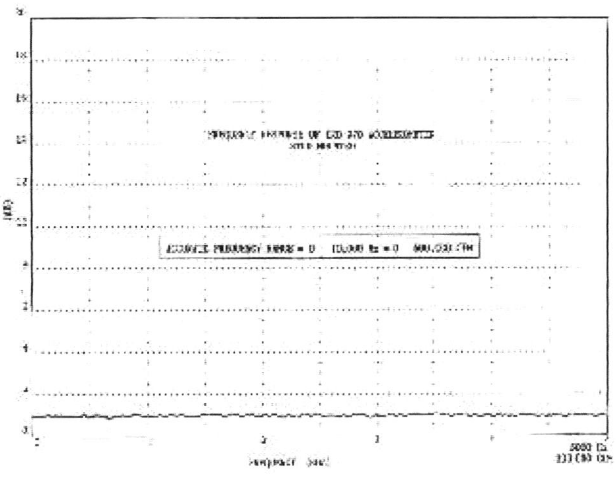

Figure 6, IRD 970 frequency response stud mounted

Centrifugal Compressor: The above example clearly shows how the sensor mount can affect the measured data. But how much variation is actually encountered under normal day-to-day measurement circumstances due to the different sensors and typical mounts?

Data was collected on the centrifugal compressor shown in Figure 9. The acceleration and velocity spectra collected with the various sensors and mounts are shown in Figures 10 and 11. Even though these are seemingly low frequency measurements, i.e., well within the calibrated range of the sensors, there are major differences in the spectral data in both acceleration and velocity. This is due to the effects of the mount. For the 2-rail magnet, twist mount, and stinger, they are significant. For example, the data in the 3000 to 4000 Hz range on the 2-rail magnet measurements are about twice what they are when stud mounted, that is about 100% error! This clearly shows that the mounting methods typically used in day-to-day data collection can have a large effect on the amplitude accuracy of routinely collected data.

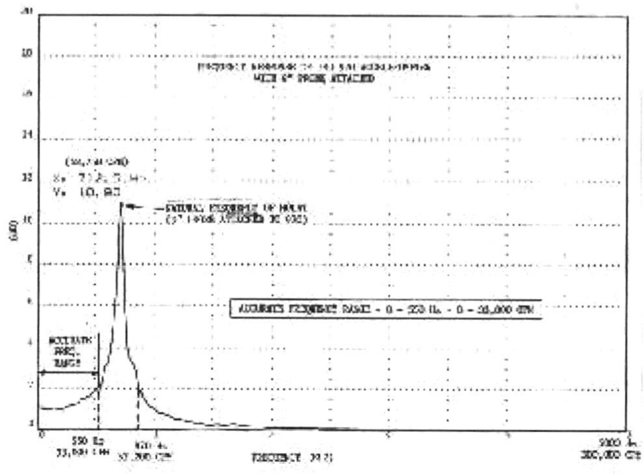

Figure 7. IRD 970 frequency response with 9" aluminum stinger

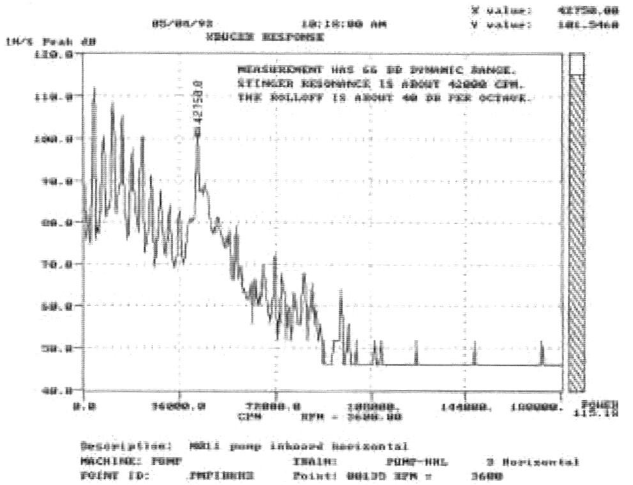

Figure 8. Log amplitude plot of pump data

Figure 9. Centrifugal compressor

Combustion Air Fan: It is obvious in the above example that the frequency response of the sensor mount can change the measurements even when the data is within the specified range of the sensor. In High Frequency Energy (HFE) readings, data is routinely analyzed that falls outside of the specified range of the sensor. Additionally, the methods, algorithms, filtering, and bandwidths used for these types of analyses differ from instrument to instrument. To complicate matters further, HFE is not a well defined measure like acceleration, velocity, and displacement. There is no actual defined physical quantity against which to measure accuracy. Because of this, amplitudes are arbitrary and based on all of the variables mentioned.

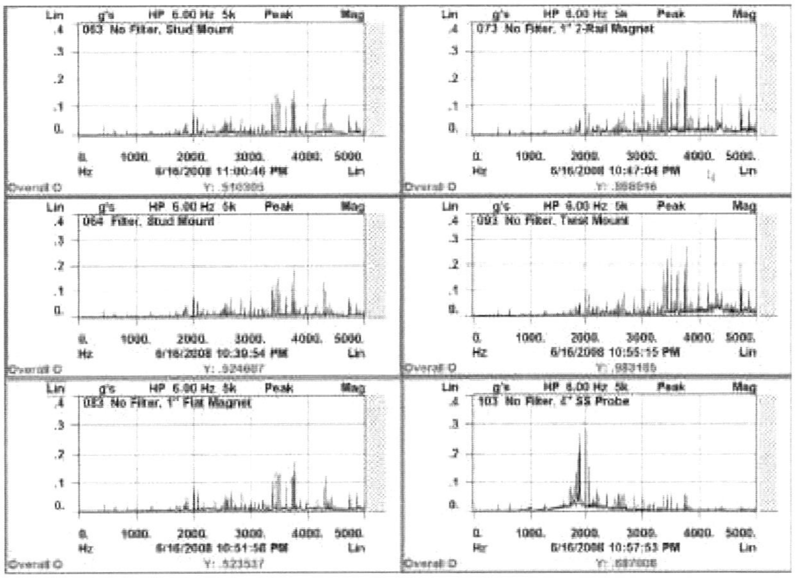

Figure 10. Acceleration spectra to 5000 Hz have significant differences in the spectral amplitudes

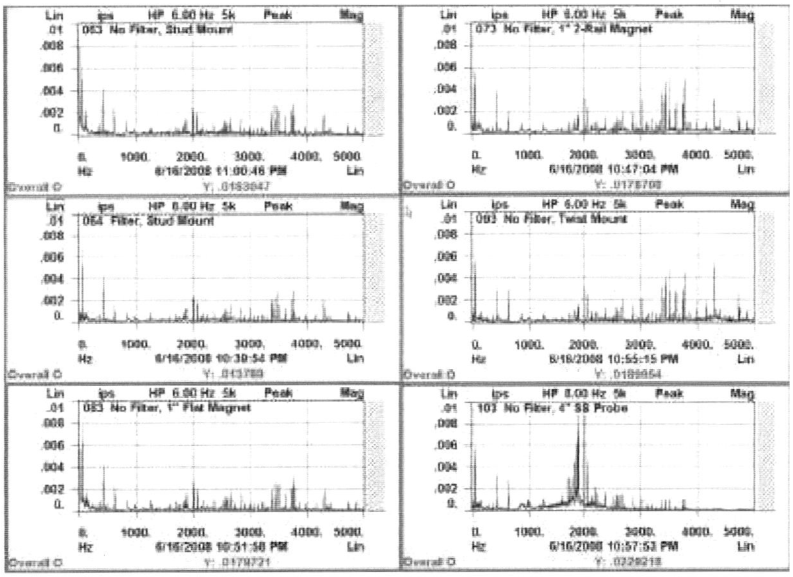

Figure 11. In velocity spectra, where an analyst would typically expect consistent data, there are also significant differences

The data shown in Table 2 summarizes the Overall Spike Energy™ readings (an HFE measurement) taken on the combustion air fan with an EntekIRD dataPAC™ 1500. The measurements were made with 1 kHz, 2 kHz, and 5 kHz High Pass (HP) Corner filters. The Low Pass Corner frequency of the data collector is fixed at 65 kHz. Thus, the frequencies measured will be well outside of the specified range of the sensor, can be at or near the sensor resonance, and can even be above it. The table is sorted by Overall Spike Energy™ (gSE units) from highest to lowest. There is a 46 to 1 difference between the highest and lowest readings! In general, the sensor with the highest frequency response and best mount has the highest values. That does not mean, however, that the other readings are wrong. They are just different.

Since HFE readings are not absolute, it is impossible to generate general severity charts based on their values. This could only be attempted for a given sensor, mount, and instrument that uses the same method, algorithm, filtering, and bandwidth all the time. In order to use the method effectively, the measurements must be made consistently and trended. Even with that, there can be amplitude variations as the generated frequencies shift relative to the sensor and mounting resonances. Though long term use and experience, an analyst may be able to determine severity guidelines for their equipment. However, be cautious, use HFE as a trigger to look at other vibration data. It is not a good idea to pull machinery from service or tear it down based totally on an HFE measurement.

					Overall Spike Energy™ (gSE)		
Pos	Sensor	Filter	Mount	Surface	1 kHz HPF	2 kHz HPF	5 kHz HPF
083	603A01	None	Flat Mag	Glue Base	6.95	6.54	6.46
063	603A01	None	Stud	Glue Base	6.85	6.36	6.45
023	603A01	None	Flat Mag	Bare	6.72	6.26	6.35
013	603A01	None	2 Pole Mag	Bare	3.79	3.50	3.49
073	603A01	None	2 Pole Mag	Glue Base	3.77	3.53	3.44
074	603C01	2-Pole	2 Pole Mag	Glue Base	2.85	2.63	2.51
014	603C01	2-Pole	2 Pole Mag	Bare	2.52	2.32	2.21
084	603C01	2-Pole	Flat mag	Glue Base	2.24	2.05	2.02
024	603C01	2-Pole	Flat Mag	Bare	2.11	1.88	1.81
064	603C01	2-Pole	Stud	Glue Base	1.91	1.65	1.58
103	603A01	None	Stinger	Bare	0.64	0.42	0.40
104	603C01	2-Pole	Stinger	Bare	0.39	0.22	0.14

Table 2, Overall Spike Energy™ for various sensors and mounts collected on the combustion air fan

Motor Generator Set: The following measurements were made on the drive motor, Figure 12, of an MG set. Spike Energy™ Spectrum (demodulated spectrum) measurements were taken with several accelerometers and mounting methods. The purpose was to see how sensors with varying frequency responses and mounting methods affected the HF demodulation. It is suspected there is a bad bearing (BPFO) as indicated by the 63 Hz spectral component. All combinations of sensors and mounts used clearly identified the fault at 63 Hz.

Figure 12. MG set drive

This was surprising when using the probe. It would have been thought that the HF content would be totally lost (mechanically filtered) but was not; It was just attenuated (note the different scales on the plots).

A sample of the data collected on the motor is shown in Figures 13 through 16. The 4-plots from top to bottom are the demodulated (gSE) spectrum, velocity spectrum, acceleration spectrum to 75 kHz to see HF response, and the time waveform with a 75 kHz BW. While the bearing fault is clearly identified in the demodulated spectrum, it is not being picked up in the velocity spectrum. Thus, while the amplitudes of HFE measurements are arbitrary, the demodulated spectrum is an excellent tool for identifying faults at early stages of development.

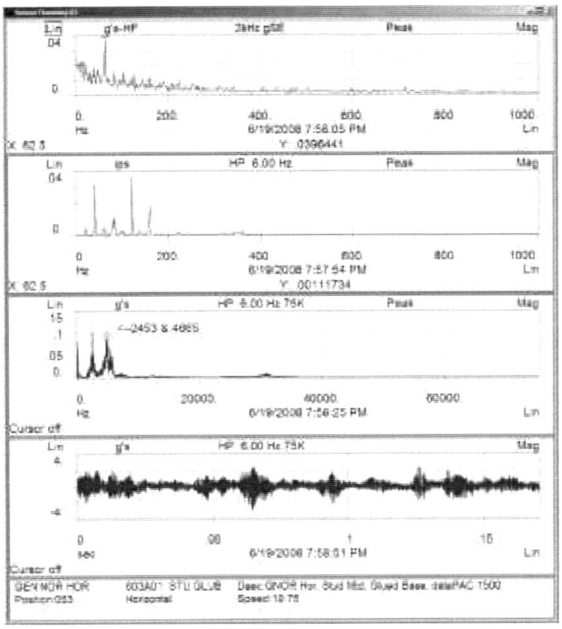

Figure 13, Unfiltered sensor with stud mount

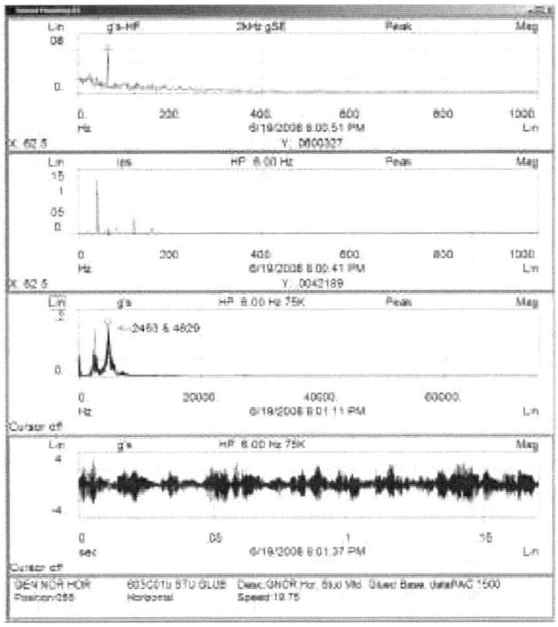

Figure 14, Filtered sensor with stud

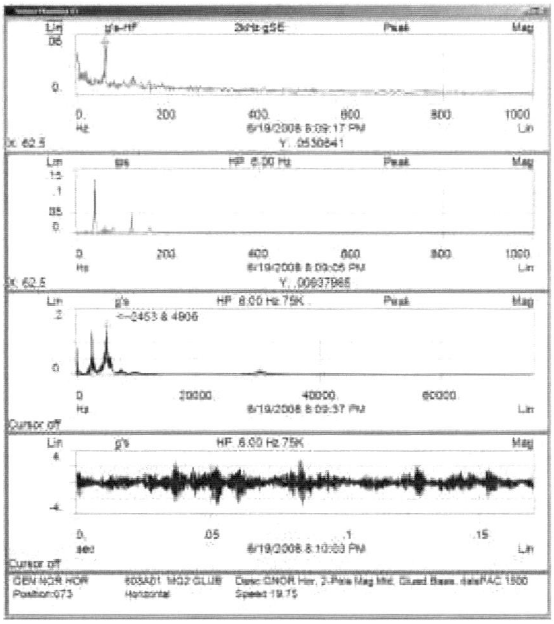

Figure 15, Unfiltered sensor with 2-rail magnet

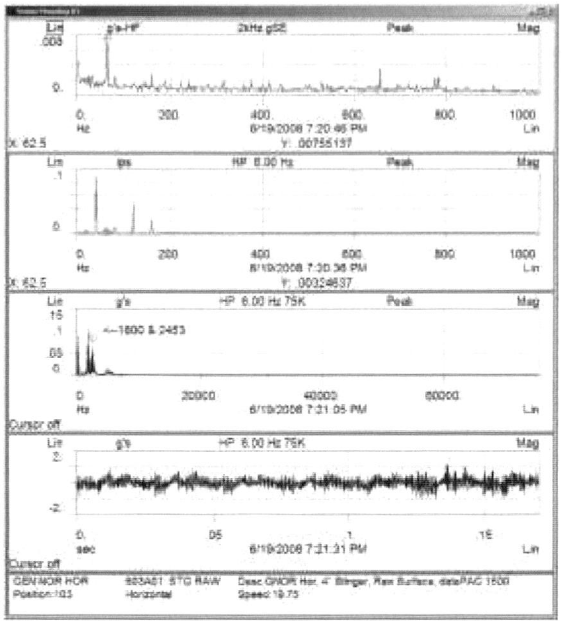

Figure 16, Unfiltered sensor with stinger

Demodulation and Bearings: Demodulation or enveloping is a signal processing technique that extracts fault frequencies from the higher frequency modulated vibration response that is generally caused by impulsive forces. The general processing steps used in vibration analysis to compute demodulated spectra are: high-pass filter, rectify, envelope, and take the FFT. However, the techniques used to accomplish this can and do vary widely.

The test setup shown in Figure 17 was used for the following tests. A SpectaQuest Rotor Kit with a slightly faulted inner race bearing, BPFI, was used to generate impacts. An EntekIRD dataPAC 1500 was used to measure time, spectral, and demodulated data.

Both an IMI 603C01 low cost industrial accelerometer and a 622B01 precision industrial accelerometer, with better high frequency response, were stud mounted and used for data collection, Table 3. In order to study the effects of filtering on time domain peaks and the demodulation process, the sensor output was run through a Wavetek Model 753A Brickwall Filter, having both a HP and LP filter with -115 dB per octave roll off. In order to simulate the full demodulation process, a half-wave rectifier was also used in conjunction with the filter. The dataPAC 1500 was used to measure the true peak acceleration and RMS acceleration, from which Crest Factor (CF) was calculated. The demodulated time waveform and demodulated spectra were also measured.

The rotor kit speed was about 1730 RPM, resulting in a BPFI of about 142.7 Hz. A rough check of the pulse width generated by the bearing fault over the speed range of the rotor kit, showed values from about 21 to 42 μs. Based on published data[2], these contact times appear to be reasonable and typical.

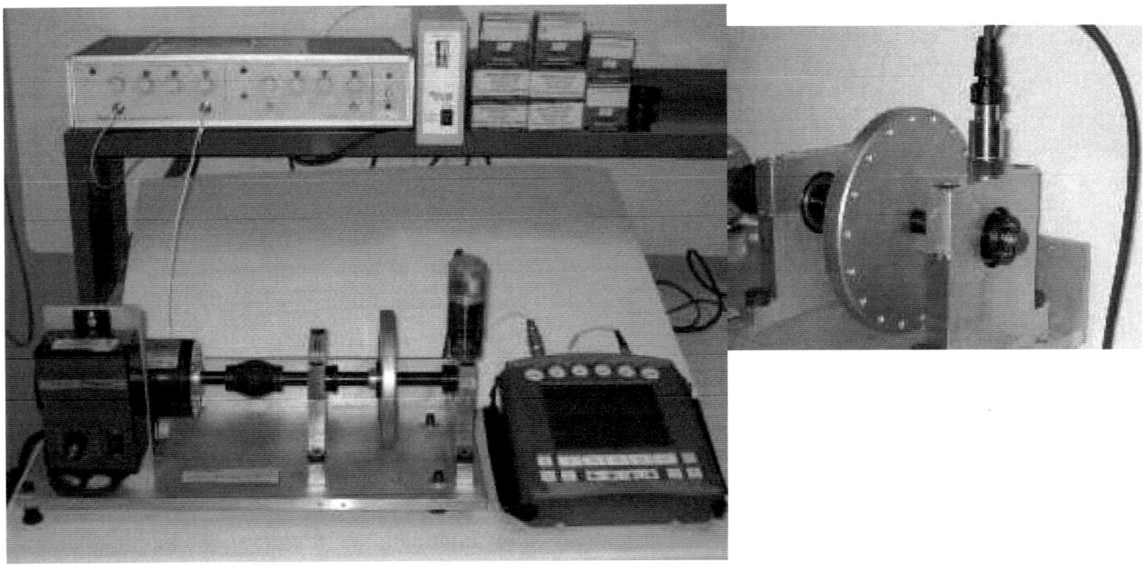

Figure 17: Rotor kit test setup with inlay photo of the 622B01 accelerometer

Model Number	±3 dB Frequency Range		Resonant Frequency	Internal Filter	Comments
	Low Freq	High Freq			
603C01	0.5 Hz	10 kHz	25 kHz	2-pole	Low cost
622B01	0.2 Hz	15 kHz	35 kHz	1-pole	Precision

Table 3. Comparison of the 603C01 and 622B01 accelerometers

Peak, RMS, and Crest Factor: Since the first step in the demodulation process is to HP filter the data (sometimes bandpass) it is revealing to see the filter affect on the amplitudes. Five measurements were collected at each of four HP filter settings for both sensors. The cutoff frequencies were chosen because they are the most typically used for vibration demodulation measurements. The average of the readings, are shown in the Table 4. A typical time waveform measurement used to generate this table is shown in Figure 18. Both the "RMS Amplitude" and the "Peak Amplitude" (true peak) are displayed on the plot. The crest factor is computed as the peak / RMS.

603C01	1000 Hz External HPF		603C01	2000 Hz External HPF		603C01	5000 Hz External HPF	
g rms	g peak	CF	g rms	g peak	CF	g rms	g peak	CF
0.373	3.157	8.46	0.281	2.496	8.9	0.146	1.785	12.26
622B01	1000 Hz External HPF		622B01	2000 Hz External HPF		622B01	5000 Hz External HPF	
g rms	g peak	CF	g rms	g peak	CF	g rms	g peak	CF
0.417	4.889	11.76	0.339	4.585	13.57	0.228	4.236	18.41

Table 4. Peak, RMS, and Crest Factor at various HP filter settings

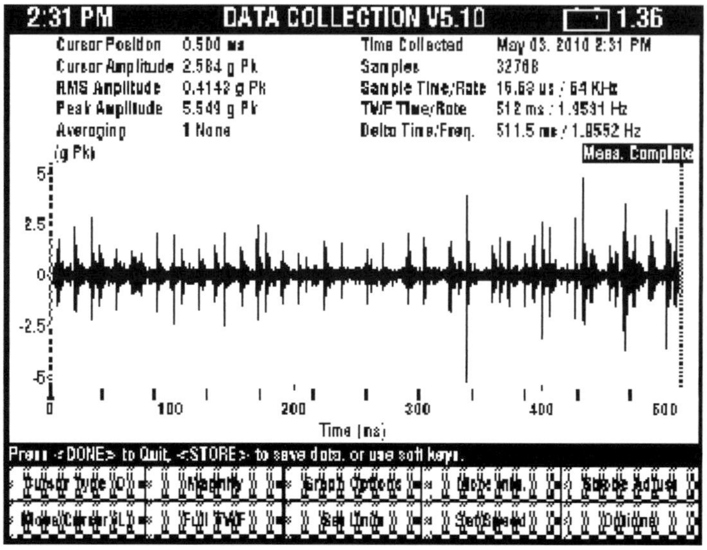

Figure 18, Typical time waveform measurement made for looking at HP filter effects

Amplitudes: Since there can be a lot of variation in the peak readings with a short data collection time, the average data is more consistent. As the HP filter cutoff frequency increases, the peak and RMS values decrease. Also, since the 622B01 has better HF response than the 603C01, its values are higher as would be expected. Further, as the HP filter cutoff frequency gets higher, the spread between the two sensor readings gets wider. This also makes sense because as more and more low frequency is filtered out, the HF accelerometer will have better response. The impacts are definitely enhanced through the filtering.

Crest Factor: A very interesting part of the analysis is in the Crest Factor. This may be a good indicator of an impending fault involving an impact, as in a bearing or gear. As more and more of the lower frequencies are filtered out, the CF gets larger. This implies that HP filtering of the data in conjunction with the CF may provide an earlier warning of an impulsive fault.

Step-by-Step Demodulation: As stated above, the basic steps in demodulation are: HP filter, rectify, envelope, and take the FFT. There are a lot of techniques for doing this particularly in the enveloping step of the analysis. In this example, enveloping is achieved by passing the signal through a LP filter. The following plots, Figures 19 through 22, show the processed signal after each step in the process using external filters and a rectifier. The final two plots, Figures 23 and 24, show the demodulated time waveform and demodulated spectra as calculated by the data collector.

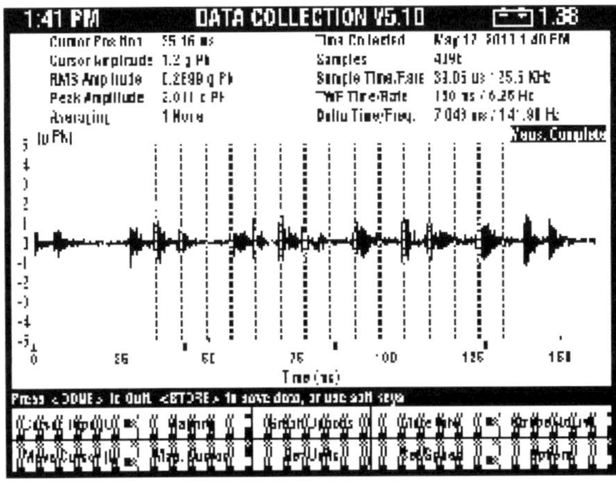

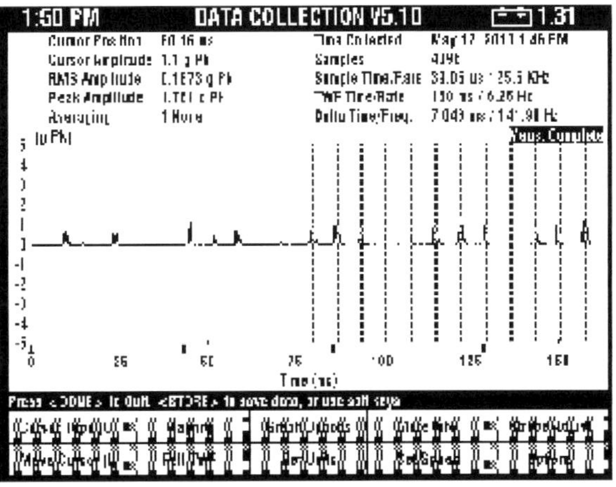

Figure 19. Step 1, HPF

Figure 20. Step 2, Rectify

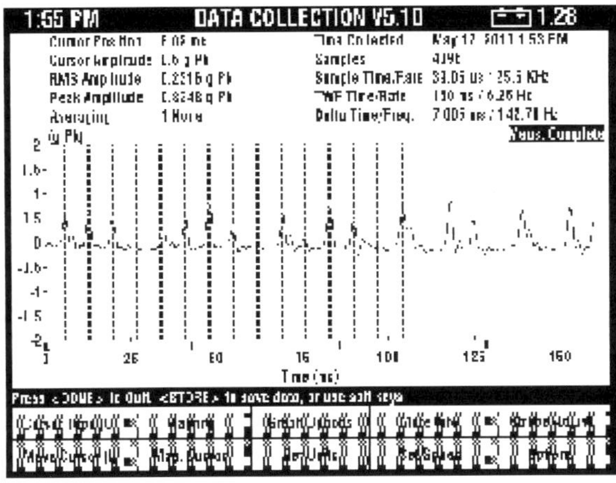

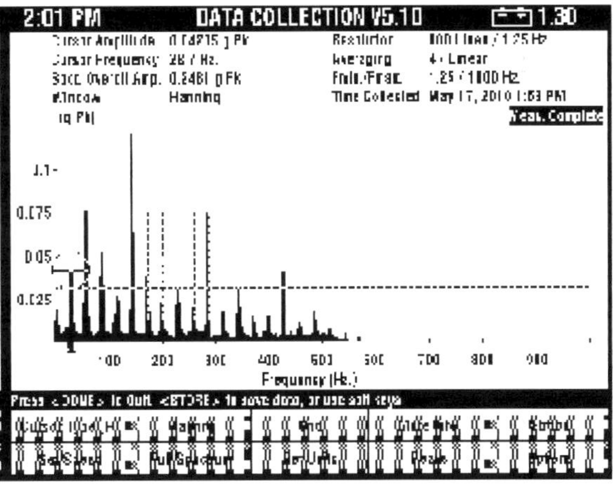

Figure 21. Step 21, LPF or Envelope yields the demodulated time waveform

Figure 22. Step 4, FFT to obtain the demodulated spectrum. The high peak is

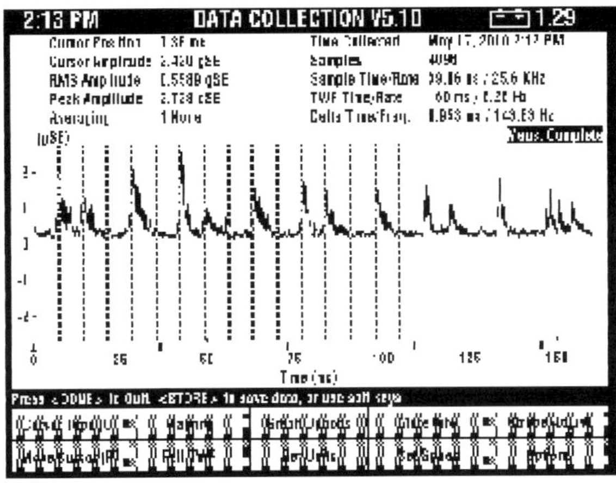

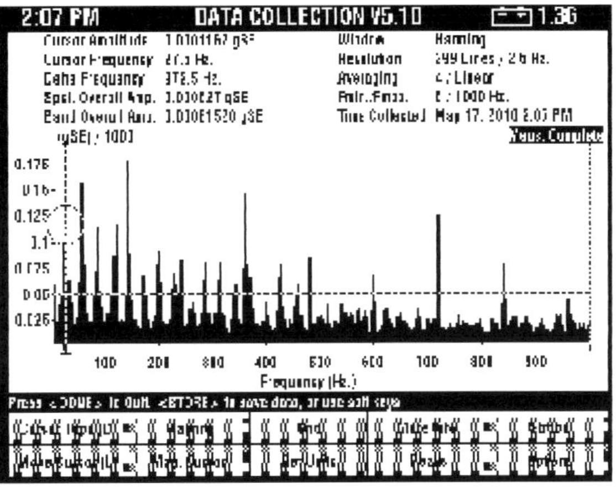

Figure 23. Demodulated time waveform produced by the data collector.

Figure 24. Demodulated spectrum (Spike Energy™ Spectrum) produced by the data

Conclusions: In order to make accurate measurements, the frequency range must not only be within the specified range of the accelerometer but must also be within the flat range of the mount. Knowing the range of the mount is a challenge and can only be determined through testing. When sophisticated calibration equipment is not available, it is possible to get an estimation of frequency response. Mount a sensor on a machine having high frequency content using various mounts. Comparing the results, as was done in the centrifugal compressor example, can provide useful information on about the sensor and mounting responses.

True high-frequency measurement such as HFE and demodulation are not physical measures so the amplitudes are arbitrary. Data must be collected consistently and trended to make use of the amplitudes. These trends should be used as an indicator to look at other vibrations data before diagnosing a problem. Demodulated time waveforms and spectra can often reveal impending faults, particularly in cases of impacting, much earlier that conventional analysis. They are a powerful complementary tool for the vibration analyst.

References and Trademarks:

1. Wilson, Jon S. *Sensor Technology Handbook*. Burlington: Elsevier, Inc., 2005

2. Robinson, James C. and James W. Walker, and James E. Berry. "Description of PeakVUE and Illustration of its Wide Array of Applications in Fault Detection and Problem Severity Assessment." Emerson Process Management reliability Conference: 2001

3. Xu, Ming. "Spike Energy™ Measurement and Case Histories." EntekIRD International Corporation. 1999.

PeakVUE is a registered trademark of Computational Systems Incorporated.

Spike Energy and dataPAC are trademarks of EntekIRD International Corporation

IMPROVING THE SAFETY MANAGEMENT SYSTEM THROUGH HFDM

Eric Bechhoefer
GPMS Inc.
40 Ridge Rd
Cornwall, VT 05753
Telephone: (937) 256-2285
eric@gpms-vt.com

Michael Augustin
mike@ivhm-inc.com
Owner
IVHM Inc.
Dallas, TX, USA

Abstract: Helicopter Flight Data Monitoring (HFDM) is an integral part of the service provider's safety management system. By capturing operational information about aircraft activities, the service provider can: identify safety hazards, initiate remedial action to maintain safety performance, facilitate monitoring and assessment of technical interactions between the crew and the aircraft, and facilitate continuous improvement of the safety management system. We wish to improve HFDM by providing aircraft metrics via automation of data download and reporting. Automation is achieved by formalizing the concept of a flight operation, exceedance monitoring and improving the HFDM architectural design to allow for the seamless movement of data. In the extreme, this model for HFDM provides protection of data even in the event of a mishap that would usually only available from crash survivable memory. This paper discusses the formalized concept of a flight operation, how regime recognition is used to support the function of an operation in order to improve the robustness of a HFDM program automated downloading and processing of data.

Key words: HFDM; regime recognition; HUMS; exceedance monitoring; automated reporting

Introduction: HFDM programs provide the data needed to proactively correct a safety issue [1]. However, a HFDM program can only be succed when the data is available for analysis. Data availability assumes that the data from a flight is downloaded, and then subsequently analyzed. In the advent of a mishap, the data from the flight might be lost. This data is invaluable for an accident investigation. HFDM systems do not require crash survivable memory or Aircraft Communication Addressing and Reporting Systems (ACARS), which could protect data, but adds considerable cost and weigh to a HFDM installation. Further more, promulgation of a HFDM system is cost sensitive, as it is not mandated on most helicopters.

Given this environment, a robust HFDM requirements were developed to:

- Assign data files associated with a flight operation such that at the end of the flight, those files can be downloaded for analysis in an automated way, and
- Provide automated reporting of events that can be identified as safety hazards which can be used as the basis for aggregation or trending, and
- Assist in triggering additional actions, such as crew contact, and

- Is low cost and weight, which is essential in order to make systems more prevalent in the industry. Cost and weight are drivers for the business case in most helicopter organizations, and
- Provide crash survivable data by transferring the data automatically in the event of a mishap.

The initial step to facility reporting requires the automated processing and downloading of appropriate data, which leads to the concept of an aircraft flight operation.

Aircraft Operation Concept: An aircraft operation is concerned with defining how data is assigned to a rotor turn or flight(s) and when it is appropriate to download that data. From power on, to rotor turn, through takeoffs/landings, mishap or hard landing, parameter and condition monitoring data needs to be associated with that operation (e.g. a flight). The operation is a fundamental concept and facilitates display and reporting of flight data, allowing:

- The ability to replay a flight from recorded aircraft states,
- Determine when component exceedances have occurred (over temp, over torque, etc), and
- Allow for a given operations to be associated with the data needed for the calculation of a rotor track and balance solution.

The definition of an operation must be robust enough that errors (loss of power, data not being downloaded) do not deter the ability to recover data from current or past operations. Fundamentally, an operation is associated with a flight or rotor turn. For example, if power is applied to the aircraft and then removed without the aircraft being started; those files that were created in anticipation of a flight are not an operation and are deleted.

Because data is typically downloaded at the end of every operation, there is continuity of data that represent both technical (aircraft system) and non-technical (pilot) parameters, which can be used by the safety management system or in an accident investigation. The flight operation consist of files associated with:

- Exceedances (operational limits defined by the original equipment manufacturer or service provider),
- Flight parameter data (latitude, longitude, altitude, velocities, accelerations, engine performance/torque/N1 (compressor RPM)/n2 (turbine RPM), airspeed, etc.),
- Mechanical diagnostics condition of rotating equipment/transmission, and Rotor Track and Balance (RTB) data, and
- Error logs.

Depending on the aircraft state at the end of the operations (e.g. normal flight vs. requirement for inspection to do an exceedance, or in the rare case of and accident/mishap), there will be different priorities on the data to be downloaded and how that data is downloaded. For example, for a normal flight, there is a period of time associated with aircraft shutdown. During shutdown, the pilot/crew chief is monitoring engine TOT temperature to ensure that lubrication will not coke. This is usually a two-minute period in which the operational data can easily be downloaded via Wi-Fi to tablet computer, smart phone/other portable device the pilot may have, or a local Wi-Fi base station. This data would include aircraft parameters data, condition indicators for mechanical diagnostics/RTB, and exceedance. Wi-Fi is the preferred download method as there is no cost associated with the bandwidth required for download. Wi-Fi can also establish a virtual private network (VPN) to ensure the security of the data. Alternatively, if no Wi-Fi is available and the aircraft state is nominal (e.g. no hard landing/inspection requirement or mishap), the files can be held locally in flash memory until Wi-Fi is available.

However, if the operation terminates with an operation exceedance, such as a hard landing, engine over temp/over torque, it is important that the parameter data (aircraft state) is downloaded by whatever means is available, such as local cellular modem using LTE (100 Mb/s download), or UMTS/HSPA (3G 7.2 Mb/s) or GSM/GPSR (2.5G 85.6 kb/s). If in the event of a mishap which causes loss of power, the HFDM should incorporate holdup power (such as an ultra-capacitor), and its own inertial measurement unit with GPS. An ultra-capacitor, in the event of power loss, can power the cellular modem. Considerations for an Inertial Navigation Unit (INU) and GPS should be made. This allows the HFDM unit to determine aircraft state independently of the aircraft avionics.

The advantage of this type of system is that: Implementing a cellular modem, Adding an ultra-capacitor holdup power, and an Internal INU/GPS for aircraft state does not greatly affect the bill of material cost or the system weight. The INU and GPS is desirable in a HFDM design as it reduces the number of aircraft interfaces needed to acquire parameter. Reducing aircraft interfaces reduces the cost and weight needed to for integration into the aircraft and greatly reduces reapplication cost. For retrofit, many aircraft have no Altitude and Heading Reference System (AHRS), so will require an INU anyway.

Because movement of data via cellular is rare (exceedances/hard landing/mishap do not happen often), the cost of the data plan associated with moving this data is low. Yet, because the data is transferred after the determination of a mishap/hard landing (within milliseconds), the data is protected even in the event of a mishap. A mishap may involve fire that could damage the HFDM persistent memory. By quickly downloading the data when a mishap is detected, the information needed for the investigation is protected. Additionally, the parameter data include positional information that can be used in locating the aircraft.

Most Type 27 Aircraft (defined by the FAA as airworthiness standards as rotorcraft up to 7,000 lb Maximum Takeoff Weight and 9 or fewer passengers) rarely fly over water for extended periods and will usually be within cellular connectivity. In the cases where the aircraft will operate in remote locations or over water, the HFDM can be equipped with a satellite modem for data transfer. Again, because the use of satellite data will be rare, the cost of the data plan will be low. As noted, the use of satellite or cellular modem only occurs when the operation is terminated due to a mishap/hard landing, or other regime that are of interest to the operator/owner of the helicopter.

Defining an Aircraft Operation: An operation is concerned with those conditions where a pilot/crew is in command of the aircraft and covers the data needed to reconstruct the pilots'/crews flight operation. The operation starts on rotor turn/power up, and reports

- The rotor turn time,
- The flight time,
- Flight parameters,
- Regime data,
- Exceedances,
- Mechanical diagnostics/RTB acquisition data (if equipped).

The operation ends on shutdown or due to a mishap/hard landing or similar condition. The type termination for the operation will define the data download priority.

The operation consists of files identified by date-time group and aircraft side number, which contain the data generated over the flight or rotor turn (ground operation). The flight operation consists of operational states that follow a set of defined functionality and rules needed to support HFDM, such as:

- Integral to the hardware file structure, the status of a flight operation files are associated by the files data structure location within persistent memory of the HFDM, such as:
    a. An active directory, which holds the current flight operation files,
    b. A transfer directory: those operations that are available for download, and
    c. An archive directory, which are those operations that have been downloaded off of the aircraft for analysis/visualization and reporting.
- If on power up there is an active operation (e.g. files in the active directory), these files are closed and then are scheduled for data transfer. An error is logged, as this suggest that the HFDM failed to close and transfer those files during the last active operation.
- A new operation is started when there is a valid time (from GPS). New flight operations files are created in the active directory.
- A major function of HFDM is to accrue time, such as power on time, rotor turn time and flight time.
    a. Rotor turn time is accrued when the main rotor RPM is greater than 10%.
    b. Flight time is time associated with flight regimes, typically torque > 30% (aircraft dependent). An important function of the regime recognition algorithm is to determine when and how the aircraft is flying.
- It is important to apply power to the radios (wireless communication) only when in appropriate regimes, such as: power on, rotor not turning, on shut down or when a mishap/hard landing is detected. A wireless connection need only be made when there are operations available in the Transfer directory.
- Rules for prioritizing which communication channel to use and which operational files to download.

a. If Impact/Hard Landing is determined and no Wi-Fi connectivity (e.g. no external Wi-Fi detected), then cellular or satellite modem is used.
b. For regimes which are associated with a mishaps, the parameter/aircraft state data is downloaded first. This may require just a few seconds, as data throughput of 3 Mb/s are typically observed.

- An operation is terminated when
  a. There is an active operation open from power up.
  b. The incremented rotor turn time is greater than zero, and the Regime allows active wireless communication, or a Mishap/Impact or Hard landing is detected.
- If an operations file transfers are complete, the operations files are move to the archive directory.

Incorporating these rules in software can be difficult. To reduce the complexity of the programing task, this can be implemented into a finite state machine. A sate machine is an abstracted construct that allows only a finite number of sates at any given time. The state machine transitions from one finite state to another state based on external inputs. The model of this state machine can be seen in Figure 1.

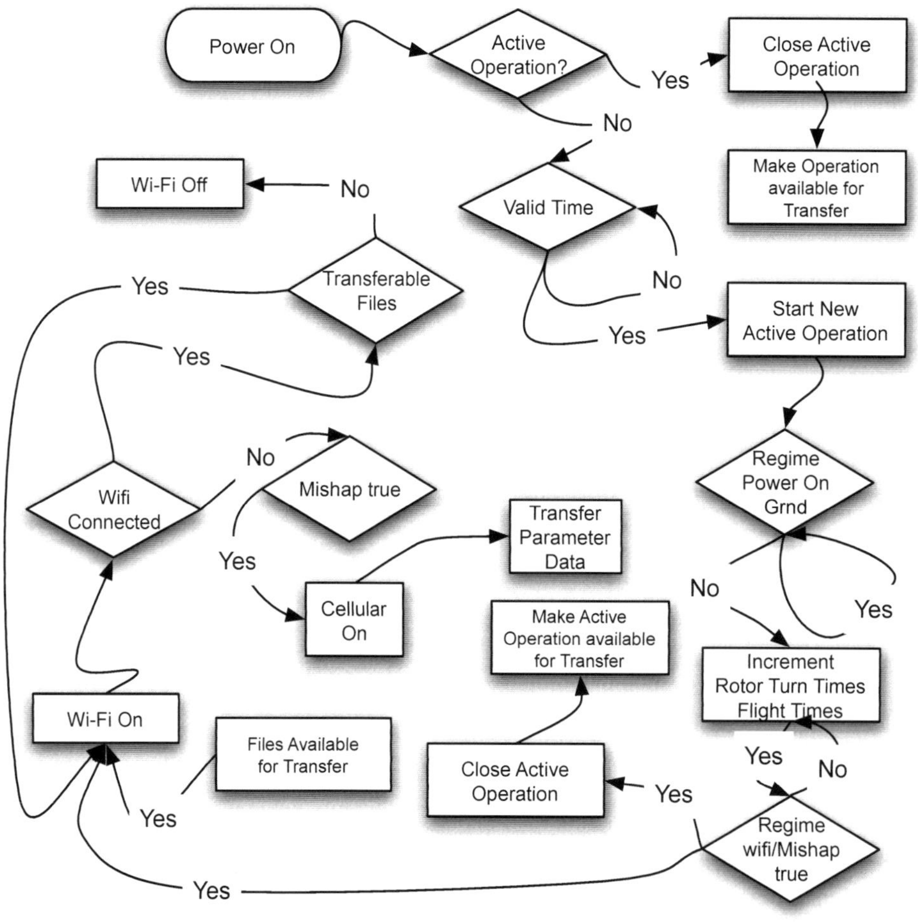

Figure 1 Operation Finite State Machine

For example, when power is applied, the state machine checks to see if there are files in the Active Operation directory. If false, the machine checks for a valid time. Once time is true, and new operation is stated, the regime being power on ground, etc.

Regime Recognition: An input into the state machine is regime. When to acquire data for mechanical diagnostics/RTB, determining when the aircraft is shutting down, or a mishap has occurred are all depended on knowing the aircraft regime. The general terms and requirements for Regime Recognition (RR) were developed in the FAA Advisory Circular "Airworthiness Approval of Rotorcraft Health Usage Monitoring Systems (HUMS)" [2]. RR in this application was defined more for loads monitoring rather than for aircraft state awareness. Similarly, in the "Aeronautical Design Standard Handbook for CBM Systems for US Army Aircraft Systems" [3], and in "Validation of FAA AC-29-2C MG-15 for Structural Usage Credits" [4], the primary usage of RR is for load monitoring.

The goal of RR used in this way is to extend the hours of life-limited structures or components. Life limited parts assume a worst-case usage spectrum. In practice, this results in an overly conservative life due to lack of real world usage knowledge. RR allows creation of the actual usage spectrum, which can then be used to re-life or update the life remaining on the part. While it is anticipated that RR for this application can be used for structural loads, the primary requirement for the project was to identify aircraft state. As noted, aircraft state supports a number of important HFDM related functions, including:

- Generation of flight statistics for HFDM reporting,
- Identifying and reporting high risk behavior: high angle of bank, rough approach, etc.,
- Determining the end of an operation for downloading of flight data, and
- Determine if the aircraft has had a hard landing/impact or crash.
- Provides an immediate time history of the operation,
- Determining when a mechanical diagnostics acquisition can be taken, and
- Associating a regime (ground, hover, 90, 120, 140 kts, etc.) for RTB acquisitions

To support these HFDM function, an easily configurable RR system allows:

- Identify when the rotor is turning, for the calculation of total rotor turn time for the given operation,
- Identify when the aircraft is flying, for the calculation of total flight time for the given operation,
- Determine when the aircraft in not flying or in ground operation, so that RF communications (Wi-Fi/Cellular Modem) can be powered to download data,
- Determine if the aircraft has suffered an event (mishap/hard landing) requiring an inspection/immediate action of the crew/maintenance team, and
- Determine when acquisitions for rotor track and balance/mechanical diagnostics can be performed.

RR Algorithm The RR application is based on mathematically rigorous methodology that is optimal in that sense that for a given probability of false alarm (e.g. stating the aircraft is in one regime, when it is in fact another regime), it maximizes the probability of correct classification. This algorithm is computationally simple to implement and configure. The algorithm is easily extended to encompass new regimes and/or new parameters. The RR algorithm uses a maximum likelihood estimator (MLE, see [5]) which assumes that input parameters are noisy. The algorithm weights the validity of a parameter by the information the parameter conveys (the inverse of variance is information). The output of the algorithm is the regime that is most likely, as a function of the *a priori* (e.g. configuration) parameter mean value and variance. This is conceptually different from algorithms using if/then/else search trees, which in effect are idealized by assuming no measurement or environmental noise.

The MLE is a quadratic classifier, which can accommodate nonlinear decision spaces. It takes the current set of measurements and calculates a normalized distance between the current measurements and a notional set of regimes. The regime that is closest statistically to the measurement is most likely. In this implementation, the MLE is structured as a multiple dimension hypothesis test, in which the aircraft parameters are used to test the hypothesis that the aircraft is in a given regime.

For example, one defines $P(H_i|\mathbf{z})$ as the probability that H_i was the true regime given a measured observation, **z**. The correct hypothesis is the one corresponding to the largest probability of the *m* possible regimes. The decision rule will be to choose H_o if: $P(H_o|\mathbf{z}) > P(H_1|\mathbf{z}), P(H_2|\mathbf{z}),\ldots P(H_m|\mathbf{z})$, else choose the greatest $P(H_i|\mathbf{z})$. The null hypothesis $P(H_o|z)$, in this test, represent power on, rotor not turning.

In testing, this RR algorithm used six input parameters: yaw rate (degrees/second), Z acceleration (-1 G is normal to the earth), N2 (turbine) RPM, main rotor RPM, torque, and indicated airspeed. As noted, the null hypothesis was power on, rotor not turning. The 22 alternative hypothesis were:

Power On Aircraft, Rotors Turning Flight Idle (65%)	Regime 2
Power On Aircraft, Rotors Turning Flight RPM (100%)	Regime 3
Shutdown (Main Rotor > 10%, N2<10%)	Regime 4
Hover	Regime 5
Left Hover Turn	Regime 6
Right Hover Turn	Regime 7
Level Flight up to 0.3 Vne (39 Knots)	Regime 8
Level Flight up between 0.3 and 0.4 Vne	Regime 9
Level Flight up Between 0.4 and 0.5 Vne (60 Knots)	Regime 10
Level Flight up Between 0.5 and 0.6 Vne	Regime 11
Level Flight up Between 0.6 and 0.7 Vne	Regime 12
Level Flight up Between 0.7 and 0.8 Vne (100 Knots)	Regime 13
Level Flight up Between 0.8 and 0.9 Vne	Regime 14
Level Flight up Between 0.9 and 1.0 Vne	Regime 15
Decent	Regime 16

Climb	Regime 17
Autorotation	Regime 18
Level Left Turn:	Regime 19
Level Right Turn:	Regime 20
Pullout: 0.8 Gs	Regime 21
Pullouts: 3.0 Gs	Regime 22
Pushover: 0.2 Gs	Regime 23

Flight Test Example: A low cost/light weight HFDM system was installed on a Bell 206 Jet Ranger in the fourth quarter of 2016. The installation was conducted under an FAA 337 field approval and cleared for flight operations in the first quarter of 2017. Initial flight tests were done to: establish threshold for mechanical diagnostics, check RTB coefficients, and ensure the HFDM automated operation generation and downloading were functional. During the initial flight, 31 HFDM parameter data was collected, of which six parameters were used for regime recognition. An example of the flight path is given in Figure 2. The aircraft hovers, climbs, circles the airfield and lands.

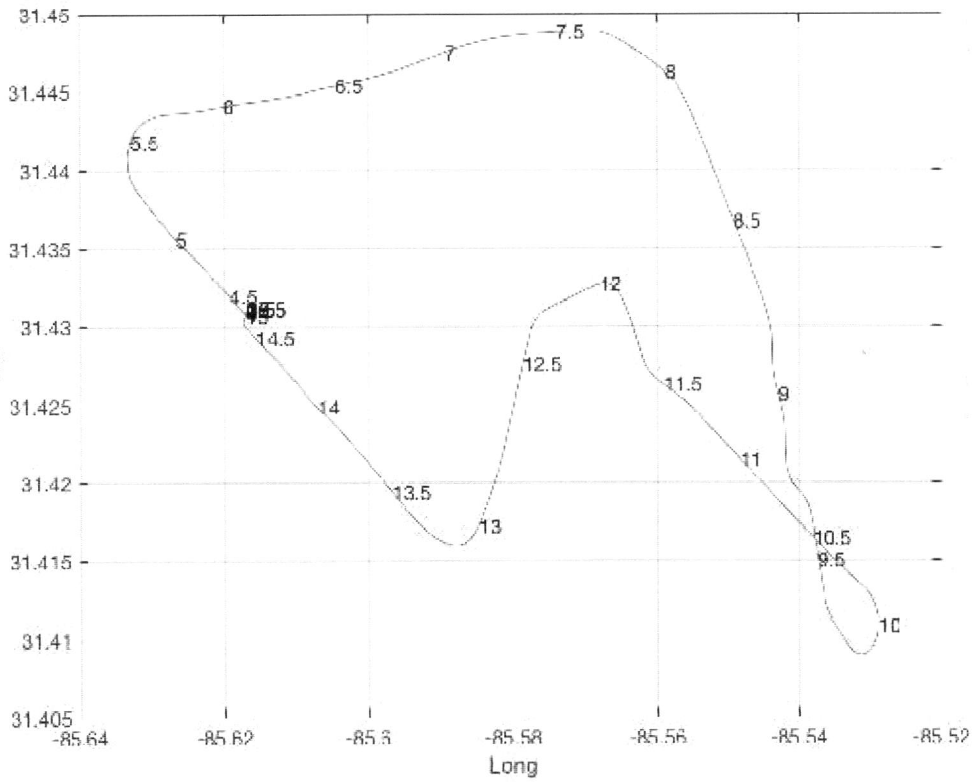

Figure 2 Example Test Flight

The times in minutes was plotted on the flight path to allow correlation with the flight parameters and regime indexes, Figure 3. The initial regime is 2, flight idle, transitioning to ground at 2:20. Note that the regime briefly goes to 17, indicating clime, then to regime 8 (low speed forward flight), before settling regime 3, ground 100%. From time 3:20 till 4:10, the aircraft is in hover (regime 5, 6, 7), and then toggles between hover and climb until 4:20. The aircraft climbs until time 6. From time 6 until time 14, the aircraft is in level flight, 70 to 100 Knots (regime 11, 12, and 13), performing right and left hand turns (see Figure 2). At time 14, the aircraft descends, hovers, goes to idle, and performs a shut down at time 17:25. The operation was successfully download and available for data analysis when the pilot returned to flight operations for debrief.

As noted, the function of this RR implementation is to provide input to the state machine. The state machine then determines when actions can be performed, such as: turn on the Wi-Fi, increment flight time, or perform and acquisition. As such, each regime is coded for what actions are be taken. This coding uses an integer representation of binary mask. For example, regime 2 (power on aircraft, rotor turning at 65%) increments rotor turn time, and allows Wi-Fi. The binary mask to represent this is 2. For regime 3 (ground, 100% main rotor RPM), rotor turn time is increments, and both mechanical diagnostics and RTB acquisitions are allowed, but Wi-Fi is off. The mask is 26. For hover (regime 5), which increments rotor turn time, flight time and allows mechanical diagnostics/RTB, the mask is 30. But for regimes 6 and 7 (left/right hand turn hover), only rotor turn time and flight time are incremented – the mask is 6.

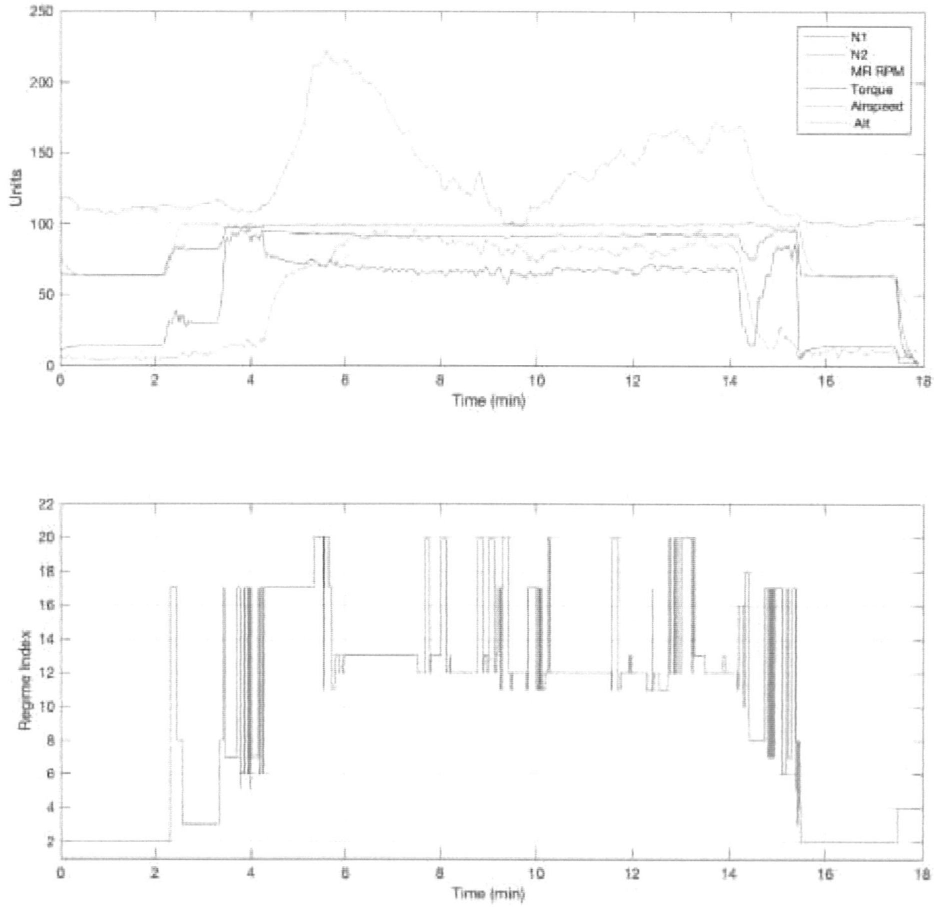

Figure 3 Flight Parameters and Regime Index

Other Functions: Flight Manual Exceedances. Automated exceedance processing is critical function of a HFMD system. Inflight occurrences require flight crew to submit a safety report and flight crews are sometimes not aware, or unwilling, to report an error. Exceedance monitoring is the most reliable method for monitoring flight crew compliance with operating limits and standard operating procedures. It captures all occurrences that take place during a flight, even those events in which the crew is unaware. Exceedance monitoring records quantifiable data that can be used to view trends. In that context, the crew's participation is a major component to provide the contextual information that is lacking in the data – to explain the issues around the event and allow insight into implementing control measures.

The HFDM exceedance monitoring function uses a hybrid system of internally derived inertial inputs, and data from a limited number of aircraft systems, both analog and digital. These parameters feed a configurable analysis engine designed to test for the operating limits as outlined by the equipment manufactures documentation. The analysis engine itself is based on the concept of simple and condition exceedance.

A simple exceendance is true if a measured parameter is greater than a given threshold, such as "the maximum operating pressure altitude is 20,000 ft". A condition exceedance is based on a simple exceedance that is conditioned on one or more other flight parameters. For example, "Vne 80 knots, when in takeoff power range of 85 to 100% torque". This exceedance is true if the condition of torque is greater than 85% is met while the IAS is greater than 80 knots. More than one exceedance can occur at a time. For example, if the torque is 101% for 6 seconds, and the airspeed is 84 IAS, both the maximum continuous power transient and the takeoff power range Vne would be exceeded.

Observation: During test flights, the HFDM system triggered the creation of new operation based on RR at aircraft startup, and correctly acquired mechanical diagnostic data in ground, hover and forward flight regime. The system also detected shutdown and automatically powered the Wi-Fi modem, downloaded data. This data was then automatically processed and available viewing by the pilot after the post flight debrief.

The RR incorrectly selected the climb regime when transitioning from Power On Aircraft, Rotors Turning Flight Idle (65%), regime 2, to Power On Aircraft, Rotors Turning Flight RPM (100%), regime 3. During the transition, the torque was higher than would be expected for regime 3. Its likely that adding an additional parameter, vertical rate (which available from the INU) will improve the discrimination of the RR algorithm for detecting climb and decent.

Operationally, it has been found that at some locations the Wi-Fi access points firewall blocks the transmission of data from the helicopter to the cloud based storage server. This is another reason for implementing cellular modem, which facilitate IoT by bypassing any firewall or other local security.

Conclusion: HFDM enhanced the safety and reliability of helicopters. The expense and weight of system makes it difficult for most Type 27 helicopter operators to make the business case and justify the installation. By reducing the cost/weight and automating the download and analysis, the business case is easier to make. This should expand the number of aircraft using HFDM. Overall, automation designed into the test system performed to expectation. Some additional configuration, such as adding vertical rate as a parameter to the RR algorithm, will improve the ability to discriminate during regime transitions. It is hoped that operators will find that the enhanced safety and reliability of HFDM equipped aircraft will justify the expense of installation.

References:

[1] Hurst, M, Cramp, T., Helicopter Flight Data Monitoring Industry Best Practices, Global HFDM Steering Group, 2012.

[2] FAA Advisory Circular AC-29-2C MG-15, "Airworthiness Approval of Rotorcraft Health Usage Monitoring Systems (HUMS)," February 12, 2003

[3] "Aeronautical Design Standard Handbook for Condition Based Maintenance Systems for US Army Aircraft," U.S. Army ADS-79-HDBK, Appendix F, "Fatigue Life Management," February 17, 2009.

[4] Hong, C., Algera, D., Green, D., DiCampli, E., and Chin, H., "Validation of FAA AC-29- 2C MG-15 for Usage Monitoring Credits," FAA report DOT/FAA/TC-15/11, to be published in 2015.

[5] Fukunaga, K., Introduction to. Statistical. Pattern Recognition, Academic Press Professional, Inc. San Diego, CA, USA, 1990

Bibliography:

[1] Eric Bechhoefer is the president of GPMS, Inc., a company focused on the development of low cost condition monitoring systems. Dr. Bechhoefer is the author of over 100+ juried papers on condition monitoring and prognostics health management, and holds 23 patents, and 5 patent applications in the field of CBM.

[2] Mike Augustin is the President and Principle Investigator at Integrated Vehicle Health Management (IVHM), Inc. Mike is an expert in advanced sensors, wired and wireless system architectures, RFID, system requirements, Condition Based Maintenance (CBM) application support, Maintenance Credit development and information fusion. Prior to IVHM, Mike was a staff engineer at Bell Helicopter, where he was responsible for the development and certification of HUMS, flight data recorder systems, and global data fusion of HUMS/CBM data.

LOCAL FAULT DIAGNOSIS OF NON-STATIONARY GEARBOX BASED ON ORDER ENVELOPE ANALYSIS

Liming Wang, Yimin Shao, Pan Sun, Fang Guo and Jing Liu
State Key Laboratory of Mechanical Transmission
Shazheng Street 174
Shapingba District
Chongqing, 400044, CHINA
ymshao@cqu.edu.cn

Chaokun Gu and Ben Zhou
Ningbo Baoxin Stainless Steel Co.,Ltd
Xiapu town
Beilun District
Ningbo, 315807, CHINA

Abstract: Tooth root crack is a common fault in gear system, it's of significance to detect the crack fault during the operation process of the gearbox. However, the gearbox usually working in a non-stationary condition, namely speed or load are time varying, which increases the difficulty of fault diagnosis since the statistic features and spectrum vary by time. In this paper, an order envelope analysis method is proposed for fault diagnosis of a two stage gearbox with local crack fault under non-stationary working conditions. Firstly, an accelerometer and an encoder are mounted on the gearbox to acquire non-stationary vibration data and rotating speed, respectively. After that, the angle domain signals are derived from the non-stationary vibration data by interpolation algorithm. Then envelope analysis is employed for the angle domain signals to detect the crack characteristic frequency components in envelope spectrum. Finally, the proposed approach is assessed by seeded tooth root crack fault in different working conditions, results show that the new method can effectively detect the tooth cracks in various non-stationary working conditions.

Key words: Fault diagnosis; non-stationary; order envelope analysis; tooth crack

Introduction: Gearboxes are the most important mechanisms in industrial machinery, wind turbines, and our daily lives to transmit power and produce high rotational speed changes and/or change the direction of motion. However, due to heavy duty and tough working conditions some local gear faults, i.e. tooth root crack and spalling, are easily observed in gearbox [1, 2]. It is indicated that gear faults account for 80% of transmission break-down [3]. Thus, research on fault diagnosis of gearbox has attracted a lot of attention from scholars around the world. In practical cases, gearbox working in a non-stationary condition, namely speed or load are time varying, traditional statistical analysis [4] and spectral analysis [5] suffer degradation in detecting the gear fault. Some scholars tried to use some new method to deal with this problem. For instance, Feng presented a time–frequency (TF) analysis method to reveal the constituent frequency components of non-stationary signals and their time-varying features for planetary gearbox monitoring [6]. Saidi introduced a joint method of bi-spectrum and empirical mode decomposition (EMD) to analyze non-stationary vibration signals' behavior for bearing failures detection [7]. He proposed a novel nonlinear time–frequency feature based on a time–frequency manifold (TFM) technique to overcome the effects of noise and condition variance issues in sampling signals [8]. Above all, though EMD and TF methods can extract fault features in non-stationary working conditions for fault diagnosis, both the two method are time consuming and require users have a good knowledge of vibration signals. Order tracking is considered as a classic and effective technique of the non-stationary vibration analysis for rotating machinery, which can overcome the disadvantage of the spectral smearing of signal in the order domain [9]. Thus, order tracking technique will be employed in this paper. The rest of the paper is organized as follows: section 2 describes the fault diagnosis method based on order envelope analysis. In section 3, experimental setup are introduced. Test results are shown and discussed in section 4. Conclusions of this paper are drawn in section 5.

Fault diagnosis method based on order envelope analysis: Spectral analysis method is usually used in fault diagnosis of gearbox, however, most gearboxes commonly working in a non-stationary condition, i.e. time-varying rotating speed and load. Therefore, frequency values in spectrum derived from the non-stationary time domain signal will be time-varying too and this phenomenon is named as "frequency smearing", as shown in Figure 1, which make is difficult to distinguish the characteristic frequency value in spectrum directly no mention to fault diagnosis.

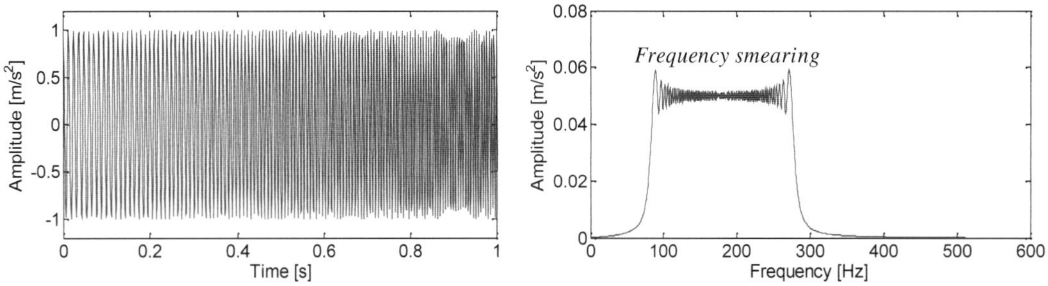

Figure 1: Time domain signal and spectrum in time-varying rotating speed condition

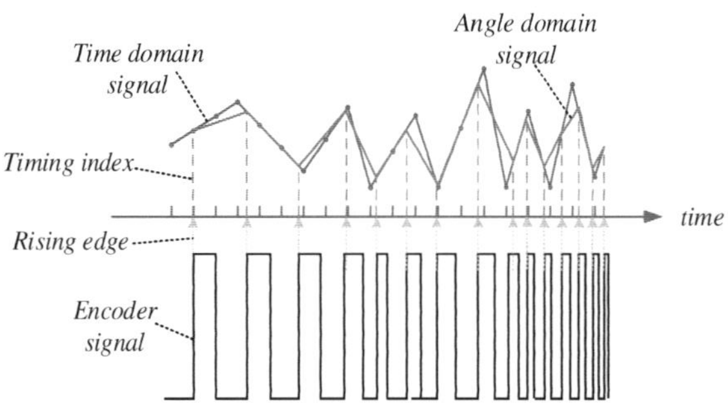

Figure 2: Resampling process of the angle domain signal

In order to overcome the disadvantage of spectrum-based method for fault diagnosis of gearbox under non-stationary working conditions, order tracking method is usually employed, where vibration signals are sampled at constant increments of shaft angle instead of at constant time interval, and encoder is used to provide trigger impulse at each shaft angle. However, many useful information in the signal may be lost if the angle domain sampling rate is too low or there are multi-stages in the gearbox. Therefore, time domain vibration signal as well as encoder signal are acquired simultaneously, then angle domain signal is derived by resampling from time domain signal, as shown in Figure 2. Firstly, the time indexes of the rising edges of the encoder signal are obtained as,

$$TI^{\theta} = [t_1^{\theta}, t_2^{\theta}, ...t_i^{\theta}, ..., t_M^{\theta}] \tag{1}$$

Then angle domain signal y can be derived from time domain signal x by Lagrange interpolation method,

$$y(i) = x(t_j) + \frac{x(t_{j+1}) - x(t_j)}{t_{j+1} - t_j}(t_i^{\theta} - t_j),\ t_j < t_i^{\theta} < t_{j+1} \tag{2}$$

When there is a local fault in a gear pair, impulsive shock is expected to be observed in every rotation, and these shocks will be modulated by meshing frequency, which leads to complex sidebands in frequency spectrum. Therefore, after obtaining the angle domain signals envelope analysis is done to extract the impulsive shocks caused by local fault. First Hilbert transformation of the angle domain signal is obtained,

$$\hat{y}(\theta) = -\frac{1}{\pi}\int_{-\infty}^{+\infty} \frac{y(\theta - \tau_{\theta})}{\tau_{\theta}} d\tau_{\theta} \tag{3}$$

Then the analytic signal of $y_a(\theta)$ and the envelope signal $A(\theta)$ are obtained,

$$y_a(\theta) = y(\theta) + j\hat{y}(\theta) = A(\theta)e^{j\phi(\theta)},\quad j = \sqrt{-1} \tag{4}$$

where, $A(\theta) = \sqrt{y^2(\theta) + \hat{y}^2(\theta)}$ and $\phi(\theta) = \arctan(\hat{y}(\theta)/y(\theta))$.

And finally order Envelope spectrum can be derived after FFT of the envelope signal,

$$S_{oea}(o) = \int_{-\infty}^{\infty} A(\theta)e^{-j2\pi o t} d\theta \tag{5}$$

The proposed signal processing procedure is shown in Figure 3. First of all, raw signal including acceleration signal and encoder signal is acquired, then based on Lagrange interpolation method time domain signals is converted to angle domain signal to reduce the non-stationarity caused by speed fluctuation. After that Hilbert transformation and FFT are used to extract the order envelope spectrum to identify the fault characteristic frequencies.

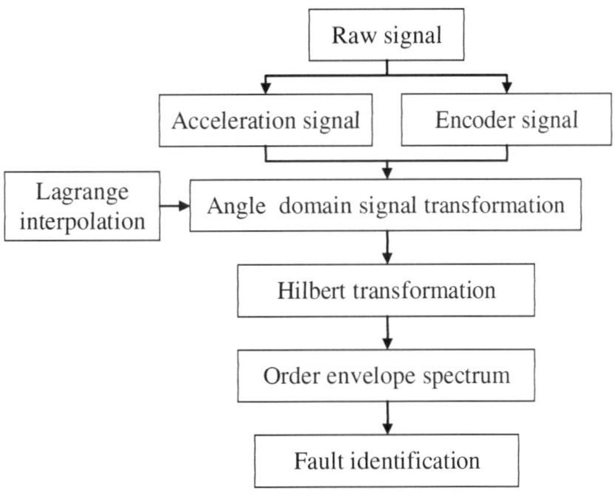

Figure 3: Fault diagnosis procedure based on order envelope analysis

Experimental descriptions: Figure 4 shows the gearbox test setup, which consists of a two-stage gearbox, a motor, a magnetic powder brake, several flexible couplings and control unit. The teeth number of the first stage and second stage gear pair are 23/39 and 25/53 respectively, Table 1 give the design parameters of gears. The rotating speed of the 3-phase induction motor is controlled by the frequency converter and the load of magnetic powder brake is controlled by a load controller, which allow the tested gear to operate under various speeds and various loads. A tooth root crack is seeded in the second-stage gear pair as shown in Figure 4, the manufacture parameter of the crack is shown in Figure 5(a), where α_c is 75° and q_0 equals to 1.8 mm.

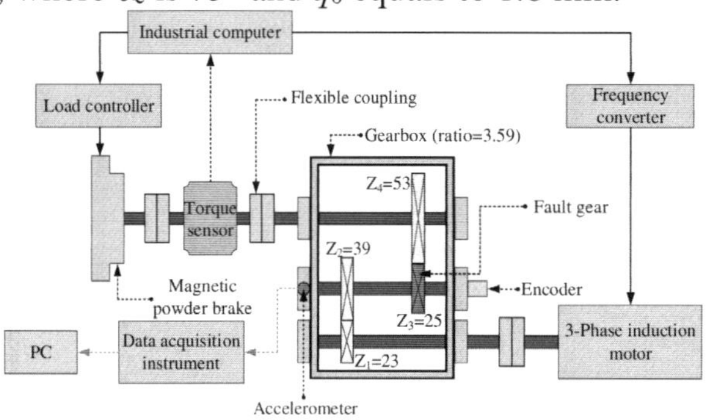

Figure 4: Schematic of the test system

Table 1: The parameters of gears

Parameter	Value	Parameter	Value
Module	3	Pressure angle	20°
Tip clearance	0.25	Addendum	1
Tooth width (mm)	60		

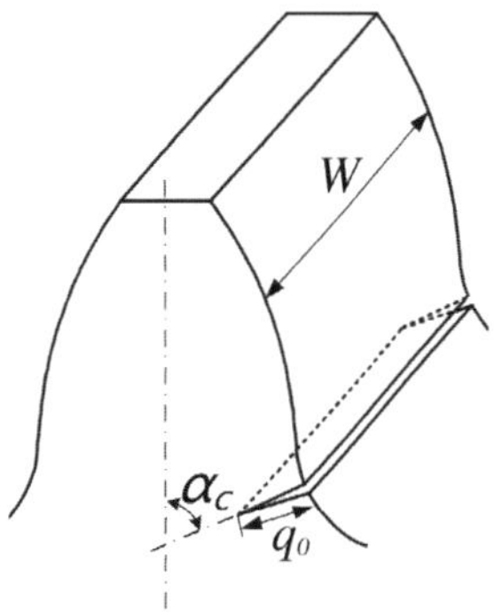

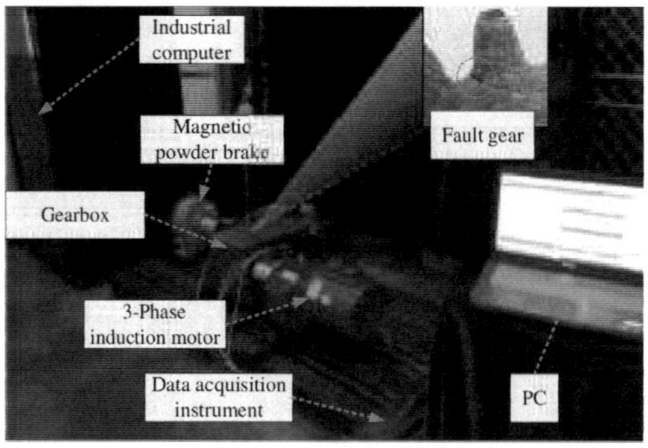

(a) (b)

Figure 5: (a) Schematic diagram of tooth root crack and (b) photograph of the test system

One accelerometer is mounted on the bearing casing of the gearbox and an encoder (360 p/r) is mounted at the end of the middle shaft, a LMS data acquisition system (DAQ) and a PC with the data acquisition software are used to acquire the vibration data and encoder signal for further processing, Figure 5(b) shows the photograph of the data collection process. Vibration data is acquired under 4 loading conditions: 0 Nm, 20 Nm, 40 Nm and 60 Nm. Rotating speed ranges from 500 rpm to 700 rpm then to 900 rpm, the signal length for each dataset is 6 seconds. For each loading condition, there are 15 datasets collected and totally 120 datasets collected for both normal and fault situation. In addition, a speed-up and speed-down dataset is collected as an example for order envelope analysis. The sampling frequency is 5120 Hz. Table 2 gives more detailed description of various datasets.

Table 2: Detailed description of various datasets

Fault severity	Motor speed (rpm)	Load (Nm)	Time length (s)
Normal	500-700-900	0/20/40/60	6
Cracked	500-700-900	0/20/40/60	6
	500-700-500	0	25

Results and discussions: The speed-up and speed-down dataset is analyzed by the proposed method as an example. Figure 6 shows the time domain signal and its zoomed plot, a clear amplitude increase and decrease trend can be observed as the rotating speed go up and down, as shown in Figure 6(a), and some impulse shocks can be found in Figure 6(b), the time interval between the two marked shocks is 0.155s which almost equals to the rotating speed period of the middle shaft at 700 rpm. Figure 7 shows the corresponding spectrum of the time domain signal after FFT, the spectrum density concentrates around 506.3 Hz as shown in Figure 7(a), however it is neither the meshing frequency components at 500 rpm nor at 700 rpm, because the meshing frequencies of the gearbox are 191.7 Hz (first gear pair) and 122.9 Hz (second gear pair) at 500 rpm respectively and 268.3 Hz (first gear pair) and 172 Hz (second gear pair) at 700 rpm respectively. And it is also found that there exist "frequency smearing" phenomenon in the spectrum, as shown in Figure 7(b). The envelope spectrum of the dataset is shown in Figure 8, and the characteristic frequency cannot easily be distinguished due to meshing modulation.

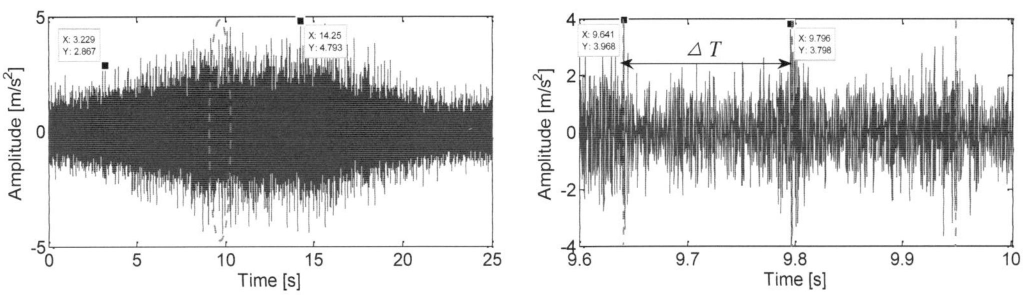

(a) (b)

Figure 6: (a) Time domain signal and (b) its zoomed plot

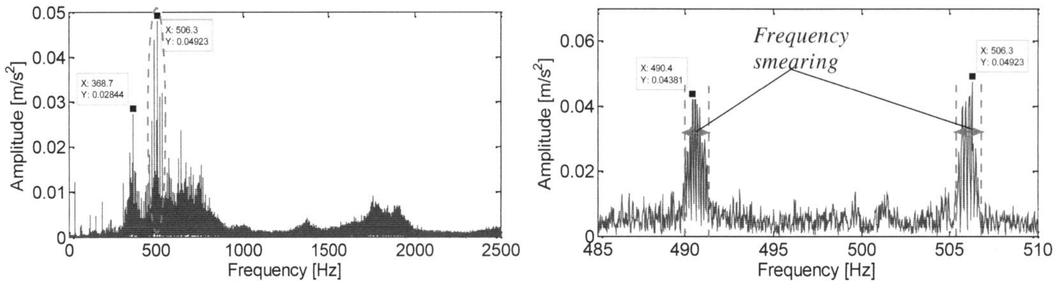

(a) (b)

Figure 7: (a) Spectrum and (b) its zoomed plot from 485 Hz to 510 Hz

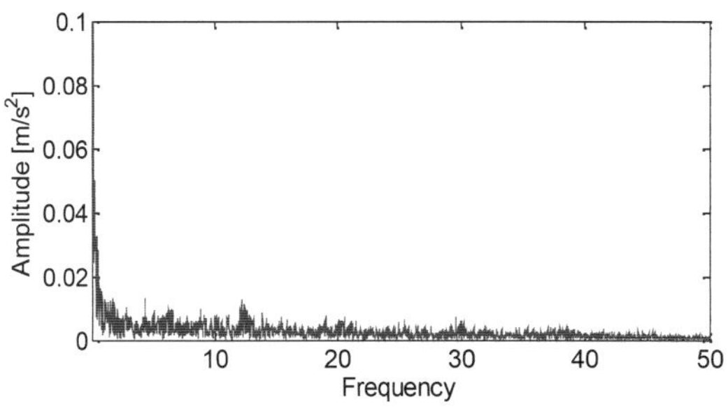

Figure 8: Envelope spectrum

After converting the time domain signal to angle domain, FFT is employed to derive the order spectrum as shown in Figure 9(a), where 1 order equals to the rotating frequency of the middle shaft. It can be observed that the spectrum density concentrates at 78 order, 64 order and 50 order, 78 order is the 2nd harmonics of the meshing frequency of the first gear pair, and 50 order is the 2nd harmonics of the meshing frequency of the second gear pair, but there is no corresponding frequency for 64 order which may resulting from the coupling of the first and second gear pair meshing. The modulation sidebands can be observed in Figure 9(b), where 78 ± 1 order are obvious, so after the envelope analysis the order envelope spectrum can be obtained as shown in Figure 10, distinct spectral components of order 1 and its harmonics can be found compared to Figure 8.

Aiming to monitor the health condition of the gearbox, a health indicator is defined as the summation of the amplitudes near the defect orders and its harmonics based on order envelope spectrum (*SDOE*), and the vibration signal acquired from the faulty gearbox is expected to have a higher *SDOE*.

$$SDOE = \sum_{n=1}^{N} \sum_{o=n-bw}^{n+bw} S_{oea}(o) \tag{6}$$

where, S_{oea} () denotes the amplitude of the oth spectrum line of the order envelope spectrum, n indicates the 1st order of the fault frequency, N is the number of harmonics, the value of N is set to 3 in this study, bw is the order band width around the fault orders which is chosen as 0.1.

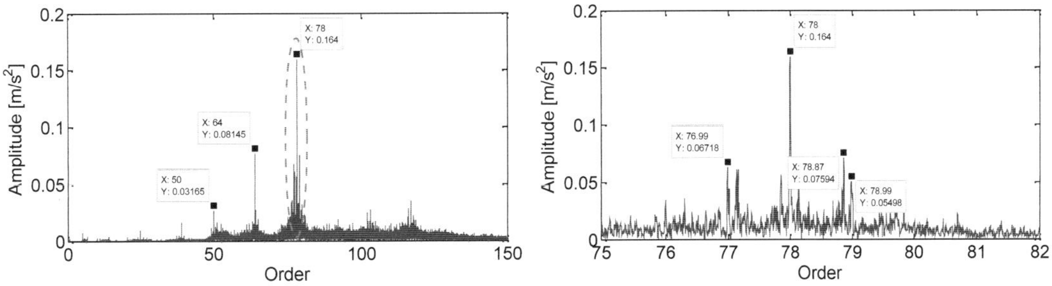

(a) (b)

Figure 9: (a) Order spectrum and (b) its zoomed plot from 75 order to 82 order

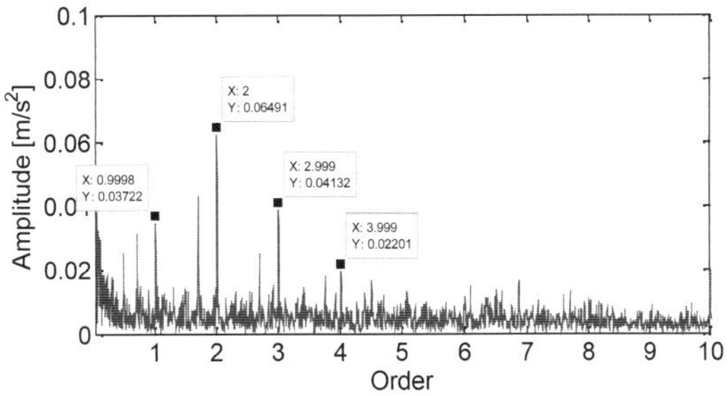

Figure 10: Order envelope spectrum

Comparisons of *SDOE* between normal and cracked situation are shown in Figure 11 (a) to (d) under various loading conditions, where x label denotes the dataset No. and y label is the amplitude of *SDOE*. It can be observed that the *SDOE* indicator can distinguish the cracked situation from normal situation in most times, however there is a large variation for the *SDOE* of normal situation in 20 Nm loading condition, as shown in Figure 12, there is an increasing trend for both normal and cracked *SDOE* indicator. And the *SDOE* of cracked situation always yields higher mean value than normal situation. So conclusion can be drawn that *SDOE* is a useful indicator in detecting early tooth root crack.

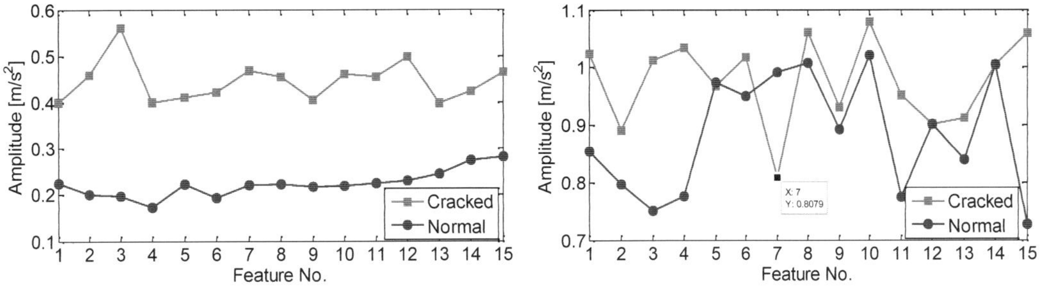

(a) (b)

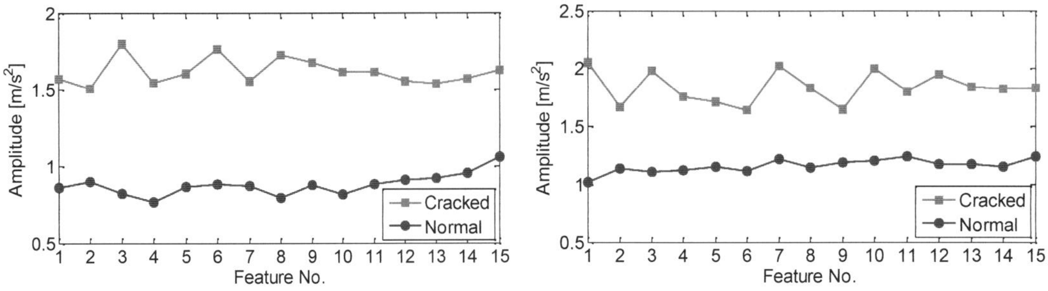

(c) (d)

Figure 11: Comparisons of *SDOE* between normal and cracked situation under (a) 0 Nm (b) 20 Nm (c) 40 Nm and (d) 60 Nm conditions

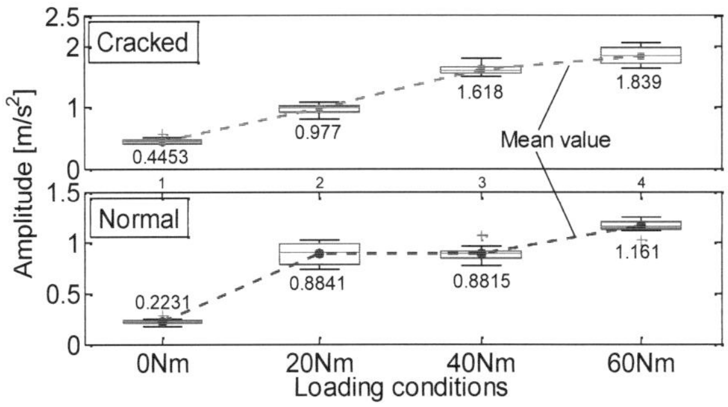

Figure 12: Boxplot comparisons between normal and cracked situation

Conclusion: This paper propose an order envelope analysis method for fault diagnosis of a two stage gearbox with local crack fault under non-stationary working conditions. First the time domain vibration signal is converted to angle domain to reduce the influence of non-stationarity caused by time-varying working condition. Then envelope analysis is applied to the angle domain signal to detect the crack characteristic frequency components. After that a useful health indicator is proposed to detect the crack fault. It is concluded that the order envelope analysis is effective in eliminating the non-stationarity of the signal, useful frequency components can be found; and the proposed health indicator *SDOE* see its success in detecting crack fault under various working conditions.

Acknowledgement: The authors are grateful for the financial support provided by Ningbo Baoxin Stainless Steel Co., Ltd. and the National Natural Science Foundation of China under Contract No. 51475053.

References:

[1] Ma R, Chen Y. Research on the dynamic mechanism of the gear system with local crack and spalling failure. Engineering Failure Analysis, 2012, 26: 12-20.

[2] Chen Z, Shao Y. Dynamic simulation of spur gear with tooth root crack propagating along tooth width and crack depth. Engineering Failure Analysis, 2011, 18(8): 2149-2164.

[3] Li Z, Yan X, Yuan C, et al. Virtual prototype and experimental research on gear multi-fault diagnosis using wavelet-autoregressive model and principal component analysis method. Mechanical Systems and Signal Processing, 2011, 25(7): 2589-2607.
[4] Samanta B, Al-Balushi K R. Artificial neural network based fault diagnostics of rolling element bearings using time-domain features. Mechanical systems and signal processing, 2003, 17(2): 317-328.
[5] Yang Y, Yu D, Cheng J. A fault diagnosis approach for roller bearing based on IMF envelope spectrum and SVM. Measurement, 2007, 40(9): 943-950.
[6] Feng Z, Liang M. Fault diagnosis of wind turbine planetary gearbox under nonstationary conditions via adaptive optimal kernel time–frequency analysis. Renewable Energy, 2014, 66: 468-477.
[7] Saidi L, Ali J B, Fnaiech F. Bi-spectrum based-EMD applied to the non-stationary vibration signals for bearing faults diagnosis. ISA transactions, 2014, 53(5): 1650-1660.
[8] He Q. Time–frequency manifold for nonlinear feature extraction in machinery fault diagnosis. Mechanical Systems and Signal Processing, 2013, 35(1): 200-218.
[9] Zhao M, Lin J, Wang X, et al. A tacho-less order tracking technique for large speed variations. Mechanical Systems and Signal Processing, 2013, 40(1): 76-90.

Misalignment Fault Detection in Dual-rotor System Based on Time Frequency Techniques

Nan-fei Wang, Dong-xiang Jiang*, Te Han
State Key Laboratory of Control and Simulation of Power System and Generation Equipment, Department of Thermal Engineering,
Tsinghua University, Beijing, 100084, China

Abstract. In order to improve the energy efficiency and compact structure, the dual-rotor structure including low-pressure rotor and high-pressure rotor has been widely used in aero-engine. Misalignment is one of the most common faults in dual-rotor system, which will causes malfunctions. It is a significant task for rotor dynamics personnel to monitor and defect faults in dual-rotor system. In the paper, the dual-rotor vibration signals are applied to solve the fault identification problem by utilizing time frequency techniques. Numerical simulations are carried out through finite element analysis of dual-rotor system with misalignment fault. Two signal processing tools namely Short Time Fourier Transform (STFT) and Continuous Wavelet Transform are used to detect the misalignment fault and compared to evaluate their diagnosis performance. The effect of addition of Signal to Noise (SNR) on three frequency techniques is presented. Experiments are carried out to obtain the vibration data of dual-rotor test rig and the results from the work show that the technique can be used for the monitoring of misalignment, which will have applications in the condition monitoring and maintenance of various types of rotating machinery.

1 Introduction

Misalignment is a situation in which the input shaft and the output shaft are not in the identical centreline. It is very challenging to attain a perfectly aligned condition between two shafts in industrial environment. Even if an accurate alignment is secured, it cannot not be continued for a long duration due to many external effects, like base foundation disturbance [1, 2]. Shaft misalignment is a frequently encountered fault observed in large rotor bearing systems like steam turbine shafts, gas turbines shafts and aircraft rotating components, and it is a significant source of vibration. High levels of misalignment may lead to fatigue cracks or rotor to stator rubbing. Therefore, misalignment detection and monitoring is an important objective for successful operation in the industrial scenario. Precision detection of shaft misalignment faults can cut the operation cost and facilitate long and safe operation.

Due to the complexity of the misalignment fault, many comprehensive researches on the misalignment experiments and misalignment dynamics mechanism have been carried out based on the vibration information of a misalignment rotor system. The vibration response of the rotor provides knowledge on operating conditions that are connected to the fault characteristics. However, due to the confusing spectral characteristics of vibration, vibration based misalignment detection lead to less reliability [1]. It was practically proved that vibration magnitude varies significantly at 2× and 4× frequencies of operating frequency of the rotating member as the misalignment increases [3]. Also it is reported that the frequencies of forces and moments from misaligned shafts are even multiples of the operating speed of the motor [2, 4]. Analysis of other faults with analytical models of rotating systems reported by Hamzaoui et al. [5] also indicated that misaligned shafts generates vibrations such that the magnitude of the 2× operating harmonic frequency component is sensitive to the fault. Sekhar and Prabhu [6] analyzed shaft misalignment in rotor bearing system with a flexible coupling using higher-order finite element model by applying theoretically modeled forces and moments. The effect of harmonic and flexural mode shapes on vibration magnitude is reported. Lee and Lee [7] introduced a dynamic model for shaft misalignment in a rotor-bearing system with roller bearing connected by flexible coupling. A model-based evaluation of defects in frequency domain was reported by [8-10]. Sinha et al. [11] introduced a technique to diagnose the shaft misalignment from a shutdown data in which the shaft misalignment at couplings is assumed to developed steady forces and moments.

In the above literature, the researchers pay more attention to the single-rotor system, which has significant difference with actual aero-engine. The arrangement of dual rotors is a common configuration utilized in practise to obtain the higher thrust to weight ratio [12]. Up to now, the effects of misalignment on the dynamic responses of the dual-rotor system have not been investigated in detail. Hence, it is of crucial importance to investigate the misalignment fault mechanism of dual-rotor system for safe and steady running of aero-engine.

The objective of the study is to detect misalignment faults existing in a dual-rotor system. In the study both simulations and experimental investigations of misalignment fault detection are considered. Use of CWT algorithm is highlighted along with a comparative study using the STFT approach. The vibration equations of dual-rotor system are computed by numerical method and the results are investigated by the time waveform, frequency spectrum and time-frequency maps. Besides, the experiments with unbalance-misalignment is carried out on a dual-rotor test rig and the vibration displacement signals of the low pressure rotor are analyzed in detail.

2 Overview of time frequency techniques

In the following sub-sections, the time frequency techniques used in the present study for fault diagnosis are discussed briefly. Using time-frequency techniques, a view on rotor fault diagnosis is reported by Feng et al. [13] and Yan et al. [14].

2.1 Short Time Fourier Transform (STFT)

STFT is also called windowed Fourier transform, it is used to examine the features of a response with respect to time. For a vibration response $y(t)$, with $w(\tau - t)$ as the window function centered at time t, the response $y(t)$ using this window function is $y(t)w(\tau - t)$. Shifting the function $w(\tau - t)$ along the time axis of the response and employing the Fourier analysis to every window components lead to the following integral [13]:

$$S_y(t,f) = \int_{-\infty}^{+\infty} y(\tau)g(\tau - t)e^{-2\pi f\tau}d \qquad (1)$$

2.2 Continuous Wavelet Transform (CWT)

Continuous Wavelet Transform uses basis functions with translation and dilation parameters instead of simple complex exponentials unlike Fourier transform [13]. It adds a dilation parameter along with a time translation parameter to the analysis. Hence, it is best suited for time-frequency analysis, and is effective in analysing non-stationary signals like run-up vibration data of the rotor. CWT of a time signal $y(t)$ is obtained by solving the following convolution integral:

$$C_y(t,u) = \frac{1}{\sqrt{u}}\int_{-\infty}^{+\infty} y(\tau)\psi\left(\frac{t-\tau}{u}\right)d \qquad (2)$$

In the present work Complex Morlet (CMOR) wavelet function is chosen as the base wavelet

3 Simulation experiment

In order to evaluate the performance of STFT and CWT, a signal s that contains frequency components of f_1=500Hz and f_2=1000Hz was constructed. The specific expression of signal s is defined as follows:

$$s = \begin{cases} \sin(2\pi f_1 t) + \sin(2\pi f_2 t) & (t \in [0,0.256]) \\ s + 5 & (t = 0.190) \\ s + 5 & (t = 0.196) \end{cases} \qquad (3)$$

The time-frequency spectrum of STFT, time waveform and spectral analysis of the constructed signal are presented in Figure 1. The time-frequency spectrum of wavelet is illustrated in Figure 1. It can be observed from the expression and the time waveform of the signal that it has two peaks at $t = 0.190$ and $t = 0.196$, and two frequency components corresponding to f_1 and f_2 can be separated accurately. However, the time-frequency resolution of the STFT is fixed when the window function and its length are chosen. The continuous wavelet transform (CWT) is a well-developed tool employing wavelets as the basis to overcome the limitations of the STFT, adding a scale variable in addition to the time variable. Therefore, by comparing the time-frequency spectrum of Figure 1 and Figure 2, it can be found that the time-frequency map of CWT has better performance.

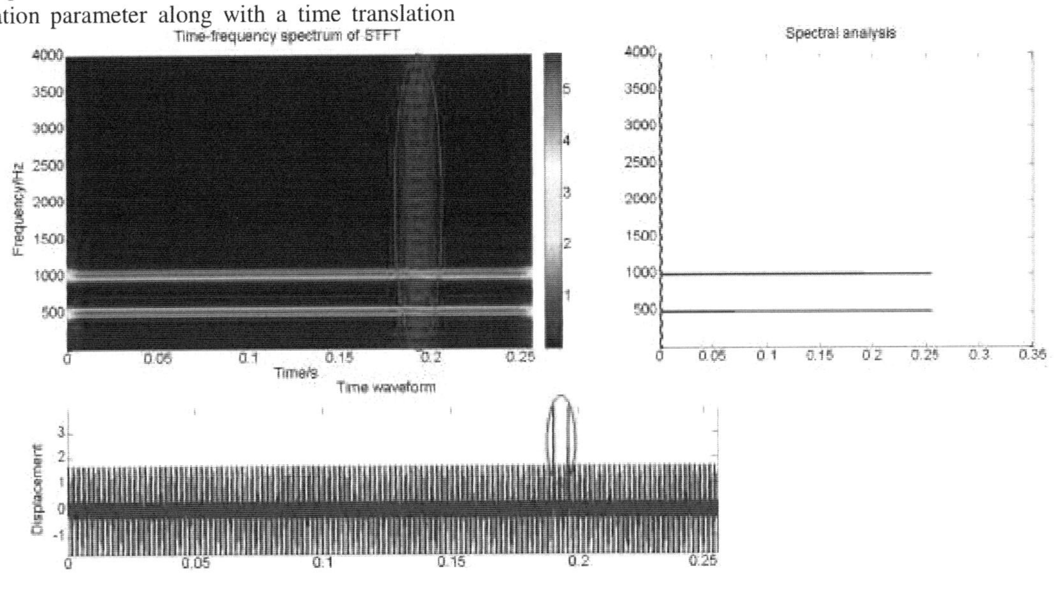

Figure 1 Time-frequency spectrum, time waveform and spectral analysis of the constructed signal

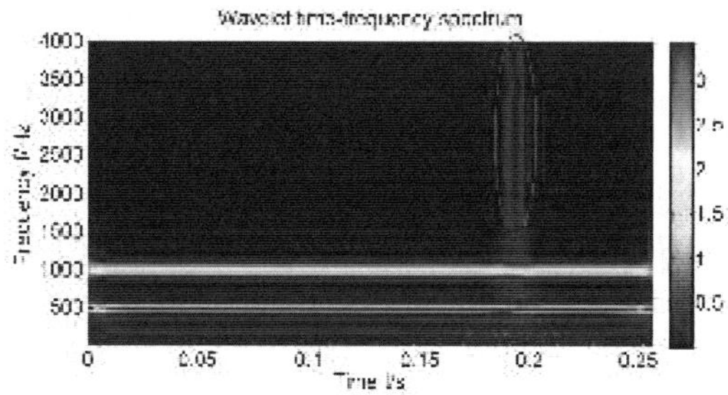

Figure 2 Time-frequency spectrum of wavelet

4 Dual-rotor system model with fault

The dual-rotor system with misalignment fault is shown in Figure 3, which consists of a low pressure rotor (rotor 1, lumped mass point 1 to 5) with a lower rotating speed and a high pressure rotor (rotor 2, lumped mass point 6 to 9) with a high rotating speed. There are four bearings included in the model, in which bearing 3 denotes the inter-shaft bearing. The disks 1-4, located, respectively at lumped mass points 2, 4, 7 and 9, represent the concentrations of low-pressure compressor disk, low-pressure turbine disk, high-pressure turbine disk and high-pressure compressor disk. The coupling misalignment fault including parallel and angular misalignment is modelled. It is assumed that the lumped masses at every lumped mass point are m_i ($i = 1 \cdots 9$), the corresponding damping coefficients of which are c_i ($i = 1 \cdots 9$), respectively. k_i ($i = 1 \cdots 7$) are the stiffness coefficients of shaft segments, respectively. In order to research the dynamic characteristics of the dual-rotor model, the following assumptions are employed:

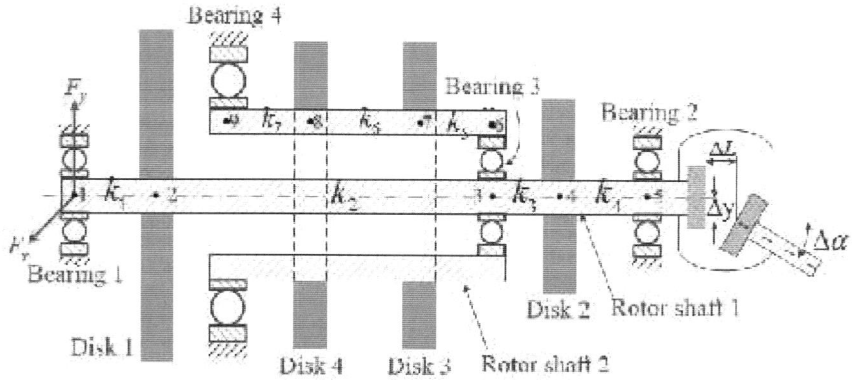

Figure 3 Mechanical model of the dual-rotor system

3.1 Misalignment modelling

The model in the study mainly considers the coupling integrated misalignment, which includes two cases of angle misalignment and parallel misalignment. When the master and slave rotor axis connected through the coupling exists misalignment, coupling housing is forced to make circular motion around its center by the limit of the two halve-coupling rotating around their respective axes. Its trajectory can be described by the following formulas:

$$\begin{cases} x = \Delta E \sin\theta = \frac{1}{2}(\Delta y + \Delta L \tan(\Delta\alpha/2)) \sin 2\theta \\ y = \Delta E \quad -\frac{1}{2}\Delta E = \frac{1}{2}(\Delta y + \Delta L \tan(\Delta\alpha/2)) \cos 2\theta \end{cases} \tag{4}$$

where ΔE is the equivalent amount of misalignment, which is determined by the coupling spacing ΔL, parallel misalignment amount Δy and misalignment angle $\Delta\alpha$, $\Delta E = \Delta y + \Delta L \cdot \tan(\Delta\alpha/2)$.

Through Newton's second law of motion and acceleration of couplings can obtain equivalent misalignment force:

$$\begin{cases} F_x = m_L \cdot (\Delta y + \Delta L \cdot \tan(\Delta\alpha/2)) \cdot \omega_1^2 \cdot \sin(2\omega_1 t) \\ F_y = m_L \cdot (\Delta y + \Delta L \cdot \tan(\Delta\alpha/2)) \cdot \omega_1^2 \cdot \cos(2\omega_1 t) \end{cases} \tag{5}$$

where $\omega_1$is the rotating speed, m_0 denotes the mass of the coupling housing.

3.2 Equation of motion

Based on the above analysis and mechanical model of the dual-rotor system shown in Figure 3, using the principle of force equilibrium and Newton's law of motion, with the external forces exerted on the dual rotors that include the imbalance force, the bearing restoring force, viscous damping force and gravity, the system's governing equations of motion can be obtained

$$
\begin{cases}
m_1\ddot{x}_1 + c_1\dot{x}_1 + k_1(x_1 - x_2) = F_{x1} \\
m_1\ddot{y}_1 + c_1\dot{y}_1 + k_1(y_1 - y_2) = F_{y1} - m_1 g \\
m_2\ddot{x}_2 + c_2\dot{x}_2 + k_1(x_2 - x_1) + k_2(x_2 - x_3) = m_2 e_1 \omega_1 \cos(\omega_1 t) \\
m_2\ddot{y}_2 + c_2\dot{y}_2 + k_1(y_2 - y_1) + k_2(y_2 - y_3) = m_2 e_1 \omega_1 \sin(\omega_1 t) - m_2 g \\
m_3\ddot{x}_3 + c_3\dot{x}_3 + k_2(x_3 - x_2) + k_3(x_3 - x_4) = -F_x \\
m_3\ddot{y}_3 + c_3\dot{y}_3 + k_2(y_3 - y_2) + k_3(y_3 - y_4) = -F_y \\
m_4\ddot{x}_4 + c_4\dot{x}_4 + k_3(x_4 - x_3) + k_4(x_4 - x_5) = m_4 e_2 \omega_1 \cos(\omega_1 t) \\
m_4\ddot{y}_4 + c_4\dot{y}_4 + k_3(y_4 - y_3) + k_4(y_4 - y_5) = m_4 e_2 \omega_1 \sin(\omega_1 t) - m_4 g \\
m_5\ddot{x}_5 + c_5\dot{x}_5 + k_4(x_5 - x_4) = F_{x2} + F_x \\
m_5\ddot{y}_5 + c_5\dot{y}_5 + k_4(y_5 - y_4) = F_{y2} + F_y - m_1 g \\
m_6\ddot{x}_6 + c_6\dot{x}_6 + k_5(x_6 - x_7) = -F_x \\
m_6\ddot{y}_6 + c_6\dot{y}_6 + k_5(y_6 - y_7) = -F_{y} - m_6 g \\
m_7\ddot{x}_7 + c_7\dot{x}_7 + k_5(x_7 - x_6) + k_6(x_7 - x_8) = m_7 e_3 \omega_2 \cos(\omega_2 t) + F_x \\
m_7\ddot{y}_7 + c_7\dot{y}_7 + k_5(y_7 - y_6) + k_6(y_7 - y_8) = m_7 e_3 \omega_2 \sin(\omega_2 t) + F_y - m_7 g \\
m_8\ddot{x}_8 + c_8\dot{x}_8 + k_6(x_8 - x_7) + k_7(x_8 - x_9) = m_8 e_4 \omega_2 \cos(\omega_2 t) \\
m_8\ddot{y}_8 + c_8\dot{y}_8 + k_6(y_8 - y_7) + k_7(y_8 - y_9) = m_8 e_4 \omega_2 \sin(\omega_2 t) - m_8 g \\
m_9\ddot{x}_9 + c_9\dot{x}_9 + k_7(x_9 - x_8) = F_{x4} \\
m_9\ddot{y}_9 + c_9\dot{y}_9 + k_7(y_9 - y_8) = F_{y4} - m_9 g
\end{cases}
\tag{6}
$$

where e_i $(i = 1 \cdots 4)$ are defined as the mass eccentricities of different disks, respectively; $(x_i, y_i)(i = 1 \cdots 9)$ denotes the vibration displacements of different lumped mass points, respectively; F_{xi} and F_{yi} represent the force components on the rotor from ith bearing $(i = 1,2,4)$, respectively; F_x and F_y are the force components exerted on the inner rotor due to misalignment faults.

It can be clearly observed from Figure 3 that (x_R, y_R) is (x_5, y_5), and (x_R, y_R) is (x_7, y_7). Therefore, the Equations (14)~(17) can be rewritten in a matrix form as

$$
\mathbf{M}\ddot{X} + \mathbf{C}\dot{X} + \mathbf{K}X = F(t) \tag{7}
$$

where $X = \{x_1, y_1, \cdots x_9, y_9, x_w, y_w, x_b, y_b\}$ is the displacement vector corresponding to each mass. **M, C** and $\mathbf{K}$ are the mass, damping and stiffness matrices, respectively. $F(t)$ stands for the extra force vector.

5 Results and discussion

Based on the actual dual-rotor test rig, we pay more attention to co-rotation. In the case of $\omega_1 = 2000rpm$, $\omega_2 = 3000rpm$ $(\kappa = 1.5)$, lateral vibration plot and corresponding spectrum of lumped mass point 5 under unbalance-misalignment coupling faults are exhibited in Figure 4 and Figure 5 (e_2=5mm, $\Delta\alpha = \pi/18$, $\Delta y = 5mm$). It can be found from Figure 4 and Figure 5 that the rotational frequencies of inner and outer rotors and double frequency of inner rotor are very prominent, which is in accordance with literature [15]. Meanwhile, the CWT has better performance than STFT on extracting the fault characteristics.

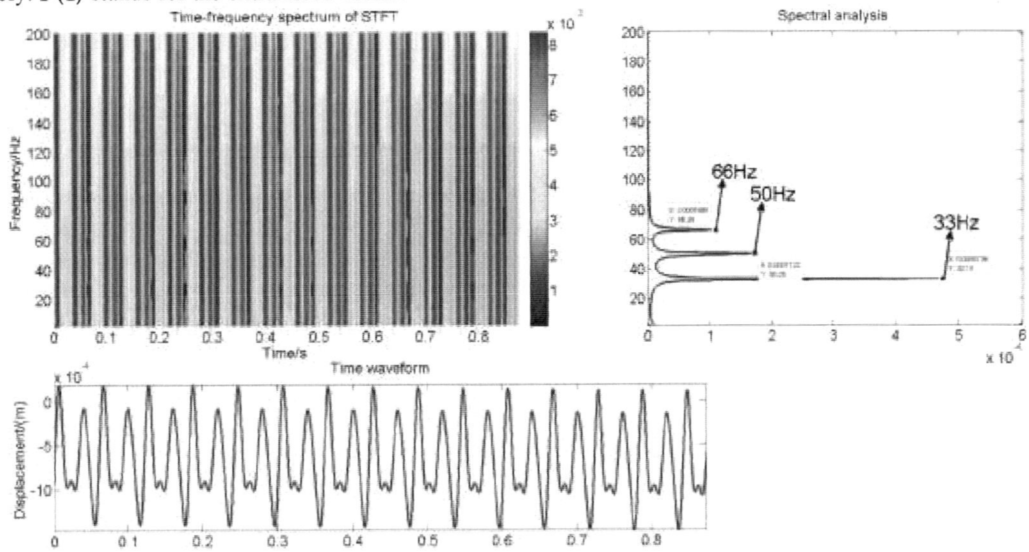

Figure 4 Time-frequency spectrum, time waveform and spectral analysis of the simulation results

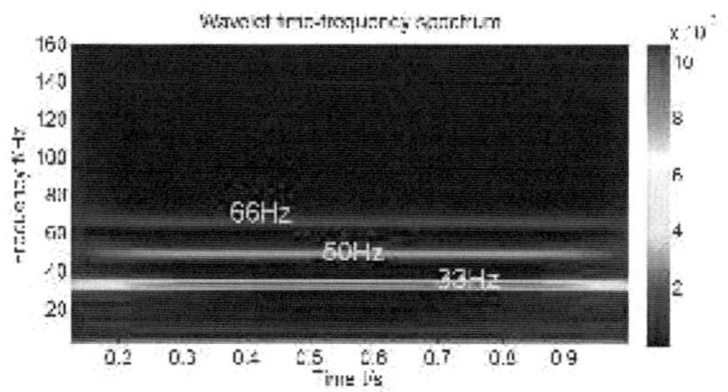

Figure 5 Time-frequency spectrum of wavelet

6 Experimental verification

Experiment are carried out on the dual-rotor system with a centrally mounted disk. In the section the experiment study to detect misalignment from vibration signals is presented. The test dual-rotor setup is as shown in Figure 6.

Figure 6 dual-rotor test rig

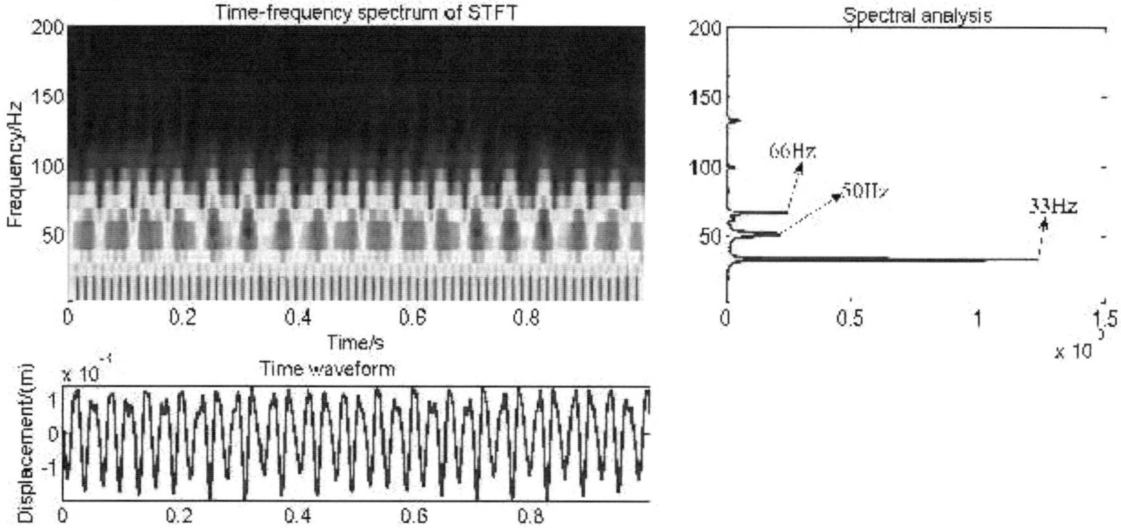

Figure 7 Time-frequency spectrum, time waveform and spectral analysis of the experimental signal

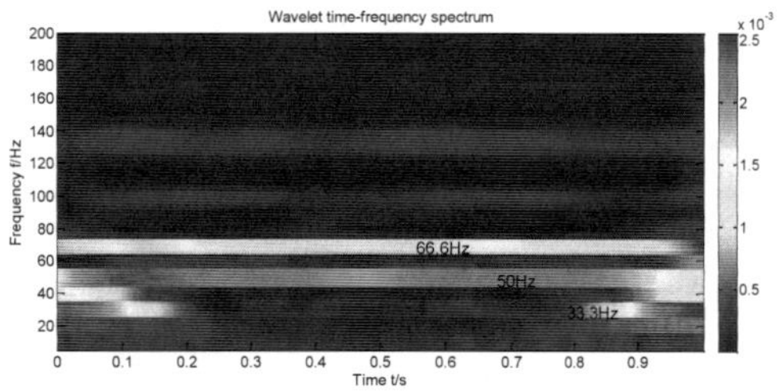

Figure 8 Time-frequency spectrum of wavelet

It can be observed from Figure 7 and Figure 8 that there are three frequency components, in which 33 Hz is the rotation frequency of inner rotor and has the maximum amplitude, while the other two frequency components, respectively, are the rotation frequency of outer rotor and double frequency of inner rotor, which are consistent with the numerical results. According to rotor vibration theory, prominent rotation frequency means the eccentricity of rotor, apparently, the eccentricity results from the imbalance of the screw. Meanwhile, 66 Hz is the double frequency of inner rotor, which demonstrates the misalignment existence of inner rotor.

7 Conclusion

A dynamic model for researching the unbalance-misalignment coupling faults of dual-rotor system with inter-shaft bearing is established in the paper. The vibration characteristics of outer and inner rotors are revealed when the dual-rotor with unbalance-misalignment coupling faults by numerical analysis and experiments. Since the time-frequency resolution of STFT is fixed when the window function and its length are chosen. the time-frequency map of CWT has better performance on extracting misalignment fault characteristics.

Acknowledgements

The support for this research under National Natural Science Foundations of China (No. 11572167) is gratefully acknowledged. The authors are grateful to the editor and anonymous reviewers for their valuable comments.

References

1. J. Piotrowski, Shaft Alignment Handbook, Dekker, New York, 1986.
2. M. Xu, R.D. Marangoni, Vibration analysis of a motor-flexible coupling-rotor system subject to misalignment and unbalance, Part I: Theoretical model and analysis, J. Sound Vib. 176 (1994) 663–679.
3. D.L. Dewell, L.D. Mitchell, Detection of a misaligned disk coupling using spectrum analysis, Trans. ASME J. Vib. Acoust. Stress Reliab. Des. 106 (1984) 9 - 16.
4. M. Xu, R.D. Marangoni, Vibration analysis of a motor-flexible coupling-rotor system subject to misalignment and unbalance, Part II: Experimental validation, J. Sound Vib. 176 (1994) 681–691.
5. N. Hamzaoui, C. Boisson, C. Lesueur, Vibro-acoustic analysis and identification of defects in rotating machinery, Part I: Theoretical model, J. Sound Vib. 216 (1998) 553–570.
6. A.S. Sekhar, B.S. Prabhu, Effects of coupling misalignment on vibrations of rotating machinery, J. Sound Vib. 185 (1995) 655–671.
7. Y.-S. Lee, C.-W. Lee, Modelling and vibration analysis of misaligned rotor-ball bearing systems, J. Sound Vib. 224 (1999) 17–32.
8. N. Bachschmid, P. Pennacchi, A. Vania, Identification of multiple faults in rotor systems, J. Sound Vib. 254 (2002) 327–366.
9. P. Pennacchi, A. Vania, Diagnosis and model based identification of a coupling misalignment, Shock Vib. 12 (2005) 293–308.
10. P. Pennacchi, N. Bachschmid, A. Vania, G.A. Zanetta, L. Gregori. Use of modal representation for the supporting structure in model-based fault identification of large rotating machinery: Part I – Theoretical remarks, Mech. Syst. Signal Process. 20 (2006) 662–681.
11. J.K. Sinha, A.W. Lees, M.I. Friswell, Estimating unbalance and misalignment of a flexible rotating machine from a single rundown, J. Sound Vib. 272 (2004) 967–989.
12. Fei Z X, Tong S G, Wei C. Investigation of the dynamic characteristics of a dual rotor system and its start-up simulation based on finite element method [J]. Journal of Zhejiang University-SCIENCE A, 2013, 14(4):268-280.
13. Feng Z, Liang M, Chu F. Recent advances in time–frequency analysis methods for machinery fault diagnosis: A review with application examples [J]. Mechanical Systems & Signal Processing, 2013, 38(1):165-205.
14. Yan R, Gao R X, Chen X. Wavelets for fault diagnosis of rotary machines: A review with applications[J]. Signal Processing, 2014, 96(5):1–15.
15. Sinha J K. Higher Order Spectra for Crack and Misalignment Identification in the Shaft of a Rotating Machine [J]. Structural Health Monitoring, 2007, 6(4):325-334

NOVEL MICRO, WIRELESS, MEMS-BASED CONDITION MONITORING SYSTEM FOR MODERN MACHINE TOOLS WITH LIMITED ACCESS

Wojciech Staszewski, Adam Jablonski and Tomasz Barszcz
AGH University of Science and Technology
Department of Mechanical Engineering
Aleja Adama Mickiewicza 30
30-059 Kraków, Polska
Telephone: (48) 509-805-483
ajab@agh.edu.pl

Abstract: Currently available condition monitoring systems (CMS) offer many types of tools, including stationary systems, portable on-site instrumentation and finally wireless, autonomous systems. However, for a particular group of modern machine tools with limited access, like lathes, milling or grinding machines, none of this tools could be used due to limited access, space constraints and cabling restrictions. Firstly, stationary CMS use wires connecting vibration sensors with data acquisition unit (DAQ), which are prohibited by both, safety and topological reasons because of sealed door. Secondly, portable systems could not be used, because modern mining machinery is entirely closed, and a diagnostic engineer is not allowed to be present inside the machine housing while operating. Moreover, some of machines perform open lubrication, which makes supervised monitoring impossible due to constant lubricant splashes. Tertiary, currently available leading wireless sensors are characterized by relatively significant size, ca. 1.5 by 4 inch, and close to 10 oz. weight; therefore, introducing significant volume and mass to the machine spindle. Moreover, the rapid characteristics of the spindle movement would generate a significant inertia to the sensor causing extra vibrations and possible detriment to sensor mounting. For these reasons, modern machine tools are generally not equipped with external CMS systems.
Recently developed MEMS technology made it possible to design a novel, small-size unit, which is capable to work autonomously in environments with limited accessibility, where volume and weight matter. In contrast to commonly used piezoelectric accelerometers, MEMS vibration sensors need much less operational power, which is a major concern in wireless designs. Moreover, latest MEMS sensors are characterized by comparable frequency response to accelerometers. The short range wireless communication can be applied to avoid connecting cabling. The paper shows a prototype of a micro-size unit for data acquisition, data processing, data storage, and transfer. The prototype is evaluated on industrial lathe.

Key words: Condition monitoring system; diagnostics; microelectromechanical sensors; micro data acquisition system; portable machine health management tools; wireless vibration sensors

Introduction: Nowadays, in the highly industrialized age, the purpose of monitoring the conditions of an engineering systems, especially machinery has become a concern that requires great affection. A Condition Monitoring System (CMS) that bases on vibration monitoring may play a very important role in pushing out the maximum potential of all kinds of machinery by minimizing the downtime. As it has been widely shown in the literature, for instance in [1-3] the maintenance that bases on CMS in compare to traditional, scheduled or corrective maintenance, more accurately fits in many aspects only with few disadvantages. Unfortunately, the design and installation cost of condition monitoring system is substantial in comparison to other maintenance approaches; therefore, usually it is only used for a huge machinery, like turbo generators or wind turbines which downtime is extremely expensive The entire purpose is to show a potential for more affordable option for constructing a simple data logger, which might be eventually turned into a basic CMS, offering limited functionality that might lead into establishment of preventive maintenance approach, as well as provide a way to introduce such a system for machines for which it is currently impossible.

Condition Monitoring System: Condition monitoring (CM) can be shortly described as a process of monitoring a parameter of condition in machinery, like vibration, in order to identify a significant change in the range of this parameter which usually may indicate a progression of ongoing fault. As this process is a major component of well-known predictive maintenance, the use of CM allows to schedule a maintenance or undertake other actions to prevent further failure as well as avoid its consequences. Condition Monitoring has a benefit in compare to other techniques (breakdown and preventive (or scheduled) maintenance) – deterioration of the crucial parameters that would shorten a typical lifespan of a component can be addressed before they could develop into a major failure [4], causing an unplanned downtime. Condition monitoring techniques are usually used on rotating machinery and other equipment like pumps, electric motors, internal combustion engines, presses, while periodic inspection using non-destructive testing techniques are used for stationary plant equipment such as steam boilers, piping and heat exchangers. The advantages that have been offered by the application of predictive maintenance techniques have in a past few years led to the increase in the development of a number of methods for condition monitoring/fault-diagnostic systems [5]. After detection of possible failure (significant change of monitored parameter), mostly the diagnostic staff is responsible for taking the necessary actions in order to verify the condition of the machinery and later on act accordingly to prevent from machine failure. Those activities may require further data measurement in order to gather more data that might be necessary in order to make appropriate decision regarding machines condition, but at the point where the failure starts to develop it is necessary to notify the engineers that the condition of machinery is changing. The schematic of a modern condition monitoring system is presented in the Figure 1. Nowadays, the new methods of modern systems theory show that use of the mathematical process and signal models, identification and estimation methods can provide better, more accurate and faster available information regarding machine condition. With use of theme it is possible to establish new, advanced methods of fault detection and diagnosis in order to detect even small faults that occur quite early and to diagnose their origins in order to settle the primary cause [1].

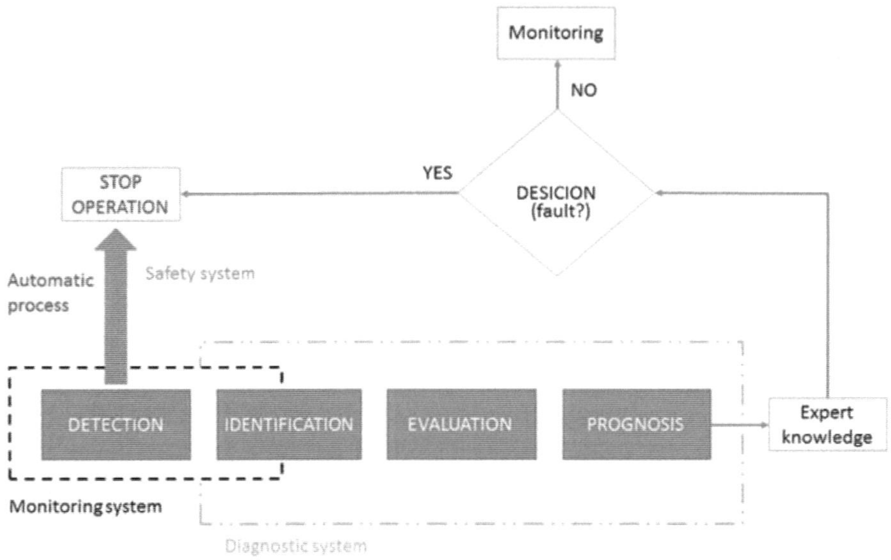

Figure 1: Scheme of modern condition monitoring system.

The process of automatic supervision of machines state is mostly realized by the limit checking (or threshold checking) of some variables that have been established as crucial for failure identification, e.g. force, speed, pressure, liquid level, temperature. Usually the alarms are raised when the limit value for a specific parameter is exceeded – before that a warning indicator may indicate, that condition is changing [3]. This process enables the introduction of automatic protection system, which in the moment of failure detection (exceeding the threshold value) will power down the machine in order to prevent from further damage, if it would harm workers or could cause permanent damage to the machinery.

Microelectomechanical Systems (MEMS) – it is compelling: As it was mentioned before, the condition monitoring systems are usually introduced for the machines that are huge in size and usually cost much, where the implementation and deployment of continuous monitoring may provide significantly lower service cost and drastically decrease maintenance time. The low cost system based on MEMS accelerometers such as the one considered in this paper could be the solution for smaller machines. The Micro Electro-Mechanical System (MEMS) based condition monitoring systems could easily reduce the installation cost per each node from $1000's to $100's [6]. The MEMS based sensors offer the capability to interconnect various interfaces used nowadays in the industrial applications and simplify implementation for considered systems. The MEMS technology is compelling, because of its attributes, such as size, weight, power, cost and high levels of functionality [7], but unfortunately their presence on the market has been limited because of relatively low level of performance in terms of noise density and resonant frequencies – in compare to other solutions, like piezoelectric accelerometers. High frequency MEMS accelerometers have been available on the market since few years, offering resonant frequencies as high as 22 kHz and Full Scale Ranges up to ±500g – unfortunately with the presence of high noise levels. Contrarily, available low noise

MEMS sensors have low resonant frequency operation, used for some condition monitoring systems, where its application requires very low frequency operation, which of course limits the possibilities of diagnostic evaluation [8].
The investment and further development in MEMS process technology has advanced to the point where the improvements in performance obtained via technology development are enough to make MEMS a new available option for a wider range of CMS applications. MEMS sensors are nowadays also tolerant to the shock conditions, with stable sensitivity after subjecting to 1000's g's of shock, or vibration to 10's of kHz. The accelerometers with embedded signal conditioning generally offer a full electrostatic self-test of the moving element and signal conditioning circuits.
In order to affirm the thesis, that MEMS accelerometers could be easily used for own-made condition monitoring system the test case has been established based on a measurement of a simple domestic fan with an office clip attached to one of its blades serving as an imbalance. Two configurations have been used in order to gather measurement data: (i) the professional ACQ unit "Vibmonitor" equipped with 24bit ADC and industrial piezoelectric accelerometer IMI 61A02, (ii) proposed system with 12bit ADC and ADXL001-70 MEMS accelerometer. To point out, the main imbalance component is well preserved in MEMS measurement. However, the representation of higher components is worse than in case of piezoelectric accelerometer (used in a professional system), because of additional damping since the latest development of MEMS sensors is available only as a raw electronic board; therefore, in order to preserve high spectral bandwidth, it is necessary to develop a housing with relatively slow damping.

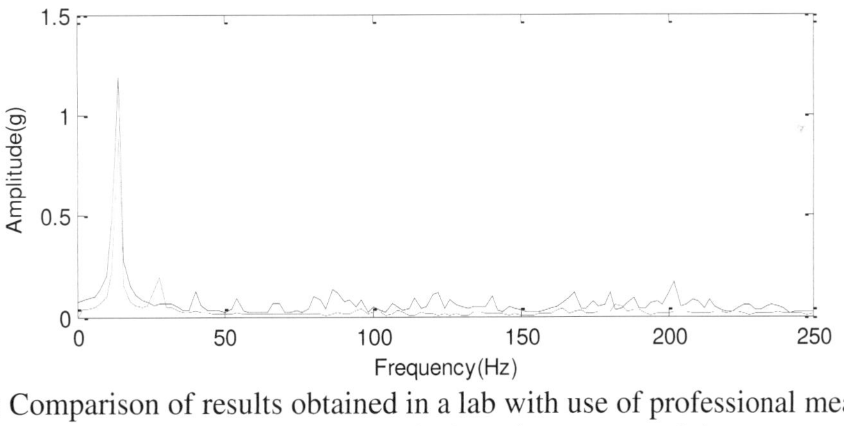

Figure 2: Comparison of results obtained in a lab with use of professional measurement system and own-made single-point sensor and daq.

Single-point sensor and data acquisition unit: For many years, in order to achieve very accurate measurement of vibration signals of individual components of the machine in the industrial conditions, the maintenance engineers were obligated to use very expensive equipment. Those appliances very often exceed the budget for the maintenance process and there-fore only for a few cases those solutions have been used. That is why the authors came to a conclusion that there is an enormous market to create low-cost architecture of data-acquisition device that could work in the industrial environment. The authors have been listening requirements and requests from companies and truly understood the need for this kind of cheap and accurate device. At the beginning it was necessary to focus primarily on the selection of appropriate architecture of data-acquisition device. Microprocessor STM32 caught authors' attention during the long process of selection the main heart and brain of whole system. STM32 exceeds many different solutions and microcontrollers due to its powerful tools, embedded possibilities and obviously very low price. The research was based on STM32 version F401re which is working on ARM M4 Cortex and the maximum system core clock is 84Mhz which seems to be fine for own-made systems. The key features of ARM Cortex M4 are presented in Table 1.

Table 1: ARM Cortex M4 features.

Parameters
up to 96 Kbyte of SRAM,
1.7 to 3.6V application supply and I/Os,
Power consumption of run mode: 146microA/MHz,
1 x 12 bit, 2.4 MSPS A/D Converter,
General-purpose DMA,
Up to 10 timers,
RTC

For the sake of the measurements, the STM32f401re NUCLEO board was used that bases on ARM Cortex M4 considering the following criteria (comparing to ARM Cortex M7):

- At the moment when the development was started the price of chosen M4 based solution was around half the price of M7 one and the idea behind the single-point sensor and daq is to cut costs as much as possible, while preserving functionality and decent accuracy,
- The main considered parameter (speed) of the ADCs present on both M4 and M7 is comparable (~2.4MSPS),
- The other parameters of ADCs have not been considered as it was out of the scope of this paper,
- Despite the fact that M7 has bigger flash, the idea is to either store the measurement data on SD card or (after further development) send the data on-line, using wireless protocols and therefore it was not crucial for this project.

In the Figure 3 the conceptual design of a single-point measurement and data acquisition unit is presented.

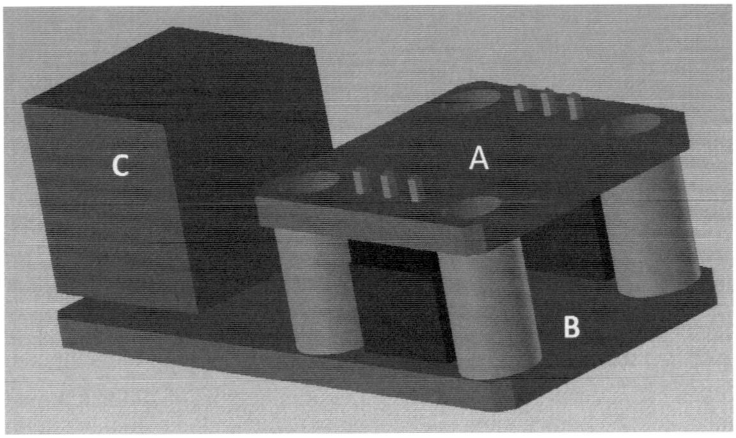

Figure 3: Scheme of single-point sensor and daq device: A – accelerometer, B – processing unit, C – connectivity.

Development of portable, wireless MEMS-based sensor: A typical low energy (LE) application involves sending the sensor data from a LE peripheral device (such as a smart monitor) to a LE Central device (such as a smartphone, or central DAQ). The architecture on the system level of such a device for needed applications has three major functional blocks, as shown in the Figure 4: measurement, data and event processing, and connectivity. Each considered block supplements to the average current of the overall system.

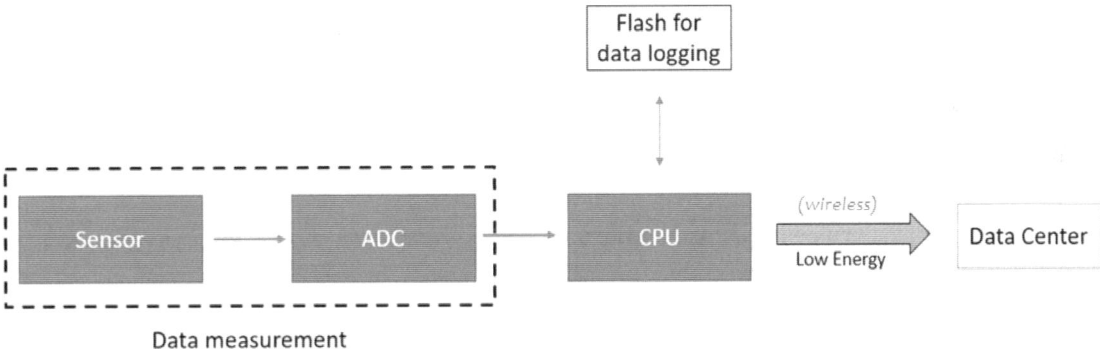

Figure 4: System Block Diagram for wireless sensor.

The measurement (sensing) block is composed of a sensor (MEMS accelerometer) and transducer hardware that converts the sensor information (digital signal) into an analog, electrical signal. A transducer is followed by a circuit that is used for data filtration and amplification (signal-conditioning circuit). The conditioned analog signal is converted into digital data for processing by an analog-to-digital converter (ADC). The main, key factors that regulate the overall current consumption in the sensing operation are: (a) signal conditioning and analog-to-digital conversion - the ADC consume significant current, (b) sampling rate - necessary to have enough samples for accurate processing of the data, (c) scan rate - the rate to detect an activity, which allows to save power when

there is no sensing activity in the system. The data acquired from the sensor requires processing in order to get a meaningful information. It may also be necessary to compress the data to reduce the bandwidth required to transfer the data to the central device (over-the-air data transfer). Furthermore, a LE device might use more than one sensing unit. In such a case it is necessary to detect pending activity in multiple sensors, process the data from multiple sensing points, and send the measurement data over the LE link. Each sensor's scan rate can be synchronous (with some limitations) and asynchronous to the others and the LE link. The knowledge how to efficiently manage those activities might be a crucial factor that contributes to the overall current consumption and leading into increase in the possible lifetime of the device on single battery.
The LE communication enables to send the measurement data from a sensor to a central device through the notifications or send read requests from the central device to the sensing device. The time taken to process the notifications and send the measurement data to the central device efficiently will at the end determine the overall current consumption.

Wireless Low Energy – increasing battery life resulting in extending operational time for each device: It is apparent that battery life can be increased by using mainly the following two approaches that lead to reduction the average current consumption: (a) reduce the active mode current: The data measurement and data processing functions contribute almost equally to the overall system current consumption and therefore, it should be a priority to reduce the current consumption in active operations - sensing, processing, and transmission, (b) reduce the active time: Reduce the time spent for active operations in CPU or RF resulting in longer idle state of the system. The actual current consumption in these idle (or sleep) states is as important as the current consumed during active operations. The overall current consumption could be limited by using the low-power modes as often and for as long as possible. The device can the following states, distinguished by the current usage:

(i) The Active state stand for that all the scanning and processing activities are carried out at the highest required rate.

(ii) The Idle state saves power when the device in not being used for an idle period (an interval, after which the device goes into idle mode).

(iii) The system and LE connection are in the Deep-Sleep mode for most of the time.

(iv) When the battery voltage drops below the operable voltage, the system stops all the activities and shuts down.

The final application should be designed in a manner to allow the CPU to operate at lower clock frequencies in order to reduce the current usage. The power-optimal algorithms and implementations should be implemented and tuned for the CPU architecture to reduce the actual processing power required for sensor data processing and not to exceed the needs. The algorithm can put the unused peripherals and clocks into the available low-power modes permanently. In addition, peripherals and clocks that are not required often or that are required only in specific usage modes should be put into their lowest power modes, wherever possible. Therefore, the application should have an algorithm that will allow to change the current operation mode as often as possible in order to reduce the overall current usage and to extend the battery lifetime.

The application should align the sensing and data processing operations to the beginning or end of LE connection events. The application can then complete all the processing and put the system into the Deep-Sleep mode along with the BLESS until the next LE connection event. This is more power efficient than processing asynchronous to the LE events, because it allows the system to be in the Deep-Sleep mode for longer times.
If no sensor use or activity is detected for some time, the device can reduce the rate at which it scans for the activity. In addition, it can increase the connection interval of the LE link and enter low-power modes to reduce the system current. While this may introduce latency in reacting when an activity occurs, the overall system power can be significantly reduced.

The sampling rate of the ADC in the device should be reduced to the minimum frequency required to effectively reproduce the analog signal being sampled. Based on the sampling frequency, the ADC configurations that allow low-power operation should be used. In addition, the channel resolution can also be reduced to decrease the conversion time and thus save power.

Conclusion: The conducted study on the use of low cost MEMS-based accelerometer solutions for the purpose of building cheap, portable and power-efficient condition monitoring system that could be applied in the conditions where typical systems cannot be introduced shows an approach of own-made system, suitable for different scenarios. The presented concept of a device that can support multiple systems and LE low-power modes that enable to design a highly power-efficient solution. Wireless low energy brings the wireless connectivity to low cost, small battery operated devices that require months or even years of battery lifetime. It has been designed to address the needs of energy efficiency and simplicity of design when employing it to develop connected products. The battery life of a LE application depends on the system-level current consumption. In most applications, sensing and processing can consume more power than the LE operations, so attention must be paid to reduce the overall system current.

The concept of creating own-made condition monitoring system that bases on portable, wireless sensors embedded with wireless low energy shows the potential of this technology. Not only could the presented realization be the solution for smaller machines, for which the cost of CMS installation would exceed the budget for its maintenance but the use of wireless, portable, MEMS based sensors would allow to reduce the installation cost per each monitoring node by the factor of 10 (or even more while thinking about the mass production).

Acknowledgements:
The work presented in this paper was supported by funding from the research project co-financed by the KIC InnoEnergy Project Agreement number 32_2014_IP110_XSENSOR

References:

[1] Blanke, M., Kinnaert, M., Lunze, J., Staroswiecki, M., Diagnosis and fault tolerant control. Springer, Berlin, 2006, 2nd edition.

[2] Hameed Z., Ahn S.H., Cho Y.M., Practical aspects of a condition monitoring system for a wind turbine with emphasis on its design, system architecture, testing and installation. Renewable Energy vol. 5 issue 5:879-894, 2011, doi:10.1016/.renene.2009.10.031.

[3] Isermann R., Fault-Diagnosis Applications, Blackwell, 2011, London Book chapter.

[4] Milind A. Siddhpura, Arti M. Siddhpura, S.K. Bhave, Vibration as a parameter for monitoring the health of precision machine tools, ICDM-08, Coimbatore, India, February 01-02, 2008.

[5] Wang W., Tse P.W., Lee J., Remote machine maintenance system through In-ternet and mobile communication, International Journal of Advanced Manufacturing technology 31:783-789, 2007, doi:10.1007/s00170-005-0236-1.

[6] Albarbar A., Badri A., Jyoti K., Starr S., Performance evaluation of MEMS ac-celerometers, Measurement 42, 790–795, 2008, doi:10.1016/j.measurement.2008.12.002.

[7] Nagel DJ, Zaghloul ME., MEMS: micro technology, mega impact, IEEE Circuits and Dev Magazine vol. 17 issue 2, 2001, doi:10.1109/101.920875.

[8] Ratcliffe C., Heider D., Crane R., Krauthauser C., Yoon M.K., Gillespie J.W., Investigation into the use of low cost MEMS accelerometers for vibration based damage detection, Composite Structures 82, p. 61–70, 2007, doi:10.1016/j.compstruct.2006.11.012.

Bibliography:

[1] A. Muszyńska, Rotordynamics, Taylor & Francis Group, LLC, Minden (2005), USA.

[2] R.J. Patton, P.M. Frank, R.N. Clark (eds.), Issues of fault diagnosis for dynamic systems, Springer-Verlag, London (2000).

[3] M. Strączkiewicz, A. Jabłoński, T. Barszcz, Detection and classification of alarm threshold violation in condition monitoring systems working in highly varying operational conditions, Journal of Physics, Conference Series (2015), vol, 628, s. 012087-012087-B.

PECULIAR CASES OF PINION VIBRATION IN PARALELL SHAFT DOUBLE HELICAL GEAR UNITS – IDENTIFICATION AND MITIGATION

Mantosh Bhattacharya
Petrofac International UAE
+971501469853
mantosh.b@petrofac.com

Abstract – Speed increasing parallel shaft double helical gear units are commonly used in centrifugal compressors and high energy pump applications. Such turbo-gears are API 613 compliant (Titled as -Special Purpose Gear Units for Petroleum, Chemical, and Gas Industry Services) and have horizontal offset transmission line, with journal bearings on the high speed shaft and low speed shaft. When a full load full speed complete unit test is carried out, pinion dynamic behavior is found to be different than what observed during full speed no load condition as mandated in API 613 . During this test, gear box high speed pinion may show up with high vibration or combination of both during certain load and speed combination in spite of proper balancing and alignment. Since full load full speed test is kept as an option in API 613 , hence these type of vibration may not be detected during a no load test in gear box manufacturer's premises. These types of vibration patterns are not explicitly addressed in API 613 .
The first objective of this paper is to cover analytical, design and diagnostic aspects which can be helpful to mitigate above issues before it is towed out from manufacturer's premises.
The second objective of the paper is to suggest installation /innovation / design modification on high speed pinion as a pre-emptive approach which can save time to mitigate if encountered during any certain speed –load combination.

Keywords – Sub -synchronous , relief , super-synchronous , tuning , nodes, FEM

Introduction-

Normal Vibration phenomena of a gear box includes 1x vibration, hunting tooth error , impacting and large side bands across the gear mesh frequencies . These can be easily identified by various techniques and mitigation action can be chalked out . Sometimes , baffling vibration phenomena such as sub- synchronous / super- synchronous vibration occurs which are not well cataloged in vibration analysis text books and international standards such as API 613 and ISO 6336.

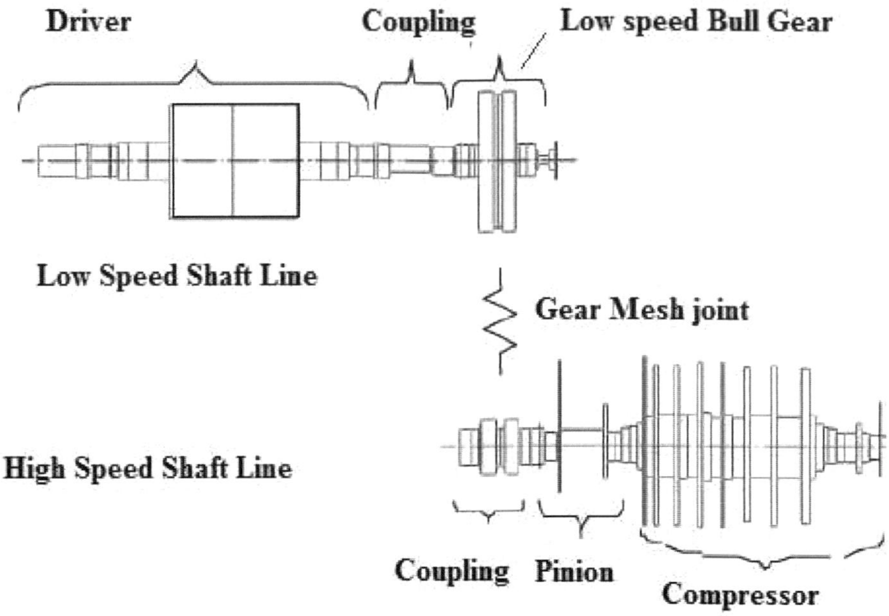

Fig -1 - Torsional Model of a turbo compressor train

Mandates by API 613 5th edition for design and testing -

As per API 613 latest edition, the design of rotor-bearing systems, consideration is given to all potential sources of periodic forcing phenomena (excitation)
The design aspect also covers resonances of structural support systems that may affect the rotor vibration amplitude that occurs within the specified operating speed range or the specified separation margins[1]

As mandated in API 613, three lateral critical speed modes are generally of concern. The frequencies at which these modes occur vary as a function of the transmitted load, primarily due to the resulting stiffness change of the bearing-oil film due to varying stiffness.
API addresses that sometimes it is not possible to operate the gear rotor below the first critical speed. When a gear rotor must operate above the first critical speed, the rotor should be designed with the first critical speed about 60% of operating speed. First critical speeds below 60%of running speed are prone to create bearing instability or oil whip and if running much above 60% are too close to running speed.
2. Gear wheels that operate between 40% – 50% of pinion speed (2:1to 2.5:1 gear ratio) can excite instability in the pinion bearings. Therefore, if gear ratios between 2:1 and 2.5:1 are required special consideration should be given to performing a pinion stability analysis. Instability phenomena at bearings can be caused by excitations having frequency around the half of the shaft speed. This is the reason why the API requires the stability verification when the gear ratio is around 2 to 2.5.
Gear Box are tested as mechanical running tests at OEM (Original Equipment Manufacturer)works as per API 613 and following measurements are taken -
a. Inlet oil temperature and pressure.

b. Oil flow.
c. Outlet oil (drain) temperature.
d. Shaft vibration, frequency, and amplitude, both filtered (synchronous) and unfiltered
e. Bearing pad temperatures.
f. Casing vibration

During API 613 compliant mechanical run test (MRT) of gear box, following criteria are explicitly followed –

1. Steady state shaft and casing vibration - (cl.2.6.6.5 & Table 9)
2. If the amplitude of any discrete, nonsynchronous vibration exceeds 20% of the pinion shaft (or 40% for shafts running below 2000 rpm) of the allowable vibration as defined in 2.6.6.5, the purchaser and the vendor shall mutually agree on requirements for any additional testing and on the gear unit's suitability for shipment.(cl. 4.3.2.2.9)
3. Unbalance response test, if specified –(cl.2.6.3.1)

Dynamics of a pinion in parallel shaft double helical gear unit

The dynamics of high speed pinion of such turbo-gear is significantly different than a centrifugal compressor / turbine / pump rotor. The bearing load on a single casing multistage compressor / pump is approximately half of rotor weight while for a pinion due to torque force developed in a meshing gear, can be several times of its own weight .Due to this , bearings of a parallel shaft industrial gear are heavily loaded at particular direction compared to bearings of compressors. The load direction is based on meshing direction of two gear elements. When these gearboxes are tested with no load (5-10% of actual load during MRT) in OEM works, they are to meet acceptance criteria of casing and rotor vibration as mandated by API. During this test, error profile, pitch line error, hunting tooth issues, gear mesh frequencies and its sidebands are identified with a spectral range of 1500 Hz.

The stiffness and damping characteristics of the bearings are functions of the radial load applied to them.

Depending on the tooth geometry the radial gear load will typically be at a 20 degree angle from the vertical, known as the pressure angle. With a centrally located bull gear, a pinion may be 'upward driven' or 'downward driven'. The gear load and the weight of the rotor shall be oriented approximately 20 degrees apart on a downward driven pinion. On the other hand, upward driven pinions have the gear load and weight vectors oriented approximately 160 (180 minus 20) degrees apart. [7] This creates a larger variation in the resultant load angle in off design conditions i.e. during no load MRT. However , during full load full speed test the dynamic behavior normally simulates actual site running condition.

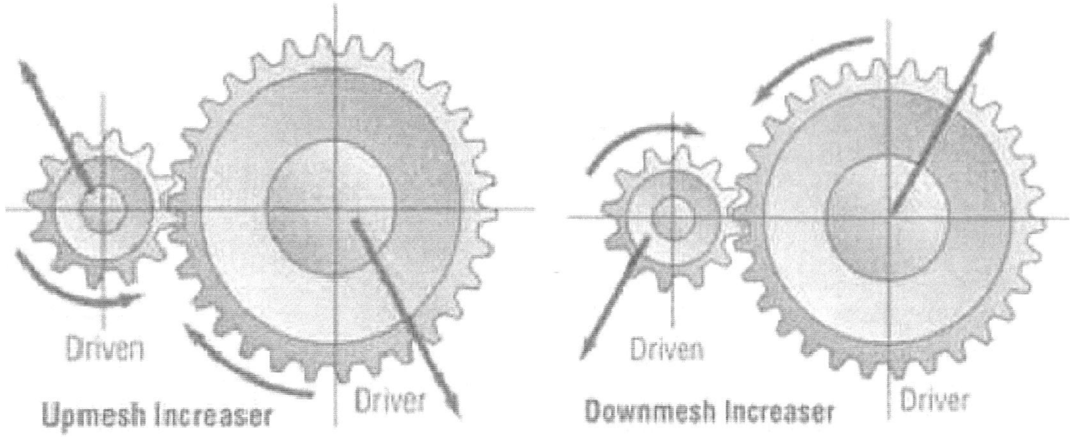

Fig 2 Gear mesh force vectors

Type of Bearings used in high speed pinion of turbo gears -
Bearings play important part in damping the vibrations and sometimes a source of vibration too . Hence, this part gives a brief account of type of bearings used in gear box .

- Cylindrical 2-lobe
- Offset half
- Double offset
- Pressure dam
- Lemon bore
- Tilting pad journal

Axially split journal bearings (Fixed Geometry) are used for the standard design in the gear shafts . If higher speeds are involved (shaft circumferential velocity exceeding 90 m/sec), radial tilting pad bearings are normally used .
For a constant direction of rotation with high load operation (70% -100% load) ,lemon bore bearings are used as a normal practice**.**
If the gear box is to operate at lower speed during driver barring and subsequent change in load direction, pressure dam bearings is a good choice.
The concern for oil film excitation of the rotor lowest natural frequency is eliminated by the use of the tilting pad journal bearings. Tri- pad tilting pad journal bearing has higher load operation capability(P, v) i.e. lower Tmax. (Pad temperature) .Now a days, to reduce power loss, evacuated direct lubrication type journal bearings is used.
For high speed pinion bearings, directed lubrication is used to reduce pad temperatures in high speed, heavily loaded bearings. These bearing designs utilize several design techniques to reduce pad temperatures, including the use of:
• Copper alloy pad materials
• Offset pivots
• Directed lubrication (running evacuated)
• "Spray Bar Blockers" (to reduce hot oil carryover)[5]
• By-pass cooling where extra oil is introduced behind the pad to increase heat transfer.
Evacuated bearing design equipped with spray bar blocker with bypass cooling) are normally used for high speed high load application

Since the loaded pads require the least amount of oil, due to their thin oil films, they typically run with full films while the unloaded top pads run with various degrees of starvation. So , for a pinion ,during various load and speed combination , location of partially unloaded pads may change. On prima facie , SFD (Squeeze film damper) bearings can be thought of a better choice but they cannot be used in such gear units as at high load the damper can settle and rotor may not operate in desired path.

Sub-synchronous vibration in pinion

Sometimes , the tilting pad bearings demonstrate chaotic behavior with a comparatively raised noise floor. It is often found the sub-synchronous vibration issue are at very low frequency and spread all over up to 0.5X region of waterfall spectrum. This type of problems takes long time to identify and mitigate.

This type of sub-synchronous vibration (SSV) has low amplitude at random frequency . The concern from OEM original equipment manufacturers and end user is that this vibration, although small, fails to meet overall API vibration limits. See fig 3 below -

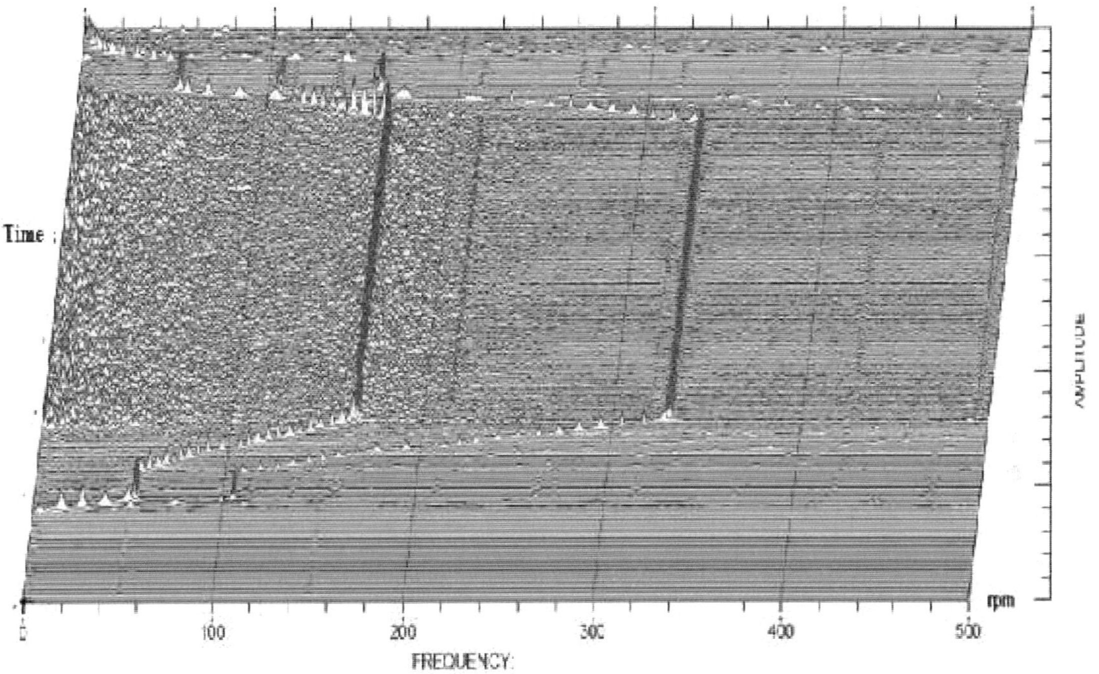

Fig 3 – Low frequency Sub-synchronous vibration

Pad Flutter is phenomena occurring to unloaded pads stationed opposite to load vector and not finding an equilibrium position. This means when the oil supply is not adequate on a relatively unloaded pad as stated earlier , the pad tends to swing in nonlinear mode creating a chaotic behavior and later stage they start spragging (infrequent hitting the journal).[4]

TPJB (Tilting pad journal bearing) particularly a three pad journal bearing, because of their geometrical and nonlinear fluid film characteristics, show high chances of the onset of chaos but are still selected when journal load and rotor peripheral velocity is too high to be accommodated by regular 4,5 pad design. If this is a case , a detailed analysis to be carried out during design phase itself and subsequent remedial action should be ready . During a loaded condition, based on direction of gear mesh as described above, unloaded top or bottom pads and require more oil to satisfy full film requirement. The shortfall of oil may lead to pad flutter.

To further confirm and identify the starved pad all pad temperatures may be recorded and compared with other pads during internal test carried out by gear box manufacturer .

The most effective means of controlling pad fluttering has been explored to be relieving the leading edge of each pad. The depth of the relief must be sufficient to ensure that a convergent wedge is formed at the leading edge of the pad as it approaches the journal.

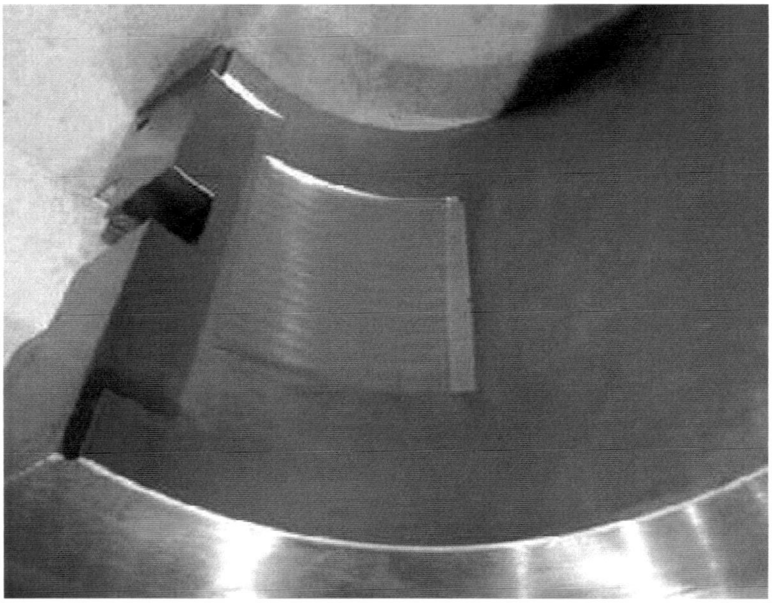

Fig 4 - Relief created in pad to avoid sub synchronous vibration

Standard tilt pad bearings achieve low cross-coupling through rocking or sliding motion. Flexure Pivot tilt pad radial bearings achieve the same low cross-coupling and high stability through flexure and rotation of the center post while eliminating any pivot wear and high contact stress. The rotational stiffness increases the pad natural frequency and eliminates pad flutter and spragging which are encountered on the unloaded pads in standard type tilt pad bearing. Although this design is used in integrally geared compressor rotor but it is yet to be used on a parallel speed increaser double helical gear box pinion for its high speed and high load capability.

Pinion vibration during start up

A high vibration spike on pinion is observed sometimes during the start up which gets registered in spectrum but does not trigger alarm / trip of machine. The reason is due to

high inertia of bull gear which literally lifts the pinion during meshing . This is considered as benign vibration till it is within 70% of pad and journal clearance.
A common phenomena in speed increasers during API 613 no load test in pinion vibration recordings that exhibit higher vibration levels than the allowable limits set in API 613, clause2.6.6.5.
These vibration values of the pinion vary within a certain range. The above mentioned behavior is due to a slight instability of one of the rotors which can occur when the tangential force (driven torque) is equal to the pinion rotor shaft weight. Once there is load on the gear unit (< approx. 10%),there will be a defined force, which will move the pinion to its normal defined operating position(upward and out). At this point the instability will immediately disappear and the amplitude of the vibration will drop below the maximum allowable values. If this phenomena occurs during the test , it is advised to conduct at least a part load full speed test to assure stability of pinion during its operating speed regime .
In this case the energy associated with the start-up is low enough whereby no harm can occur.

Another phenomena commonly experienced in these type of gear units higher than normal vibration on hot starts. In this case higher vibration levels may be experienced on both rotors after a brief shutdown (hot start). When the gearbox is shut down lube oil continues to spray into the local area of the standstill mesh . The cooling effect of the oil brings the localized temperature down causing a relative thermal contraction of the steel in a localized area relative to the rest of the rotors gear teeth. This cooled area temporarily generates a thermally induced relative transient transmission error .This behavior is experienced for a short while and then gradually diminishes until the rotors are thermally stable to uniform load. In such cases a due diligence is to be taken to employ vibration trip multiplier during startup.

Pinion Super synchronous vibration

In last few years oil and gas industry has encountered. gearbox pinion high radial vibrations at a particular frequency that was well away from the gear mesh frequency, but which was excited by the pinion's harmonics at 7x , 8x ,9× , 10× or 11× running speed.
ISO 6336-1:2006 has identified super synchronous vibration due to tooth mesh frequency only . When excitation frequencies (such as tooth meshing frequency and its harmonics) coincide or nearly coincide with a natural frequency of vibration of the gearing system, the resonant forced vibration may cause high dynamic tooth loading. When the magnitude of internal dynamic load at a speed involving resonance becomes large, operation near this speed should be avoided

As the super-synchronous vibration on pinion is identified, investigations are systematically carried out by comparing the uncoupled and coupled natural frequencies and their mode shapes with varying gear mesh stiffness, taking into account rotating speeds, and by comparing the strain energies of the lateral and torsional vibration modes. The results show that some modes may generate coupled lateral and torsional mode characteristics when the gear mesh stiffness increases over a certain value or a dominant mode may change from torsional to a lateral or vice versa .

The parameter excitation is normally generated the pitch error during the gear teeth manufacturing within API accepted manufacturing tolerance. So , 7x, 8x or 9x and so on could be a harmonic multiple inside the pitch error deviation around the circumference. The harmonics can be found by an FFT-analysis of the pitch error

Sample Case study -

With variable speed drive turbo compressor train, turbo gear pinion showed high super-synchronous (7x) vibration at NDE side in certain speed load range which was inside the compressor operating map. It was found that pinion 4th lateral critical speed was excited at 7 times of its operating speed. The shaft circumferential velocity was below just below 100 m/sec and was equipped with tilting pad bearing. It was taken into consideration that train 10th torsional critical is coinciding with pinion 4th critical and mode was amplified.. Changing of coupling and its overhung to shift the 10th torsional critical did not improve the 7x component of vibration.

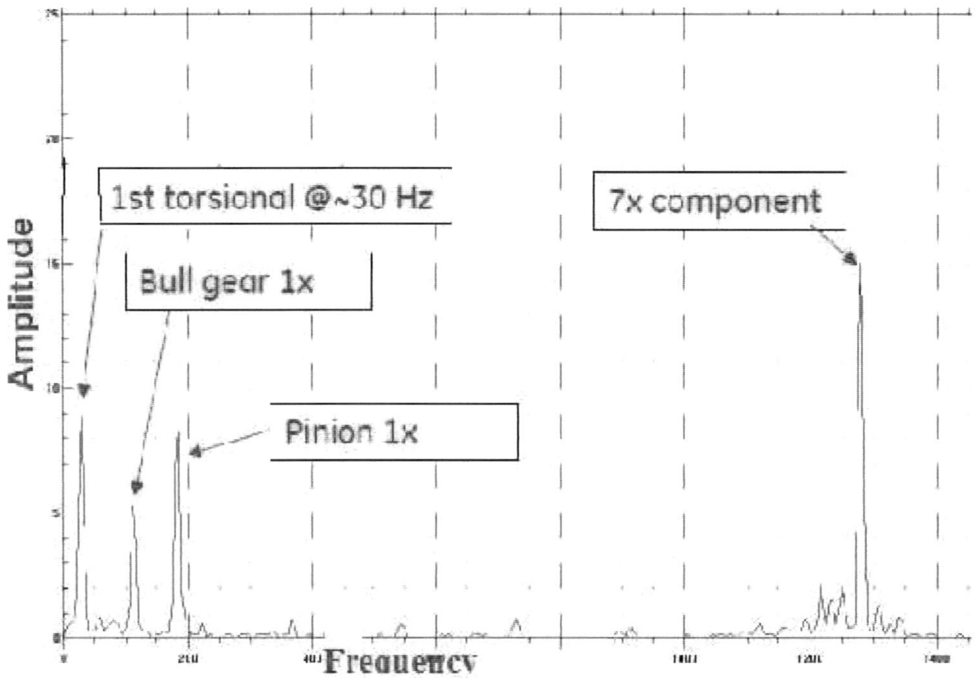

Fig 5 – Pinion super- synchronous vibration

Later on, the bearings were changed to fixed profile (lemon bore) bearings with much significant improvement, with lowered the 7X component but not below API criteria. With a revised rotor dynamic analysis of data gathered at test bed, in was observed that horizontal stiffness of pinion NDE bearing at that particular speed load range was very different than vertical. It was also seen in 4th damped eigenvalue mode shape plot that NDE bearing was at its nodal point .To shift the pinion critical speed and damp 4th mode after analysis of 4th damped eigenvalue plot, bearing damping was increased and as the 4 the mode was very near to nodal point, bearing span was changed by @12 mm.

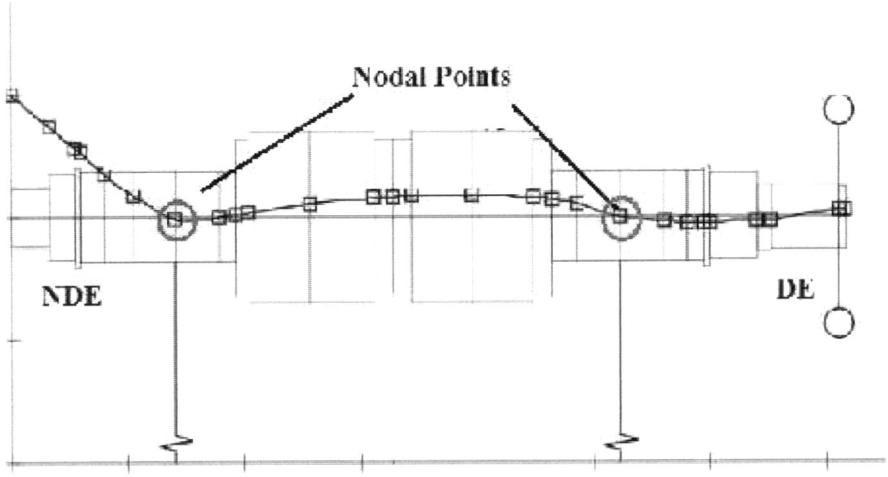

Fig 6 - Nodal points of a mode shape

Pinion vibration came down to meet API criteria to its full operating range.
In other cases, pinion was tuned where it was not possible to change the bearing type due shaft peripheral velocity and journal load (high PV) or location of nodes in mode shape [7]
After analysis of plots, pinion was over tuned by adding mass at NDE side (care was taken the mass does not exceed more that 7% of pinion weight)
In a case the pinion was under tuned by removing the mass from NDE side.

Role of torsional –lateral coupling in pinion vibration –
It is known that two dynamic systems influence each other, if connected. In the case of rotating machinery, lateral and torsional vibrations are usually analyzed separately. Such analysis is allowed in the case when the coupling effect is weak. When the coupling is strong, coupled vibrations is to be analyzed . It is to be noted that in the case of lightly damped structures, there are ranges of unstable rotor speeds. The unstable speed ranges are at frequencies for which the natural frequencies of both dynamical systems intersect with each other on the Campbell diagram.

At certain speed and load, sharp increase in vibration of pinion may be noticed but as the load and speed are increased, the vibration may be disappeared. In such case this can be concluded that a torsional Eigenmode excited by low loaded pinion bearings. When a torsional and lateral coupled system is modeled by a FE based software, it is found that at certain speed both modes exist and coupled at gear mesh point. By undertaking a stability analysis of various load and speed combination as shown below, bearing geometry and type can be changed to stabilize the pinions in its full operating regime

Transmission error (TE) is considered to be an important excitation mechanism for gear vibration during certain load/ interaction with torsional Eigen mode of train.
The causes of transmission error are -

- Cumulative pitch-line error
- Contact deformations (hertzian) in the gear mesh

- Gear teeth bending deflections
- Gear blank deflections
- Shaft deflections
- Bearing and gearbox casing flexibility

Normally practiced evaluation method.

Since the couplings with a very low lateral stiffness are used, it is understood that the rotational frequency from another component such as compressor rotor / pinion in the train will not trigger a lateral-resonant speed. Regarding the possibility of torsional and lateral critical speed affecting each other , they are drawn in different color lines denoting torsional and lateral natural frequencies are drawn in horizontal axis and possible excitation frequencies are drawn .Then they are checked for separation margins to one another

Little known resonant condition and frequencies -

Gear blank resonance- The gear blanks of high-speed, lightweight gearing may have natural frequencies within the operating speed range. If the gear blank is excited by a frequency which is close to one of its natural frequencies, the resonant deflections may cause high dynamic tooth loads.

Ghost Frequency -In addition to the tooth mesh frequency, there are sometimes "ghost" or "phantom" frequencies appear in spectrum, as a result of cyclic errors in the master worm wheel drive in the grinding machine used to finish cut the gears. These errors are imprinted onto the gear teeth as a helix undulation on each tooth and they are sometimes an important source of transmission error.
The ghost frequency usually shows up as the product of the number of teeth on the cutting mechanism and the running speed of the rotor and is caused from the imperfections of the gear hobbing tool that cut the gear at the time of manufacturing. Every gearbox may have a fault frequency set up for the ghost frequency, not to alert us of problems but to show us what that unknown frequency is.

Mitigation strategies

It is understood that all issues encountered during operation , cannot be pre- identified in a mechanical run no load complete unit test. Owing to various reasons such as -project completion schedule , work loads on OEM shop and test bed availability a full load full speed string test cannot be executed. In this case OEM's proactive approach to mitigate such type of vibratory phenomena plays a key role to successfully commission and hand over the train to end user. Below paragraphs endeavor to draw a road map for such actions .

Before discussing the mitigation strategy , it is imperative to mention that base plate of gear box should be free from any twist in all planes . If there is any shaft driven main lube oil pump , it should be supported adequately ensuring no stress on casing . Return oil drain pipe (which is often a big diameter pipe) shall be fitted without any residual stress on gear box casing and casing shall be properly dowelled after ensuring that there is no soft foot present in installation. In recent years , this issue has baffled condition monitoring and rotor-dynamic engineers to solve gear box vibration problem.

Sub synchronous vibration -Most commonly used design tools do not consider the pad inertia as a standard feature. The inherent stability and self-alignment ability are the most important advantages tilting pad journal bearings have. An essential requirement for rotor-bearing system analysis is the calculation of the oil film force. Reynolds Equation is considered to be acceptable for evaluation of oil film force based on database. The database is based on number of pads , orientation , pad angular extension and length to diameter ratio.
The swing of bearing pads must be taken into account for detailed dynamic analysis of rotor bearing system.
One of proactive mitigation strategy may be to increase oil flow than critical oil flow to load all pads and keep them in dynamic equilibrium.

Super synchronous vibration-To detect the traces of peaks at high harmonics , the Fast Fourier Transform spectrum should be accordingly set up . Normally vibration peaks up to 1500 Hz are measured and identified in no load test.
At lower load at 70% speed, as the meshing stiffness is lower, pinion can excite at higher critical mode. Hence , a detailed asynchronous response analysis to be done addressing forcing functions at 7x,8x,9x and 10 x operating speeds to determine the pinion and bearing system sensitivity to fourth and higher lateral mode as shown in fig 7a . The forcing function shall address the cumulative pitch error just below API 613 acceptance limit .Root locus plot of pinion for various loads and speeds should be drawn to see the stability of the system.

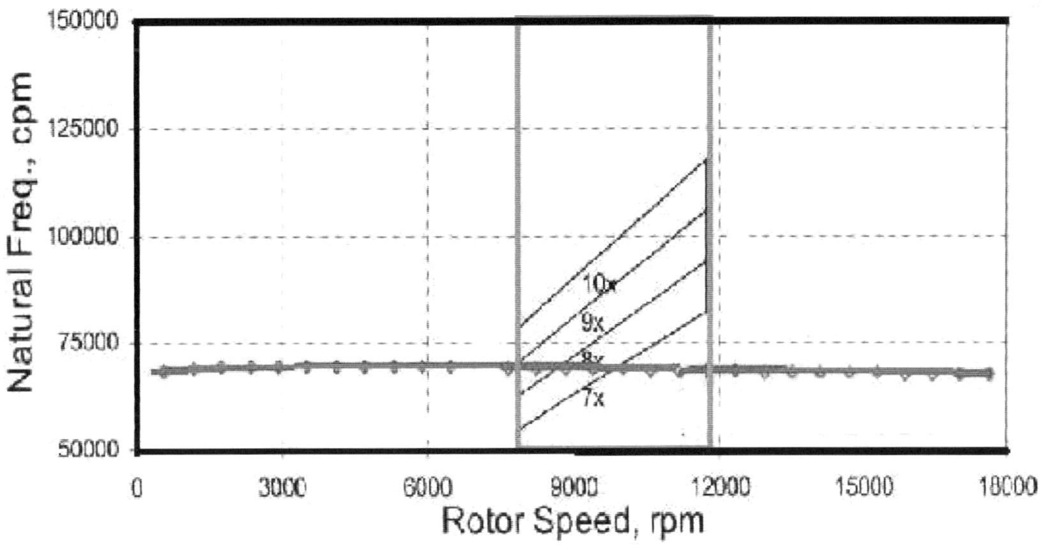

Fig 7 - Identification of corresponding harmonics and 4th lateral frequency

Root Locus plot -
Load force vectors imposed on the pinion journal bearings vary due to load variations and speed changes for a centrifugal compressor driven by variable speed drive motor or two shaft gas / steam turbine. In most of the cases the pinion load in on upward direction due to up mesh. The horizontal load vector is due to separating forces generated at mesh. In mechanical run test, load vectors of such magnitude are not experienced. The resulting bearing load vector during start up is downward and as the speed and torque is increased, it tends to go upward. So it is imperative to draw a root locus plot for shop condition and various load – speed condition.

Prediction of possible torsional coupling with a realistic rotor-dynamic analysis should be carried out by using Finite Element model of bull gear and pinion.. Bearing pad radial movements should be modeled in the analysis.FFT(Fast Fourier Transform)-analysis of the pitch error can be fruitful during no load test; a log based spectrum plot can be generated. The gear box sensitivity analysis can be done based on the pitch error with variables of meshing stiffness and torsional -lateral coupling.[6]

The complete train system is to be analyzed in order to define the optimum compromise to avoid the excitation mechanism (in speed harmonics were known), or to increase the system modal damping. Gear Box OEM along with bearing designer should keep the preliminary design of alternative geometry bearing ready if found that pinion nth lateral critical speed is excited at higher harmonics of pinion and mode can be damped.
In addition to above OEM should ensure to have wider space in bearing housing during design and fabrication so that bearing can be shifted to avoid nodal location in Eigen modes if required . Oil supply passages in gear housing must be designed to meet this modification (if required to be carried out at site). [6]

Fig 8 - Gear Box bearing housing

Identification of defect frequency if any during mechanical run test (ghost frequency) can be done and can be stamped in installation and maintenance manual .
Forecasting of tuning of pinion -Addition / removal of 7-10 % of pinion's weight at the non-drive end.

.

Proactive design approach to solve the issue at site with minimum time -
Sensitivity analysis of effect of coupling overhung on pinion Eigen frequencies and modes should be carried out as a part of OEM internal test .Before and after installing the bull gear and pinion on the casing a bump test may be carried out and same can be verified with historical data of modal parameters.
If FLFS (Full load full speed) test of complete unit is not an exercised option then proactive approach shall be very beneficial -
1. Tuning of pinion to shift the Eigenmodes is a proven solution . But this type of activity is carried out when after a prolonged and detailed study reveals the excitation of higher mode .Hence as a proactive approach OEM can design the pinion in such way that it is easy to remove weight or add weight to tune at site . The pinion should be drilled with a proper centering device up to location of NDE bearing location . It shall be then closed with a stepped plug before assembly . If the requirement of tuning the pinion is needed , this pug can be removed to shift the natural frequency of pinion.
2.Additionally , a set of two segmented ring / disc can be machined and kept as spare which can be installed during diagnostic tests if issue of super-synchronous vibration arises . The rings can be mounted as interference fit using coupling hub mounting device or induction heater used to mount antifriction bearing . The rings / discs can be joined at NDE end with fatigue proof bolts . The total mass forcasted to tune the pinion can be easily splits into two rings or discs to fine tune it. While designing the rings for mounting , it is important that these can be easily removed by hydraulic or mechanical puller device if required .

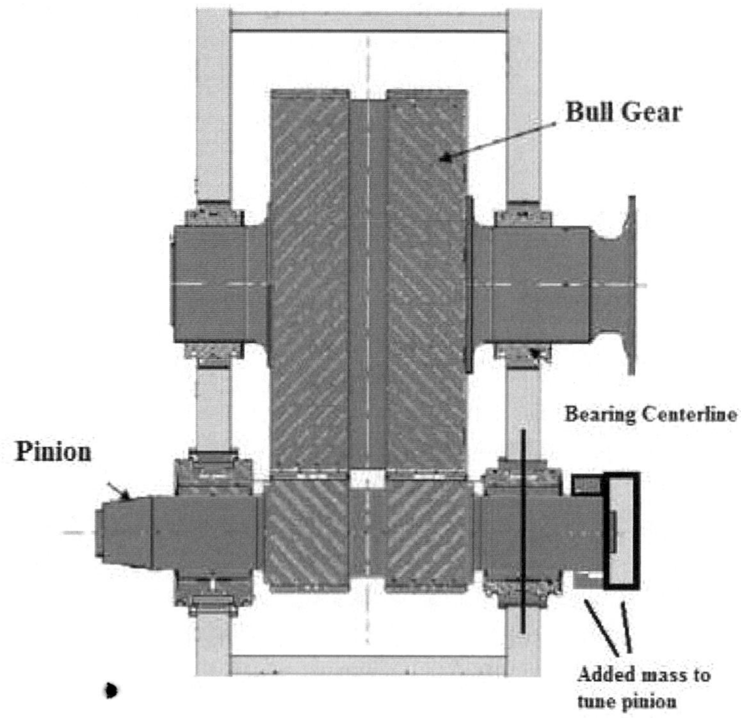

Fig -9 - Tuning of pinion by adding mass

After installing the added mass on NDE side of pinion , in-situ balancing shall be carried out .A balancing machine shall be available at site which can be hooked with machinery diagnostic system .
For both cases a loss of interference study shall be done using a FE analysis .

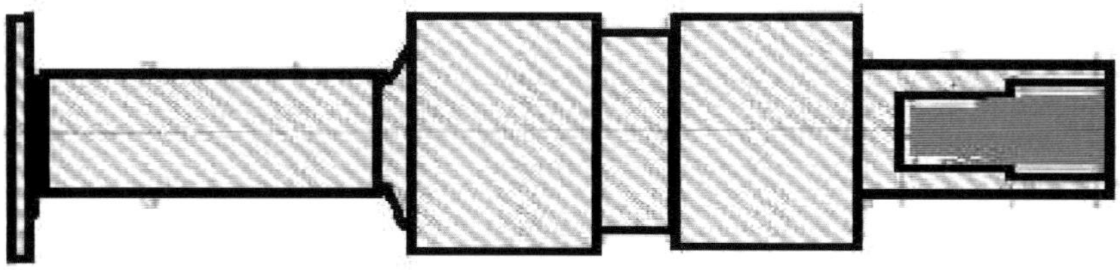

Fig10- Tuning of pinion by removing mass (temporary plug inserted)

Technical support at site -

1. Availability of personnel from OEM with a master vibration analyst (Cat III or IV).
2. Vibration analysis tool and software
3. On line availability of rotor-dynamics experts from OEM and active participation of End user to come to a quick decision .

Conclusion

From instances described above , it is evident that no load mechanical run or no load complete unit test may not accurately predict the probability of sub synchronous and super synchronous vibration . Of course the nonlinearity associated with tooth contact can give rise to instability producing super and sub harmonic responses, and even chaos, but it may not then correspond to actual operating conditions where the load is sufficient to maintain tooth contact.

Instead of focusing just on the fourth lateral vibration mode(as has been the common phenomena as reported in various technical papers), the alignment of any pinion torsional modes with lightly damped lateral ones should be avoided. [6]It is important to synthesize the bearing support dynamics , pad movement , pure lateral and pure torsional mode to create a reliable solution. Coupling modifications may become a less costly, and less time-intensive, and may be explored if the torsional mode high participation is confirmed. Plausible mitigation action suggested above should be deeply reviewed by OEM in terms of risks and rewards (saving time in not conducting a full load string test , analyzing the issue , finding ways to tune the pinion / damping the vibration at pinion lateral mode)

References -

1. API Standard 613 -Special Purpose Gear Units for Petroleum, Chemical and Gas Industry Services- Fifth edition.
2. AGMA 6011-Specification for High Speed Helical Gear Units
3. ISO (2006) Calculation of load capacity of spur and helical gears—part 1: basic principles ,introduction and general influence factors
4. The effect of starvation of the dynamic behavior of tilting pad bearings –John K. Whalen, Vaclav Cerny,Minhui He and Vaclav Polreich- 44 th Turbomachinery symposium TAMU
5. Tilting Pad Journal Bearings with Spray-Bar Blockers and By-Pass Cooling for High Speed, High Load Applications,"- Dr. J.C.Nicholas, Proceedings of the 32nd Turbomachinery Symposium, Texas A&M University, College Station, Texas, 2003.
6. Analysis of Super-Synchronous, Sub-mesh Frequency Vibrations in a Speed Increaser Gearbox- C. Hunter Cloud, James M. Byrne, Minhui He, José A. Vázquez, Timothy J. Hattenbach - Proceedings of the 9th IFToMM International Conference on Rotor Dynamics
7. Case Histories of High Pinion Vibration When Eight Times Running Speed Coincides with the Pinion's Fourth Natural Frequency- Edmund A. Memmott- CMVA/ACVM 24thMachinery Dynamics Seminar, October 2006

Bibliography –

1. Artec Technical Documents-Deformation of Teeth and Rotors, Bearings for High Speed Gears

2. Practical Uses of Advanced Rotor Dynamics Tools to Ensure Trouble-free Operation of a Gearbox -Manuel Marin, -38th Turbomachinery symposium Case study
3. Analysis of a tilting pad journal bearing to avoid pad fluttering- Hargreavesa, and. Fillon
4. Coupled bending-torsional vibration of geared rotors –Rao, J. S. / Chang, J. R. / Shiau, T. N. et al.
5. Gearbox Spectral Components Presentation- Timothy Erwin
6. Cepstrum Analysis and Gearbox Fault Diagnosis- B&K application notes-R.B. Randall
7. Dudley – Handbook of practical gear design
8. Case Study – Load Dependent Critical Speed-Mr. Gregory L. Elliott,Mr. James F. McCraw, Mr. Thomas P. Shoup
9. Analysis of Systems with Complex Gears-Frédéric Gaulard , Alexander Kosenkov , Dr. Joachim Schmied ,Delta JS
10. Dynamic analysis of a multi-mesh helical gear train-Kahraman A
11. Modal and whirling analysis of coupled lateral and torsional vibration of herringbone gear -Siyu Chen, Jinyuan Tang, Changjiang Zhou, Zehua Hu
12. Chaotic Dynamics of Rotor Supported on tilting pad journal bearings (TPJB);-Issam Abu-Mahfouz, Proceedings of WTC2005 World Tribology Congress III September 12-16, 2005, Washington, D.C.,
13. Finite Element Analysis of Flywheel Considering Interference Fit- Lijesh K.P-, Indian Institute of Technology, Delhi , India
14. Changing the Effective Mass to Control Resonance Problems-Richard Smith S&V /May 2001
15. Dispersion of critical rotational speeds of gearbox: effect of bearings stiffness-F. Mayeux, E. Rigaud, J. Perret-Liaudet
16. Simulation of the dynamic behaviour of a geared transmission on hydrodynamic journal bearings-Romain Fargere, Jean-Francois Sigrist, Raphael Bosio, Christian Menard, Philippe Velex

PERFORMANCE MONITORING OF SOCCER PLAYERS USING PHYSIOLOGICAL SIGNALS AND PREDICTIVE ANALYTICS

Sampurna Ravindranathan, Hossein Davari Ardakani, Andrew Pimental, and Jay Lee
NSF I/UCRC for Intelligent Maintenance Systems (IMS)
University of Cincinnati
560 Baldwin Hall
Cincinnati, OH 45221-0072
Telephone: (513)556-3412
hossein.ardakani@uc.edu

Robert Mangine. PT, ATC
Assistant AD, Head Athletic Trainer University of Cincinnati
Nova Care
Cincinnati Oh

Joseph F. Clark
Professor of Neurology
University of Cincinnati
MSB 7110
Cincinnati, OH 45267-0536
Telephone: (513)556-7085

Abstract: With the rapid advancement in wearable sensor technologies and predictive analytics, college and professional sports teams are facing new opportunities of leveraging such technologies and at the same time challenges of maintaining their competency. In professional sports today, the margin between winning and losing is narrow although its impact can be massive. Effective use of predictive analytics along with the implementation of advanced sensors can bring about a deeper understanding of players' physical condition and performance potential throughout individual games or the entire season, and help sports medicine personnel get the most from available resources while keeping players healthy and minimizing their risks of injury. This paper presents a predictive analytics framework for analyzing and predicting soccer players' performance data. The data consists of GPS and physiological measurements, collected from female soccer players during both practices and games using Zephyr Bioharness device. The proposed framework consists of data cleaning, filtering, visualizations and analytics modules to provide deeper insights into the data. The preprocessing modules automatically remove outliers using intelligent tools and determine first half, second half and potential overtime RISKS based on data patterns. Furthermore, comparison-based metrics have been developed to analyze the performance of players from different aspects including their activity level, fitness and consistency. For instance, Kolmogorov-Smirnov (KS) test was utilized to extract performance metrics based on players' Heart Rate and Speed, or a Neural Network-based approach was utilized to analyze the Heart Rate recovery rate of the players and quantify their recovery rate, which is important for effective play. At the end, different visualization tools were used to combine players' running patterns and speed profiles, along with various metrics. Potentially clinically relevant trends related to objective performance parameters could be observed for players during individual games and the entire season which can provide the athletic training staff with a better understanding of player's performance and inclines or declines in their performance. Future work could validate the methodology for return to play, safe play and pull from play decision tools for the coaching and training staff. The tools used for analyzing biological signals can also be expanded for other applications involving human performance monitoring.

Key words: Sports analytics; physiological signal analysis; human performance monitoring; soccer; wearable sensors;

1. Introduction

Wearable sensor technologies are facilitating the collection of high-quality data from the human body in a variety of applications including sports, workers in harsh and remote environments and military. Not only monitoring the critical physiological signals can help protect workers, soldiers or athletes, but the implementation of predictive analytics on such parameters can bring further insight and vision into the performance of individuals, which can be used for improving their performance and minimize risks associated with injuries, and physical and mental fatigue. Soccer is one the sports in which sports analytics plays a significant role and there is still a substantial opportunity for implementing predictive analytics for improving the outcome. A great example of such implementation is Leicester City, which is one of the most advanced in the English premier league soccer to use data analytics, and has been using wearable technology and analytic tools in the past 10 years [1]. There is a growing trend in professional soccer towards using metrics and analytics.

Despite the development of analytic tools for analyzing physiological signals, performance analysis in Soccer has been done with mostly descriptive game statistics like the number of goals, passes, shots [2, 3] or event-based data [4]. For example, the uniqueness and consistency in player in-game movements is quantified to identify potential replacements to a given player [4]. Ball pass origins, destinations and spatiotemporal data has been used extensively to develop a conceptual field map that help predict the likelihood of the passes leading to shots [5]. Another dimension to this analysis is to find a relation between a player's physiological performance and on field performance. The fitness and conditioning of players is of great importance during a game as well as for the duration of the season. A comparison of professional and amateur players indicates that elite players have higher measures of maximum volume of oxygen (VO_2) and lower heart rates [6]. Other works, indicate the significance of heart rate (HR) along with VO_2 levels, vertical jump height, weight etc., to classify players based on those parameters as well as their proficiency and field position [7,8,9]. Apart from raw

HR, monitoring heart rate recovery (HRR) after a sub maximal running test has been used to predict changes in physical performance of players [10].

Various studies have briefly analyzed and established the importance of parameters like HR, HRR, speed profiles, player running patters and fatigue levels during a game. It is challenging to incorporate all these measures to quantify or monitor players' performance. Our work proposes a novel approach to both visualize and quantify these measures to assist sports medicine personnel to make decisions.

2. Data

The data was collected using Zephyr Bioharness from 21 players of a soccer team during both practices and games for an entire season. The key variables used in this study were GPS and heart rate. The data was collected at the sampling rate of 1 Hz. Table 1 shows a complete list of variables used in this study.

Table 1. List of variables used in the study with units

Parameters	Units
Time	Seconds
Heart Rate	Beats per minute
Breath Rate	Breaths per minute
Speed	Mile per hour
Peak Acceleration	G Force
GPS (Latitude, Longitude Co-ordinates)	Degrees Minutes Seconds

3. Methods

The methodology for analyzing the data was developed with the oversight of the sports medical staff to ensure un-encumbered performance during games and practice, to ensure the applicability of such analysis. The methodology consisted of data preprocessing, player activity profile extraction and visualization, similarity-based player performance metric development, and heart rate recovery modeling. This section briefly explains the methodologies used for this study.

3.1. Data Preprocessing: The data collect from bio sensors tend to have large amounts of noisy data. This noise comes from down time, players going out of range for the sensors, local EM parameters etc. The players also typically wear the sensors well before the game begins and when they aren't actively playing on field. To ensure that such extraneous data doesn't bias the models we developed preprocessing methods to account for the non-game noise. This process consisted of removing outliers and out-of-field GPS samples. A simple k-means algorithm was implemented to estimate game start and end times of each game based on player speed profiles and time.

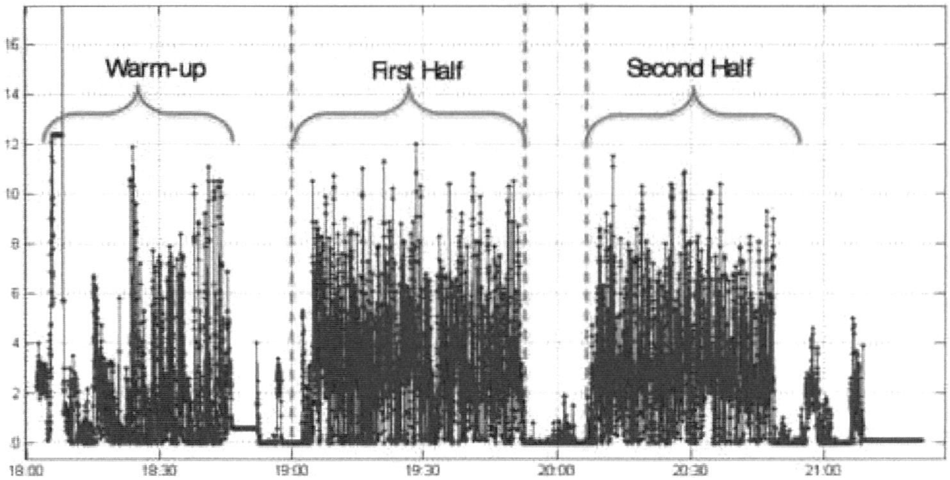

Figure 1. An example of a player's speed profile during one game; horizontal axis shows the time and vertical axes shows the speed in miles per hour

3.2. Player Activity Profile: Players' activity profiles were established based on analyzing the GPS measurements from two different perspectives: their speed profiles and its consistency throughout the game, and from game to game, and the movement vectors and movement patterns during the game. This was further extended to observe the relationship between player HR and distance profiles. Speed and HR are two indicators of how active a player is on the field and how easily they get fatigued. Figure 2 shows the movement pattern during one half of a game with intensity being represented by the color gradient white (low speed) to red (high speed).

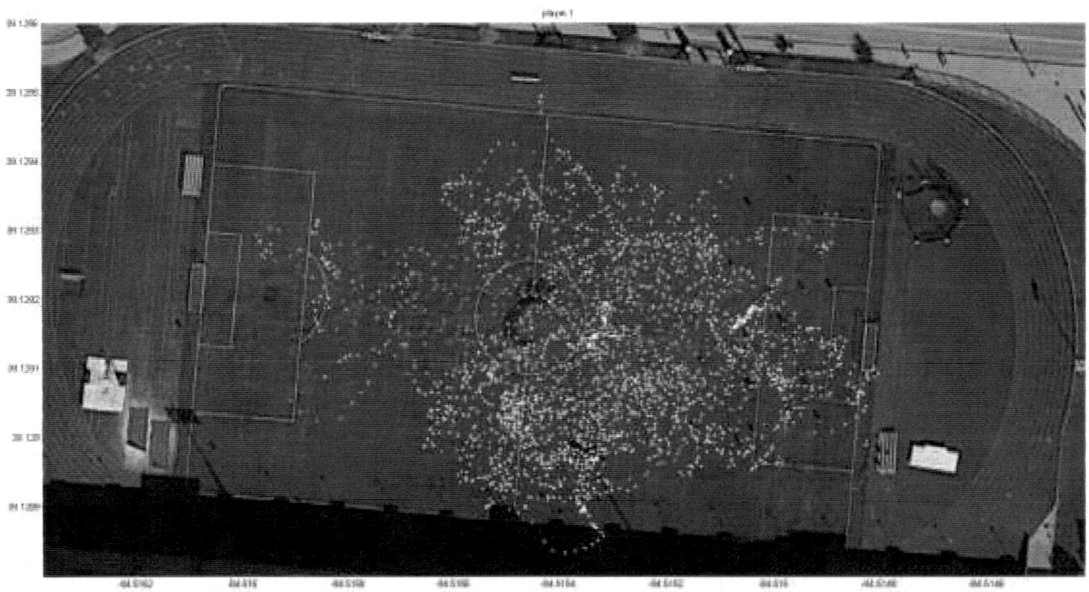

Figure 2. An example of a player's movement pattern during one half of a game; the color of each point represents the speed of the player at that point, red being the highest speed and white the lowest.

To quantify these speed data per game over the season, speed distributions were computed at 10-minute time windows during the game as shown in Figure 3. Three different speed zones were identified including 0-4 mph (low), 4-8 mph (medium) and 8-16 (high), and raw speeds were categorized into respective zones based on their values. The relationship between a players' percentage of maximum HR (%HRmax) exerted and distance traversed during a game has been visualized as shown in Figure 4.

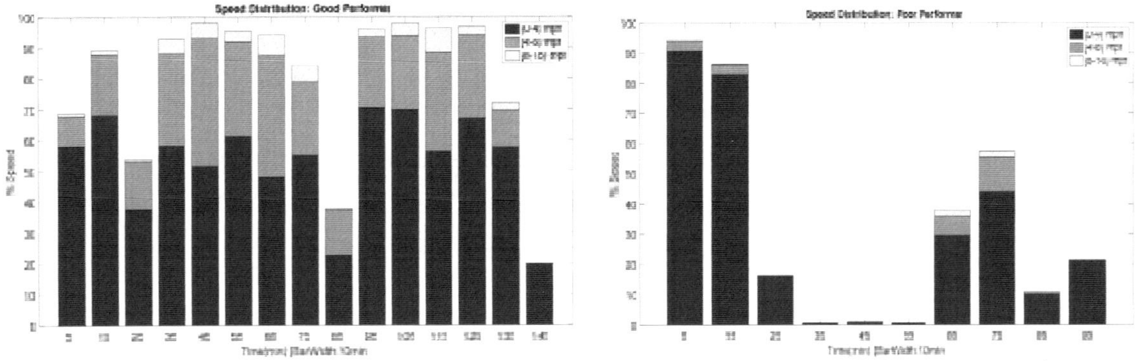

Figure 3. An example of a player's movement pattern during one half of a game; the color of each point represents the speed of the player at that point, red being the highest speed and white the lowest.

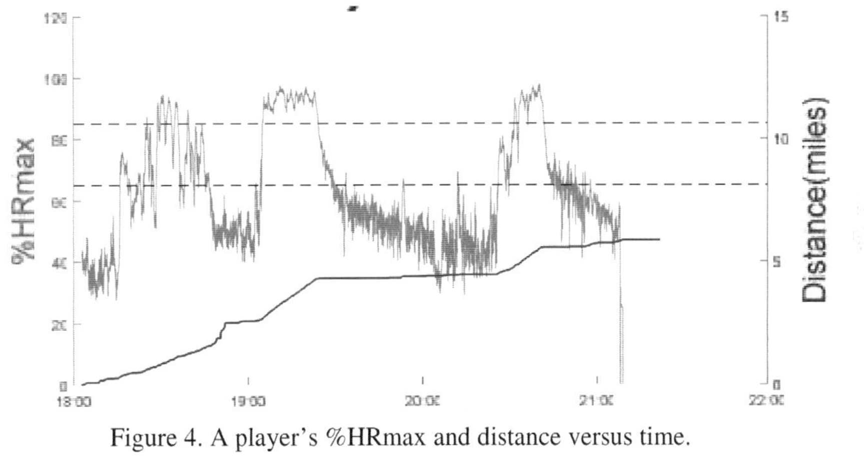

Figure 4. A player's %HRmax and distance versus time.

3.3. Similarity-based Player Performance Metric Development

The similarity-based performance metric development is based on comparing the measured Heart Rate and Speed parameters during the games. The ratio of Heart Rate to Speed (HR/S) at each instance was calculated. Kolmogorov-Smirnov (KS) test was used to benchmark each player's performance at a desired period of time with the rest of the data (or a specific player to benchmark against). KS test, explained in [11], utilizes the Cumulative Distribution Function (CDF) to determine whether a target distribution is statistically similar to a reference distribution. In this case, the data from the rest of the players is considered as the reference distribution. In order to quantify the similarity (or difference) between the two distributions, a distance metric D between the two distribution is defined. Mathematically, it is defined as:

$$D = \underset{-\infty < x < \infty}{m} |T(x) - R(x)| \qquad (1)$$

where $T(x)$ is the target distribution and $R(x)$ is the reference distribution. As mentioned, the parameter HR/S was considered as the input for calculating the metric. Figure 5 shows an example of the distribution functions during 11 games for a player with higher performance than the team average. As it can be observed from Figure 5, the ECDF for Player 21 is at the left side of the ECDF of the rest of the team, meaning that this player has a lower heart rate compared to the rest of the team. Figure 6 shows the metric for HR/S which is equivalent to the maximum distance between the two distributions in each of the subplots in Figure 5. The performance consistency of Player 21 can be observed from the metric values. Figures 7 and 8 show the ECDF's and the HR/S metric for another player with less performance consistency.

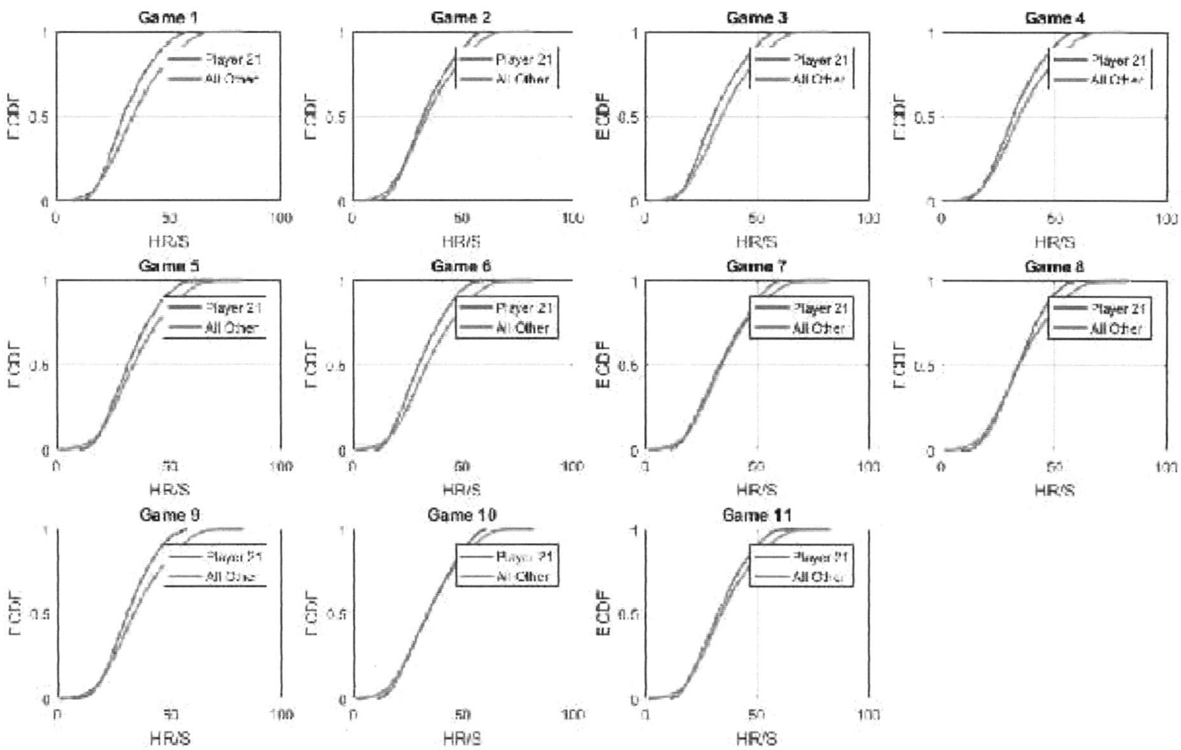

Figure 5. Empirical Cumulative Distribution Function of Heart Rate / Speed metric for a player during 11 games.

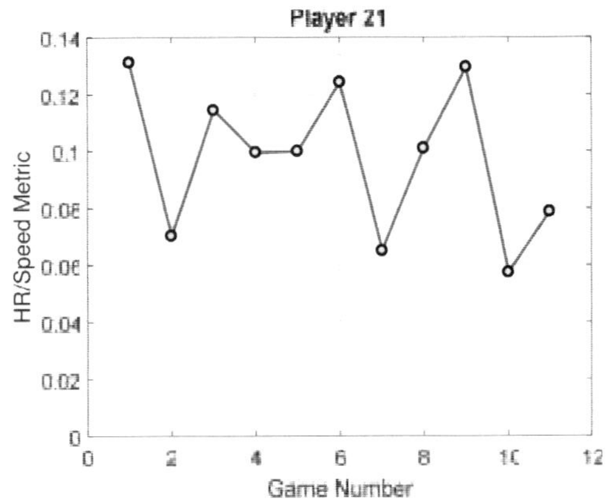

Figure 6. Similarity-based Heart Rate / Speed metric for a player during 11 games.

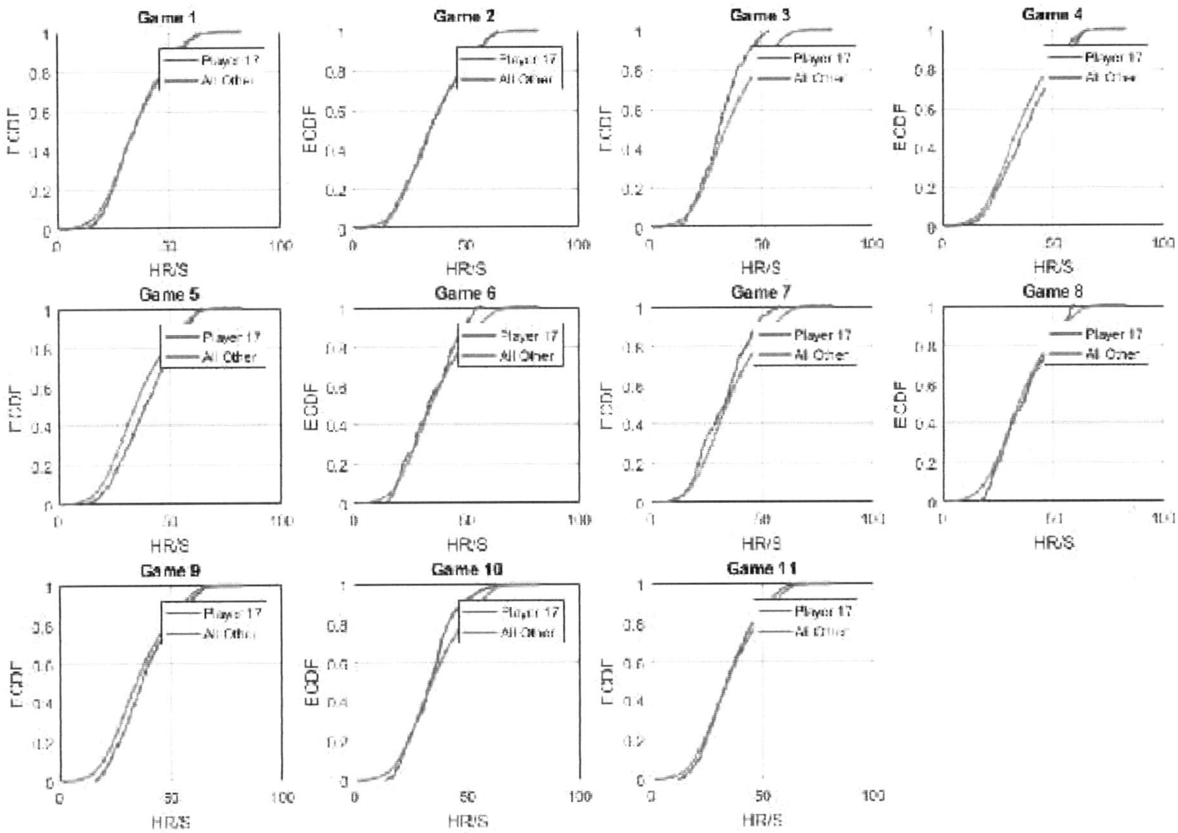

Figure 7. Empirical Cumulative Distribution Function of Heart Rate / Speed metric for a player during 11 games.

3.4. Heart Rate Recovery Modeling

The heart rate recovery model uses a neural network to understand how players compare under similar levels of strain in respect to their heart rate recovery. Through examining how player's actual data compares to a modeled player, a set of scores was then acquired. Figure 9 gives a description of the model detailed below.

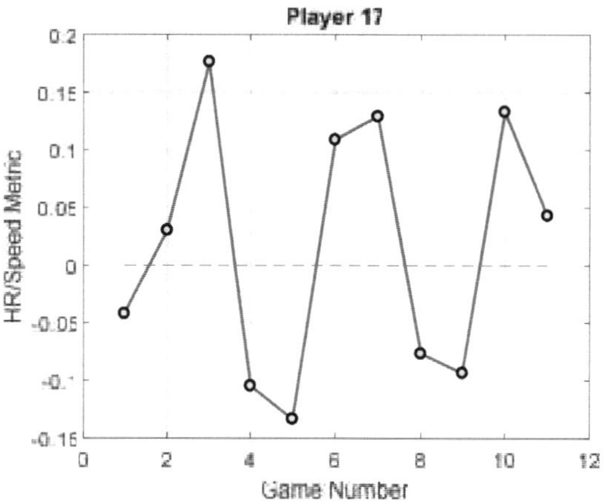

Figure 8. Similarity-based Heart Rate / Speed metric for a player during 11 games.

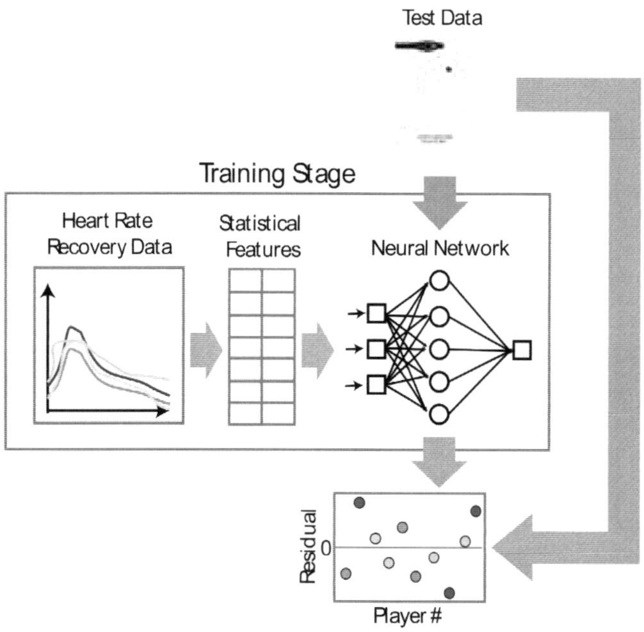

Figure 9. Heart Rate Recovery Model.

Subsets of preprocessed data were extracted to represent the moments in which players are in a state of recovery after a high speed moment. To do this, the speed data for a single player at a time was first grouped into sections of continuous time to remove issues with improper data acquisition. Then, for each section of a significant length, the data was passed into a windowed mean algorithm with a window size of 10 to smooth the data to show the pattern of play for the player. These windowed mean speeds were then passed into a filter that set all values less than a limit of 1.5 mile per hour to zero. This then created a set of speed spikes in which the players exhibited a higher speed for an amount time that then dropped below the threshold. The locations of these speed spikes were then recorded at the end of the spike to either the next speed spike or 150 seconds later depending on whether there was a nearby speed spike after the current one or not. This Heart Rate Recovery data was used to generate the statistical featured for the neural network.

The features fed into the neural network were the current heart rate, the mean speeds for 5-50 seconds at 5 second intervals, the current speed, the current distance traveled since the start of the game, the time since the start of the half on the field, the current activity and the respiration rate data. The most influential features in predicting the target value were mean speeds for the last 50 second and the last 5 seconds. The target value of the neural network is the change in heart rate per a second of time.

To train the neural network, the players labeled by the coach as top performers and bottom performers were used. Then, the final score was found by then examining the difference between the predicted values for the change in heart rate over a second compared to the actual change in heart rate over a second of time. This further allowed examining the rate of recovery of the players compared to a shown good player.

4. Results and Discussion

Data preprocessing: The data processing section was developed to be autonomous and versatile for different situations and data settings. It was established based on common sense and occasional errors observed in the recorded data. The beginning and end of the game, half time etc. were also determined by means of a k-means clustering based approach and proved to work successfully by considering the speed profile of all the players.

Metric development: Metrics can be very informative and indicative for coaches and training staff to get a better understanding of each individual player and the roster as a whole, along with increases or decreases in their abilities, strength and readiness. For this study, two different approaches are introduced to produce metrics mainly based on heart rate and speed. Although the metrics need improvements and standardization, they provided helpful insight for the training staff in terms of trends observed throughout the season, and relative performance of each player compared to the rest of the roster. In a number of players, a consistent trend was observed in HR/S metric throughout the season indicating either improvement or decline in performance. This metric is currently showing promise in terms of quantifying players' performance but as part of the future work, it needs to be standardized and validated using cases where the performance and readiness level of a player is known i.e. identified by the coaching and training staff for validation purposes.

Overall, distribution comparison techniques provide a simple but powerful tool to quantify the similarities and differences between players' physiological signature and performance.

A neural network-based approach was also proposed to develop a data-driven model which learns based on the observed heart rate recovery rates. In sports medicine, rapid heart rate recovery is considered a sign of good conditioning and lack of fatigue, so this parameter could be important or helpful to the sports medical professional. Further, fatigue in the fit individual is thought to be associated with an increased risk of injury, therefore this modeling system could be used to keep athletes safe by removing them from play and reduce risk of injury. The model was developed to predict the heart rate recovery patterns based on the given heart rate values from prior samples. Ideally, the model would need to be trained based on data from a collection of players who are fit and in their best condition so that the model can give each individual player a score based on comparing them to the best.

Visualization: In sports predictive analytics, data visualization is key. The data and the information extracted from it need to be visualized in a way that makes it easier for the coach and training staff to make informed decisions. This may be more significant when coaches monitor real-time data and analytics to make decisions on substitutions or change tactics. In this study, the visualizations above were developed based on regular feedback from coaching staff, to provide the information in the most convenient way. For example, Visualizing the players' movement patterns during the game (Figure 2) helped in analyzing a player's positions during the game along with improving cardiovascular performance in training. Figure 3, captures the difference in speed profiles of a Good and Poor performer. It is observed that a Good performer has spent greater duration in mid/ high speed zones compared to a player with lower performance. In Figure 4 a strong correlation between distance traversed and %HRmax is observed, also good performers do not cross 80-90 %HRmax typically. The figure also helps in identifying warm-up, game halves and half-time zones. High activity zones are when %HRmax and distance increase.

Players' positions: As each position in the soccer field requires specific sets of skills and physical capabilities, considering the position of each player can significantly improve the outcome of the analysis and make the metrics more informative.

5. Conclusion and Future Work

As the wearable sensor technology is helping us capture more data from the field, there is a great opportunity for sports teams to develop advanced Predictive Analytics to monitor the performance of athletes, optimize their training programs for maximized physical readiness while minimizing the injury risks, and win more games. This work provides a collection of simple but effective tools for analyzing physiological and game data, by integrating the knowledge and experience of University of Cincinnati's Sports Medicine team, along with the experience of researchers at Center for Intelligent Maintenance Systems (IMS) at the University of Cincinnati in developing predictive analytic tools. The future work consists of enhancing the analysis and more particularly metrics, and adding more tools to this collection for analyzing other variables being measured. Moreover, as next step, these algorithms will be integrated into a web-based predictive analytics platform that provides the coaching team, training staff and the athletes' access to the results of the analysis at any time for making optimum decisions.

6. References

[1] Foxy Leicester City FC won Premiership with Data Analytics, Simon Creasey 16 May 2016 10:45. http://www.computerweekly.com/news/450296302/Foxy-Leicester-City-FC-won-Premiership-with-data-analytics

[2] Anderson, Chris, and David Sally. The numbers game: why everything you know about football is wrong. Penguin UK, 2013.

[3] Duch, Jordi, Joshua S. Waitzman, and Luís A. Nunes Amaral. "Quantifying the performance of individual players in a team activity." PloS one 5.6 (2010): e10937.

[4] Gyarmati, László, and Mohamed Hefeeda. "Analyzing In-Game Movements of Soccer Players at Scale." arXiv preprint arXiv:1603.05583 (2016).

[5] Brooks, Joel, Matthew Kerr, and John Guttag. "Developing a Data-Driven Player Ranking in Soccer Using Predictive Model Weights." Proceedings of the 22nd ACM SIGKDD International Conference on Knowledge Discovery and Data Mining. ACM, 2016.

[6] OSTOJIC, SERGEJ M. "Elite and nonelite soccer players: preseasonal physical and physiological characteristics." Research in Sports Medicine 12.2 (2004): 143-150.

[7] Little, Thomas, and Alun G. Williams. "Measures of exercise intensity during soccer training drills with professional soccer players." The Journal of Strength & Conditioning Research 21.2 (2007): 367-371.

[8] Ostojić, Sergej M. "Physical and physiological characteristics of elite Serbian soccer players." Facta universitatis-series: Physical Education and Sport 1.7 (2000): 23-29.

[9] Sporis, Goran, et al. "Fitness profiling in soccer: physical and physiologic characteristics of elite players." The Journal of Strength & Conditioning Research 23.7 (2009): 1947-1953.

[10] Buchheit, M., et al. "Monitoring changes in physical performance with heart rate measures in young soccer players." European journal of applied physiology 112.2 (2012): 711-723.

[11] Cong, Feiyun, Jin Chen, and Yuna Pan. "Kolmogorov-Smirnov test for rolling bearing performance degradation assessment and prognosis." Journal of vibration and control 17.9 (2011): 1337-1347.

PHM/IVHM: CHECKPOINT, RESTART, AND OTHER DESIGN CONSIDERATIONS

James P. Hofmeister, Douglas L. Goodman, and Ferenc Szidarovszky
Ridgetop Group, Inc.
3580 West Ina Road
Suite 200
Tucson, AZ 85741
Telephone: (520)-742-3300 x107
hoffy@ridgetopgroup.com

Abstract: An often overlooked set of services associated with Prognostics Health Monitoring and/or Management (PHM)/Integrated Vehicle Health Management (IVHM) are those associated with Checkpoint and Restart (Save and Restore), which is necessary to save and restore operational states. The design of a framework for a PHM system includes consideration of services and support for resource management, such as the following: (1) "What are the considerations for checkpoint/restart?" (2) "How often should nodes be sampled?" (3) "What are the accuracy and precision requirements? (4) "What are the prognostic distance and horizon requirements?" And (5) "How much noise filtering and mitigation is needed to meet requirements?" The answers to these questions and others are especially important for a PHM system using condition-based data (CBD) to support condition-based maintenance (CBM) solutions. In this paper we present concepts and considerations to provide the reader with basic tools and knowledge as a basis to design the framework of a PHM system.[1]

Key words: PHM systems; framework for checkpoint (save) and restart (restore); sampling rate; remaining useful life; prognostic accuracy and precision; prognostic distance; prognostic horizon; noise filtering and mitigation.

Introduction: An often overlooked set of services associated with Prognostics Health Monitoring (Management) are those associated with Checkpoint (Save) and Restart (Restore). We often spend a great deal of time in the design and selection of approaches and algorithms: examples include model-driven versus data-driven versus hybrid; physics of failure and reliability modeling versus statistical modeling versus; and distributions, acceleration factors, and likelihood ratio test (LRT), maximum likelihood estimation (MLE) test, and mean square error (MSE) [1] [2].

We also need to consider how to handle interrupted and resumed operation of subordinate systems, assemblies, and components: (1) "what are the considerations for checkpoint/restart?" Other considerations include, for example, the following: (2) "How

---

often should nodes be sampled?" (3) "What are the accuracy and precision requirements? (4) "What are the prognostic distance and horizon requirements?" And (5) "How much noise filtering and mitigation is needed to meet requirements? The answers to these questions and others are especially important for a PHM system using condition-based data (CBD) for condition-based maintenance (CBM) solutions. This paper uses a hypothetical situation to illustrate checkpoint/restart and other considerations to design a complex PHM system [3].

Hypothetical Example: You are asked to prognostic enable the solenoid to start a turbojet engine; you are informed it has been determined the failure mode is a gradual increase in coil resistance; and you are informed it is not permissible to make direct resistance or impedance measurements – the starter solenoid cannot be removed unless it is replaced.

The customer specifications for the prognostic system include the following: (1) failure is defined to occur when the nominal resistance of the coil of the solenoid changes by 5%; (2) failure must be detected at least 72 flight hours prior to failure; (3) prediction estimates shall have a one-hour precision; and (4) the relative accuracy (α) requirements are (A) within 25% at or before 75% remaining health, (B) within 10% at or before 50% remaining health, and (C) within 5% at or before 25% remaining health.

Reliability Information: You determine (after research, discussions and meetings) some starter solenoids begin to degrade after as little as 200 flight hours, and by 1,000 flight hours all have begun to degrade. After onset of degradation: MTTF (Mean Time to Failure) is 1,200 hours, the smallest TTF is 400 hours, and the largest TTF is 2,700 hours – see the SoH plots in Figure 1. After due consideration, you decide to use CBD to support Condition-based Maintenance (CBM) for the PHM system because you believe reliability-based modeling, usage-base modeling, and/or data-driven methods and approaches will not meet the accuracy and precision requirements of the customer.

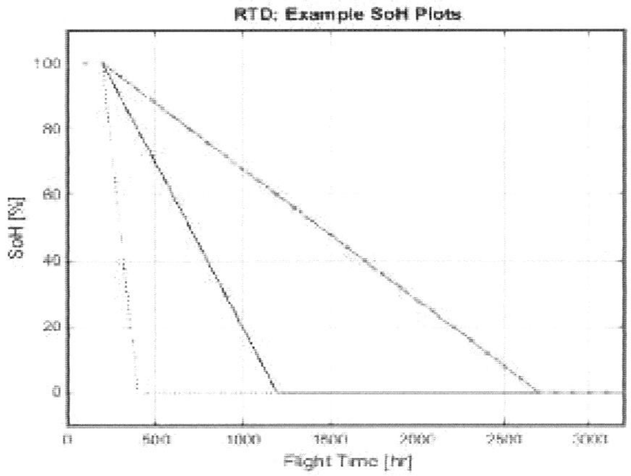

Figure 1: Example of TTF Values Relative to the Time of Onset of Degradation.

Design Approach: You choose to use the design approach shown in Figure 2, in which the framework of the PHM system comprises the following: sensor, feature-vector, prediction, and fault-management [4], which is not shown because you decide to defer that part of the design.

Control and Data Flow: The approach leads you to design a generic PHM system having a data- and control-flow architecture similar to that shown in Figure 3 (after [4]). Data is collected by one or more sensors in a Sensor Framework from one or more monitored nodes; the data comprises one or more leading indicators of failure from which Feature Data (FD) such as voltage, current, and temperature is extracted and input to a Feature Data Framework. Specialized processing such as data fusion, data transforms, and domain transforms are used to produce a Degradation Progression Signature (DPS) and then a Functional Failure Signature (FFS). Each data point in an FFS is passed to a Prediction Framework to produce prognostic information in the form of estimates of State-of-Health (SoH), Remaining Useful Life (RUL), and Prognostic Horizon (PH). Because of the precision and accuracy specifications, you select a sampling rate of twice per hour.

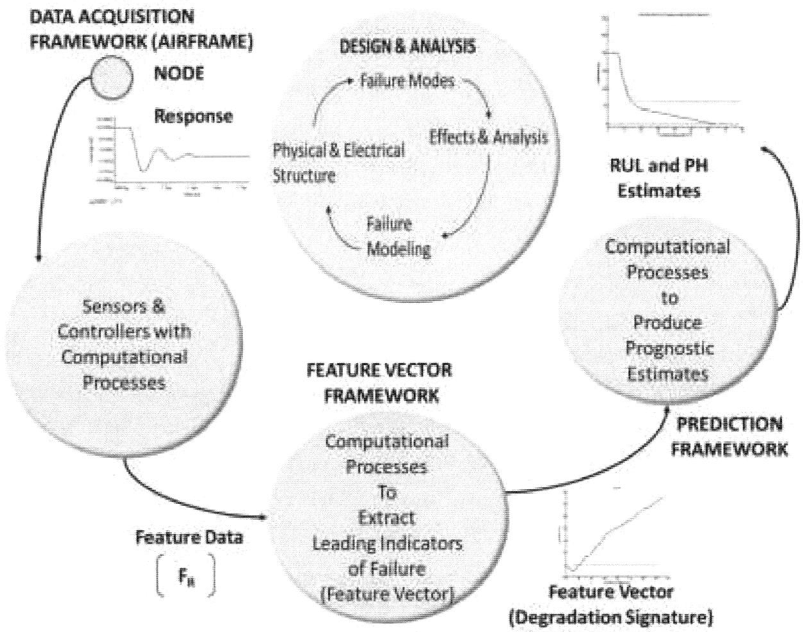

Figure 2: Diagram of an Approach to Designing a PHM Framework.

Prognostic Information: You present your high-level design to customer engineers and you realize those engineers are expecting you to present them with a solution based on their understanding of reliability: Reliability $R(t)$ of a prognostic target is the probability that the prognostic target will operate satisfactorily for a required period of time, and reliability is related to lifetime λ and mean time between failures (MTBF) by the following [5]:

$M = (NT)/F$ Where N = number of test units, T = test time, and F = number of test failures (1)

$\lambda = 1/M$ Lifetime (2)

$R(t) = e^{(-t)} = e^{(-t/M)}$ Reliability (3)

Performance Metrics: Because your customer is aware of certain performance metrics as defined by NASA [6], you need to explain your PHM system does not use reliability-based modeling; and your need to define and/or explain your prognostic terminology. In addition to RUL and SoH, your terminology includes the following (see Table 1):

Prognostic Horizon (PH) = Estimated EOL = T_S + RUL (4)

Relative Accuracy (α–λ) = (PH / TEOL) (5)

RULα = Estimated TTF - Tα = RUL at Tα (6)

Prognostic Efficiency (χ) = (RULα / PD) (7)

Where T_S = time when data is sampled

Tα = time when estimated EOL is within specified relative accuracy

PD = prognostic distance = PH when degradation is first detected

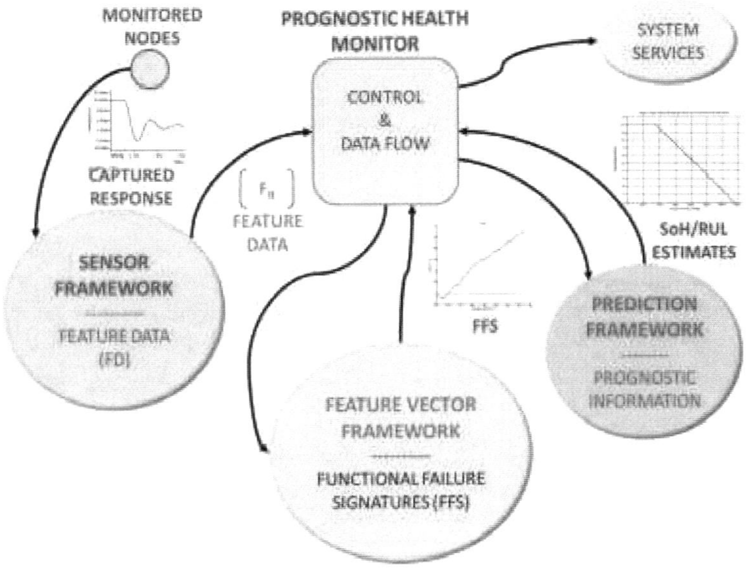

Figure 3: Diagram of a Generic PHM System.

Table 1: Comparison of Terminology and Definitions

Term	NASA Definition	PHM System Definition
EOL	End-of-Life, time of failure	Same
Prognostic Horizon (PH)	EOL – time when RUL estimates are within α accuracy	T_S + RUL
Error Margin (α)	Percent	Same
Relative Accuracy (α–λ)	(α–λ) = (estimated RUL) / (EOL – time of data sample)	(α–λ) = (PH / TEOL)
Prognostic Distance (PD)	Not defined	PD = EOL – time when degradation first detected
Convergence Efficiency (χ)	Not defined	χ = (RULα / PD) When RUL within α accuracy

Rationale for Terminology: The original performance metrics are deficient in two significant ways: initial estimate error and near EOL errors. At the onset of degradation, a

PHM system must make an initial RUL estimate: in our example, a reasonable estimate would be an RUL of 1,200 hours (mean TTF). For a solenoid fails in 400 or 2,700 hours, there is an initial estimate error: an underestimation error of 67% and an overestimation of error of 125%. The NASA-defined relative accuracy results in near EOL errors when estimates occur within a few sampling periods of failure as shown in Figure 4, which is explained as follows: "it may be prudent to evaluate PH on an error band that is limited in extent on the time x-axis by the time instant t_{EoUP}, which denotes the End-of-Useful-Predictions (EoUP), ... these predictions are of little or no use practically." [6]

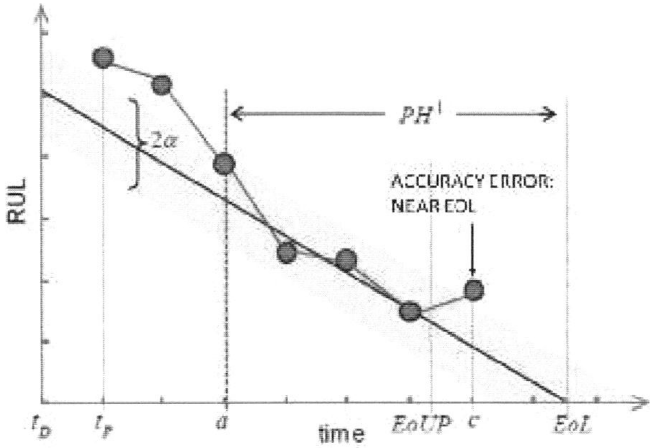

Figure 4: Example of Accuracy Error Near EOL (from [6]).

It is preferable to use a relative accuracy definition that is meaningful, is correct, and does not exhibit such "anomalous" behavior.

Examples of Data and Metrics: To simplify processing, data amplitudes should be normalized by dividing by a nominal value ($D_{NORMALIZED} = D_{MEASURED}/D_{NOMINAL}$) as exemplified by Figure 5. Data should also be made relative to the time when degradation is detected (see Figure 6). The plots in Figure 7 are for ideal PH, ideal RUL, ideal SoH, and ideal PH accuracy; the plots in Figure 8 are examples of initial-estimate errors; and Figure 9 shows examples of non-ideality in FFS transfer curves: signal conditioning such as noise filtering and mitigation is necessary to sufficiently reduce non-ideality. [7]

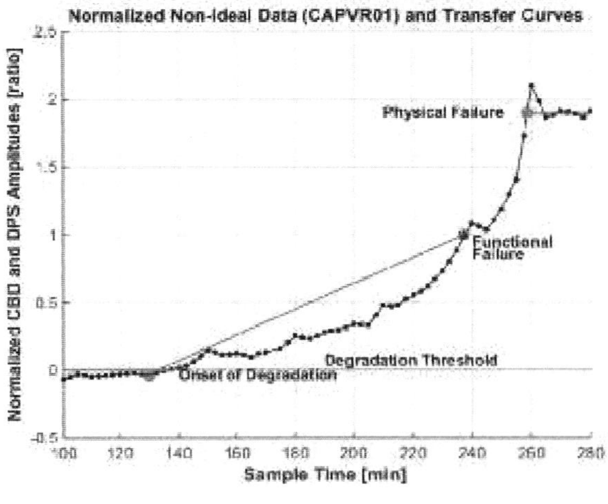

Figure 5: Example of Normalized and Transformed CBD.

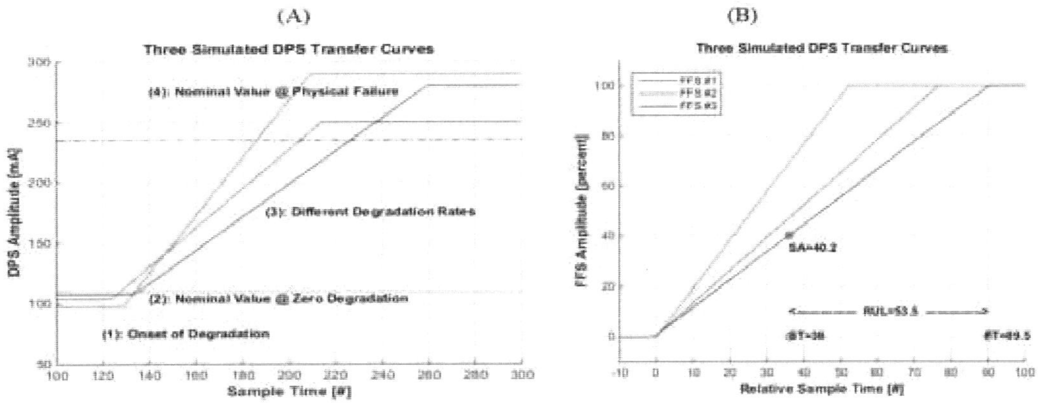

Figure 6: Example of DPS (A) and FFS (B) Transfer Curves.

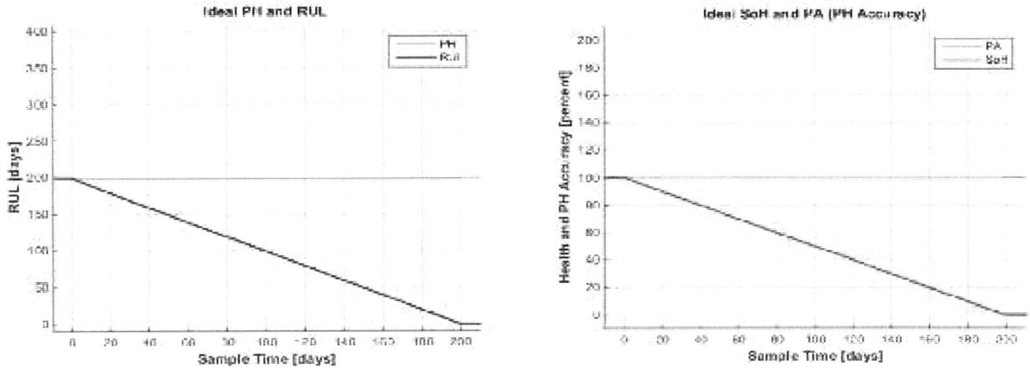

Figure 7: Examples of Ideal Prognostic Information – PH, RUL, SoH, and PH Accuracy.

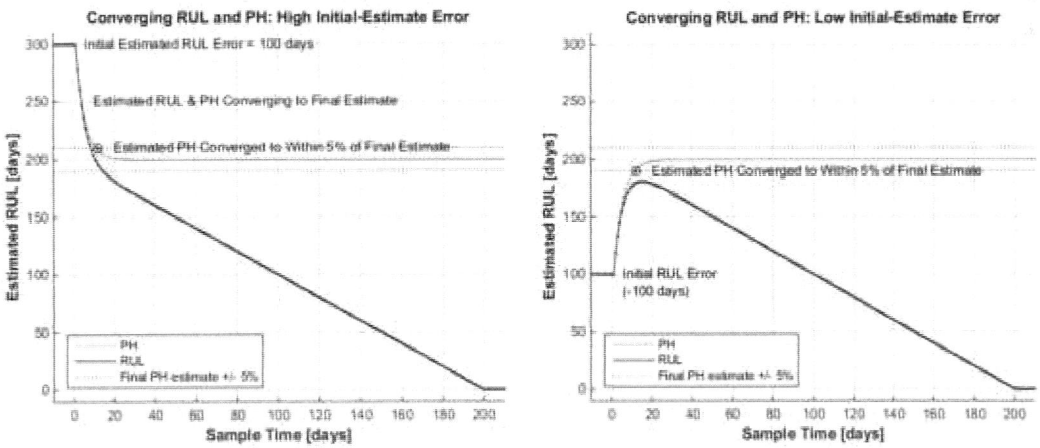

Figure 8: Example of Initial-Estimate Errors and Convergence

FSS Non-ideality: An ideal FSS, straight line, is not achievable in practice due primarily two causes: (1) a change in the failure mode and (2) distortion due to noise which is defined as any variability in CBD not caused by degradation. The first could be addressed by employing a multi-function DPS transform – but experimentation reveals such complexity is not needed. The latter is addressed by employing sufficient noise and distortion filtering and mitigation: see Figure 9. [7]

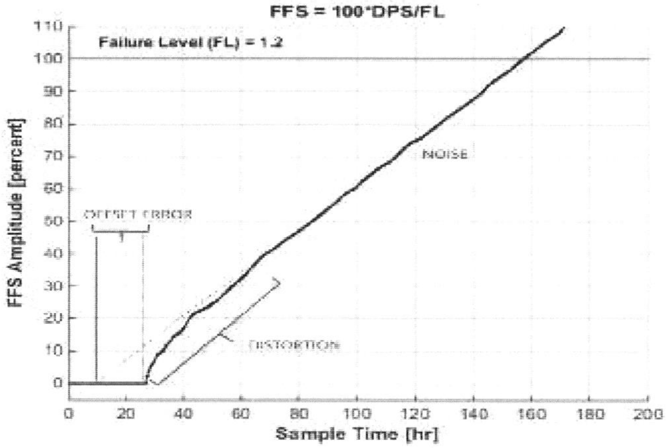

Figure 9: Examples of Non-ideality in FFS Transfer Curves

Experimental FADEC Test Data: You are provided with simulated measurement data obtained from a test bed for a Full Authority Digital Engine Controller (FADEC): see Figure 10. The PHM system performs data conditioning, data fusion, and data transforms with results as shown in Figure 11 and Figure 12 (A).[7]

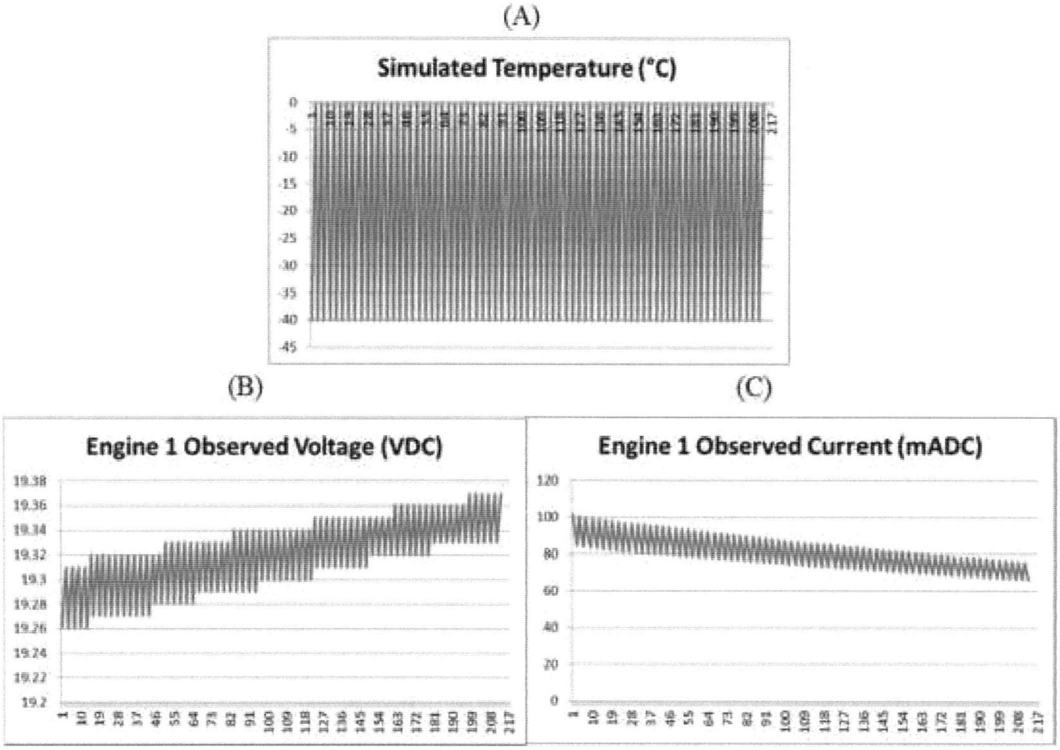

Figure 10: Starter Solenoid Measurements - Temperature, Voltage, and Current.

Prognostic Information – Results: The FFS data is input to your Prognostic Framework with results shown in Figure 12 (B) and (C). The performance is evaluated as "far exceeds requirements" as follows:

- Prognostic Horizon: within 25% at SOH = 99.5%
- Prognostic Horizon: within 10% at SOH = 98.8%
- Prognostic Horizon: < 5% at SOH = 79.4%

(A) (B)

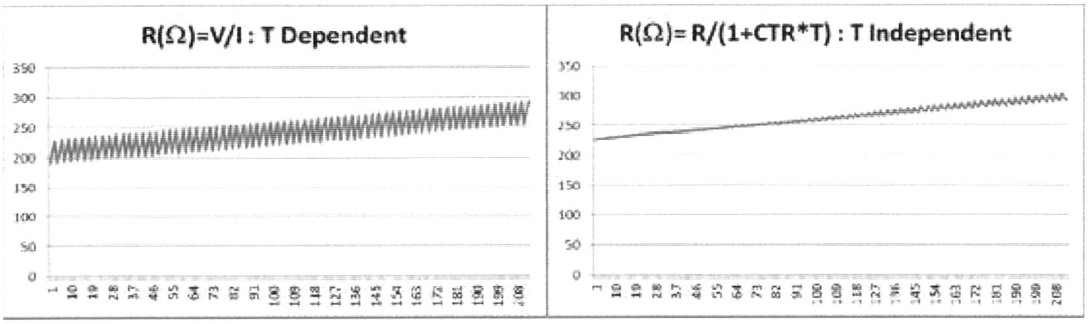

Figure 11: Test Data After Fusing (A) and Transforming (B)

Alpha Test – Revised PHM System: The PHM system is alpha tested and fails because the customer powered off your PHM system between simulated flights: no provisions were made for Check Point/Restart of the system. A redesigned PHM system is necessary: see in Figure 13.

(A)

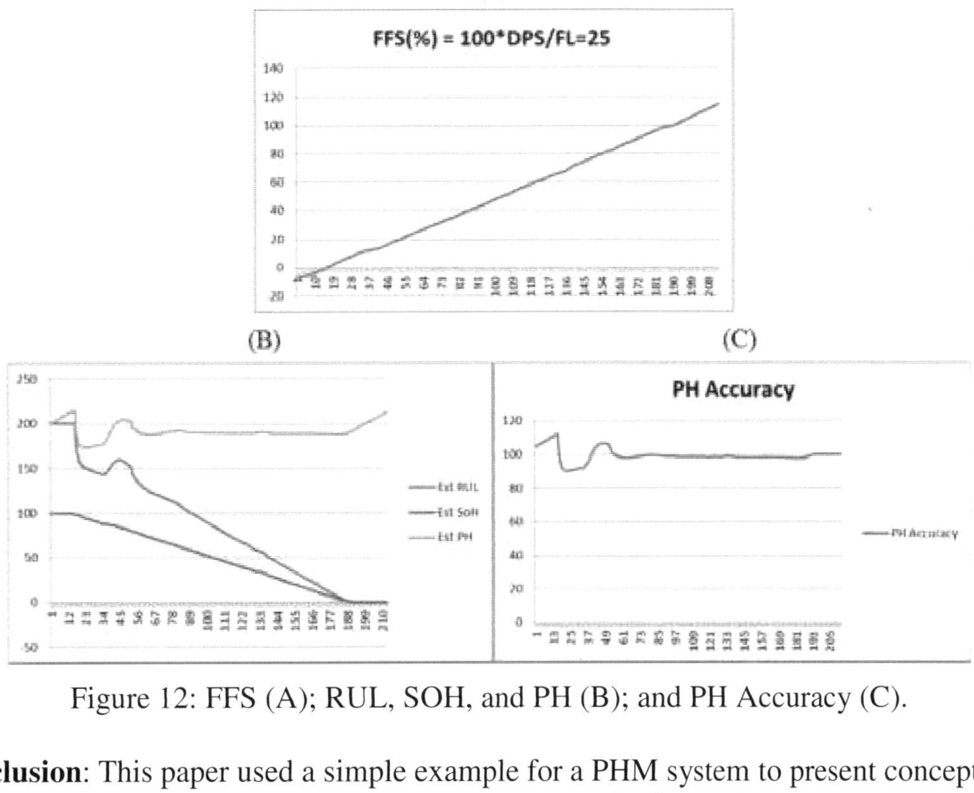

Figure 12: FFS (A); RUL, SOH, and PH (B); and PH Accuracy (C).

Conclusion: This paper used a simple example for a PHM system to present concepts and considerations associated with Prognostics Health Monitoring and/or Management (PHM). Considerations included sampling rate; prognostic information and performance metrics such as RUL, SoH, PH, PD, ($\alpha-\lambda$), and χ; data conditioning, including noise and distortion, fusion, and transforming; and Checkpoint/Restart services. [7]

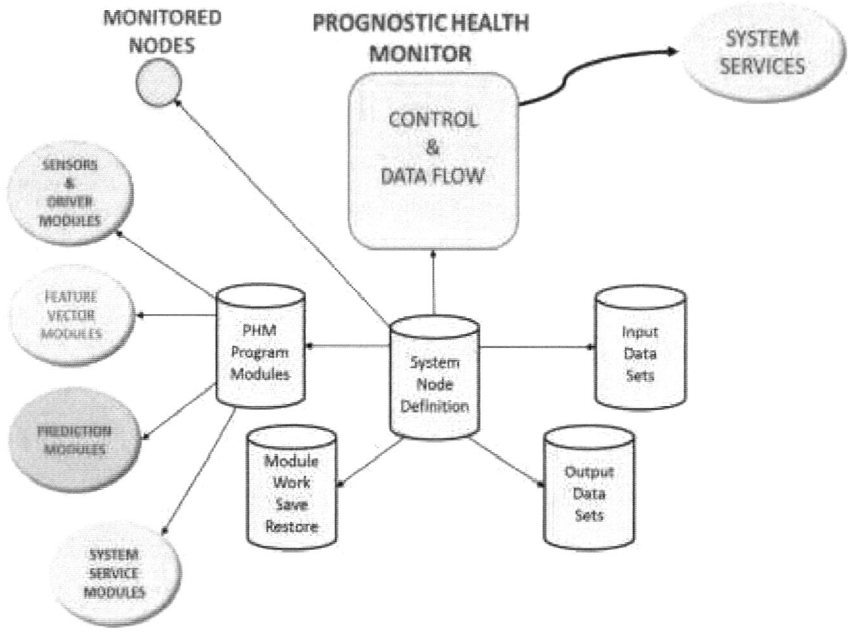

Figure 13: Revised PHM System with Save/Restore (Checkpoint/Restart).

References:

[1] M. Pecht; *Prognostics and Health Management of Electronics*, ISBN 978-0-470-27802-4. Hoboken, New Jersey: John Wiley & Sons, Inc., 2008.

[2] S. Kumar and M. Pecht, "Modeling Approaches for Prognostics and Health Management of Electronics," *International Journal of Performability Engineering*, Vol. 6, No. 5, pp. 467-476, Sep. 2010.

[3] James Hofmeister, Robert Wagoner, and Douglas Goodman," Prognostic Health Management (PHM) of Electrical Systems using Conditioned-based Data for Anomaly and Prognostic Reasoning," *Chemical Engineering Transactions*, Vol. 33, pp. 992-996, 2013.

[4] Prognostics Health Management for Electronics. (Accessed 2015, Nov.) Auburn University, Auburn, Alabama. [Online]. Available: http://cave.auburn.edu/rsrch-thrusts/prognostic-health-management-for-electronics.html.

[5] S. Speaks. (Accessed 2005, Aug.). Reliability and MTBF Overview. VICORP. Andover, MD. [Online]. Available: www.vicorpower.com/documents/quality/Rel_MTBF.pdf.

[6] A. Saxena, J. Celaya, B. Saha, S. Saha, and K. Goebel, "On Applying the Prognostic performance Metrics," *Annual Conference of the Prognostics and Health Management Society (PHM09),* CA, 2009.

[7] James Hofmeister, Douglas Goodman, and Robert Wagoner, "Advanced Anomaly Detection Method for Condition Monitoring of Complex Equipment and Systems," *2016 Machine Failure Prevention Technology,* Dayton, OH, 24-26 May 2016.

PRACTICAL ROTOR DYNAMICS FOR USERS

Thomas J. Walter
Halfmoon, NY 12065-6233
Telephone: (518) 557-2211
twalter02@gmail.com

Abstract:

Following a brief overview of the history of the development of key aspects of rotor dynamic technology, this paper describes ways to estimate critical speeds, and cipher them from resonances and other vibration phenomena. The influence of various bearing types on a rotor's synchronous and non-synchronous dynamics is reviewed. Target audiences for this material are those most likely to be making and interpreting vibration measurements - in other words, machinery users and maintainers.

Key words: Bearings; critical speed maps; damping; diagnostics; health management; prognostics; rotor dynamics; vibration assessment

Introduction: In many cases, the in-field vibration analyst has minimal design information (shaft dimensions, bearing properties, predicted critical speeds, damping, etc.) regarding the machine being assessed. Nonetheless, s/he must make determinations regarding the "health" of the equipment and in many cases, recommend and/or implement a "fix" if the machine behavior is changing or faulty.

This paper recalls some of the things the in-field analyst might consider to help determine a machine's operational condition. The intention of this short review is to provide insight/memory jogs regarding test methods to acquire information that can help get the job done. Technology advances in the past 50 years make the work more accurate and easier.

The Past Enables the Present: Acknowledging this conference's theme "Fifty years of failure prevention technology innovation," this section summarizes some of the enabling advances that benefit today's rotor dynamic technology. In actuality, some developments have been more than 50 years in the making, as described below. Looking back, advancements have closely paralleled the progress made in digital computing; principally increases in computing power and decreases in cost and size.

<u>Analysis Methods:</u> The study of a rotor's dynamic behavior began in the late 1800's (1-3). Early theoretical and experimental contributors included William J. M. Rankine (1869 - first performed an analysis of a spinning shaft), Gustaf de Laval (1889 - ran a steam turbine to supercritical speeds), and Stanley Dunkerley (1895 - published an experimental paper describing supercritical speeds). In 1916, W. Kerr published a paper showing experimental evidence of a second critical speed, and in 1919 Henry Jeffcott

published a paper now considered classic in which he confirmed the existence of stable supercritical speeds. History suggests August Föppl in 1895 described work similar to Jeffcott's, but it was unfortunately published in a German civil engineering journal that went unnoticed by the rotating machinery technical community at the time.

Through the mid 1900's, much work was done in the area of instabilities and modeling techniques culminating in the works of H. Holzer, Nils Myklestad and M. A. Prohl which led to the transfer matrix method (TMM) for analyzing rotors. The approach involves assumptions and variation of a frequency until the boundary conditions are met. The most prevalent analysis technique used today is the finite element method (FEM), and there are many PC-based programs commercially available. The FEM method assumes a mode shape in equations of motion that meet the boundary conditions and iterates the assumed mode shape until it does not change, thus obtaining a frequency that relates to the calculated mode shape.

In parallel with this activity, groundbreaking work was being accomplished to assess the dynamic performance of bearings, and the resulting development of bearing dynamic properties that are used in rotor dynamic computer programs (4-7). Much of the early work focused on fluid film bearings, the principal style used in ever-larger turbomachinery being developed at the time.

<u>Sensors:</u> Of the three principal ways to assess machinery vibration (displacement, velocity and acceleration), in the author's opinion it is the development of the eddy current displacement sensor that over the past 50 years has had the most important role in development of rotor dynamic technology. The ability to directly and accurately measure shaft motion without contact over wide speed ranges has provided practitioners with information needed to verify or dispute predicted performance. Among the early developers was Donald E. Bently (1924-2012). In today's marketplace, eddy current "proximity" sensors are available as commodities from multiple suppliers.

<u>Instrumentation:</u> From a field measurement perspective, this is where the most obvious advances have occurred. Improvements have enabled easier and more accurate acquisition and interpretation of rotor dynamic data. Fifty years ago, vibration instruments were analog devices, and the input signal was passed through filters of various widths. Analyzers were large, bulky, heavy devices that were called portable if their cases had handles! Recording devices were multichannel analog tape recorders that were even bigger and weightier. To enable machinery reliable enough for months-long voyages and quick refits during crew changeovers, the U.S. Navy submarine fleet spurred development of crew-usable equipment, and had the most comprehensive machinery vibration recording equipment and onboard machinery health monitoring methodology. The digital revolution enabled evolution to minicomputers, microcomputers, personal computers to today's hand held devices. Dynamic range for these instruments has increased from 30-40 dB to 120 dB or more, enabling easier and far more accurate and concise data acquisition and processing.

Using Transient Vibration to Assess Rotor Dynamics: For the in-field vibration analyst, measuring a machine's transient vibration provides a wealth of rotor dynamic information. Benefiting from the type of advances outlined in the previous section, these measurements can be accomplished more easily and with better accuracy than ever. Available from many text and vibration books and course notes, information in this section has been principally extracted from references 2 and 8.

Observing transient vibration can help identify natural frequencies, confirm system resonances, obtain damping values, and obtain modal information if needed to correction of resonance. There are several ways to obtain the data needed. One of these is illustrated below. Other methods (9-12) include measurements made as machine speed sweeps through a rotor critical speed, the half power method, and the phase angle method.

This example uses the shaft's lateral motion (ring-down) to a transient event, perhaps an impact on the machine's housing. The machine may be operating or non-operating during these measurements. Depending on the style of bearings the machine has (see a following section), results may be improved if the machine is running. In most instances, structural and other non-rotor-bearing resonances may not be too noticeable in ring-down tests. One must of course be aware of any sensor mounting resonances that may be encountered, since shaft observing sensors will observe those as well.

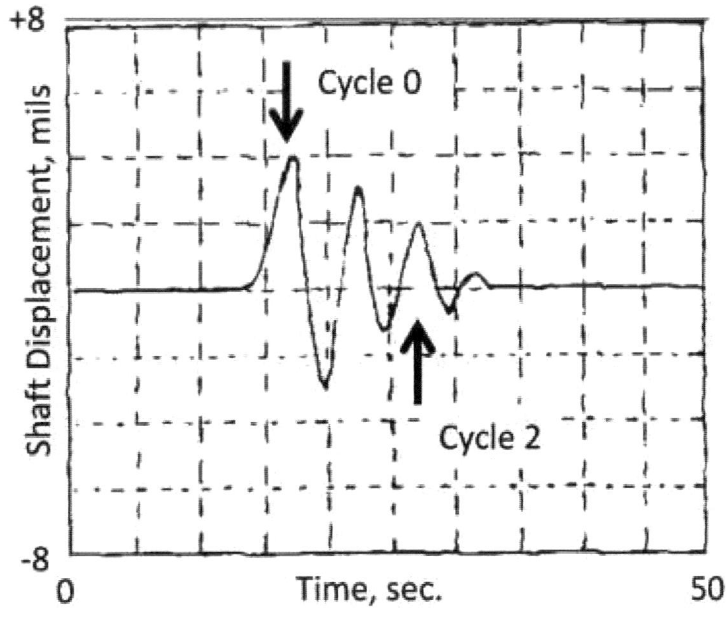

Figure 1: Rotor Response to Impact

Classical derivations (9, 10) provide the following forms for logarithmic decrement (used to determine system damping from measurements), damping ratio, amplification factor, period of vibration, and natural frequency:

$$Logdec = \delta = \frac{1}{n} Ln \frac{X0}{Xn} \qquad = \frac{1}{2} Ln \frac{4}{2} = 0.3466 \tag{1}$$

$$\xi = \frac{1}{2\pi}\delta = \frac{.3466}{6.28} = 0.055 \tag{2}$$

$$Q = \frac{1}{2\xi} = \frac{1}{2*0.055} = 9.1 \tag{3}$$

$$T_d = 5m\sec = 0.005\sec \tag{4}$$

$$f_d = \frac{1cycle}{.005\sec} = 200Hz \tag{5}$$

Based on the above equations and Figure 1 data, the logarithmic decrement is 0.3466 (equation 1). The damping ratio (2) is 0.055a small but oftentimes sufficient amount. The amplification factor (3) is 9.1, a tolerable value in many machines. The measured period (4) is 0.005 seconds and the natural frequency (5) is 200 Hz.

Forced Harmonic Vibrations – A Key to Understanding Rotor Dynamics: Harmonic vibration is found in most real world machines, with mass unbalance (see a following section) being the most common. Although other types of excitation are always present, knowing how the system responds to harmonic excitation provides insight into more complex behavior too. Most of the material presented in this section comes from reference 2.

Based on classical derivations, Figure 2 presents the motion of a machine which can be modeled as a single degree of freedom system. The Magnification Factor depicts the factor by which the zero frequency deflection must be multiplied to determine the amplitude X, based on how near or far the frequency is from the natural frequency, where $\frac{\omega}{\omega_n} = 1$.

The figure shows that amplitude and phase in the vicinity of a natural frequency are functions only of the frequency ratio and the damping factor. As shown, damping has a large influence on amplitude and phase in this region, but little or no influence elsewhere. In other words, damping is important only when the forcing frequency is close to the natural frequency.

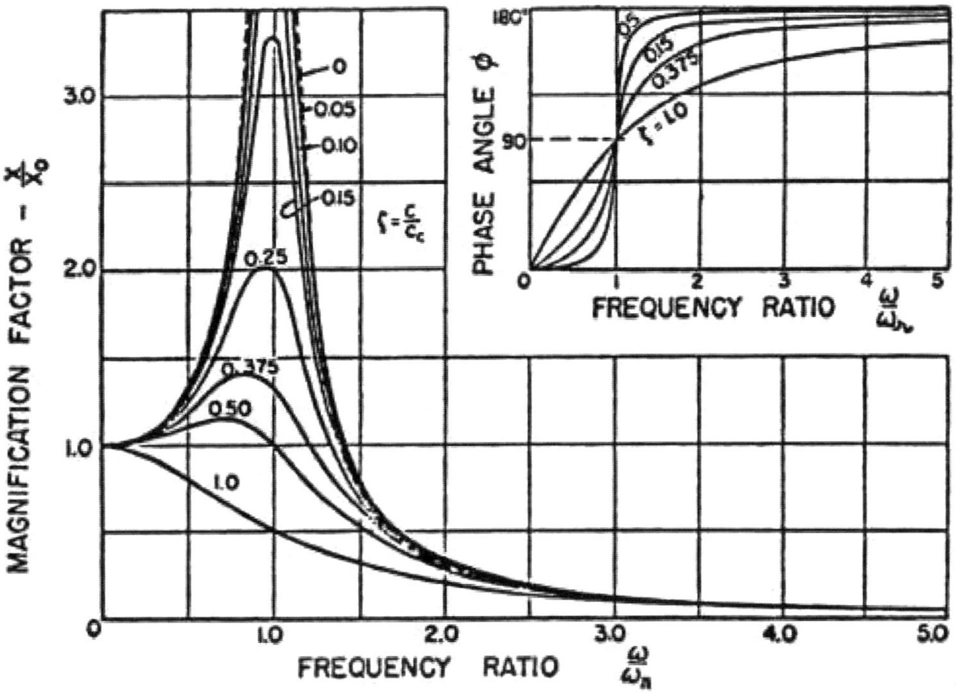

Figure 2: Forced Vibration of a Viscously Damped System

Shafts and Rotors: It is important for in-field vibration analysts to understand the concept of rigid and flexible rotors. Figures 3 and 4 show some basic concepts that lead to the critical speed map of Figure 5, the key take-away from this section. Information in this section was principally extracted from references 2 and 11.

A rotor consists of a shaft with an assembly of disks mounted on it (Figure 3a) or a massive shaft with varied changes in geometry (Figure 3b). A rigid rotor operates substantially below its first bending critical speed. In other words, its flexural vibration is inconsequential compared to bouncing and rocking on the bearings. A rigid rotor can be balanced at an arbitrary speed in two planes and remain in balance at all operating speeds. A flexible rotor, Figure 4, operates close to or beyond its first bending critical speed and dynamic effects influence relative rotor deformation.

Rotor flexibility depends on the relative stiffness between the bearings and the rotor and its operating speed. Figure 5 shows a critical speed map. This map shows the effect of the relative flexibility of the rotor and bearings on system natural frequencies.

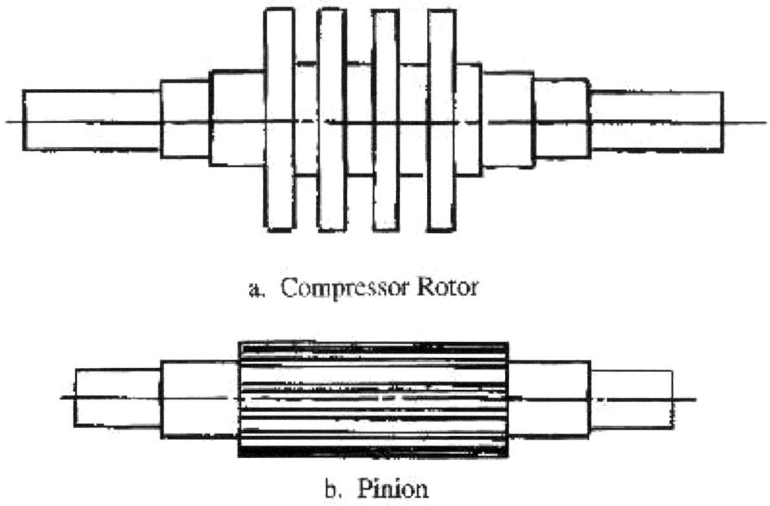

Figure 3: Rotor Geometry

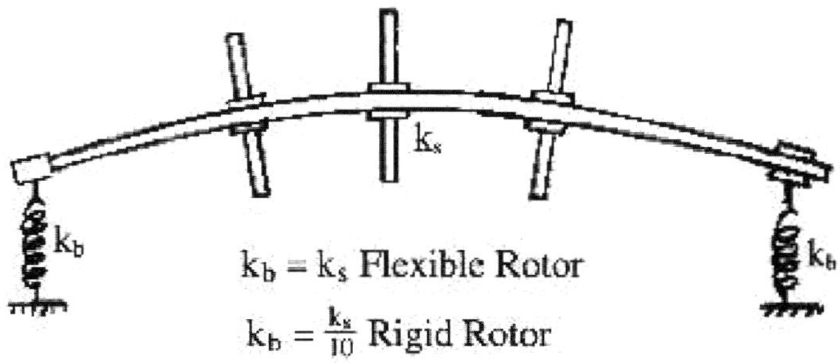

Figure 4: Rotor-Bearing Flexibility

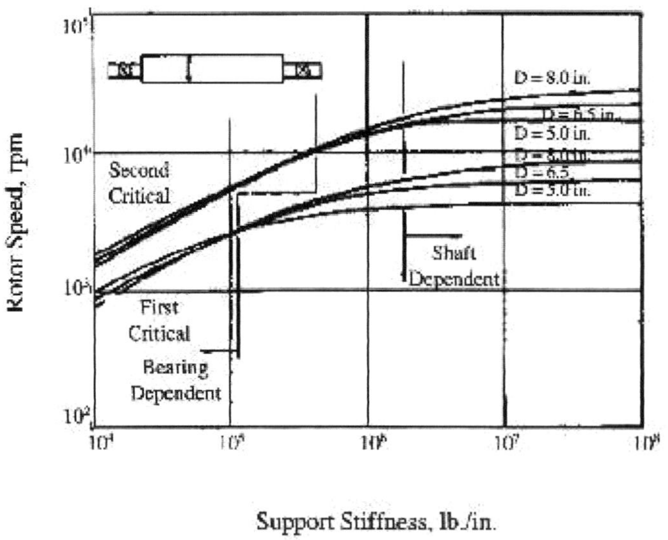

Figure 5: Critical Speed Map Showing Rigid and Flexible Shaft Sectors (11)

Mass Unbalance Cause a Principal Force in Machinery: The classical mass unbalance effects result in a once-per-revolution frequency with amplitude proportional to the rotor mass, mass eccentricity, and rotor speed squared (equation 6). Most of the material presented in this section comes from references 2 and 12.

$$F_c = \frac{w}{g}\omega^2 a \tag{6}$$

Where a = eccentricity, in.
w = weight, lb.
ω = speed, rad./sec.
g = gravitational constant, in./sec.

Mass unbalance arises from the lack of coincidence of the mass center with the spin axis of the rotor (Figure 6). As it is accelerated, this mass eccentricity causes a centrifugal force on the rotor. The force due to mass unbalance increases as the speed squared. In addition, it can be seen (Figure 2) that the amplitude will be a function of shaft and support stiffness, damping, and mass; i.e., the natural frequency (critical speed). Extracted from reference 12, typical causes of mass unbalance are shown in Table 1.

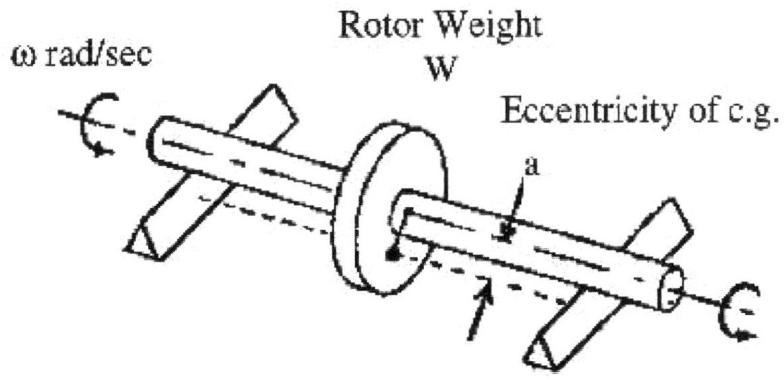

Figure 6: Variation of a rotor's Mass Center

Table 1: Typical Causes of Rotor Unbalance

CAUSES OF UNBALANCE	OBSERVABLE SIGNS	FREQUENCY OF VIBRATION
Disk or component eccentric on shaft	Detectable runout on slow rotation, e.g. runs to bottom on knife edges	one/rev
Dimensional inaccuracies	Measurable lack of symmetry	one/rev
Eccentric machining or forming inaccuracies	Detectable runout	one/rev
Oblique angled component	Detectable angular runout Measure with dial gage on knife edges	one/rev
Bent shaft Distorted assembly Stress relaxation with time	Detectable runout on slow rotation, often heavy vibration during rotation	one/rev
Section of blade or vane broken off	Observable bearing vibration during operation	one/rev; possible process pulsations
Eccentric accumulation of process dirt on blade	Bearing vibration	one/rev
Differential thermal expansion	Shaft bends and throws c.g. out. Source of heavy vibration	one/rev
Non-homogeneous component structure Sub-surface voids in casting	Rotor machined concentric Bearing vibration during operation. C.G. runs to bottom on knife edges	one/rev
Non-uniform process erosion	Bearing vibration	one/rev
Loose bolt or component slip	Vibration reappears after balancing due to component angular movement	one/rev Possible magnitude and phase changes
Trapped fluid inside rotor, possible condensing/ vaporizing with process cycle	Vibration reappears after balancing Apparent c.g. angular movement occurs	one/rev Possible magnitude and phase changes
Ball bearing wear	Bearing vibration Eccentric orbit with possible multi-loops	one, two, or higher per rev

Bearings & Their Effect on Rotor Dynamics: A critical speed map (Figure 5) shows how the effective stiffness and damping of a machine's suspension affects its vibration overall. This section briefly describes the principal styles of machinery bearings and compares their relative attributes.

- Rolling Element Bearings: These bearings come in a wide variety of styles and sizes, including Deep Groove, Angular Contact, Four-point Contact, Self-aligning, and even Linear Sleeve and Linear Spline. Rolling element bearings are used in wide speed ranges from low to very high. Compared to other bearings, they have low friction (they are sometimes called antifriction bearings!), high stiffness, and low damping.

- Plain Bearings: These are the simplest of bearings. They involve just two surfaces: one moving, one stationary. They are often compact, lightweight, and inexpensive. Nonetheless, they are capable of very high loads. Two general categories include Integral, where the bearings are built into an object, and Bushing, where the bearing is an independent object that is inserted into a housing. Plain bearings use one of three categories of lubrication systems: Class I – lubricant applied from an external source; Class II – lubricant contained within the bearing walls; Class III – the bearing material is the lubricant.

- Fluid Bearings: For these bearings, fluid lubrication separates the two surfaces through fluid dynamic effects. The fluid can be liquid or gas. There are two categories of fluid bearings: Hydrostatic bearings, where external pressure provides the load capacity, and hydrodynamic bearings that use the relative motion between surfaces to generate the load carrying capacity. Hydrostatic bearings are often used when machinery speeds are low. Hydrodynamic bearings are more suitable for higher speeds. Gas lubrication is better suited to light loads and high speeds with lower losses, while liquid lube enables higher loads and more effective cooling.

- Magnetic Bearings: These bearings support machinery without physical contact. They can be used over a wide speed range, and have very low friction, and no mechanical wear. Compared to rolling element or oil lubed bearings, magnetic bearings have lower stiffness. There are two categories of magnetic bearings: Permanent magnet (PM), which are passive and use the repulsive nature of north and south pole magnets, and electromagnet (EM), which can carry higher loads but need shaft position sensors and a controller that adjusts magnetic field strength on-the-fly. PM bearings have low stiffness and little damping. EM bearings can include sophisticated controllers that adjust the bearing's effective stiffness and damping. Auxiliary bearings for overload or controller power loss are often needed for EM bearings.

Conclusions: Although the in-field vibration analyst's access to design-related background information is often limited, proper selection and application of sensors, instrumentation, and some straightforward tests can provide meaningful insight towards understanding the vibration of the machine at hand. It is important to understand that the machine's rotors and bearings form a dynamic system, and that "fixes" for one type of machine may not apply universally. The capabilities available to assist today's practitioner build upon the developments of many theorists and practitioners in decades past.

Acknowledgements: The author wishes to acknowledge Dr. Ron Eshleman, Technical Director of the Vibration Institute, for his encouragement to prepare this subject matter. As cited, portions of the material herein have been extracted from the notes the Vibration Institute uses in its Practical Rotor Dynamics and Modeling© course. This course is part of the Institute's Category VI Vibration Analyst training program. Appreciation is also expressed to Mr. Peter Chapman and Mechanical Solutions, Inc., former colleague and employer, respectively, for background and bearings-related information.

References:

[1] Nelson. F. A. "Rotordynamics without Equations" published in International Journal of COMADEM, vol. 10, no. 3, July 2007, ISSN 1363-7681.
[2] Eshleman, R. L. "Practical Rotor Dynamics & Modeling© – Course Notes" Vibration Institute, 2014.
[3] Wikipedia, https://en.wikipedia.org/wiki/Rotordynamics.
[4] Wilcock, D. F. and Booser, E. R. "Bearing Design and Application." McGraw-Hill, New York 1957.
[5] Lund, J. W. and Sternlicht, B. "Rotor Bearing Dynamics with Emphasis on Attenuation." ASME Paper No. 61-WA-68.
[6] Lund, J. W. "Stability and Damped Critical Speeds of a Flexible Rotor in Fluid-Film Bearings." ASME Paper No. 73-DET-103.
[7] McHugh, J. D. Principals of Turbomachinery Bearings." Proceedings of the Eighth Turbomachinery Symposium, 1979.
[8] Eshleman, R. L. "Vibration Specialist III – Course Notes" Vibration Institute, June 2000.
[9] Den Hartog, J. P. "Mechanical Vibrations." Fourth edition. McGraw-Hill, 1956.
[10] Thomson, W. T. "Theory of Vibration with Applications." Prentice-Hall, 1972.
[11] Jackson, C. "Turbomachines: How to Avoid Operating Problem," Hydrocarbon Processing, November 1979.
[12] Rieger, N. F., and Crofoot, J., "Vibrations of Rotating Machinery," Vibration Institute, 1977.

Bibliography:

[1] Chapman, P. A. "Overview of Modern Bearing Technologies", Mechanical Solutions Inc. presentation 11P8, presented at Vibration Institute Hudson-Mohawk Chapter's Vibo-Rama, March 10, 2011, Albany, NY.

[2] Deery, J. “The ‘Real’ History of Real-Time Spectrum Analyzers - A 50-Year Trip Down Memory Lane.” Sound and Vibration, January 2007.

[3] Dimarogonas, A. D. “A Brief History of Rotordynamics.” In book Rotordynamics ’92 pp.1-10. January 1992; ISBN: 978-3-540-19754-6 DOI:10.1.

[4] Eisenmann, Robert C., Sr. P.E. and Eisenmann, Robert C., Jr. Machinery Malfunction Diagnosis and Correction - Vibration Analysis and Troubleshooting for the Process Industries PTR Prentice Hall; 1st edition (November 26, 1997). ISBN-13: 978-0132409469; ISBN-10: 0132409461

[5] Nelson, F. A. "A Brief History of Early Rotor Dynamics." Sound and Vibration, June 2003.

PREDICTION OF SENSOR SYSTEM RELIABILITY

Shan Guan, Christopher Taylor, and Narasi Sridhar
Materials Technology Development
DNV GL (USA)
Dublin, OH 43017
Telephone: (614) 397-1062
shan.guan@dnvgl.com

Abstract
This paper summarized influencing factors to the sensor system reliability used in Oil and Gas industry. A management plan based on criticalities of influencing factors to the overall system is proposed. Prediction of sensor system reliability will be especially useful in the situation where sensor systems can degrade over time in service. A modeling approach has been carried out in this paper to combine the Bayesian network modeling and "Analytical Redundancy relations" methodology for assessing sensor reliability in a digital downhole application.

Key words: Sensor; Reliability; FMECA; Bayesian network; Digital downhole.

Introduction:

A sensor is a device that measures a physical quantity by generating a functionally related output which can be read by an observer or by an electronic instrument. Sensors can be categorized in many different ways based on the mechanism by which they transform a particular input into an output, for example as physical, chemical, biochemical, and electrochemical sensors. Most sensors produce an electrical output for ease of transmission, storage, and read out. "Sensor" is often used interchangeably with the term "sensing element". However, most state-of-the-art sensors consist of multiple components. For example, "smart sensors" or "sensor systems" use built-in compute resources to perform predefined functions upon the detection of specific input and then process data before passing it on. A working sensor system is a composite of four distinct parts shown as a block diagram in Figure 1 including 1) A sensing element, which produces an electrical response when the device is stimulated by external factors; 2) A signal conditioning element that modifies and processes the electrical signal to be understood properly by the receiver; 3) A sensor interface that allows the device to acquire, store, and communicate with an external interface, and; 4) A power system to provide external power resources or to harvest energy.

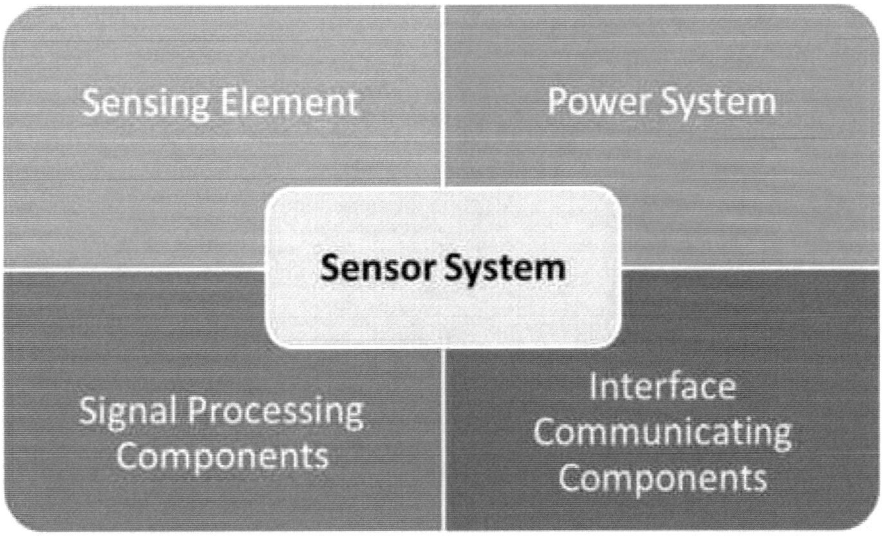

Figure 1. A schematic of a sensor system

Sensor system reliability can be defined as the ability of a sensor system to perform its required functions under stated conditions for a specified period. For this reason, the reliability of a sensor system will be strongly dependent upon its age, context and application. Assessing sensor system reliability is a major challenge in developing new sensors and sensor selection. It is not uncommon that a sensor system possesses a high reliability in one application but becomes unreliable in another situation. Additionally, reliability requirements depend upon how and where the sensors are applied. To ensure long-term performance, embedded or permanent sensors that are difficult to calibrate will need to have a higher reliability than temporary or manually operated sensors. A high reliability is especially critical for the following application scenarios as illustrated in Figure 2 including 1) Complex and smart sensor systems (e.g. Sensor Fusion); 2) Highly integrated miniature sensor systems (e.g. MEMS/NEMS sensors); 3) Long term monitoring requirements (e.g. Condition Monitoring); 4) High Consequence /Mission Critical applications (e.g. Leak Detection or Aviation); 5) Application requires dynamic response (e.g. Dynamic Positioning System); and 6) Harsh and extreme working environments (e.g. High Temperature High Pressure in Oil &Gas fields). Any equipment including sensors will eventually fail regardless of how superior the design is and how well it is maintained. During the lifecycle of a sensor system, its failure rates tend to follow the 'bathtub' curve. To reduce cost, a typical strategy that sensor users adopt is to extend the time of "wear-out" period as long as possible. This could be realized through a well-defined and well-executed maintenance plan. In some situations, with detailed knowledge of the sensor system and enough statistical data, it is possible to estimate the remaining useful life. However, for each application scenario, the lifecycle curve may not be identical even for the same sensors manufactured in the same batch. Therefore, when predicting the remaining life of a sensor system, many factors including sensor design, materials selection, manufacturing and packaging process, maintenance and calibration, and the sensor working environment should be systematically considered.

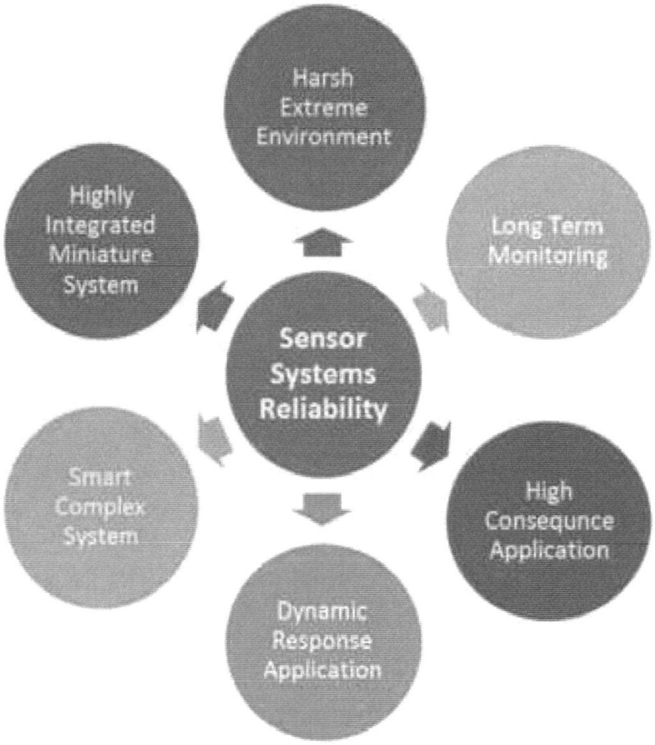

Figure 2. Application scenarios where sensor system reliability is extremely important

Factors Affecting Sensor Reliability

To evaluate, predict or improve the reliability of a sensor system, it is necessary to know the reliability of each component of the sensor system i.e. sensor element, signal processing components, power system, and interface communicating components. Additionally, design, manufacturing, packaging, installation, maintenance, and calibration will also play a role on sensor system reliability. The software used by a sensor system also affects its reliability. One method for assessing sensor system reliability is to apply Failure Mode, Effects and Criticality Analysis (FMECA), which is an extension of Failure Mode and Effects Analysis (FMEA, or just referred to as "failure mode"). In this methodology, a risk assessment is made of each failure mode to determine its criticality. Criticality is derived from an assessment of the probability that a particular failure will occur combined with the severity of the failure if it does occur (i.e. the consequence).

To carry out FMECA for a sensor system, a systematic analysis need to be performed based on the sensing mechanism, the working environments and the overall redundancy. Table 1 shows the criticality of a pressure sensor (generic) used in subsea processing, where the criticalities were divided into five categories of Very Low, Low, Medium,

High, and Very High. Factors with a criticality of Very Low or Low can be subject to corrective maintenance. Factors with a criticality of Very High must be considered for re-design or adding redundancy to decrease criticality. The factor ranked with a criticality of Medium to High is to be evaluated further, for example, to consider its effects on other components as well as on the overall system. [1]

Table 1. Failure Modes and Their Criticality Ranking for the Reliability of a Pressure Sensor System

Sensor System Components	Failure Modes	Criticality				
		Very Low	Low	Medium	High	Very High
Sensor Elements	Degradation of Sensing Materials				■	■
	Thermal Induced		■		■	
Signal Processing Components	Degradation of contacts or connections				■	■
	Thermal Induced		■		■	
	Degradation of Signal Processor		■		■	
Power System	Degradation of contacts or connections				■	■
	Faulty Electronics				■	■
	Loss of Power					■
Interface Communicating Components	Degradation of contacts or connections				■	■
	Faulty Electronics		■		■	
	Poor/Insecure Signal		■			

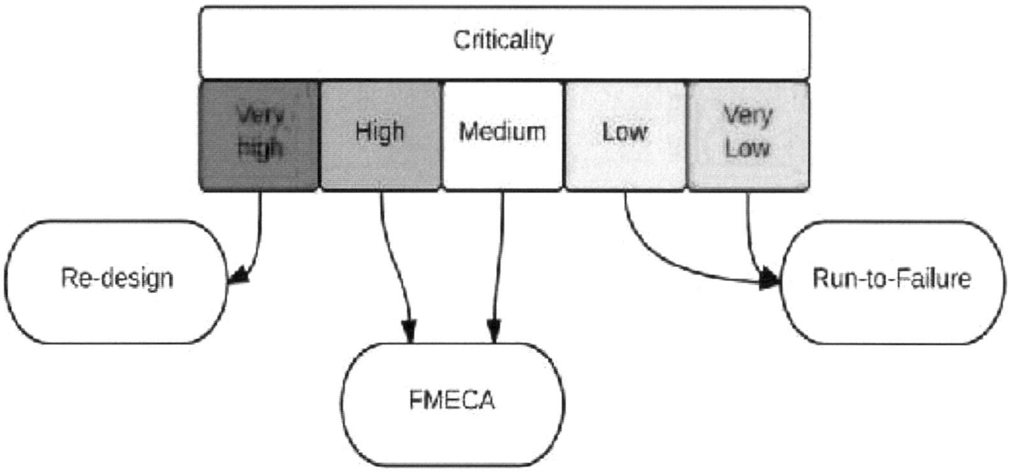

Figure 3. The criticality definition and the corresponding actions for the reliability of a sensor system

One of the most important factors affecting the reliability of a sensor system is degradation, primarily because of materials aging, corrosion, and wear. In general, three classes of sensor system degradation should be considered including 1) Degradation of the sensing element itself, e.g., in harsh environment, accelerometers' sensitivity and accuracy gradually getting worse; 2) Degradation of connectors, usually a problem that results in false positives. For example, a false positive from water-in-oil sensors can cause dry-dock of the ship unnecessarily, thus resulting in huge financial loss; and 3) Degradation of power systems – this is especially important for sensing systems operated via batteries. For example, for acoustic sensors used to sense buried pipeline leaks, it is difficult to replace the battery without excavation.

Since ageing and degradation of a sensor system are time dependent, the reliability will inevitably decrease over time and thus the potential for faults and failures will increase. Unfortunately, this issue is not always considered during the planning phase of sensor deployment. Taking the condition monitoring of ship machinery as an example, sensor degradation may not disable the monitoring system, but will likely create false positive or negative readings. False positive may lead to unnecessary actions to rectify a potential problem or, if sufficient false positives are seen, may lead to de-sensitization of staff to an event. False negatives may lead to ignoring a potential threat. Degradation can also result in imprecise sensor data in the form of drift or bias. In such cases, errors will be produced in the system condition diagnosis resulting in ineffective control. A typical example can be found in the drift of pressure sensors. While the exact mechanism of pressure sensor drift is not understood completely, typically it is believed to result from the change in response of the physical properties of the sensor materials to external environments, including the frequency of the pressure changes and the exposure to temperature extremes. An example of this phenomenon is the SIEMENS WEPS-100 Series Subsea Pressure Sensor. According to the data sheet provided by the sensor manufacturer, brand new ("out of the box") sensor yield a reading accuracy of +/- 0.35

bar, which increase to +/- 9.1 bar after 25 years of usage [2]. That is, one can experience averages of 100% drift each year! In theory, the drift can be corrected through calibration, while in special situations, such as applications in subsea oil and gas processing, it is impractical to calibrate the sensors without interference with production.

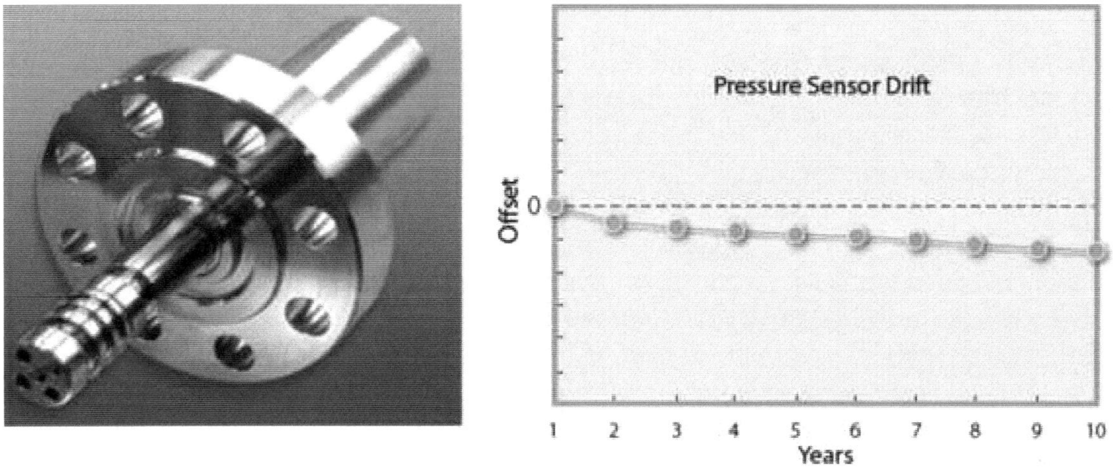

Figure 3. (left) WEPS-100 Series Subsea Pressure Sensors, [2] and (right), A schematic of pressure sensor drift.

Prediction of Sensor System Reliability

A number of algorithms have been developed to detect faulty sensor data as a screening tool prior to passing information on to decision-aid tools. These computational methods include 1) Principal Component Analysis (PCA); 2) Artificial Neural Networks (ANNs); 3) Recurrent Neural Networks (RNNs); 4) Auto-Associative Kernel Regression (AAKR); 5) Independent Principal Component Analysis; 6) Support Vector Machines (SVMs); and 7) Fuzzy Similarity, etc. As summarized by Sharma et al., four categories of detection methods can be discerned [3] including 1)Rule-based methods define heuristic rules/constraints that the sensor readings must satisfy; 2) Estimation methods define "normal" sensor behavior by leveraging spatial correlation in measurements at different sensors; 3) Time series analysis based methods compare a sensor measurement against its predicted value based on time series forecasting to determine if it is faulty; and 4) Learning-based methods infer statistically established models to identify faulty sensor readings using training data. These methods have been found to be useful in many application scenarios. However, there also exists the need to estimate the uncertainty attached to the sensor data being received. This information will be especially useful in the situation where sensor systems can degrade over time in service, and thus the data may need to be corrected or compensated before decisions can be made. From the condition based monitoring standpoint, it will be useful to predict when the reliability of a sensor system has become unacceptable, and thus action must be taken to calibrate or replace the sensors.

Assessing the probability of an unreliable sensor system failure using modeling tools is challenging for three reasons: 1) no model is accurate in all situations, 2) the input data used to run the models is never exact, and 3) the knowledge of the system is often incomplete or unclear. [4, 5] For this purpose, we introduce the concept of using Bayesian network modeling to assess the failure probability of a deployed sensor system. A Bayesian network is a probabilistic graphical model based on Bayes' theorem for combining prior knowledge and data. It can combine diverse models (i.e. mechanistic and empirical, with different sources and different programing languages) into one unified method. Therefore, the methodology makes it easy to update the overall framework when new knowledge is produced. If we learn enough information about a sensor system, including the working mechanism of sensors, its technical specifications, failure modes and working environments, it is feasible to construct a Bayesian network model for filtering and assessing unreliable sensor data. Figure 4 demonstrates the generic structure of a simplified Bayesian network that might be used to predict the reliability of sensor data under the influence of degradation and aging of the sensor components. It can be seen that the working environments, the maintenance, and calibration of the sensors will play a direct role on the reliability. A suitable algorithm, established through a consideration of the failure modes revealed through accelerated life testing and/or updateable field experience, will be able to compute the probability of unreliable sensor data. These degradation models can then be used to make decisions regarding sensor placement and timing for maintenance and/or replacement. The degradation models can also be used to help in the statistical determination of whether a constraint breaking event has occurred (i.e. distinguishing an actual fault from the noise). It is critical to be able to update the model through the technique of Bayesian inference, as sensor systems are often extended to new environments or longer service lives due to life extension programs, in which case new failure modes associated with materials ageing and other physical or chemical processes will emerge.

Figure 4. Using Bayesian network models to predict the effects of degradation on the reliability of sensor systems.

Case Study: Predicting Sensor Reliability in the Digital Downhole

The digital downhole refers to the entire instrumented drill string system, the sensor communication system, and the digital twin model [6]. As drilling operations in oil and gas have to deal with deeper wells, more aggressive environments, or more complex in geometries (think horizontal drilling and fracking), the drill string systems will need to become more "intelligent" so that operations can be conducted safely, economically and efficiently. Sensor systems provide the means for assessing the conditions downhole and reporting them to the operations managers. Sensors of significance include pressure and flow sensors, as well as sensors that can monitor temperature and chemistry of the fluid systems being encountered. The sensor data flowing back to the operations center provides the inputs needed to construct the real-time digital representation (i.e. the digital twin) of the downhole system, so that operators see an exact picture of the state of health, and can make adjustments to correct any deviations or initiate emergency procedures. Due to the aggressive nature of drilling operations, including abrasive wear from solids in the drilling fluids, high temperatures and pressures and the chemical environments themselves, the sensors will also be subject to degradation. But, how do we distinguish between faulty sensor data and a true alarm? One technique is to use Analytical Redundancy Relations that generate digital "fingerprints" for normal operations versus failure modes. [7] Analytical redundancy relations are derived from the physics that connects sensor readings from one part of the operation (say the pressure at the drilling fluid pump) to the readings at another part of the operation (say the downhole pressure- the differences should be related to the hydrostatic pressure plus losses proportional to the friction coefficient and the square of the fluid flow). The digital twin model is ideally poised to utilize such information in providing a first "health assessment" of the sensor data since it contains a complete multiphysics representation of the asset. To give an example, consider the type of setup shown in Figure 5, adapted from the representation in Willersrud et al. [7] Sensors that measure flow rates and fluid pressure are placed along the assembly, including before and after the pump, at the drill bit, within the annulus and around the choke point. These sensors report back to the operating station at which point the signals are processed to update the digital twin model for the system.

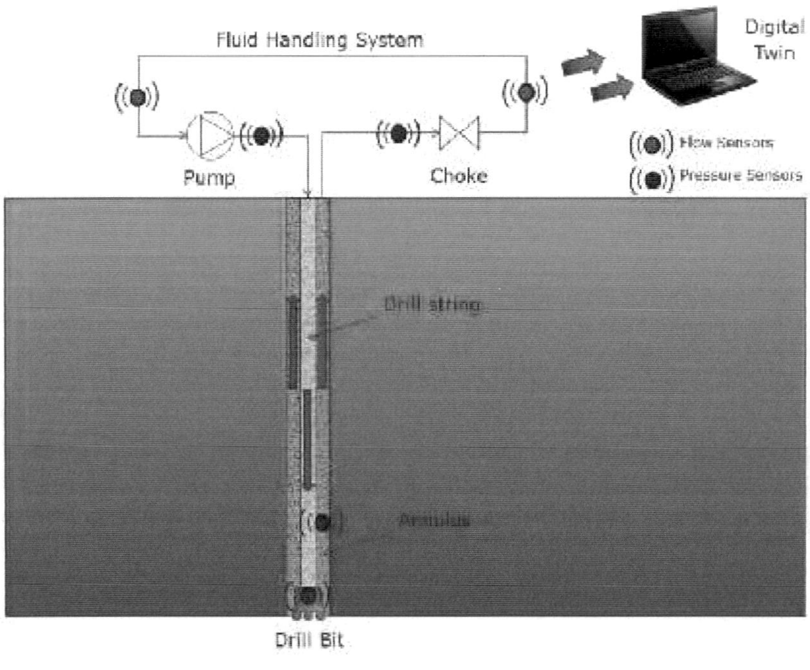

Figure 5. The Digital Downhole instrumented by pressure and flow sensors, in this example, reporting data back to update the Digital Twin at the operational station. [7]

At each of the sensor points, multiple sensors could be placed to provide redundancy. Even without this measure, however, there are "redundancy relations" that can be analytically derived. To take the example above, the pressure downhole, at the drill bit, should relate to the pressure at the pump according to the hydrostatic pressure (i.e. density x gravity x height difference) minus the losses according to the friction resisting fluid flow down the drill string (proportional to the square of the fluid flow rate). Thus, the sensors for pump pressure, downhole pressure, and fluid flow should be related according to a physical equation. This relationship is used to build a constraint in such a way that it will mathematically resolve to the value of zero under normal, fault-free conditions (within a noise threshold). Similar constraints can be derived for the other sensor signals in the system. When a fault occurs, however, one or more of these constraints will return a non-zero reading. It turns out that different failure modes will produce different responses in the non-zero constraint vectors (see Figure 6). A sensor failure at the pressure pump, for example, will break certain constraints but not others. Likewise, plugging of the bit nozzle by debris will have its own unique pattern of broken constraints. Accordingly, the constraint matrix will provide a way to fingerprint different kinds of system failures, thanks to the analytical redundancy relations that were encoded into the digital twin that handles the sensor data.

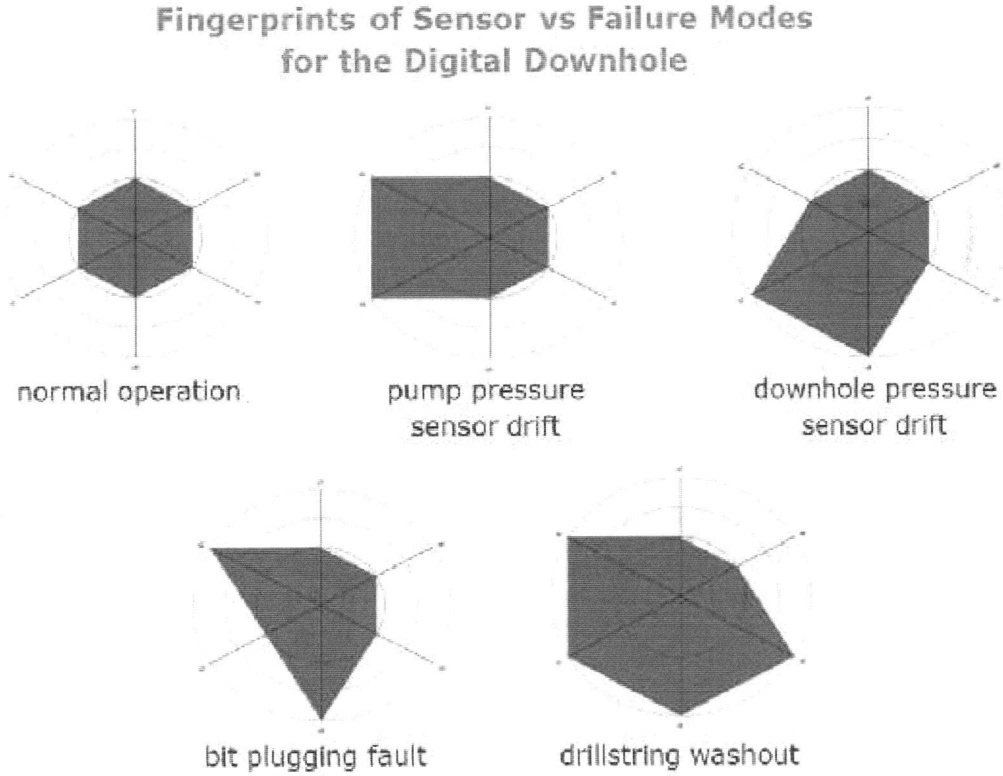

Figure 6. Each failure mode for the downhole system corresponding to a unique pattern of deviation in analytical redundancy relations can be used as fingerprints that allow fault detection and isolation within the Digital downhole system.

Like all physical components of the downhole system, the sensors are composed of materials that must meet structural, functional, and electronic/communication requirements for successful operation. In qualifying sensors for use in the Digital Downhole, the reliability of those materials must be assessed. The evolving probability of sensor part failure over time can be estimated in advance to some extent through accelerated testing. The "structured reliability assurance process" developed by Veneruso and co-workers [8] describes a system that integrates laboratory testing with continuous documentation of field data as a means for continuously improving the quantitative reliability assessment. The necessity to include field data comes about because laboratory testing can never fully replicate the range of conditions (nor the desired asset lifetimes) experienced by the materials when they are placed downhole. In interpreting data for materials health assessment through either experiment of collated field data, probability distributions provide a key way of quantifying the reliability. Most commonly, the "survival data" for materials exposed to the field are fitted to extreme value statistics, such as the Weibull distribution function. As an example, see the hypothetical case in Figure 7 based on a study of monitoring and control systems performed by Veneruso et al. [8]. An initial model, based on laboratory testing initially fits the survival probabilities well. However, due to testing constraints (such as only a short time available to perform testing), the model performs more poorly beyond a certain lifetime. At this point, the historical data can be used to update the probability of sensor survival and build a new model.

This example shows one way in which the sensor reliability assessment resulting from fingerprinting and the Digital Twin can be combined with survival probability models such as Bayesian network models. As the Digital Downhole system collects data and reports on sensor performance through the use of analytical redundancy relations, the Bayesian network models for the sensor reliability can be updated. Conversely, the updatable Bayesian network models, shown in Figure 7, can be used to internally assess the likelihood of a sensor failure. What does this mean? In performing the fingerprint analysis, as shown in Figure 6, the digital twin needs to be aware of "thresholds" that distinguish a true constraint-breaking event from the background noise. As the sensor systems age or adverse conditions are experienced, the likelihood for accepting a constraint-breaking event should increase. Hence, there will need to be a feedback between the construction of the internal models, and the decision to accept a broken constraint.

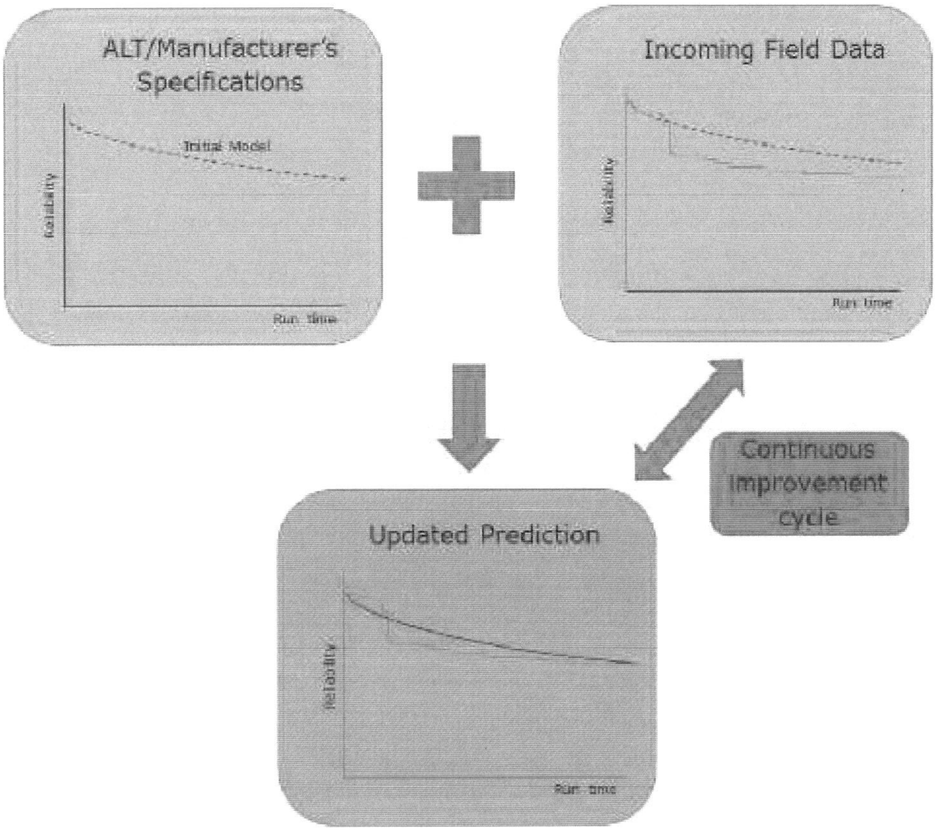

Figure 7. Updateable models for the survival probability of sensors components begin with an initial model constructed through lab-based testing, but are then updated in real-time by data coming in through "fingerprint" analysis performed by the Digital Twin.

Conclusion

Sensor system reliability is a challenging yet critical issue that almost all industrial sectors are facing. As sensor utilization in the next ten years is expected to increase at a pace of more than 10% each year, all sectors of the modern industrial world, and society in general, will increasingly rely on information obtained from sensor systems. There is an increasing trend in sensors that are embedded or left in place in operating environments. These sensor systems then degrade due to exposure to the operating environments, affecting their reliability over time. Therefore, the risks associated with sensor system reliability need to be carefully evaluated. Algorithms for checking the reliability of sensor data need be developed using the best combination of physical, statistical and probabilistic modelling tools. In this paper, influencing factors to the sensor system reliability used in Oil and Gas industry are summarized, and a management plan based on criticalities of influencing factors to the overall system is proposed. This was demonstrated through identifying failure models and criticality ranking for the reliability of a generic pressure sensor system. Prediction of sensor system reliability will be especially useful in the situation where sensor systems can degrade over time in service. A modeling approach has been carried out in this paper to combine the Bayesian network modeling and "Analytical Redundancy relations" Methodology for assessing sensor reliability in a digital downhole application.

References

[1] S. Guan, "Sensors Requirements for Condition Monitoring and Data-driven Classification: A Case Study", DNV GL Report No.: GGRUS915SGUA (91515017), May 8, 2015.
[2] Technical Data Sheet, WEPS-100 Series Subsea Pressure Sensors, SIEMENS.
8. Technical Data Sheet, MEMS accelerometer (Model 40366), Endevco Corp. (a Meggitt Group Company)
[3] A.B. Sharma, L. Golubchik, and R. Govindan, "Sensor faults: Detection methods and prevalence in real-world datasets", ACM Transactions on Sensor Networks (TOSN), 6(3), 2010.
[4] F. Ayello, S. Jain, N. Sridhar and G.H. Koch, "Quantitative Assessment of Corrosion Probability – A Bayesian Network Approach", Corrosion, vol. 70, 2014.
[5] S. Guan, F. Ayello, A. N. Sánchez, V. Khare, and N. Sridhar, "Internal Corrosion Direct Assessment using Bayesian Networks Modeling with Limited Data: A Case Study", CORROSION/2016 paper #7078, NACE International, Houston, TX, (2016).
[6] S. Guan, C. Taylor, and N. Sridhar, "Understanding Sensor System Reliability", DNV GL Position Paper, 2016-2.
[7] A.Willersrud, M. Blanke, and L. Imsland, and A. Pavlov, "Fault diagnosis of downhole drilling incidents using adaptive observers and statistical change detection", J. Process Control. 30, 2015, pp. 90-103.
[8] A.F. Veneruso, H. Kohli, and M. Webster, "Towards Truly Permanent Intelligent Completions: Lifelong System Survivability Through a Structured Reliability Assurance Process", SPE 84326, Society of Petroleum Engineers Inc., 2003.

Bibliography:

[1] Shan Guan is a Principle Consultant and Senior Project Manager at DNVGL, where his research focus was on managing the risks of pipelines and oil & gas production systems, as well as applications of sensors on Maritime, Oil and Gas and Energy industry. He obtained a Ph.D. in Mechanical Engineering in 2004, and a Masters of Electrical Engineering in 2003, both from the University of Minnesota. He obtained a M.S. and B.S. in Materials Science from Bei Hang University, in 2000 and 1997, respectively. He has published over 50 papers and has been awarded more than 10 patents in the areas of MEMS, Ink-Jet Printing, Microfabrication and Electrochemical Processing.

RELIABILITY MODELING OF COMPLEX MULTI - STATE SYSTEM BASED ON BAYESIAN NETWORKS

Fangfang Shao, Yaqiu Li and Yufeng Sun*
School of Reliability and System Engineering, Beihang University
37 Xueyuan Road
Room 537
Haidian District, Beijing
Telephone: (86) 18801270510
shaofangf@buaa.edu.cn

Abstract: Traditional reliability analysis methods are not suitable for some systes with multi-state probability. In this paper, a Bayesian Network based multi-state reliability model is built according to the state relationship between the system and its constituting components. Universal generating function is used to build the conditional probability table of the non-root nodes. Based on this, a method of computing the system failure state's probability is proposed. The algorithm of the importance of each component in the system is inferred. Combined with an example of a typical natural gas pipeline compressor zone system is given. The result shows that this method in this paper provides a simple and intuitive measure to deal with the reliability analysis of natural gas pipeline compressor zone system with multi-state property. The reliability level can be evaluated and the affection of each component to the system reliability can be confirmed effectively.

Key words: Multi-state reliability; universal generating function; Bayesian Network

1 Introduction: Traditional reliability analysis is based on the assumption that events are binary, i.e., success and completely failure, while in actual engineering, many systems are in a state of partial failure between normal and complete failure, these states should be taken into account, especially when more accurate assessment of system reliability and system optimization design are necessary. Thus, the traditional re-liability theory of two states system can't meet these requirements; Multi-State System (MSS) reliability analysis is very important and essential.

MSS reliability analysis has received extensive attention over the past years; many researchers have put forward many methods in this field [1]. Because of the similarity between the MSS and traditional two states (or failure) system in many ways, the traditional modeling and analysis methods are extended and expanded in the field of MSS [2-3]. The modeling and analysis methods of multi-state fault tree are studied, based on the extension of traditional Binary Decision Diagram (BDD), a multi - state binary decision diagram (MBDD), a binary decision diagram (LBDD) of logarithmic transformation and a Multi - state binary decision diagram Multi-valued decision diagram MMDD, etc., are formed, which have been widely used. But the expression and solution method of the block diagram in the multi-state system reliability is relatively small; Markov analysis has the defects of state space explosion and cannot be used to model the complex and bulky systems.

In this paper, a new method for reliability modeling of complex multi-state systems based on Bayesian networks (BN) is proposed. The nodes of the BN and the various states of the components of the sys-tem are determined, and the probability of each state is given [4]. Then, conditional probability distributing (CPD) is used to describe the relations among the states of the elements to express the state of the associated nodes. Finally, a BN model of multi-state complex system is established. The model can express the state and state probability of the system and components clearly, and can directly calculate the system reliability according to the multiple state probability of the component, and qualitatively analyze and quantitatively evaluate the reliability of the multi-state system. In this paper, the marine steam turbine generator system is used as an example to verify the model, and the Bayesian network inference is used to analyze the reliability and component importance of turbo-generator system.

2 Introduction of multi-state system and Bayesian Network:

2.1 Multi-state systems: As the name suggests, a multi-state system is a system with multiple operating states, these different states can be different in performance, for example, the generator set can work at different power levels; can also be different in failure states, such as system with the long-time operation, due to wear and tear, fatigue and other failure modes lead to the gradual degradation of system performance, making the sys-tem to run a different level of degradation; can also be a system with different failure modes, which has different effects on the upper level system or other, such as the two failure modes of the valve. To sum up, a multi-state system is a system with a limited number of performance levels, and its various sub-systems and components are generally multi-state.

In the context of this study, the definition of the working state of the system or component is a certain state which can achieve a certain level of performance requirements or complete certain tasks, as well, the failure state of system or component which is unable to achieve a certain level of performance requirements or to complete certain tasks.

To analyze the reliability of a multi-state system, it is first necessary to analyze the characteristics of its components. Assume that any component j of the system has k_j

kinds of different states without loss of generality, corresponding to the performance level of the component in different states, which can be represented by a set as follow:

$$\boldsymbol{g}_j = \{g_{jC}, g_{j1}, \dots, g_{jK_j-1}\} \tag{1}$$

Where g_j represents the performance level of component which is in the ith state, $i=0,1,\dots,k_j-1$.

The performance level G_j of component j at any time is a random variable, which value range is determined by $\boldsymbol{g}_j$, $G_j \in g_j$. In some cases, the performance level G_j of component j cannot be determined by a single random variable, but may be determined by the random variables of several performance parameters, so the performance level at this time should be defined as random vector $\boldsymbol{G}_j$. In the multi-state system discussed in this paper, this situation is not considered.

The state probability of component j at different performance levels can be expressed as a set:

$$\boldsymbol{p}_j = \{p_{jC}, p_{j1}, \dots, p_{jK_j-1}\} \tag{2}$$

Where $p_j = \Pr\{G_j = g_j\}$.

Similar to the definition in a two-state system, the state probability of a component of a multi-state system is the probability of being in a state at a particular task time, and it is required that these state events be mutually exclusive, i.e. component j can only be in one state, and the sum of the probabilities for each state is 1.

Equation (2) defines the probabilistic quality function of the random variable G_j, i.e. the set of performance level states and their combinations of probability (g_{ji}, p_{ji}) can form the distribution of the performance level of the component. The cumulative distribution is shown in Figure 1. The horizontal axis x represents the performance levels of the component, and the vertical axis represents the probability that the component's performance level is not less than x.

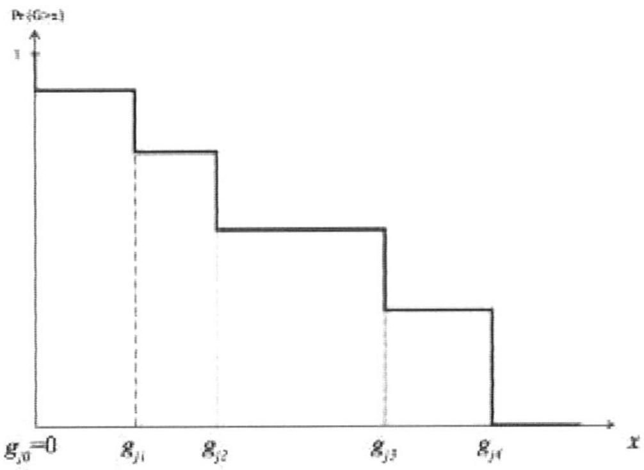

Figure 1: Cumulative Performance Distribution Curve of Multi - State System Components

For a multi-state system with n components, the performance level of the system is determined by the performance level of each component. Assuming that the whole system has K different states and g_i represents the performance level of the system at the ith state $(i \in \{0,1,\cdots,K-1\})$, then the performance level of the system is a random variable whose value range can form a set as $\{g_0, g_1, \cdots, g_{K-1}\}$.

A model of a multi-state system is defined, which includes the probability-quality function of the performance levels of the various system components (Eq. (3)) and the structural function of the system (Eq. (4)).

$$g_j, p_j, 1 \leq j \leq n \quad (3)$$

$$\varphi(G_1, K, G_n) \quad (4)$$

The universal generating function is used to describe the serial-parallel structure of multi-state component or system, and then the generating function polynomial of multi-state element is [5-7]:

$$U_j(z) = \sum_{i=0}^{K_j-1} p_j z^{g_j} \quad (5)$$

Where, z is an auxiliary variable, this formula also applies to multi-state systems.

1) Assuming n components in series connection, the generating function of the system is:

$$U_j(z) = m\left(U_0(z), U_1(z), \dots, U_{n-1}(z)\right) =$$

$$\sum_{i_0=0}^{K_0-1} \sum_{i_1=0}^{K_1-1} \dots \sum_{i_n=0}^{K_n-1} \prod_{j=0}^{K_n-1} p_{ji_j} z^{m\left(g_{0i_0}, g_{1i_1}, \dots, g_{n-1 i_{n-1}}\right)} \quad (6)$$

2) Assuming n components connected in parallel, the generating function of the system is:

$$U_j(z) = m\left(U_0(z), U_1(z), \dots, U_{n-1}(z)\right) =$$

$$\sum_{i_0=0}^{K_0-1} \sum_{i_1=0}^{K_1-1} \dots \sum_{i_n=0}^{K_n-1} \prod_{j=0}^{K_n-1} p_{ji_j} z^{m\left(g_{0i_0}, g_{1i_1}, \dots, g_{n-1 i_{n-1}}\right)} \quad (7)$$

2.2 Bayesian Network: Bayesian network (BN), also known as Belief Network, is a directed acyclic graph (DAG) with probability annotated [8]. Considering a finite discrete random variable set $U=V_1, V_2, \cdots, V_K$, where, each variable V_i can take a finite number of values. Formally, a Bayesian network is a two tuple $S=\langle G, P\rangle$. Of these, the first part G is a directed acyclic graph where the nodes mirror the random variables $V_1, V_2, \cdots, V_K$, and directed arch shows conditional dependencies between variables. It contains the following conditional independence assumption: given set of parent nodes, each variable is independent on its non-descendant nodes. The second part P is a set of parameters that describe the network distribution of the conditional probability and quantitatively represents dependence of each node with its parent nodes. Node without the parent node is called the root node; node without descendants is called a leaf node.

Based on the probability multiplication formula:

$$p(V_1, V_2, \dots, V_k) = \prod_{i=1}^{k} p(V_i | V_1, V_2, \dots, V_{i-1}) \quad (8)$$

If $A(V_i)$ represents any subset of nodes which non- V_i descendant nodes constitute, $Pa(V_i)$ represents direct parent nodes, then the assumption based on the conditional independence is:

$$p(V_i | A(V_i), P\ (V_i)) = p(V_i | P\ (V_i)) \quad (9)$$

Variable U joint probability distribution Bayesian network described can eventually be determined by the following formula only:

$$p(U) = p(V_1, V_2, \dots, V_k) = \prod_{i=1}^{k} p(V_i | P\ (V_i)) \quad (10)$$

Bayesian network construction steps are as follows:

a) Determine the corresponding relationship between variables with each node of the network.
b) Constructing a directed acyclic graph indicating conditional independence.
c) Determine conditional probability distribution parameters of nodes.

Figure 2 is a simple yet typical Bayesian network model, the joint probability of the variables in the model can be deduced by equation (10):

$$p(V_1, V_2, V_3, V_4) = \prod_{i=1}^{4} p(V_i | P\ (V_i)) = p(V_4 | V_2, V_3) p(V_3 | V_1) p(V_2 | V_1) p(V_1) \quad (11)$$

Bayesian network inference is realized by computing the probability of certain conditions, including the joint probability, the marginal probability as well as the calculation of each conditional probability.

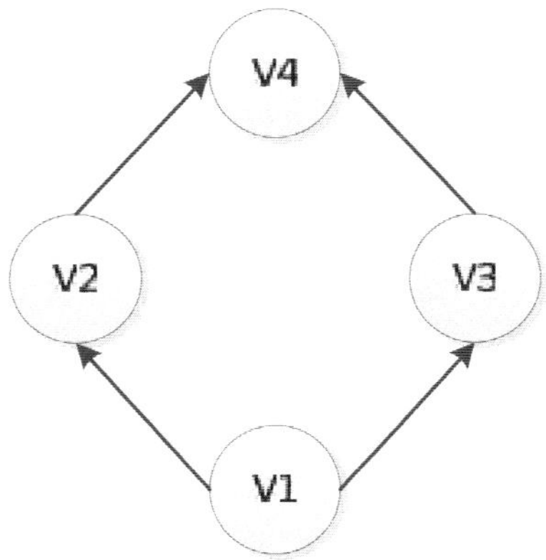

Figure 2: Bayesian network model (omitting the CPT)

Let the Evidence set of variables be E, Query be Q, under the condition of given evidence variable values $E = e$, Bayesian network inference calculates conditional probability distribution of the query variable Q. It can be described as follows:

$$p(Q | E = e) = \sum_{U-E} p(V_1, V_2, \dots, V_k) = \sum_{U-E} \prod_{i=1}^{k} p(V_i | P\ (V_i)) \quad (12)$$

3 Multi-state systems reliability modeling:

3.1 The state probability of component: Before modeling the system by the Bayesian network, the state of all the basic components is analyzed and divided, and the state probability of each state should be solved. For different situations, different methods can be used to obtain the state probability of the system.

1. State probability of non - repairable system with known reliability function
Assuming that the reliability function of the component j which changes with time t is $R_j(t)$, so that the reliability value of the component at any time t_0 can be obtained, the state probability distribution of the component at time t_0 can be considered as shown in Table 1:

Table 1: State probability distribution of non - repairable two - state system

State X	0	1
Probability	1-$R_j(t_0)$	$R_j(t_0)$

2. The state probability of the degenerate component with known degradation behavior
In a multi-state system, many components progress gradually over time and their performance is gradually declining until they fail to meet the performance requirements of the system. For such components, the performance degradation analysis theory can be used to analyze the state of the component.

For the components whose performance level is represented by a single performance parameter, we can perform a degradation test on them, the test data are distributed and fitted to obtain the degraded failure distribution of the components. For highly reliable products, an accelerated degradation test can be performed to get the test data [9].

After the degradation failure of the component is obtained, a degenerate failure probability can be obtained by setting a performance level as a failure threshold for the two-state component. For multi-state components, it is necessary to discretize this distribution, specify some state of discrimination, and set the threshold of each state, which can be the status of each state probability.

3. Solving method of state probability based on Markov Process
According to the time is continuous or discrete, the Markov process is divided into continuous-time Markov process and discrete event Markov process.

The probability of its state can be estimated by the Markov chain when the probability of transition between the states of the component is known [10]. However, the general Markov process can only solve the case where the life distribution of the component is exponential distribution, so for the case of non-exponential distribution, the semi-Markov process can be used to solve. Firstly, the process of converting between states in multi-state components is expressed as a continuous-time discrete state stochastic process, and then the integral equation of state transition is given as shown in Eq. (13).

$$\theta_{ik}(t)=\delta_{ik}\left[1-F_i(t)\right]+\sum_{l=1}^{m_j}\int_0^t q_{il}(\tau)\theta_{lk}(t-\tau)d\tau \tag{13}$$

Where

$$q_{il}(\tau)=\frac{dQ_{il}(\tau)}{d\tau},F_i(t)=\sum_{l=1}^{m_j}Q_{il}(\tau),\delta_{ik}(\tau)=\begin{cases}1, & i=k\\ 0, & i\neq k\end{cases}$$

Then, after giving the kernel matrix of the semi-Markov process and the initial conditions, we can obtain the state probability of each state [11].

3.2 Calculation of system failure probability: According to the expression of Bayesian network inference, the joint probability distribution is the product of all probabilities in the network [12] [9], which is:

$$P(X_0,X_1,\ldots,X_{n-1}) = \prod_{k=0}^{n-1} P(X_k|p\ (X_k)) \tag{14}$$

Where, *pa* (X_k) is the parent node of X_k.

Therefore, in the Bayesian network, the joint probability distribution can be used to directly compute the probability of occurrence of the top event *T*, that is, the failure probability of the system:

$$P(T=0) = \sum_{x_0,x_1,\ldots,x_i} P(X_0=x_0,X_1=x_1,\ldots,X_i=x_i,T=0) \tag{15}$$

Where X_0, X_1, ..., X_i are the bottom and middle nodes in the network, x_i is the possible state value of event *i*, 0 is the complete failure of the event; *i* is the number of non-leaf nodes in the network. According to Eq. (7) and (8), we can calculate the probability of occurrence of the top event *T*, that is, the failure probability of the system.

3.3 Calculation of structural importance: Computing the structural importance of a component is an important part of the reliability analysis. It is used to reflect the degree of change of the system caused by the slight change of the state of a component [13-14] [10-11], and to find out the weak points of the system.

The structural importance degree $I(j)$ of the element X_j can be calculated by the Bayesian network inference [15] [12], the equation as follows:

$$I(j) = \frac{1}{m-1}\sum_{l=0}^{m_j}\left[\left|P(T=0|X_j=l) - P(T=0|X_j=1)\right|\right] \tag{16}$$

Where m_j represents the fault state of the element X_j, l indicates that the element X_j is in the fault state, and m denotes the number of states of the element X_j.

The posterior probability algorithm formula in Eq. (15) is as follows:

$$P(T=0|X_j=0) = \frac{P(T=0,X_j=0)}{P(X_j=0)} \tag{17}$$

Where,

$$P(T=0|X_j=0) = \sum_{X_0,X_1,\dots,X_t} P(X_0=x_0, X_1=x_1, \dots, X_t=x_t, T=0, X_j=0) \quad (i \neq j) \tag{18}$$

$$P(X_j=0) = \sum_{X_0,X_1,\dots,X_t} P(X_0=x_0, X_1=x_1, \dots, X_t=x_t, X_j=0) \quad (i \neq j) \tag{19}$$

The posterior probability $P(T = 0 \mid X_j = 0)$ and the posterior probability $P(T = 0 \mid X_j = 1)$ are given by Eq. (14) (17) (18) (19) .The two posteriori probabilities are brought into Eq. (16) to compute the importance $I(j)$ of the component X_j.

4 CASE STUDIES: This chapter mainly uses the modeling method proposed in this paper to apply some typical systems to verify the effectiveness of the method. A typical natural gas pipeline compressor zone system is analyzed for the system reliability.

4.1 System introduction: A typical compressor area configuration diagram shown in Figure 3.

It can be seen from the figure, the natural gas from the compressor inlet tube input, respectively, through the three compressor lines, after the compressor pressurized, concentrated by the compressor outlet into the air cooler area through the air cooler cooling, and then to the downstream Piping. For each compressor line, there are inlet and outlet valves connected to the inlet and outlet piping, respectively, for the control of the function of the compressor line. The reliability diagram of the system is shown in Figure 4.

The compressor zone has three compressor lines which work together to achieve natural gas pressurization. The inlet and outlet valves in each compressor line are normally open and are closed when the pipe fails.

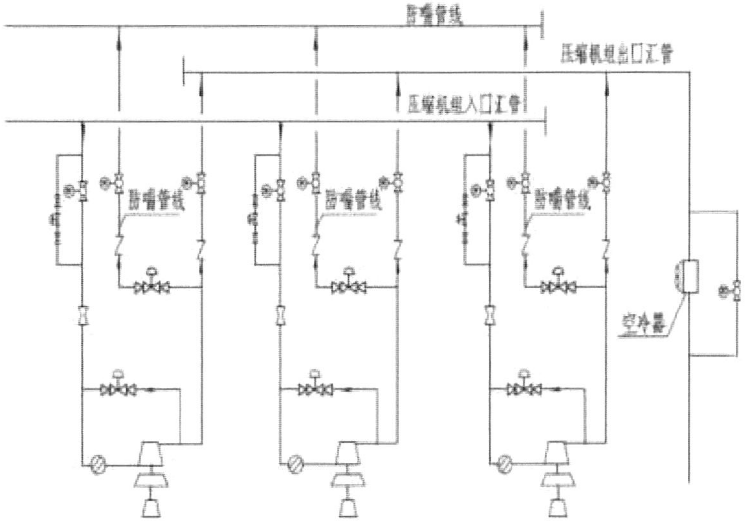

Figure 3: Typical compressor unit configuration diagram

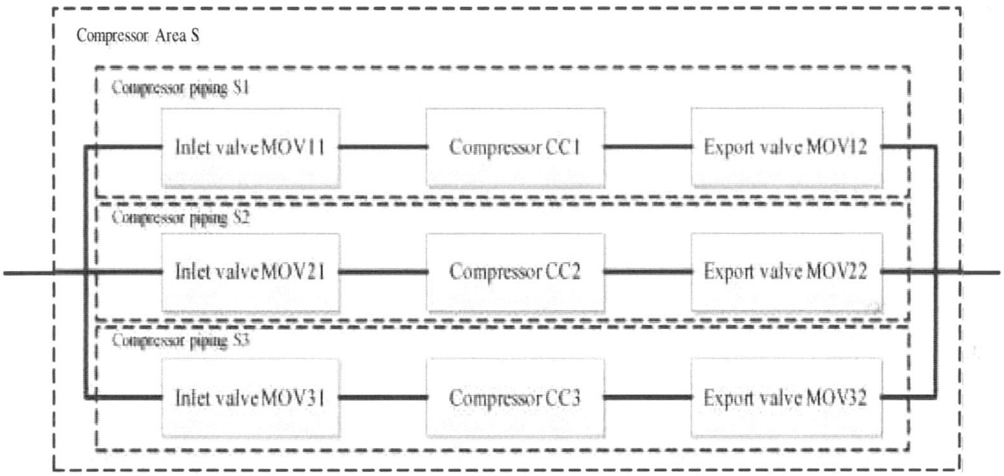

Figure 4: Typical Compressor Block Multi-State Reliability Block Diagram

4.2 Component state analysis: the compressor has two conditions by analysis, one is 3.75 million cubic meters per hour of gas transmission, and the other is 3 million cubic meters per hour, assuming its state as follows:

State 1: failure, the performance value is 0, the state probability is 0.0009.
State 2: condition 2, the performance value is 300, the state probability is 0.275.
State 3: condition 1, the performance value is 375, the state probability is 0.7241.

The universal generating function is:

$$U_{CC}(z) = 0.0009\,z^{0} + 0.275z^{300} + 0.7241z^{375} \tag{20}$$

The inlet and outlet valves have a variety of failure modes, corresponding to multiple states of the component, matching the multi-state system described in this article, after further failure mode analysis and we found that it has some failure states including leak, jam, cannot be turned off and cannot open. The different failure states have different effects on the system. From the failure of the point of view, these failure status and working conditions can be reduced to two states, so for the inlet valve and outlet valve, its state as follows:

State 1: leak, blocked, cannot be opened. The corresponding performance value is 0.
State 2: normal operation, the corresponding performance value is 1.

Assuming that the failure rate of the inlet and outlet valves is the same and the exponential distribution is obeyed, the probability of the two states is: state 1: $1-e^{-\lambda_1 t}$; state 2: $e^{-\lambda_1 t}$. Thus the general function of the inlet and outlet valves is:

$$u_{MOV}(z)=\left(1-e^{-\lambda_1 t}\right)z^1+e^{-\lambda_1 t}z^2 \qquad (21)$$

When $\lambda_1=1.9\times10^{-7}h^{-1}$ and t=87600 h, $u_{MOV}(z)=0.016506z^1+0.983494z^2$.

4.3 Reliability modeling and calculation of gas pipeline compressor area system based on Bayesian Network: Compressor line S3 circuit as a cold backup, the system dynamic operation obeys rules of m / N (w): G system. Each component life and maintenance time is subject to exponential distribution. According to the method described above to determine the failure mode of each unit and the conditional probability table shown as Table 5.

Table 2: Compressor area failure mode and data sheet

System/Component Name	Failure mode Name	Failure Rate h^{-1}	Maintenance Rate h^{-1}
Compressor area	Compressor area failure	/	/
Compressor piping	Compressor piping failure	/	/
Inlet valve	Inlet valve failure	3.64*10e-5	0.303
Export valve	Export valve failure	3.64*10e-5	0.303
Compressor	Compressor failure	2.47*10e-4	0.0641

The corresponding Bayesian Network is shown in Figure 5, set run time t = 8760 h:

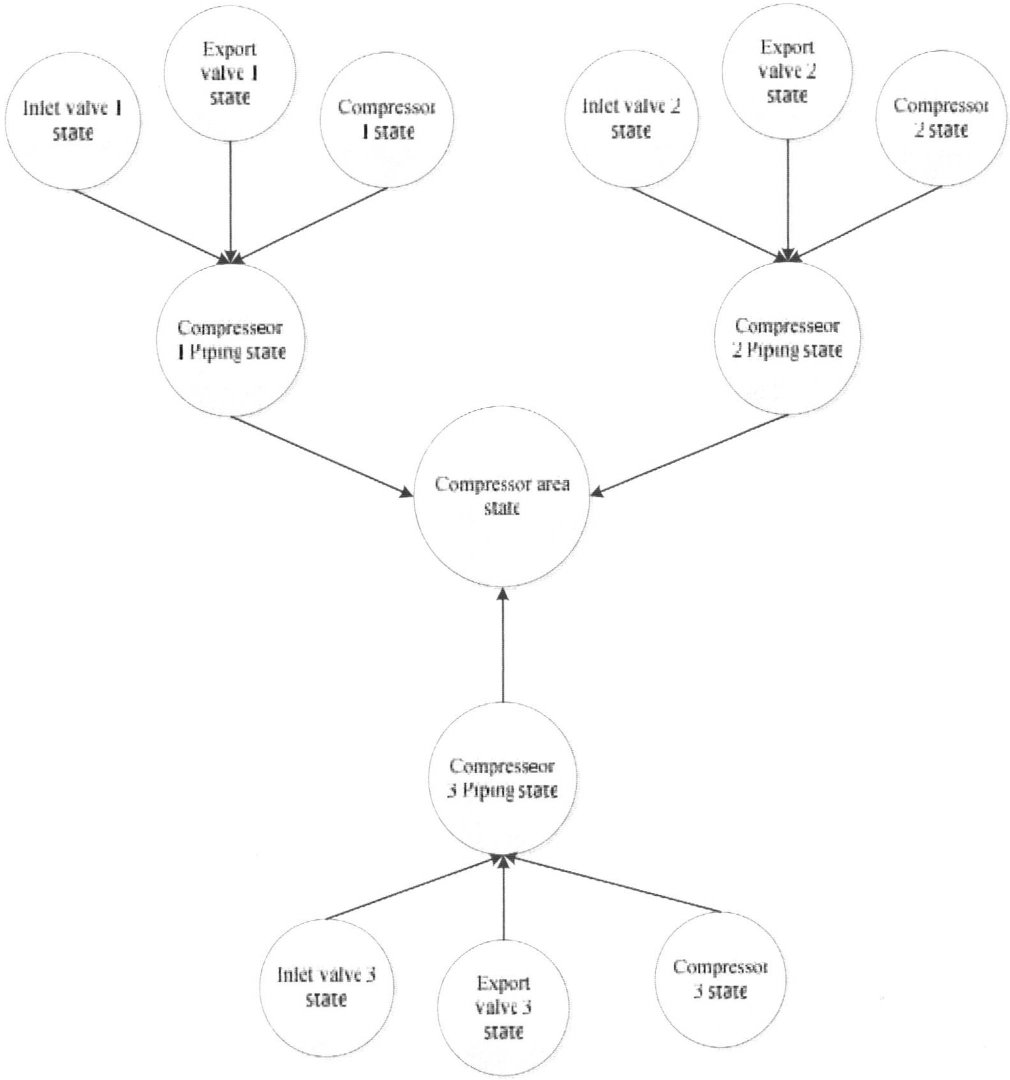

Figure 5: Bayesian network model of gas pipeline compressor area system

By using the forward predictive reasoning, the availability of compressor area system is 0.99551554. Based on the established Bayesian network model and the importance calculation method introduced in Section 3.3, the influence of the change of the bottom events on the total loss of function of the marine steam turbine generator system was obtained.

5 Conclusions: In this paper, Bayesian network and universal generating function is used to construct reliability evaluation model, and Bayesian network inference is used to get the probability of fault occurrence and the importance of each component in the system. It solves the polymorphism problem of probabilistic reliability evaluation of gas pipeline compressor area system. It is proved that this method is practical and feasible through the example of marine steam turbine generator system reliability evaluation. It can provide technical support for improving the system reliability of gas pipeline compressor area system.

References:

[1] Hirsch W M, Meisner M, Boll C. Cannibalization in multicomponent systems and the theory of reliability [J]. Naval Research Logistics Quarterly 1968, 15(3):331-360.

[2] Lisnianski A, Levitin G. Multi-state system reliability: assessment, optimization and applications [M]. World scientific. 2003.

[3] Yingkui G, Jing L. Multi-state system reliability: A new and systematic review [J]. Procedia Engineering 2012: 29:531-536.

[4] Lianwen,Zhang &Haipeng,Guo. Introduction to Bayesian Networks [M]. Beijing: Science Press. 2006.

[5] Peng,Gao &Liyang,xie. Reliability Analysis of Multi - State System Based on Improved Generating Function Method [J]. Chinese Journal of Aeronautics 2010, 31(5): 935-939.

[6] Bo,Ren. Zhenzhou,Lv &Guijie,Li. Reliability Analysis of System Life Based on Universal Generating Function[J]. Chinese Journal of Aeronautics 2013, 34(11): 2550-2556.

[7] Hoang Pham. The Universal Generating Function in Reliability Analysis and Optimization [M]. Germany: Springer. 2005.

[8] Jensen FV &Nielsen TD. "Bayesian networks and decision graphs,"2nd ed. New York: Springer. 2007.

[9] Aimin.Deng. Research on Reliability Technology of High Reliability and Long Life Products [D]. National University of Defense Science and Technology, 2006.

[10] Lisnianski A, Levitin G. Multi-state system reliability: assessment, optimization and applications[M]. World scientific, 2003.

[11] Yanlong.Shang, Qi.Cai & Xinwen.Zhao. Multi - State Reliability Analysis of Reactor Pump Unit Based on UGF and Semi - Markov Method [J]. Nuclear power engineering, 2012(01):117-123.

[12] Zhongbao,Zhou &Doudou,Dong. Application of Bayesian Networks in Reliability Analysis[J]. System Engineering Theory and Practice 2006, 6: 96-100.

[13] Timo Koski &John M Noble. Bayesian Networks an Introduction[M] UK: WILEY. 2009.

[14] Xiaowei,Yin. Wenxue,Qian &Liyang,Xie. Bayesian Network Evaluation Method for System Reliability[J]. Chinese Journal of Aeronautics 2008, 29(6): 1483-1489.

[15] Liang,Zeng &Xin,Guo. Component importance analysis of multi - state monotone correlated system[J]. Systems Engineering and Electronics 1998, 2: 72-74.

REPAIR OF SIMULATED BATTLE DAMAGED ALUMINUM USING COLD SPRAY TECHNOLOGY

Carl Paxton
Dennis Miller
Bowhead Science and Technology

Blake Barnett
Victor Champagne
Marc Pepi
US Army Research Laboratory
Weapons and Materials Research Directorate
Aberdeen Proving Ground, Maryland

Abstract

The US Army Research Laboratory was asked by the US Army Tank Automotive Research, Development and Engineering Center (TARDEC) to research the feasibility of cold-spray repairing aluminum armor panels that were shot at with 20-mm fragment simulating projectiles. The through-holes were filled using the CGT4000 cold-spray equipment. The repairs were deemed successful, since the objective was to provide protection against anticipated chemical, biological, radiological, and nuclear exposure, and to make the armor air and water tight. This result puts cold-spray technology on par with the current hull and metal component repair requirement which states, "Hull patches do not provide any additional ballistic protection; they are designed to maintain hull integrity (air/water tight)."

Introduction

Battle Damage Assessment and Repair (BDAR) is the process used to rapidly return disabled military equipment to operation by applying field expedient repairs to damaged components [1]. According to FM 4-30.31 [2], BDAR can be accomplished by bypassing components or safety devices, cannibalizing parts from like or lower priority equipment, fabricating repair parts, jury-rigging, taking shortcuts to standard maintenance, and using substitute fluids, materials, or components. References 1 and 2 do not provide remedial guidelines for ballistic repair and only list the following "fixes" with respect to repairing metal component repair:

- Metal epoxy stick (applied to damaged components as a temporary means of controlling fluid loss)
- 1:1 ratio metal polymer
- 3:1 ratio metal polymer (the best choice of polymer for engine blocks and coolant system components exposed to higher temperatures and pressures)

Welding has also been used as a means of repairing battle-damaged components; however, caution must be exercised when exposing metal components to these thermal gradients, due to the propensity of altering the prior heat treatment of the part. The US Army Tank and Automotive Research, Development and Engineering Center (TARDEC) tasked the US Army Research Laboratory (ARL) to determine the feasibility of repairing aluminum armor plates, subjected to range ballistic testing with cold-spray technology.

Cold Spray Technology

Cold spray was chosen for this repair process, because no heating of the substrate is encountered that could deleteriously affect the prior heat treatment of the armor plate. Cold spray is a mature repair technology, and the spraying of aluminum generally leads to one of the most reliable depositions of all cold-spray materials.

Cold-spray systems operate through the aerodynamic acceleration of small particles, which subsequently form a deposit through impact with a surface or with previously deposited particles [3-6]. The cold-spray process imparts supersonic velocities to metal particles by placing them in a heated nitrogen or helium gas stream that is expanded through a DeLaval converging–diverging nozzle. The powder feed is inserted at the nozzle entrance or immediately after the nozzle throat. High pressures and temperatures yield high gas velocities and high particle acceleration within the gas stream. The particles, entrained within the gas, are directed toward a surface where they embed on impact, forming a strong bond with the surface and previously deposited particles. Subsequent spray passes increase the structure thickness. The adhesion of the metal powder to the substrate, and the cohesion of the deposited material, is accomplished in the solid state.

Figure 1 is the diagram often used to illustrate this process [6]. The term "cold spray" has been used to describe this process due to the relatively low temperatures (100–500°C) of the expanded gas stream that exits the nozzle. The relatively high gas pressures used for this geometry (10–60 bar) result in the name High-Pressure Cold Spray (HPCS). The rate of powder spray (hence the rate of deposit buildup) is typically 0.05–0.2 kg/min. As the nozzle is moved over a surface, the deposited line is 5–10 mm wide, with a surface roughness, Ra, of 10–20 m [7,8].

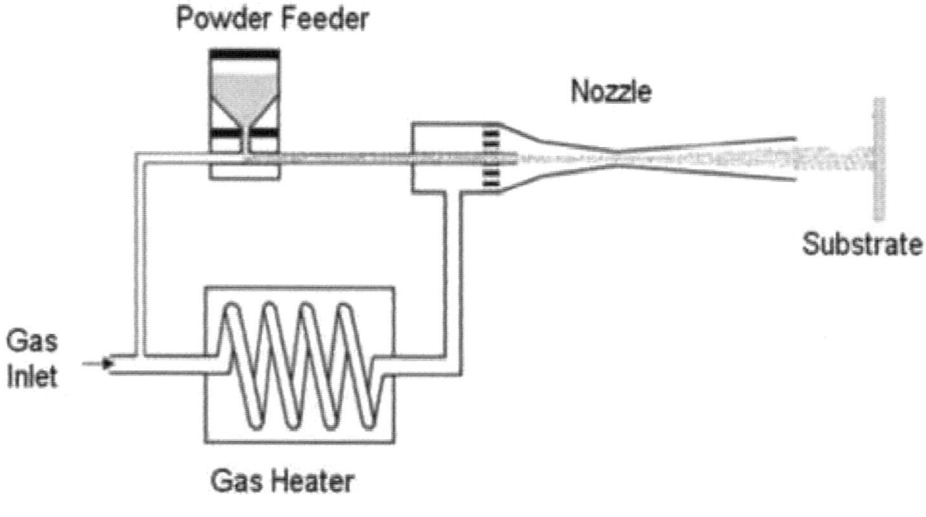

Figure 1 Schematic of a typical cold-spray assembly

Experimental

TARDEC shipped an aluminum alloy (AA) 5083 armor plate to ARL that was 24-inch × 24-inch × 0.75-inch thick (Fig. 2). It had been shot at with 20-mm fragment simulating projectiles (FSP) resulting in complete penetration through the armor. These holes were to be repaired by ARL using the cold-spray process.

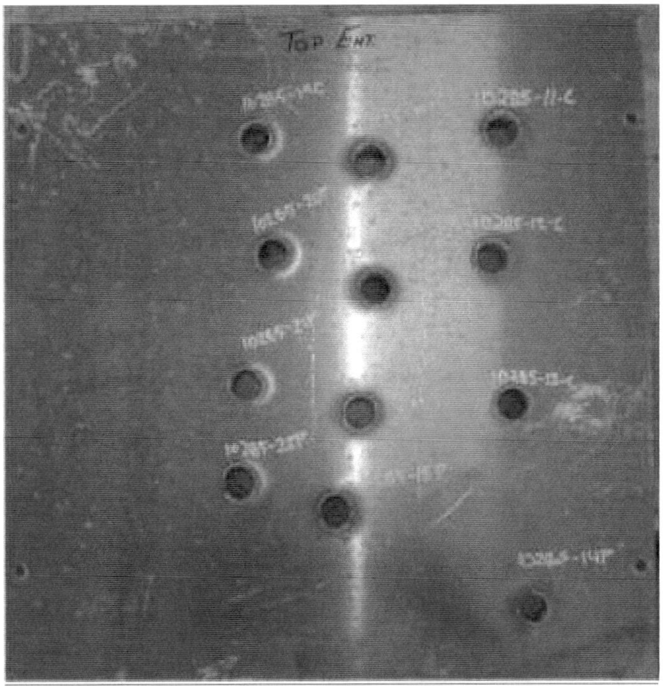

Figure 2 As-received damaged AA5083 plate prior to cold spray repairs.

ARL was tasked to identify a surface preparation protocol for repairing these through-holes, as well as a suitable aluminum alloy for cold spray. ARL began with coupon-level simulated repairs, which were subjected to microstructural evaluation to get a visual of surface adhesion and fill integrity. The test coupon is shown in Fig. 3. This simulated a chamfered hole, with a backplate, which turned out to be the process used to repair the holes in the armor plate. Figure 4 shows a cross section of the coupon after it was filled with cold spray using 80% AA6061 and 20% aluminum oxide (Al_2O_3) powder; Fig. 5 shows the cross section after metallographic preparation.

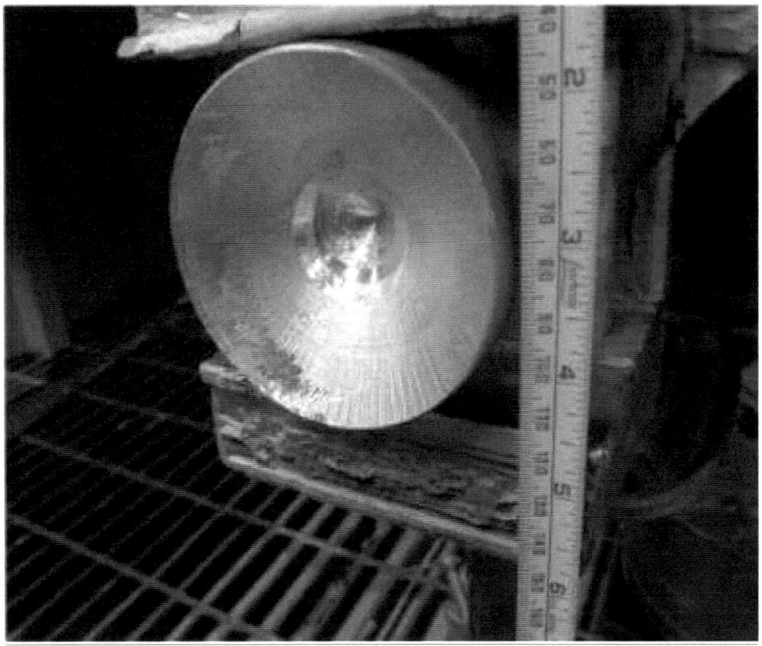

Figure 1 Test coupon utilized to determine feasibility of cold spray repair of the armor plate.

Figure 2 Cross section of test coupon showing cold spray fill (region above arrows).

Figure 3 Metallographically prepared cross section shown in Figure 4.

The CGT4000 system was used for these repairs, and the parameters used are listed in Table 1.

Table 1 Cold spray parameters used for repair

Cold Spray Parameters	Robot Programming
N_2 gas at 30 Bar Gun temperature at 500°C Powder feeder at 1 rpm	Spiral program at 2000; circle program as hole is filled; back to spiral program

There appeared to be excellent adhesion between the cold-spray fill and the substrate, and the coating showed little to no porosity/cavities.

Once it was determined that the process was adequate on coupons, an actual damaged plate was repaired via cold spray. A typical projectile through-hole is shown in Fig. 6. These holes were chamfered, as shown in Fig. 7, to allow better adherence of the cold-sprayed powder, using a countersink HS 4-3-120 tool (Fig. 8). Once the chamfering was complete, the Motoman robot that holds the cold-spray nozzle, was programmed to spray the powder perpendicular to the angled sides of the chamfer. Once robot programming was accomplished, a backplate was affixed to the aluminum armor to allow for the cold-sprayed powder to build up from the bottom. As the hole was being filled, the robot would need to be reprogrammed to accommodate the new contour of the hole. Figure 9 shows an example of a completed cold-spray filled hole. After ensuring that the fill was above the surface of the plate, the cold spray was ground down to the level of the plate (Figs. 10 and 11) using a 1-inch to 3-inch fly cutter tool, followed by a 1/2-inch end mill and surface sanded using an air Dymet sander and Scotch Brite pads. Figure 12 shows a plate in various stages of repair.

Figure 4 Typical ballistic damage noted on the as-received aluminum armor plate.

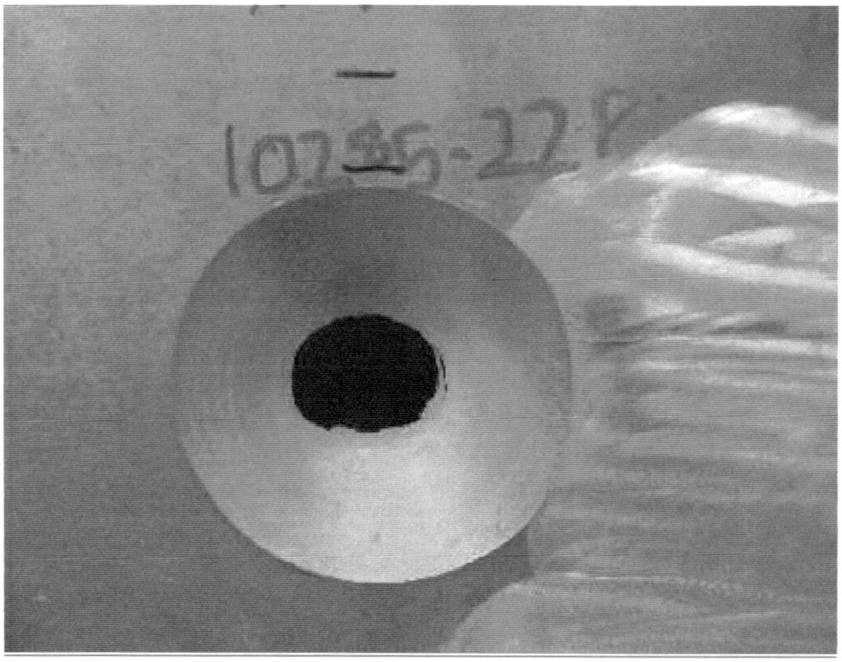

Figure 5 Chamfering of hole to allow for better cold spray adherence.

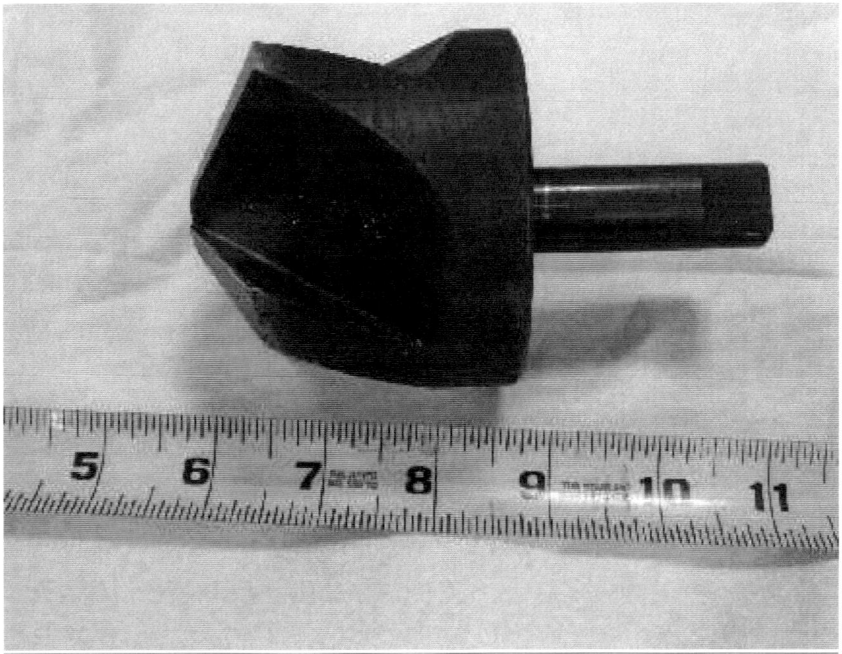

Figure 6 Tool utilized to chamfer projectile holes in preparation of cold spray fill.

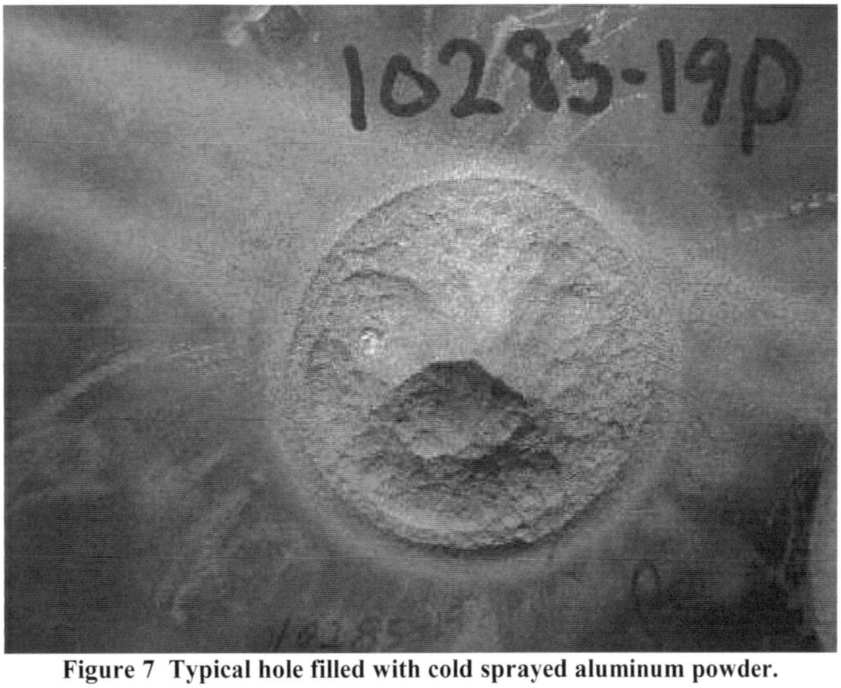

Figure 7 Typical hole filled with cold sprayed aluminum powder.

Figure 8 **Typical filled hole ground smooth with a fly cutter tool, followed by a ½-inch end mill, and surface sanded using an air Dymet® sander and Scotch Brite® pads.**

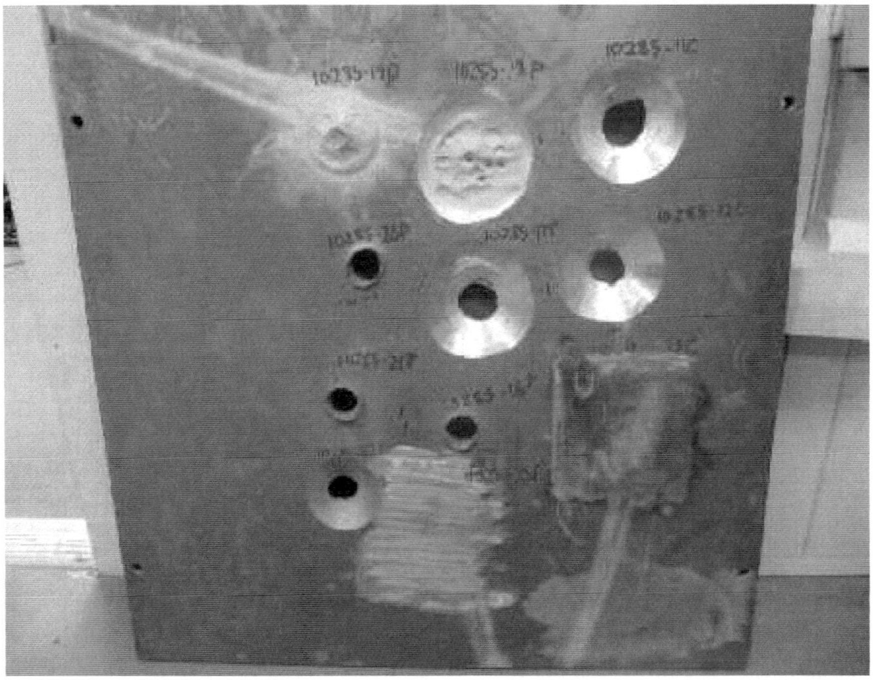

Figure 9 **Similar to Figure 10, a filled hole ground smooth with a fly cutter tool, followed by a ½-inch end mill and surface sanded using an air Dymet® sander and Scotch Brite® pads.**

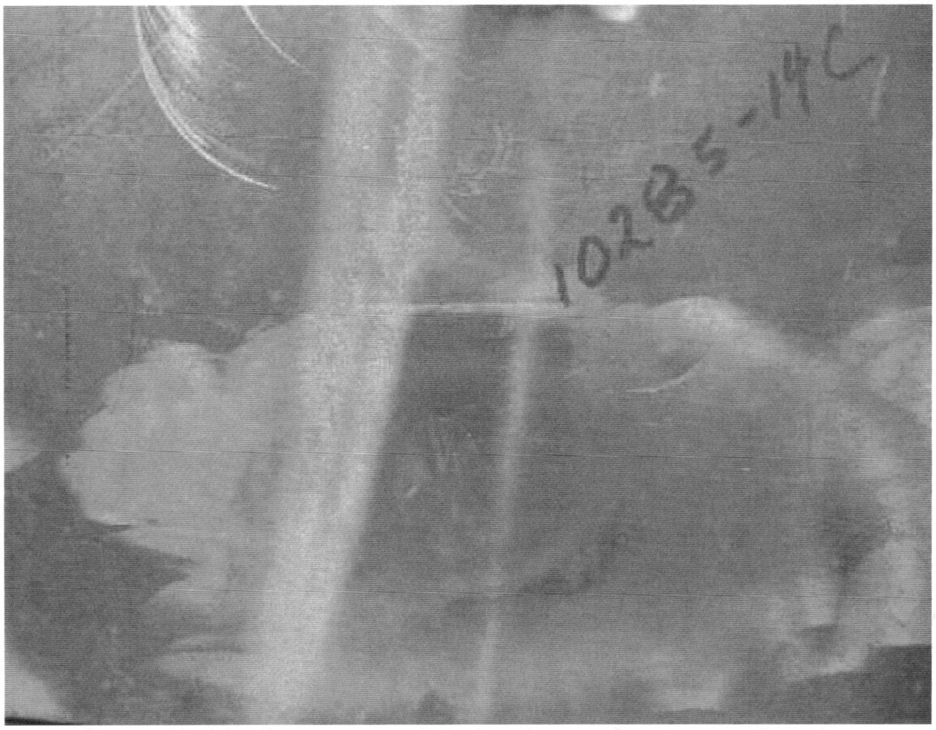

Figure 10 Aluminum armor plate showing varying degrees of repair.

Conclusions and Future Research

As far as being able to patch holes created by projectiles in aluminum armor, cold spray performed admirably. This could be deemed successful, if the object was to provide protection against anticipated chemical, biological, radiological and nuclear exposure. This result puts cold-spray technology on par with hull and metal component repair as outlined in FM 4-30.31,2 which states, "Hull patches do not provide any additional ballistic protection; they are designed to maintain hull integrity (air/water tight)."

This process could possibly be improved by first grinding off the damaged area smooth to the plate. Then, a slightly tapered cylindrical "plug" could be created from a solid rod of aluminum or of the same material as the damaged plate. The plug would need to have the larger diameter end just slightly larger than the hole in the plate. The plug would need to be driven or pounded into the hole, so that the larger end is approximately 1/8-inch below the surface of the plate. The edge of the hole would then need to be tapered, beveling the wall of the hole to 30°. This will greatly reduce the void needing to be filled with cold-sprayed powder at the inlet section of the wall and plug surface. The remaining area would simply be filled with cold spray on the plug and taper so that the coating is above the surface of the plate. The deposit could then be machined smooth with the surface of the plate as described previously, or left as coated. This would result in a much quicker repair process and be less expensive, as less powder and cold-spray operation time would be needed.

Acknowledgments
The authors wish to extend thanks to James Newman and Mike Nikodinovski, both of the US Army Tank Automotive Research, Development and Engineering Center, as the customers for this work, and their desire to employ cold spray for battle damage repair applications.

References

[1] GTA 01-14-001. Battle damage assessment & repair smart book. Aberdeen (MD): Training and Doctrine Command (TRADOC) Program Office for BDAR/R; 2012 Mar.
[2] FM 4-30.31. (FM 9-43-2); MCRP 4-11.4A (FMFRP 4-34). Recovery and battle damage assessment and repair. Washington (DC): Headquarters, Department of the Army; 2006 Sep.
[3] Alkhimov A, Kosarev V, Nesterovich N, Papyrin A, Shushpanov M: Gas-dynamic spraying method for applying a coating. United States patent US 5,302,414, 1994 Apr 12.
[4] Champagne V. Cold spray materials deposition process–fundamentals and applications. Cambridge (UK): Woodhead Publishing Limited; 2007.
[5] Schmidt T, Gartner F, Assadi H, Kreye H. Development of a generalized parameter window for cold spray deposition. Acta Materialia. 2006;54:729–742.
[6] LiW, Zhang D, Huang C, Yin S, YuM, Wang F, Liao H.Modelling of impact behaviour of cold spray particles: review. Surface Engineering. 2014;30:299–308.
[7] Champagne V, Helfritch D, Leyman P, Grendahl S, Klotz B. Interface material mixing formed by the deposition of copper on aluminum by means of the cold spray process. J Thermal Spray Tech. 2005;15:330–334.
[8] Helfritch D. Cold spray systems capabilities. Proceedings of the CSAT Conference; 2013 Jun; Villa Park, IL.

RESEARCH INTO RELIABLE, INTELLIGENT AND COST EFFECTIVE USE OF ACCELEROMETERS FOR THE CONDITION MONITORING OF ROTATING MACHINES

Rmdan Shnibha
Harouge Oil Operations- Libya
Ghanant Al_Arif
24 December Street
Telephone: (218) 911700535
rshnibh@yahoo.com

Abstract: This paper presents fundamental work being carried out on designing and building an intelligent monitoring system based on vibration measurements

The vibration sensor, e.g. the accelerometer, is the main element in the condition monitoring system and all the signal processing and decision-making depend upon its performance. Unfortunately, the high cost of accelerometers limits their usage, especially, when simple electronic circuitry interfacing is essential and the bulky and costly additional charge amplifiers are unsuitable, particularly in smart and wireless vibration measurement systems.

MEMS-based accelerometers offer an alternative solution owing to their low cost and small size. However, MEMS accelerometers are delicate and cannot stand harsh industrial environments, although no thorough investigation into their performance within such an environment has been carried out.

In this paper, three newly-developed MEMS accelerometers, from different manufacturers, were packaged in-house and their performances were examined and compared with a well-known piezoelectric accelerometer.

A 1.1 kW three phase induction motor test rig was purposely designed and built for this research to undertake vibration measurements and condition monitoring testing. MEMS accelerometers performance testing was carried out under different motor speeds and load conditions. Two types of common faults were also introduced, namely, phase imbalance, an electrical fault, and misalignment for the mechanical type of fault.

It was concluded that the performance of one of the MEMS accelerometers was improved by the packaging strategy; reasonable and comparable results were achieved.

Keywords: intelligent, condition monitoring, MEMS, induction motor

Introduction:

Condition monitoring and condition-based maintenance aim at increasing machine availability, performance, reducing consequential damage, increasing machine life, reducing spare parts inventories and reducing break-down maintenance.
A general condition system is depicted in figure 1. The key part of this model is sensing through the use of sensors that measure key attributes of machine, information-processing (monitoring, collection, and displaying) through the use of an embedded computer. The signal condition takes the transducer signal and modifies it to a desired magnitude; decision –making and action-taking are done by humans.
With an intelligent condition monitoring system, pre-warnings can identify the nature of the problem and inform maintenance personnel. A controlled or planned shutdown of the process reduces further potential problems and damage, resulting in reduced production loss and repair cost while time needed to repair and restart the process is reduced.

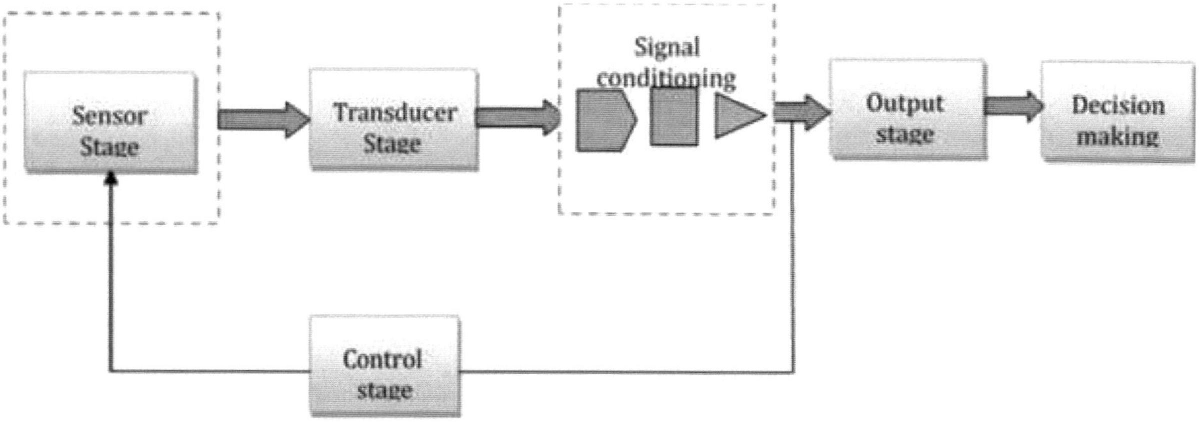

Figure 1 General condition monitoring.

As shown in figure 1 the precision selection and location of sensors is very important and the decision analysis depends upon their performances.
Typically, a smart sensor contains physical transducers, a network interface, a processor and memory core that can all be fabricated on a single die. The physical transducers can be made very small because silicon micromachining technology used to make integrated circuits is now being used to create micro-electro-mechanical systems (MEMS) that can become an integral part of a smart sensor. Thus, MEMS technologies and smart sensors are now acknowledged as a major focus of the current sensor development.
Vibration measurement is at the heart of the monitoring of rotating machines. Vibration monitoring is non-invasive but uses a number of specialised sensors, broad bandwidth and complex analysis. The precise selection and location of sensors is very important and the decision analysis depends on their performance.
Unfortunately, the high cost of the vibration sensor (e.g. the accelerometer) limits their usage, especially, when simple electronic circuitry interfacing is essential and the bulky and costly additional charge amplifiers are unsuitable. Consequently, smart and wireless

vibration measurement systems combined with specialist software and a (well paid) engineer analysing the data are required.
However, microelectromechanical system MEMS-based accelerometers offer an alternative solution owing to their low cost and small size, mobility and flexibility. They can be integrated with smart sensors to enable signal processing and self- diagnosis capabilities.
This has led to a new area of research within the field, allowing cheap sensor devices to be created which will enable the industry to move towards condition-based maintenance rather than periodic or breakdown based maintenance. This will reduce maintenance costs and avoid accidental shut-downs of critical plants. According to Maruthi et al [1], vibration analysis using a MEMS accelerometer is less expensive compared to conventional sensors (US $ 250 to US $ 25).
However, at present, the use of the MEMS accelerometers for machine condition monitoring is still limited to the testing stage in the laboratory experiments; Sabin [2] has used the MEMS accelerometer together with a conventional accelerometer for measuring the vibration of a pump during its normal operation. He found that the frequency content from both sensors was in good agreement. However, no rigorous investigation has been conducted to compare the performance of these MEMS accelerometers which are used to detect induction motor faults. Albarbar et al [3] performed a study which proved that MEMS accelerometers are of sufficient quality to be utilized for machinery health and condition monitoring purposes. He carried out a test performance of a MEMS accelerometer for different excitations and compared it with a piezoelectric accelerometer. He found that the MEMS accelerometer had good agreement with the piezoelectric accelerometer in his experiment [4]. Colin et al [5] performed a study which showed that an array of low cost MEMS transducers can determine results comparable with those obtained using high performance transducers by testing a composite vertical stabilizer (tail plane).
Induction motors are among the most important electromechanical systems, which are widely used in almost every industry. Induction motor faults such as winding faults, unbalanced stators and rotors, broken rotor bars, eccentricity and bearing faults have been studied. Several studies have shown that 30%-40% of induction motor failures are due to stator winding breakdown. Slow fluctuations in the supply voltage or even imbalance between the phases can cause operational problems such as overheating which leads to winding insulation over-stress. As a rule of thumb, for every 3.5% voltage unbalance per phase, the winding temperature increases by 25% in the phase with the highest current. Transient voltage conditions result in reduced winding life [6]. Shaft misalignment is a prevalent fault associated with rotating machines, and it occurs when the shaft of the driven machine and the shaft of the driver machine do not rotate on a common axis. Shaft misalignment is a measure of how far apart the two centrelines are from each other.
In this study, three newly developed MEMS accelerometers, from different manufacturers, were packaged in house and their performances were examined and compared with a well-known piezoelectric accelerometer. Two test rigs were conducted for sensors performance: the first was a shaker test, and the second one was the actual experiment.
This paper is organized as follows. In the second section MEMS accelerometers are presented. The proposed signal processing technique is introduced in section 3. Section 4

describes the test rig, the instrumentations and data collection. A discussion of the results and conclusions are presented in Sections 5 and 6 respectively.

2. MEMS Accelerometer

The EMS accelerometers, which include both the signal conditioning circuitry and the sensor, are fabricated together on a single monolithic chip. The output signals are analogue voltage that are proportional to acceleration. It uses sensing change in capacitance as shown in figure 2. The deflection of the internal mass changes the capacitance between the finger and the adjacent cantilever beams. The sensor has an electronic device which will convert the capacitance change due to acceleration into a voltage.

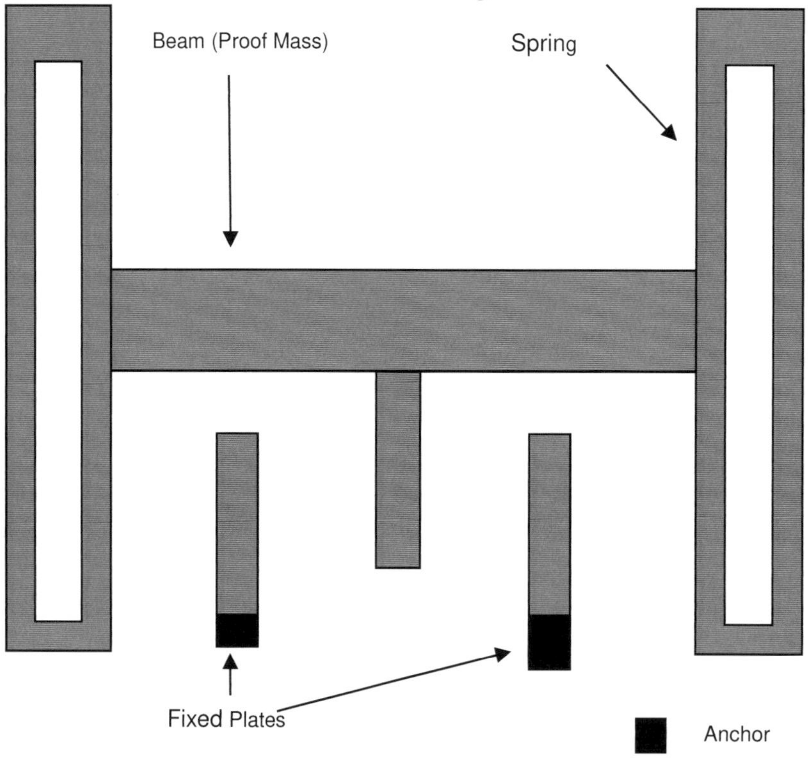

Figure 2 Capacitive MEMS accelerometer

The choice of accelerometers depends on several factors and some of them are listed below:
Static characteristics of sensors

- Input range : interval between the maximum and minimum of input
- Output range : interval between the maximum and minimum of output
- Span : the interval of output range of a measurement system Span= Omax -omin
- Zero: the system output corresponding to a zero input.
- Resolution: is the smallest detectable incremental change of input that can be detected in output signal.
- Sensitivity: is the rate of change in output corresponding to the rate of change in input.
- Repeatability: inability of a sensor to represent the same value under identical conditions.

- Deadband: range of input in which the output remains at zero.
- Hysteresis: the delay phenomenon in output due to energy dissipation.

Dynamic characteristics of sensors

A measure of a sensor's capability of the following rapid change in input:

- Delay (response time)
- Rise time
- Overshoot
- Settling time

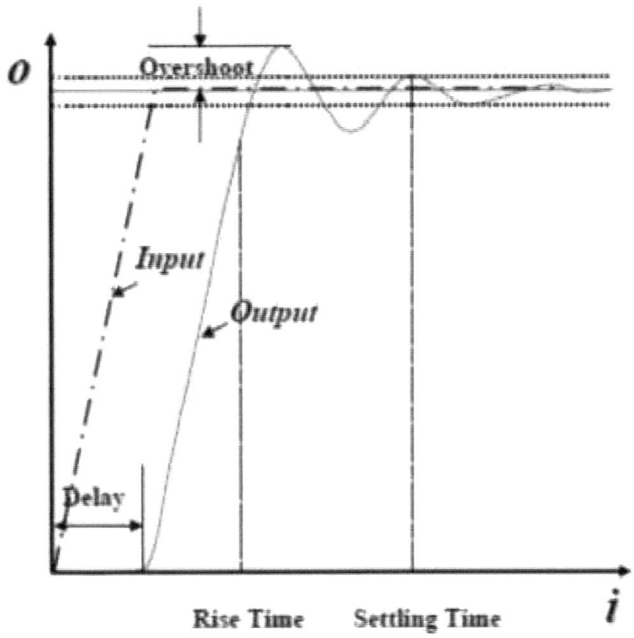

Figure 3 dynamic characteristics of sensor

There are more factors to consider when choosing the sensor such as cost, size, weight, linearity, and environment (temperature, shock, vibration, etc.).

3.1 Induction motor (Experimental) test rig.

The schematic diagram for the test rig and instrumentations is shown in Figure 4. It comprises a 1.1kW variable speed induction motor, a DC Generator and a resistor bank to act as a load and to dissipate the electrical energy. The instrumentation consists of a conventional accelerometer and three MEMS accelerometers. Conventional accelerometer sensitivity is 100mV/g. The output of the piezoelectric accelerometer is filtered and amplified via a charge amplifier. The technical specifications of the piezoelectric accelerometer and the MEMS accelerometers are briefly listed in Table 1. The model numbers and the manufacturers' names of the MEMS accelerometers used in the experiments were deliberately not mentioned, as the intention was to share experiences among several engineers and researchers involved in the area of vibration sensing and condition monitoring. The MEMS accelerometers were packaged to make them more robust for industrial use. The accelerometers' power supplies were stabilized to 5 volts using a solid state voltage regulator to prevent the power supply from affecting the sensitivity. The measured data were then transferred to the computer using 24-bit resolution data acquisition. The sampling frequency was set to 50k Hz.

The developed LabVIEW based system allows the user to monitor and store machine variables, and subsequently a MatLab code was used for further signal processing.

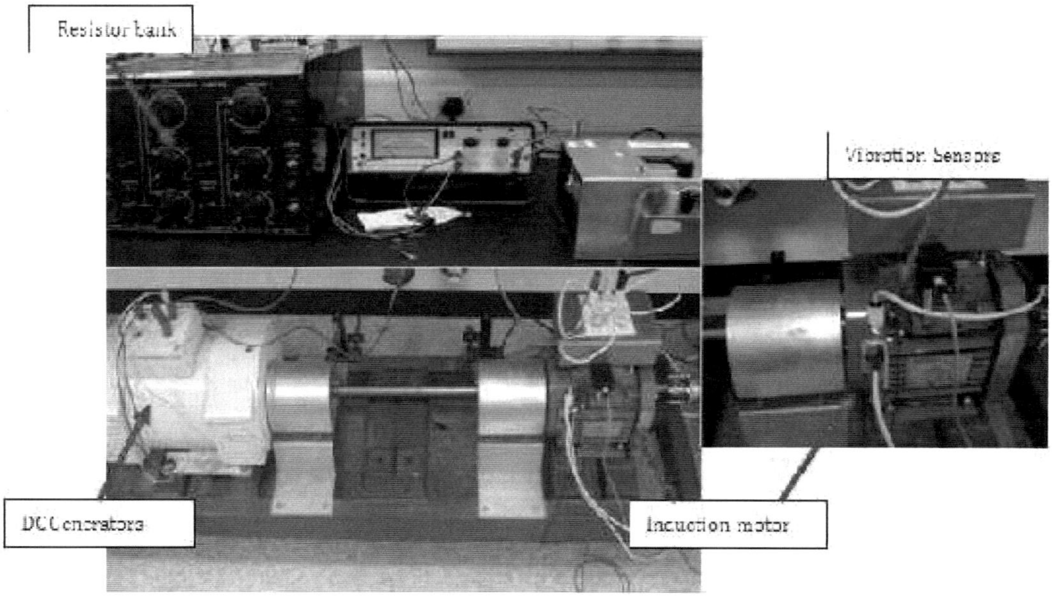

Figure 4 Schematic diagram of induction motor test rig

Table 1 Technical specifications of accelerometers.

Formatting Issue - Best Available Paper/Graphic

	Piezoelectric	MEMS 1	MEMS 2	MEMS 3
Price £	315	15	32	22
Mass	27 g	<1.0 grams	<1.0 grams	<1.0 grams
Sensitivity	100mv/g for Vs=5V	250 mV/g for Vs=5 V	250 mV/g for Vs=5 V	120 mV/g for Vs=5 V
Resolution	3mg	2mg	2mg	2 mg
Frequency response 3dB	2kHz	10 kHz	32 kHz	18 kHz
Temperature range	-40C to 200C	-40 C to + 85 C	-40 C to + 125 C	-55C to 125C

3.2. Shaker Test of MEMS sensors

The test rig was built as shown in figure 5 to test the performance of the MEMS accelerometers. The test rig contains a shaker, function generator, amplifier, data acquisition and computer. A number of test experiments conducted when the shaker was excited by sine wave at different frequencies and with different amplitude levels and responses were simultaneously measured from all accelerometers.

The time domain and frequency domain of measured signals are shown in figures 7 to 10, where the exciting frequency was set at 55 Hz and 95 Hz and at different amplitudes, 0.2 & 0.7 g.

It can be clearly shown that all the responses of the MEMS 2 are in good agreement with the Piezoelectric accelerometer.

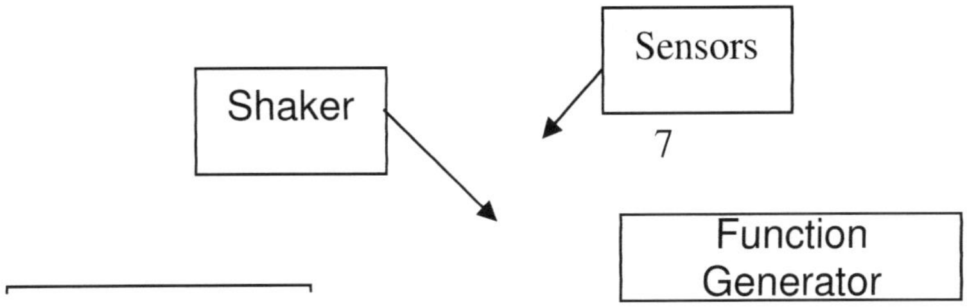

Figure 5 test rig for performance mems sensor

3.3 Packaged unit

The MEMS accelerometers were packaged as shown in figure 6 in the laboratory to assess whether MEMS accelerometers can stand harsh industrial environments are easy to use, easy to connect to machines, easy to power and can be easily protected. The package unit is cheaper compared with using a conventional sensor.

Figure 6 Packaged unit of MEMs sensors

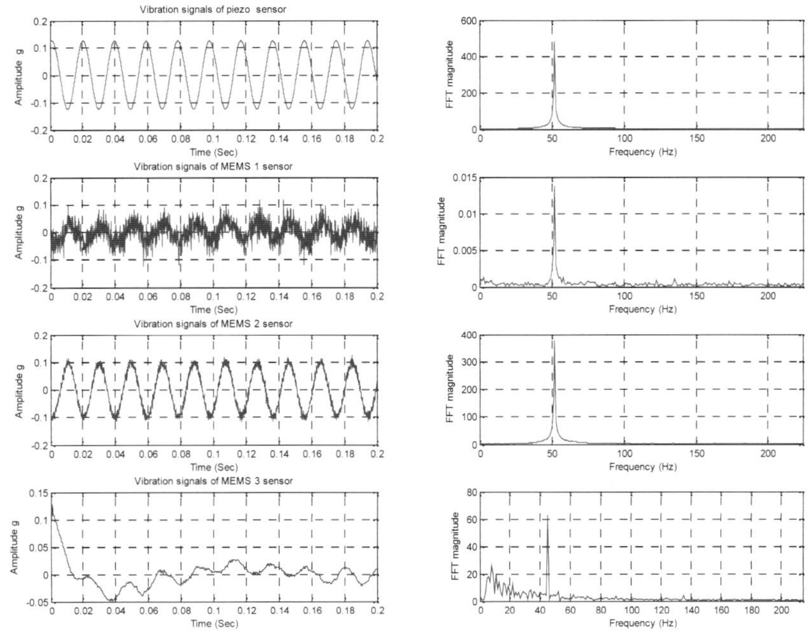

Figure 7 Measured acceleration responses by the MEMS accelerometers and the piezoelectric accelerometer at 55Hz for the excitation amplitude 0.2g

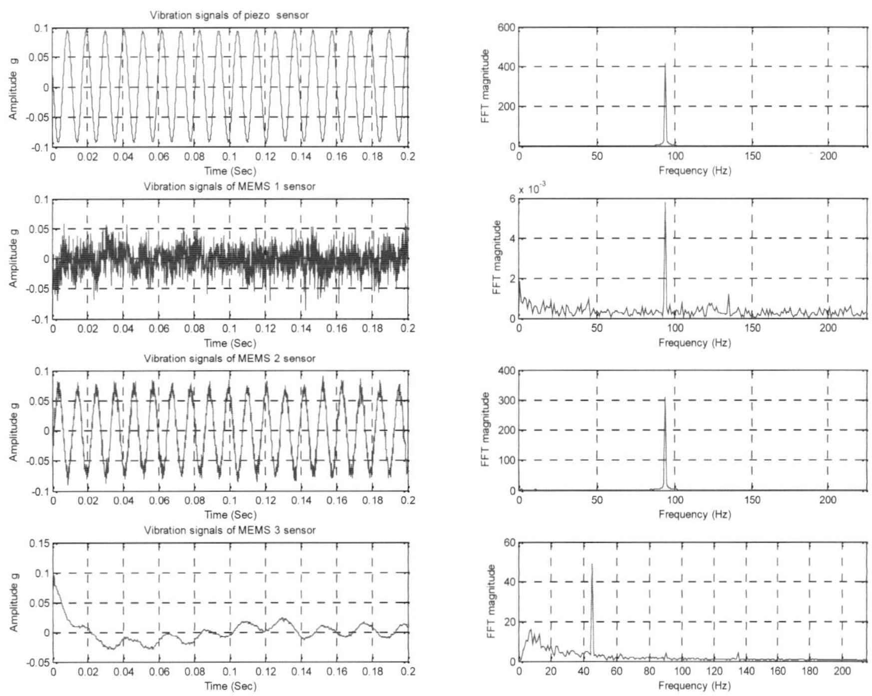

Figure 8 Measured acceleration responses by the MEMS accelerometers and the piezoelectric accelerometer at 95Hz for the excitation amplitude 0.2g

Formatting Issue - Best Available Paper/Graphic

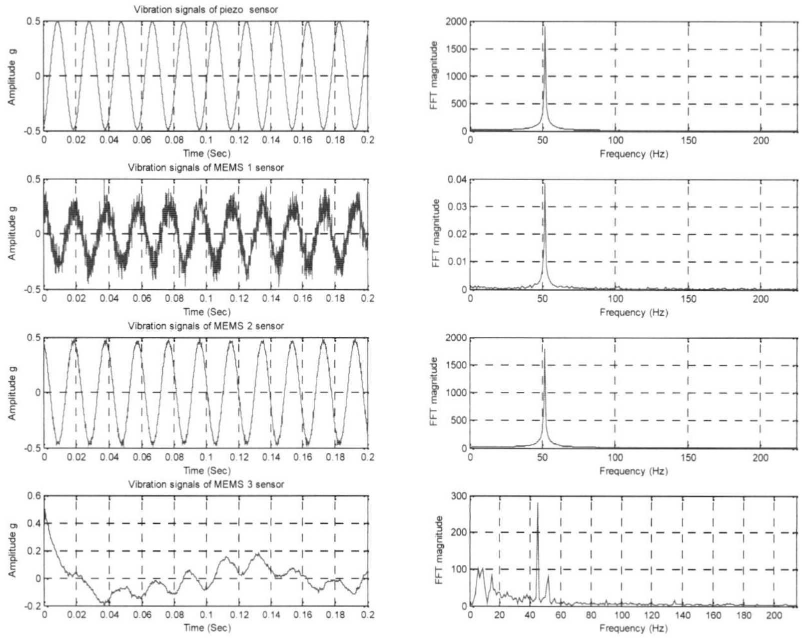

Figure 9 Measured acceleration responses by the MEMS accelerometers and the piezoelectric accelerometer at 55Hz for the excitation amplitude 0.7g

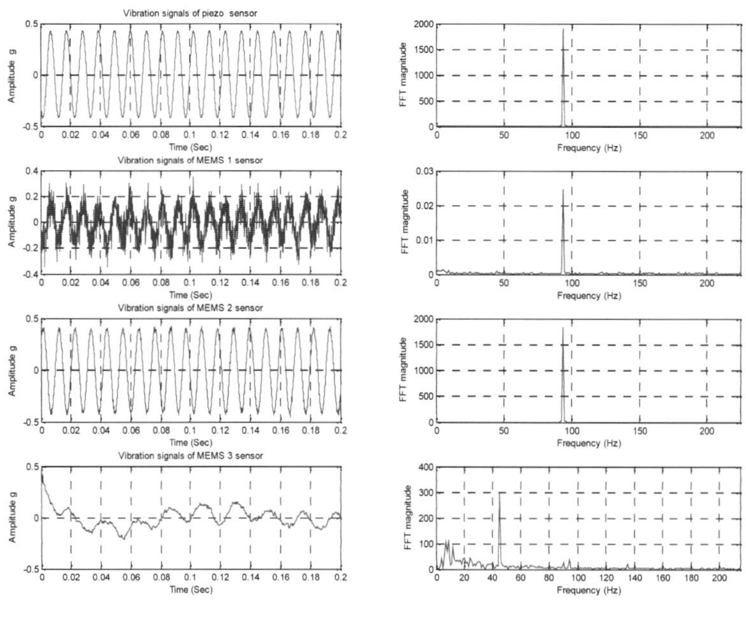

Figure 10 Measured acceleration responses by the MEMS accelerometers and the piezoelectric accelerometer at 95Hz for the excitation amplitude 0.7g

4. Experimental Result

The motor was tested at 1500 rpm and 100% load in a healthy condition and then under four different phase imbalance voltages of 5, 10, 15, 20 and 25 V; these represented 2%, 4%, 6%, 8% and 10% respectively, of the main nominal voltage level. And mechanical faults (1mm and 2mm misalignment)

A number of statistical methods were applied to the measured vibration signal (see Figures 11 and 12). The applied statistical signal processing includes RMS, STD and Kurtosis. No sufficient and reliable condition-related information can be observed. Further information about statistical signal processing including RMS, STD and Kurtosis can be found in various papers [7].

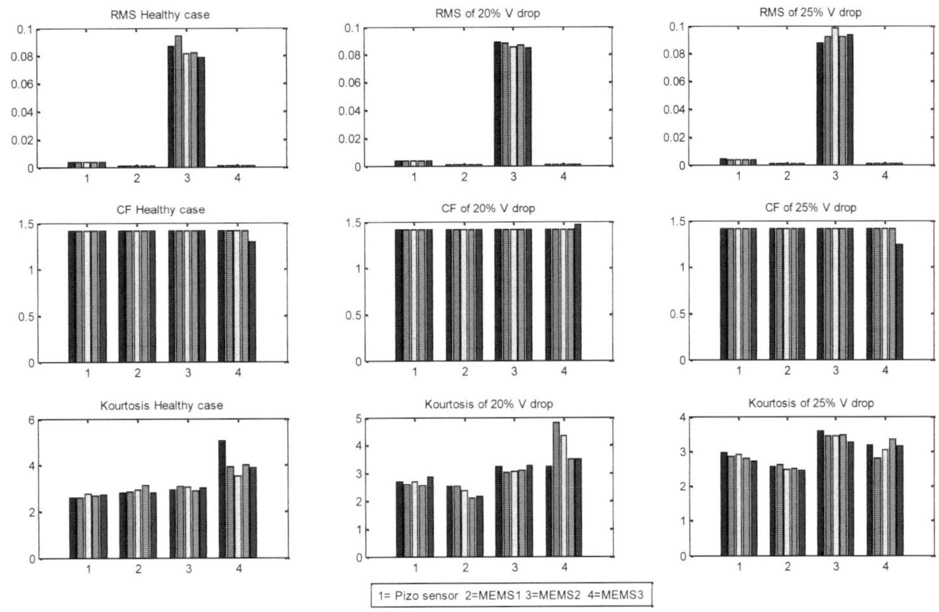

Figure 11 Time domain parameters: the motor at full load is equipped with various conditions (healthy, 20% & 25% V drop at 1500 rpm speed)

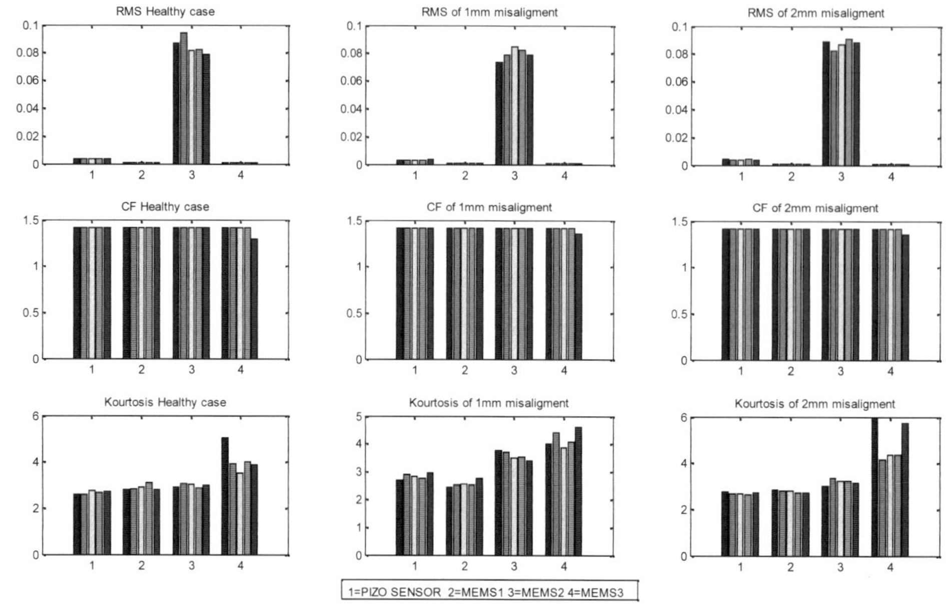

Figure 12 Time domain parameters. The motor at full load is equipped with various conditions (healthy, 1mm & 2mm misalignment at 1500 rpm speed)

The measured signals were transformed into the frequency domain, where the frequencies of their main components were identified. In Figure 13 (vibration) the first frequency component is at 25 Hz, rotor rotation speed; the second is at 50 Hz, due to the power supply or the second harmonic of rotor running speed; the third is at 100Hz, which could be the second harmonic of power supply frequency or fourth harmonic of the rotor frequency. There was a slight increase in the amplitudes of these components.

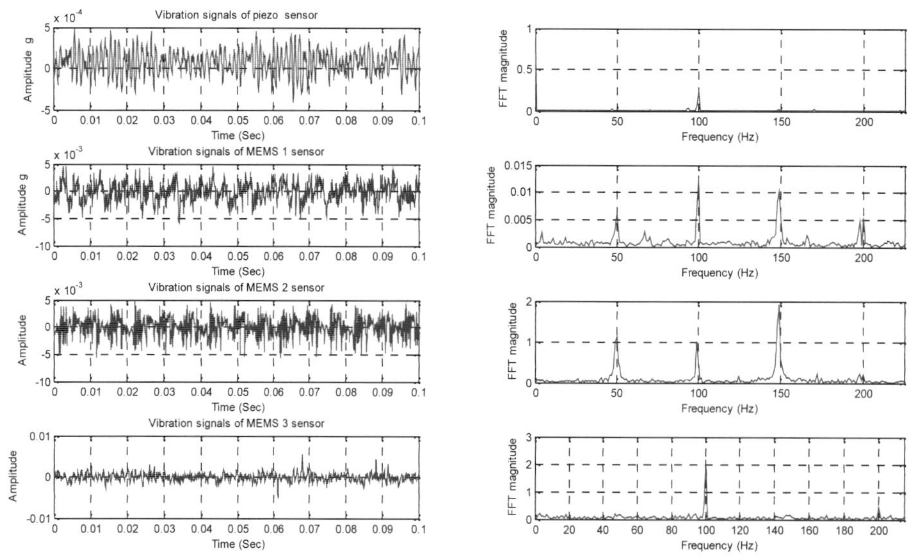

Figure 13 Time and Power spectra of the signal of the piezoelectric accelerometer and MEMS accelerometer. The motor is at full load and 1500 rpm speed

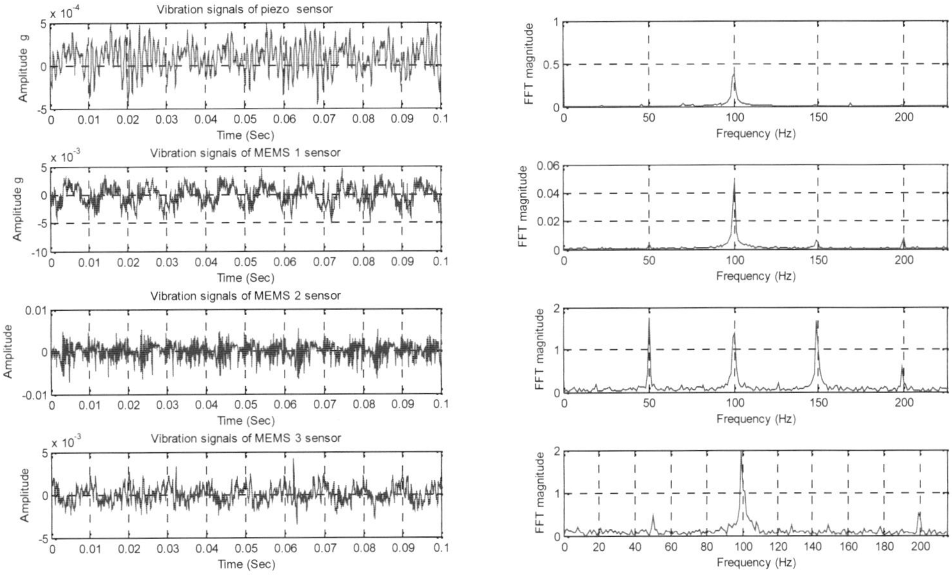

Figure 14 Time and Power spectra of the signal of the piezoelectric accelerometer and MEMS accelerometers. The motor at full load is equipped with 20% voltage drop at 1500 rpm speed.

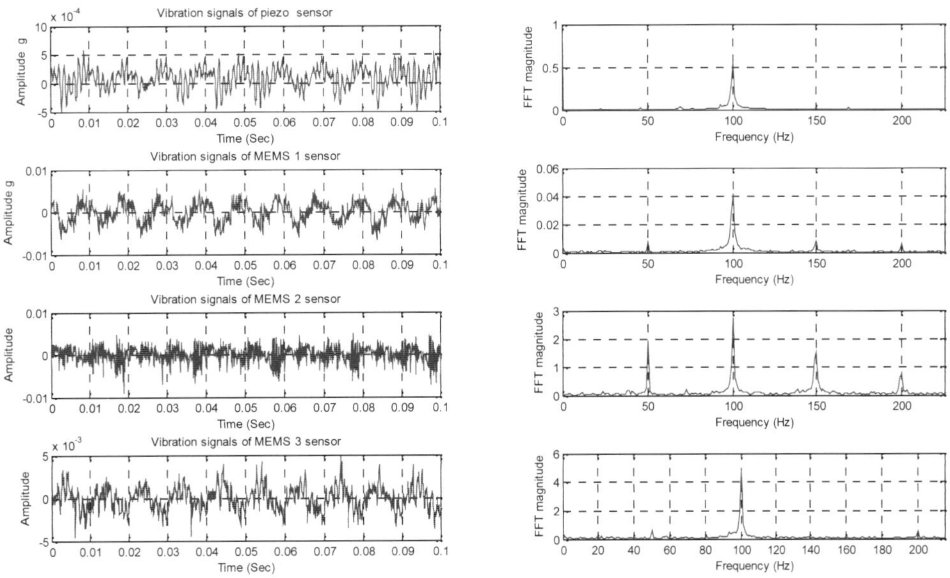

Figure 15 Time and Power spectra of the signal of the piezoelectric accelerometer and MEMS accelerometers. The motor at full load is equipped with 25% voltage drop at 1500 rpm speed.

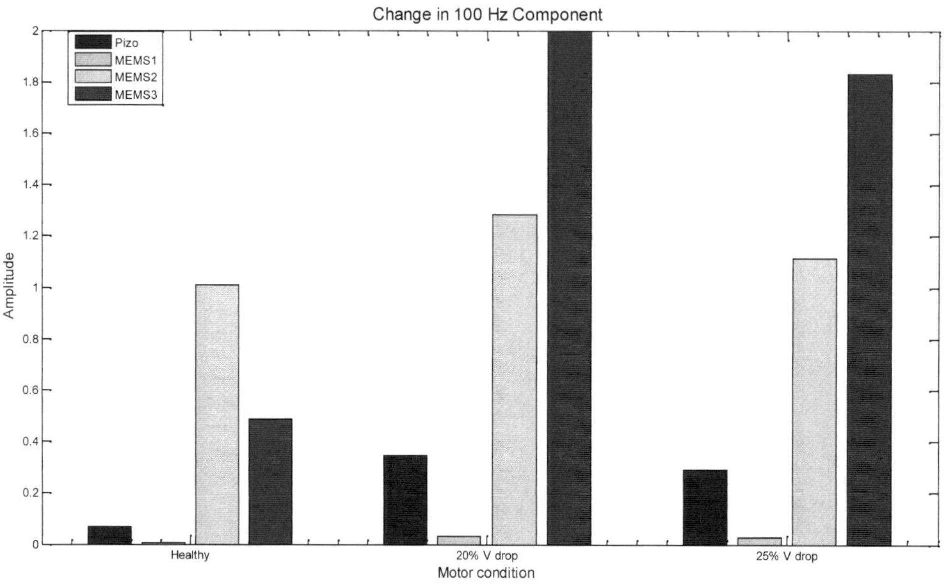

Figure 16 Changes in 100Hz components from healthy and 20% & 25% voltage drop. The motor is at full load and 1500 rpm speed.

A comprehensive comparison of the ability of each sensor, in detecting a phase imbalance fault and assessing its severity based on the change in the amplitude of the 100Hz component is shown in figure 16 (for both 20% and 25% voltage imbalance). The plots show that the MEMS2 sensor is more capable of detecting the faults.

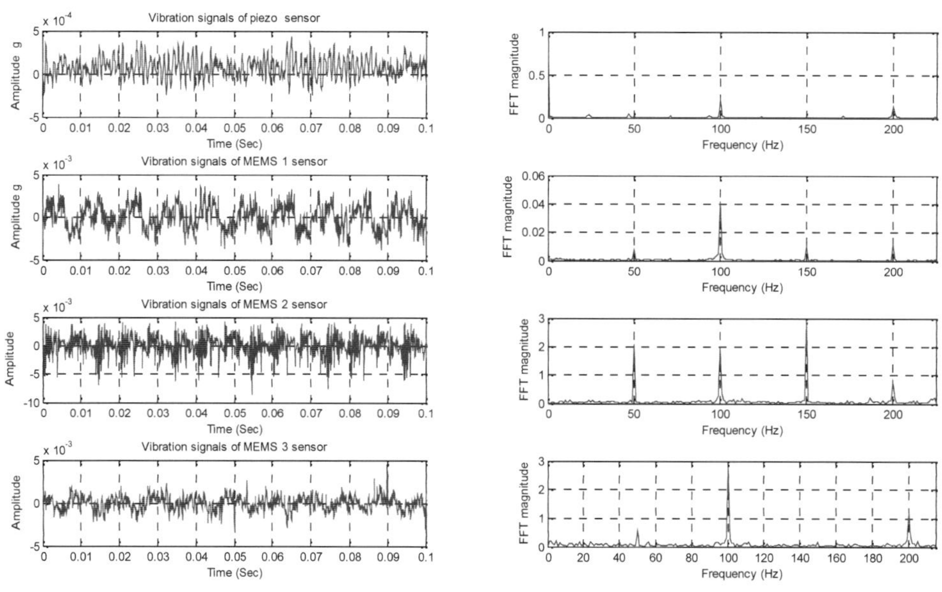

Figure 17 Time and Power spectra of the signal of the piezoelectric accelerometer and MEMS accelerometers. The motor at full load is equipped with 1mm misalignment at 1500 rpm speed.

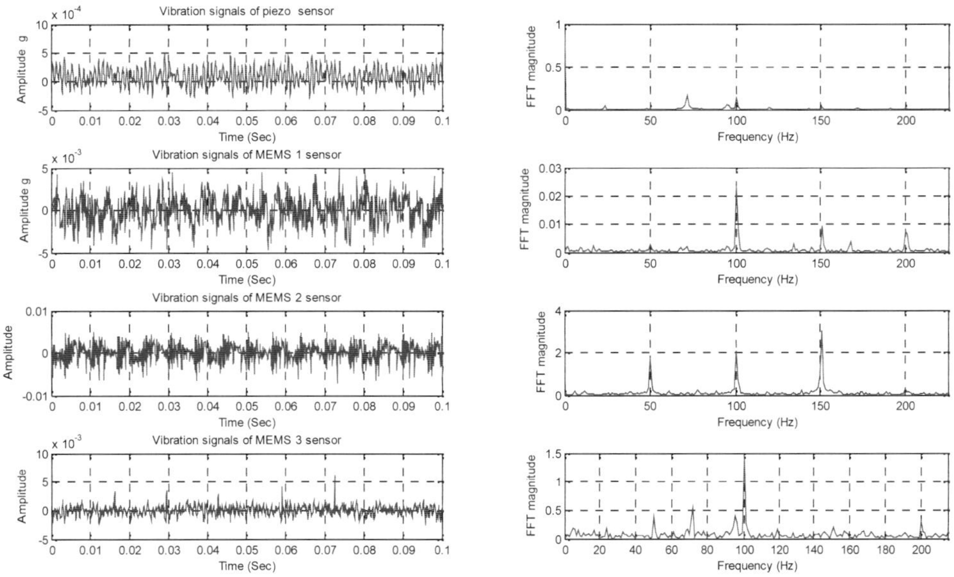

Figure 18 Time and Power spectra of the signal of the piezoelectric accelerometer and MEMS accelerometers. The motor at full load is equipped with 2mm misalignment at 1500 rpm speed.

Figure 17 and 18 show the spectra of vibration from the piezoelectric sensor and 3 different MEMS sensors for healthy operation, with 1mm misalignment and 2mm misalignment. In line with the theory, the 1st, 2nd and 3rd harmonics of rotational speed in vibration change in amplitude between healthy operation and faulty conditions. These changes are shown in figure 19.

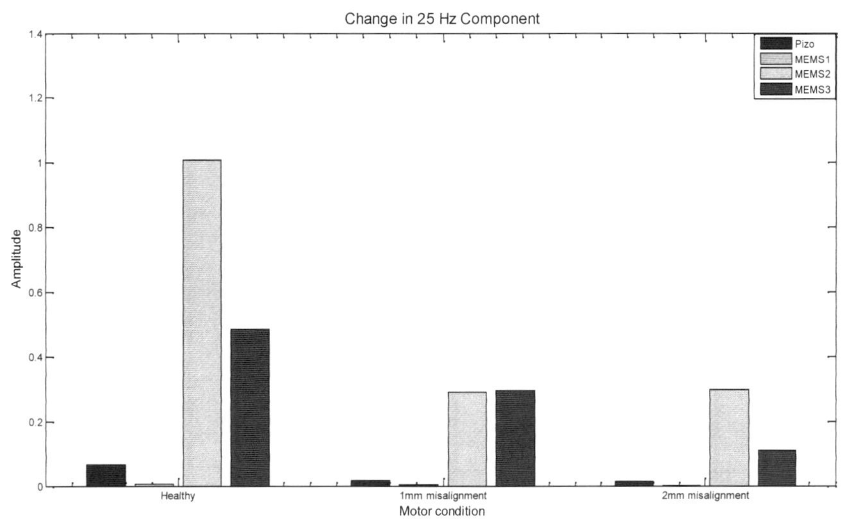

Figure 19 Change in 100Hz components from healthy and 20% & 25% voltage drop at 100. The motor is at full load and at 1500 rpm speed

A comprehensive comparison of the ability of each sensor to detect a misalignment fault and assess its severity based on the change in the amplitude of the 25Hz component (rotational speed) is shown in figure 19 (for both 1mm and 2mm misalignment). The plots show that the MEMS2 sensor is more capable of detecting the faults.

The presence of the frequency peaks is consistent in MEMS 1 and 2 responses with the reference accelerometer; however, the peaks amplitudes are not exactly the same. For MEMS 3 the frequency components and their amplitudes are not the same.

A satisfactory performance is achieved by the MEMS (2) accelerometer as expected and this was comparable and in good agreement with the conventional accelerometer. The first frequency component is at 25 Hz, rotor rotation speed. The second is at 50 Hz, due to the power supply or the second harmonic of rotor running speed. The third was at 100Hz, which could be the second harmonic of power supply frequency or fourth harmonic of the rotor as measured by both accelerometers. Line frequency and its multiples also exist in the frequency domain spectrum.

5. Conclusion and future work

The fabrication, packaging and testing of a vibration monitoring using MEMS accelerometers are presented in this study. Preliminary tests proved that the developed measurement device is capable of measuring machine vibrations for machine condition monitoring while the sensor has several advantages in terms of its compact size, low cost and high sensitivity. The results obtained from MEMS sensors are in close agreement with the conventional sensor results.

Future work will be related to building up the intelligent monitoring system in terms of developing an on-board low-power method of performance analysis and fault diagnosis using MEMS sensors, developing a low-power wireless communication system based on MEMS technology and designing, building and testing a working prototype which is programmable and has the ability to perform on- board fault detection and diagnosis and act accordingly.

References:

[1] Maruthi G.; Vittal K. Electrical fault detection in three phase squirrel cage induction motor by vibration analysis using MEMS accelerometer. IEEE pesd 2005.
[2] Sabin, E. Vibration of dry pumps. Semicond. Int. 1995, 18(8), 249-250.
[3] Albarbar,A; Samir M.; Andrew S. and Robert P.; "Suitability of MEMS Accelerometers for Condition Monitoring: An experimental study". Sensors 2008, 8, 784-799.
[3] Albarbar,A; Samir M.; Andrew S. and Robert P.; "Suitability of MEMS Accelerometers for Condition Monitoring: An experimental study". Sensors 2008, 8, 784-799.
[4] Badri A. , Sinha J. K. ,Albarbar A. ; "A Method to Calibrate the Measured Responses by MEMS Accelerometers"; International Journal for experimental mechanics
[5] Coiln R.; Drik H.;Roger C.; Carl K.; "Investigation into the use of low cost MEMS accelerometer for vibration based damage detection" . composite structure 82(2008) 61-70

[6] Siddique,G A.. Yadavan S., Singh. B., 2005, " A review of stator fault monitoring techniques of induction motors", IEEE Transactions on Energy Conversion, Vol. 20, No. 4, pp. 106-113.

[7] Shnibha R. , Albarbar A., Ibrahim G.R., Badri A. . "Three Phase Induction Motors Phase Imbalance Severity Evaluation Based on Acoustics Measurements and Energy Calculations." The 5th Libyan Arab International Conference on Electrical and Electronic Engineering LAICEEE 23-26/10/2010

Bibliography:

[1] First...
[2] Second...

REVIEW OF DATA-DRIVEN PROGNOSTICS AND HEALTH MANAGEMENT TECHNIQUES: LESSIONS LEARNED FROM PHM DATA CHALLENGE COMPETITIONS

Bin Huang, Yuan Di, Chao Jin and Jay Lee
NSF I/UCRC for Intelligent Maintenance Systems (IMS)
University of Cincinnati
560 Baldwin Hall
University of Cincinnati
Cincinnati, OH 45221
Telephone: (513) 556-4650
huangba@mail.uc.edu

Abstract: Machine learning and statistical algorithms are receiving considerable attention during the past decade in prognostics and health management (PHM). However, there is a lack of consensus and methodology on algorithm selection in different scenarios, which renders the random implementation of machine learning algorithms and inefficient development processes. PHM Data Challenge, an open data competition specialized in PHM, includes diverse issues in industrial data analytics and thus provides abundant resource for study and appropriate approach development. In this work, we first summarize the problems and datasets of PHM Data Challenge competitions. According to their objectives, the 9 problems can be classified into 3 categories, health assessment, fault classification and remaining useful life prediction. Second, common issues and unique challenges have been clearly pointed out for each problem and each category. Then, we analyze all solutions regarding what type of strategy a particular solution took, what algorithms it used and how it overcame the challenges. At last, insights in PHM solution strategies have been summarized to conclude the paper.

Key words: Diagnostics; Prognostics; Health Assessment; Data Challenge Competition; Fault Classification; Remaining Useful Life Prediction

1. Introduction: Prognostics and Health Management (PHM) is an interdisciplinary research area that aims at increasing productivity by managing operations based on assessment and prognosis of potential engineering system faults. The advancement of Industrial Internet of Things [1] and industrial big data analytics [2] has spurred huge promise of economic values through data-driven predictive modeling [3, 4] in a variety of industries. Due to its scalability, flexibility, and rapidity of deployment, data-driven PHM has clearly become a necessity for the next industrial revolution.

The PHM Society has been promoting research at the frontier of PHM by organizing the PHM Data Challenge competition nearly every year since 2008 [5]. The data in the competitions cover a wide spectrum of real-world industrial problems, from rotary machinery fault diagnosis [6] to jet engine remaining useful life prediction [7], and from time series based fault detection [8] to event-based failure prediction [9]. The proposed issues and winning algorithms each year serve as a diverse library of case studies from which we can learn about the current challenges in practice, the thinking flow of addressing these challenges, and the advantages and disadvantages of different methods.

This paper attempts to find the commonalities and insights of applying machine learning algorithms for PHM solutions based on the insights learned from the competitions. The resulting conclusions will serve as the foundation for the development of a systematic methodology in data-driven modeling and contributing to the advancement of the PHM research.

1.1. Review of Related Literature: With the rapid advancement in prognostics, quite a few research efforts provide a survey for remaining useful life (RUL) prediction as one of the most important tasks in prognostics. In [10], the author categorized the algorithms into physics-based methods, experienced-based methods and data-driven methods. In the data-driven section, four typical types of machine learning algorithms, including neural network, support vector machine, Bayesian network and hidden Markov model, were reviewed while other feature extraction techniques combined with statistical and machine learning algorithms were considered as hybrid methods. In [11], the author reviewed the data-driven methods and summarized them into two types based on whether it is possible to identify an indicator from the original raw data, direct condition monitoring approach and indirect condition monitoring data approach. Regression-based, Wiener process, Gamma process and Markovian-based methods can be used when an indicator is identified; stochastic filtering and covariate based hazard model and hidden Markov model are applicable if an indicator cannot be identified. The paper [12] reviewed the algorithms in both diagnosis and prognosis. It is noted that the author discussed prognostics based on different tasks. Other than that, it also reviewed the commonly used techniques in other steps during implementation, including data acquisition, preprocessing, feature extraction and selection and a special case when inconsistent types of data need to be processed. A systematic methodology on PHM design for rotary machine systems with a comprehensive review of algorithms was proposed in [13]. Although this paper proposed an algorithm selection method according to customer requirements and application conditions by quality function deployment (QFD) tables, its performance is largely limited by prior knowledge and subjective judgment. As we see from the above existing PHM reviews, most papers focus on discussing algorithms in terms of capability and functionality, but they rarely consider the application of these algorithms in real industrial situations. Hence this review paper will try to discuss the methods and algorithms in an application and problem-oriented fashion.

1.2. Motivation: In the past 9 years, several key issues that reflect the challenges in industrial data analytics in different industrial sectors have been generalized and released as PHM Data Challenge competitions organized by PHM Society and IEEE Reliability Society (in 2012 and 2014). The purpose of PHM Data Challenge is to gain more attention and efforts from academics and industry to address the real-world challenges. The topics of the competitions covers a wide range of areas, including aircraft engine, fuel cells, and gearbox. The topics also cover almost all the PHM tasks, which include remaining useful life (RUL) prediction, diagnosis, fault detection, health assessment, and condition monitoring. The competitions provide abundant open resource for study and research such that further improvement and extended research can be advanced. Many research works have been published based on popular datasets provided by PHM Data Challenge competitions, such as jet engine RUL prediction in 2008 [14] and gearbox diagnosis in 2009 [15, 16]. Therefore, reviewing how algorithms have been used and methods have been developed based on PHM Data Challenge datasets will surely benefit and facilitate the data-driven PHM methodology development in the future.

1.3. Objective: This study aims at providing a generalized problem-oriented PHM technique review by summarizing 9 PHM Data Challenge competitions and around thirty papers published as the winning solutions to these competitions.

The basic ideas of this review consist of three steps: 1) summarize the common issues and analytical challenges from the data competitions; 2) compare and benchmark the relevant solutions; 3) discuss the lessons learned from these competitions and solutions. By following these steps, the objective is to understand the reason why the winning algorithm outperforms others and learn the rational strategy of algorithm selection in different scenarios.

1.4. Organization: Section 2 gives an overview of Data Challenge competition, and introduces the basic information about each year's problem. Then the details of three categories of problems, health assessment, fault detection and classification and RUL prediction, are analyzed individually in Section 3, Section 4 and Section 5, respectively. Section 6 concludes the findings in each type of problem and other common issues over all data challenge problems.

2. Overview: In general, an effective PHM system is expected to have the capabilities of providing early detection and isolation of incipient faults and monitoring and predicting the progression of faults, which will further support maintenance scheduling and asset management. Such expectation is a mixed requirement of diagnostics, prognostics, health assessment and health management. Conventional diagnostics is known as a fail-and-fix process to determine the root cause of machine failure. On the contrary, prognostics is a predict-and-prevent methodology aiming to reduce downtime and maintenance cost proactively. Health assessment is mostly regarded as a critical process for detecting incipient failures before an algorithm-centered prediction process. The last step in a complete PHM system is maintenance scheduling and operation management based on outputs from prognostic modeling process. It focuses on assessing the further impact of failures and minimizing the loss through optimization.

In PHM Data Challenge competitions, a variety of topics mentioned above in the general PHM scope are covered. For the convenience of discussing different competitions, we shorten the name of each year's PHM Data Challenge competition as the following format: "ORGANIZATION'YR" to represent the competition organizer and year. For example, "PHM'08" represents the competition held by PHM Society in 2008; "IEEE'12" represents the competition held by IEEE Reliability Society in 2012. We summarize the tasks in the general PHM framework as shown in Figure 1. Three competition problems involve more than one PHM tasks. In PHM'11, a sensor health monitoring problem, given two types of dataset with different number of sensors included, it requires the detection of faulty condition for shear data and identification of all the possible combinations of multiple faulty anemometer for pair data, which are considered as health assessment and diagnostics, respectively. IEEE'14 splits the competition into two tasks: RUL Prediction and State of Health (SoH) estimation, which involves all three tasks. PHM'15 requires prediction of the failure mode, failure start time and failure end time, which is a combined task of prognostics and diagnostics, while other prognostics problem only consider the RUL prediction without the identification of failure modes.

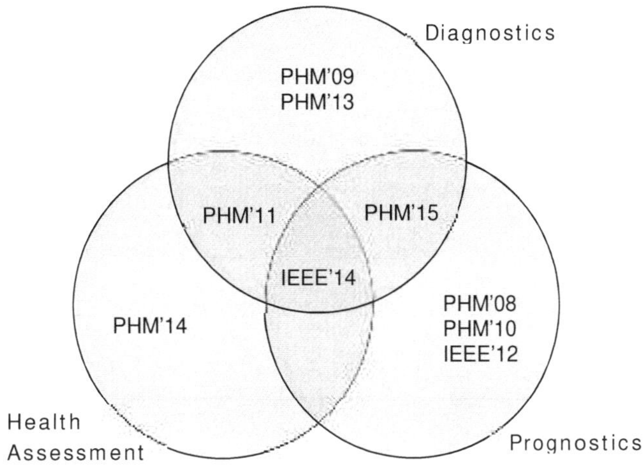

Figure 1: Tasks of PHM Data Challenge Competition Problems.

2.1. Basic Information of PHM Data Challenge Competitions: Since the tasks of competitions are sometimes multi-objective, we categorize all the competitions into three types according to the common objectives in PHM, namely, health assessment, fault classification and RUL prediction. Table 1 summarizes the basic information of PHM Data Challenge competitions and their datasets. The observations from this table are: 1) most time prediction tasks consider multi-asset and multi-regime issue without multiple failure modes, since the run-to-failure data provided is sufficient to represent the failure mode; 2) fault classification tasks focus more on the multi-failure mode issue than multi-asset and multi-regime scenarios; 3) health assessment tasks seem to focus less on regime complexity in the competitions.

3. Health Assessment: Health assessment task in the context of PHM Data Challenge competition can be defined as health condition estimation for a device or system at the time requested. Regarding its applications, health assessment can be applied either independently to real-time condition monitoring or to detecting the degradation trend with RUL prediction tasks as a pre-modeling step, as illustrated in Figure 1. Details of three health assessment problems have been listed in Table 2.

Table 1: Basic Information of PHM Data Challenge Competitions and Datasets

Task Objective		Organization & Year	Application	Asset	Failure	Regime	Data Source	Data Type	Variable Type
Health Assessment	NP	PHM 2011	Anemometer	NA	4	D	Monitoring	Short	Parameter
		PHM 2014	Unspecified Equipment	Fleet	14	NA	Maintenance & Usage	Long	Event & Parameter
	P	IEEE 2014	Fuel Cell	2	NA	S&D	Testbed	Short	Parameter
Fault Classification	NP	PHM 2009	Gearbox	NA	14	10	Testbed	Short	Waveform
		PHM 2013	Unspecified Equipment	NA	13	NA	Monitoring	Long	Event & Parameter
	P	PHM 2015	Plant Operation	Fleet	6	NA	Monitoring & Control	Long	Parameter
Remaining Useful Life Prediction		PHM 2008	Aircraft Engine	Fleet	NA	6	Simulation	RTF	Parameter
		PHM 2010	Milling Machine	6	NA	1	Monitoring & Usage	RTF	Waveform
		IEEE 2012	Bearing	17	NA	3	Testbed	RTF	Waveform
		IEEE 2014	Fuel Cell	2	NA	S&D	Testbed	RTF	Parameter

Notation:
P/NP – Prediction of failure time is required or not required.
NA – The criteria is neither clearly specified nor significant to problem.
Fleet – A large number of assets are considered.
S/D – Static or Dynamic regime is considered, respectively.
RTF – Run-to-Failure data is provided
Short/Long – Relatively short term or long term data is provided.

Table 2: Overview of Health Assessment Problems and Datasets.

	PHM'11	PHM'14	IEEE'14
Equipment	Anemometer	Unknown	Fuel Cell
Data Type	Monitoring	Maintenance Log & Usage	Testbed (Run-to-Failure)
Asset (train-test)	NA	1913-2076	1-1
Regime	Dynamic	NA	Static – Dynamic
Failure Mode	NA	14	1
Sampling Interval	Every 10min	Event	Aging: Every 30s EIS, Polarization: Weekly
Variables	Single Value	Single Value	Single Value
Variables	Pair: 16 Shear: 20 or 24	Part Consumption: 5 Usage: 3	Aging: 25 Polarization: 8 EIS: 3
Note that PHM'11 only provide a series of sample data without knowing whether they come from the same group of sensors or various assets: For pair data, 12 25-day length samples for training; 420 5-day length samples for testing. For shear data, 7 25-day length samples for training; 255 5-day length sample for testing			

Three levels of health assessment can be classified in PHM Data Challenge competitions due to their different levels of requirement: 1) abnormal detection using a feature-based indicator, which represents a base level binary health condition assessment. For example, PHM'11 only differentiates normal/abnormal or healthy/unhealthy conditions; 2) quantitative estimation of health condition using a self-defined health index, such as the failure rate of system defined different from method to method in PHM'14; 3) quantitative estimation of health condition using a special indicator, such as the diagnostic tool of electrochemical impedance spectroscopy (EIS) test being used in fuel cell (FC) SoH estimation of IEEE'14.

3.1. Challenges: Health condition always changes randomly with noise. Although the randomness or other contribution from invisible factor exists, part of the unstable behaviors can be theoretically explained or physically understood. Challenges of each health assessment problem are summarized in Table 3. In PHM'11, slow running situations are not caused by mechanical issues but temporary weather conditions, such as icing. Since they are not permanent degradation and recoverable, such cases should not be considered as true failure, and need to be filtered out during the modelling process. In PHM'14, two major changing modes of failure rate have been discovered according to the preventive maintenance policy in which the failure rate follows a bathtub curve, a case of Weibull distribution [17]. Failure rate decreases after preventive maintenance (PM) determined when a large number of parts are replaced and dramatically increases for a short period of time known as infant mortality after corrective maintenance (CM) determined when a few parts are replaced [18]. Both winning methods published this year built their models upon similar maintenance policies. In IEEE'14, a self-healing effect on FC power has been found with a local reversible degradation behavior after weekly performance or characteristic test. Therefore, how to understand and explain such phenomenon becomes a common issue for health assessment. In order to address these issues, finding suitable algorithms with deep domain knowledge is critical to develop an effective PHM solution no matter what strategies being chosen. Thoroughly understanding the background, relevant theories, underlying mechanism of the problem may help handling the uncertainty and facilitate method development.

Table 3: Summary of Challenges of Health Assessment Problems

	PHM'11	PHM'14	IEEE'14
Common Issues	• Find health indicator • Health condition degrades with uncertainty • Call for domain knowledge		
Unique Challenges	• Icing • Pattern recognition	• Implementation of maintenance policy	• IES Estimation • Little Reference

3.2. Solutions: In the competitions, the strategies took in solutions are so diverse that we cannot directly compare the solutions. Each of them represents one type of health assessment problem. Fortunately, since each solution create a health index in modeling, we discuss the solutions by following what health index was chosen and what algorithm was used.

PHM'11 – **Binary Abnormal Detection Using a Feature-Based Indicator**: Three solutions for this competition are compared. Before pattern classification, Siegel [19] and Sun [20]'s methods uses wind speed difference as health indicator and both applies power law equation and unify the data from different heights, whereas Cassity [21]'s method uses percentage difference of wind speed as health indicator and applies same wind profile theory to create wind profile models and fit in each height level. The champion solution proposed by Siegel applies an advanced algorithm Auto-Associative Neural Network (AANN) further process the health indicator for shear data other than Sum of Squared Residual (SSR) and R^2 value used by Sun and Cassity, respectively. In all three methods, the icing issue which would potentially cause false alarm is addressed by eliminating outliers. For pattern classification part, the health indicators are used as critical features and inputs of pattern recognition models. Siegel's k-mean clustering based figure of merit (FOM) method is able to partition the dataset with a bias on the samples containing lagging anemometers. Sun conducted a Euclidean similarity measurement based pattern search with a hybrid decision making method which has a close performance to the 1st place. Cassity's technique performs not as well as other two top methods do. This is probably because of the additional step it takes for initial guessing some bad sensors, as an essential step of generating labeled bad/faulty data samples for discriminant analysis which is a supervised learning method. Due to the dynamic working regime and multiple possible failure mode which is not clear to the participants in this competition, such process of using initial guess as an untrusted baseline data to detect other faulty sensors would be more likely to generate a larger error. The expertise knowledge involved in solution procedure is also summarized in the last column.

PHM'14 – **Quantitative Estimation of Health Condition Using a Self-Defined Health Index**: Similar to PHM'11, the dataset of PHM'14 also has two parts, maintenance log based part consumption data and usage measurement data. In both solutions, a deep expertise on maintenance policy has been adopted firstly, especially when defining and detecting Predictive Maintenance (PM) or Scheduled Maintenance (SM), and Corrective maintenance (CM). In spite of other statistical analysis, histogram chart is seen as the most frequently used statistical tool due to its capability of accumulating discrete data samples as frequency information that can be further converted as risk indicator. The champion's solution proposed by Rezvani [18] is quite complicated in statistic based analysis, in which two high risk time interval detection models have been individually developed for PM and CM based on Weibull Bathtub curve by only using the part consumption dataset, then used to test each asset sequentially at the requested times to determine its risk level. Instead of calculating high risk time interval sample by sample, the 2nd placed solution [22] created a high/low risk histogram-based spectrum for any type of failure mode (reason code) in time domain and in usage domain as well during the training stage, then simply test each sample whether it located in the high/low risk bins by a consensus ensemble strategy. Furthermore, unlike the champion method, this method utilizes the usage dataset revealing a linear relation of failure rate with accumulated usage, also makes a further inference from reason code to part number instead of summarizing parts information in repeated number of part numbers replaced (RNPNR). Despite of some of these part number being found to be uniquely associated to SM and CM, respectively, this solution treated them in one strategy with no difference instead of using separate strategies for PM and CM in champion's method. Overall, these two solutions are similar in methodology, technically compensate with each other and can be possibly fused into one more effective method.

PHM'15 - **Quantitative Estimation of Health Condition Using a Special Indicator**: As shown in the literature review in both papers, we see that the majority of research on FC or battery focuses on the prognostics and diagnosis, but less effort on EIS estimation which is related to the dynamic performance of FC. Basic background knowledge on FC and some expertise on EIS would be essential to solve this problem effectively. The champion of this task [23] developed a solution targeting to each issues mentioned above. The solution merged the power degradation information with Equivalent Circuit Model (ECM) for EIS estimation by considering an exponential local reversible degradation model for the degradation uncertainty issue. This requires a deep understanding of physical system of FC stacks. In the 2nd placed solution [24], not too much expertise was used in modeling process. Several different regression methods are applied to fit the real part and imaginary part independently, though much more background and domain knowledge is mentioned at the beginning. This method is very straightforward and only considered the EIS data without any degradation process. Data with current information and polarization curve hasn't been used in both solutions, but may lead to better results. Another solution published to RUL prediction of FC, anther split task of this competition using the same dataset, will be discussed later in Section 5.2.

4. Fault Classification: Finding root cause of machine failure is generally considered as the major task of diagnostics. In the content of PHM Data Challenge competition, the task of root cause finding is mostly expressed as fault classification problem that is required to determine a certain sample's fault type from a series of possible failure modes. Details of three fault classification problems have been summarized in Table 4.

Table 4: Overview of Fault Detection Problems and Datasets

	PHM'09	PHM'13	PHM'15
Equipment	Gearbox	Unknown	Plant
Data Type	Short-time Testbed	Maintenance Log & Monitoring	Monitoring
Asset (train-test)	NA	1	33-15
Regime	10	NA	Multiple
Failure Mode	14	13	6
Sampling Interval	Test	Event	Every 15min
Variables	Waveform	Single value	Single Value
Variables	3	2, 2	Monitoring: 14 Fault Event: 3

Two classification strategies can be grouped according to the data type that needs to be processed:

1) Feature-based pattern classification, in which the performance of the algorithm highly depends on the quality of features extracted from the dataset, e.g. PHM'09 and PHM'15 give a no training data situation and a no background situation for classification task, respectively, thus the former requires a powerful feature extraction methods, and the latter requires thorough data exploration process. In both situations, deep expert knowledge would be necessary.

2) Event-based inference and classification, in which major effort would be taken in revealing the complex relationship among multiple character type variables so as to classify the data samples, e.g. PHM'13 asks participants to develop a maintenance action recommender which is capable of infer the problem ID based on event type data generated from the onboard condition monitoring system.

4.1. Challenges: Despite of the types of classification technique, the challenges that each problem faces are different, mainly due to their diverse backgrounds. Challenges of each health assessment problem are summarized in Table 5. For PHM'09, because of no training data (labeled data) provided, we would only be able to mine useful information or extract features from unlabeled data, which further implies the necessity of developing more powerful methods with more expertise in signal processing for vibration data and gearbox domain knowledge. For PHM'13, three character type variables, "event", "case" and "problem type", formalizes a complex triangular relationship in which each two of them are somehow related. In addition, once each event occurs, 30 numerical parameters would be generated by device, as a snapshot of system condition. Hence the unique challenges in this competition are 1) how to reveal the triangular relationship and 2) how to appropriately combine the two sources of data and systematically use them. For PHM'15, a task consisting of both fault classification and failure time prediction, the first challenge would be determine a strategy to deal with such multi-objective task. Besides, finding high quality features without given details about the physical system is another challenge.

Table 5: Summary of Challenges of Fault Classification Problems

	PHM'09	PHM'13	PHM'15
Common Issues	• Find health indicator • Health condition degrades with uncertainty • Call for domain knowledge		
Unique Challenges	• No training data • Signal processing	• Complex Relationship among "event", "case" and "problem type" • Data Fusion	• Fault classification and time prediction • Feature extraction without knowing the mechanism of system

4.2. Solutions: The solutions to fault classification problems are discussed by following the two types of strategies mentioned above. PHM'09 and PHM'14 belong to feature based pattern classification problem. PHM'13 belongs to event based inference and classification problems.

For PHM'09, fault classification without training data, the champion Wu proposes a compact and integral solution which works in a sequential process with the best performance in this competition [25]. The idea of using many bandpass filters to decompose the whole frequency spectrum to achieve the global frequency analysis as well as local analysis of frequency component of interest simultaneously requires a deep insight of vibration signal processing skills. Al-Atat's solution is relatively complicated and module-based, but follows a more general diagnosis approach [26]. It consists of many sub-methods regarding to regime segmentation, health assessment and fault classification. These sub-methods are highly independent with each other, and thus have more flexibility being used in other applications. In this competition, all methods are used with prior expertise or for exploring the hidden relations.

For PHM'15, fault classification with downtime prediction, in order to address this multi-objective task, both two top ranked solutions take the same strategy that machine learning based fault classifiers are built in each time interval for each plant, and features are extracted based on the comprehensive data exploration. In champion's solution [27], several algorithms are used for correlation analysis for variables and components, and periodicity of failure time is also found. For 2nd place [28], features are extracted based on inference from physical interpretation, basically guessing the possible physical meaning of each correlated sensors, seasonality analysis, health condition detection by a visualization tool named empirical Probability Density Function (PDF), and a "one hour" rule being discovered by time series analysis. Both of the above two methods performs very close to each other, with score of 21015.55, 20639.71, respectively, nearly double as the 3rd place solution [29], 10221.54. The 3rd place solution also uses ensemble technique to combine Random Forest (FR) and Gradient Boosting Decision Tree (GBDT) to form a decision tree based classifier. Besides, the 3rd place solution focused on an issue of fault event overlapping. It is observed that building a multiclass classifier in each time step seems a good strategy for multi-objective task of fault classification and prediction, and preparing high quality features which requires thorough exploration is critical to the performance of a method.

Table 6 Summary of Solutions to Event Based Inference and Classification Problems

PHM'13	Score	Dataset Used	Basic Strategy	Algorithms
1st - Das	86	Event data	Matrix decomposition	ALS-NMF
2nd - Katsouros	60	Event data	Bayesian Classification	Naïve Bayesian Classifier
3rd - Kimotho Method 1	51	Event data and parametric data	Classification for event	Event-based DT + Ensemble of classifiers of BT, RF and SVM
3rd - Kimotho Method 2	48	Event data and parametric data	Classification for event	Event-based DT + SVM
2nd - Katsouros Method 2	35	Parametric data	Classification	SVM with radial based kernel
6th - Siegel Method 1	35	Event data and parametric data	Regression for problem	Event based random forest
6th - Siegel Method 2	30	Parametric data	Regression for problem	Random forest

For PHM'13, a typical fault classification problem with a dataset mixed with event data and parametric data, a benchmark of solutions to this event based inference and classification problem has been illustrated in Table 6. Parametric data is an extra dataset with 30 parameters generated by control system when a specific condition corresponding to an event code in event data is met onboard. Based on the papers and presentation slides [9] published by PHM Society, 7 solutions have been benchmarked and ranked according to their scores shown in Table 6. High performance (score>60): Both of the top two methods in this range only use event data and take their main efforts in relationship inference. The champion Das [30] introduces a collaborative filtering based method typically used for 'Recommender System' development in Information Technology (IT) field, e.g. e-commerce item recommendation [31]. The 2nd placed Bayesian classifier proposed by Katsouros [32] has much lower accuracy than the 1st placed Non-Negative Matrix Factorization (NMF) methods, but it has a straightforward inference with a big advantage in its simple structure and computationally effectiveness. Medium performance (60>score>40): The two second level methods are all based on classification for each event. Solutions in this range are generated by the 3rd place participant Kimotho, using both event data and parametric data to classify different event codes. Low performance (score<40): Other three methods given by Katsouros and Siegel either choose regression instead of classification or completely ignore the event data. Therefore we may conclude that parametric data may not be very useful as the Event ID/Code itself summaries the parameters settings for a specific condition [30]. Beyond the performance, due to the time constraint of competition, most participants in PHM'13 follows a trail-and-error strategy and gives more than one solutions to the problem, however the champion who gains more scores than others seizes the essence of the problem and introduces a highly effective method from the IT field.

5. Remaining Useful Life Prediction: RUL prediction tasks occupy nearly half of all PHM Data Challenge competitions, because of its significance of assisting maintenance scheduling and health management. It is very common to see the health assessment works as a critical step to support RUL prediction in real applications. In competitions, the objective of RUL prediction is to estimate the remaining working time of a system before its health condition deteriorates below a threshold indicating a faulty condition. In competitions, RUL prediction doesn't explicitly require to describe the health condition during degradation as long as the solution accurately estimates the time before actual failure, which make it literally differentiable from health assessment tasks in this competition. Details of four RUL prediction problems have been summarized in Table 7.

Table 7: Overview of RUL Prediction Problems and Datasets

	PHM'08	PHM'10	IEEE'12	IEEE'14
Equipment	Engine	Milling Cutter	Bearing	Fuel Cell
Data	Simulation (Run-to-Failure)	Monitoring & Usage (Run-to-Failure)	Testbed (Run-to-Failure)	Testbed (Run-to-Failure)
Asset (train-test)	218-218	3-3	6-11	1-1
Regime	6	1	3	2(Static-Dynamic)
Failure Mode	NA	NA	NA	NA
Sampling Interval	Operation Cycle - Flight	Operation Cycle - Cut	Every 10s	Every 30s
Variables	Single Value	Single Value	Waveform	Single Value
Variables	30	8	3	25

5.1. Challenges: As shown in Table 8, the common challenges of RUL prediction is finding an appropriate degradation trend which can be identified as an indicator for modeling process. However, in different systems and situations, degradation detection would meet different challenges related to number of assets and number of regimes. PHM'08 aircraft engine dataset shows a clear degradation trend with multiple regimes and a large number of units. PHM'10 milling cutter machine dataset also shows a clear degradation trend but with few units available to analysis, which causes data scarcity issue. PHM'12 bearing dataset shows a very unclear end-of-life degradation signature without an ideal gradually monotonic degradation trend. We also notice the difficulty of dealing with bear's uncertain behavior which is caused by its unbalanced dataset and multiple regimes. PHM'14 fuel cell dataset has a unique training/testing mode that training data in static regime is used as reference to estimate the test sample's RUL in dynamic regime and a special degradation mode, self-healing effect or local reversible degradation, caused by its electrochemical features.

Table 8: Summary of Challenges of Remaining Useful Life Prediction Problems

	PHM'08	PHM'10	PHM'12	IEEE'14
Common Issues	• Degradation trend detection			
Unique Challenges	• A fleet of asset • Multiple regimes	• Data scarcity	• Unclear end-of-life signature • Unbalanced dataset	• Static/dynamic regime • Reversible degradation

5.2. Solutions: A comprehensive review [14] on the papers using the C-MAPSS dataset in which PHM'08 competition dataset is included has been published. In that review, the solutions of 2nd place [33] and 3rd place [34] are classified as "Category 1 - using functional mappings between set of inputs and RUL", and champion's solution [35] are classified as "Category 2 - functional mapping between health index and RUL" as well as "Category 3 - similarity-based matching". Our review of methods provided in all competitions related to RUL prediction is basically conducted in a similar way. Over all solutions to RUL prediction problems, two strategies have been observed so far for data-driven RUL prediction methods: 1) finding a degradation indicator and predicting the trend until a threshold; 2) direct mapping from extracted features to RUL index. Since we follow a problem-centered manor of review, the solutions are discussed with their unique difficulties.

In PHM'08, RUL prediction with a fleet of assets and multiple regimes, the champion's solution proposed by Wang [35] takes the first strategy and clearly applies three techniques targeting on each of the single challenges, degradation trend fitting, asset variety and multi-regime. The other two solutions take the 2nd strategy. The 2nd place Heimes' solution [33] brings a mature tool that is specially developed for nonlinear dynamic system with capability of dealing with uncertainty and complexity. The 3rd place Peel's solution takes advantages of Kalman Filter (KF) to treat variables as time series data integrating past information and reduce noise in raw data and error in asset variety.

For PHM'10, RUL prediction with data scarcity which means only 3 assets of training samples are available for predicting another 3 testing samples whose data is only partially given, the champion's solution [36] proposed by Das follows a classic and general PHM solution workflow. The key point for his high accuracy prediction is attributed to successfully finding a critical feature, harmonic of tooth pass frequency, and employing a batch of 100 neural networks to deal with the data scarcity issue. The 2nd place solution given by Chen [37] is developed in the framework of Bayesian statistic, but not using any Bayesian inference methodology. Both of them are second category regression based approach, however, the difference is Chen proposes an individual step of estimating the initial wear state of the cutter instead of using the wear state related feature from last cut as input of regression model in Das' solution.

For PHM'12, RUL prediction with big challenges in handling an unclear degradation end-of-life signature and unbalanced dataset, the champion Sutrisno developed three independent solutions [38], two of them take the first strategy and another one take the second strategy. For the solutions with a degradation indicator, Sutrisno's Method 1 detects an ideal degradation indicator, moving-averaged spectral (MAS) kutosis with a gradual and monotonic trend, then predicts the trend by Bayesian Monte Carlo; Sutrisno's Method 3 extracts a "prognostic feature" by taking averaging of 5 maximum values of variables, eventually calculate the end-of-life time using the ratio of multiple degradation stages which can be easily observed. For the regression-based solution in Sutrisno's Method 3, also the only one in this competition, least squares-support vector regressor (LS-SVR) with features in time domain and wavelet analysis is applied for exploring the underlying relationship between features and RUL directly by the regression method, and achieved the best performance in this competition. The 2nd place solution given by Wang [39] is actually similar to Sutrisno's Method 1 in terms of its workflow, but using an health index extracted by Principal Component Analysis (PCA) and T^2 statistic. Though it applies some bearing diagnosis techniques, such as bearing characteristic features and natural frequency analysis, this method proposed does not have a strong capability of detecting the degradation trend to overcome this challenge. By comparing the strategies took in different solutions, we conclude that multi-stage degradation identification should be the key to RUL prediction problem with unclear end-of-life trend.

For IEEE'14, RUL prediction with variety of static/dynamic regime, due to its unusual dataset with only one sample in either training or testing dataset, it assumes that the variety issues between different FC stacks can be neglected, and uncertainty observed in data only comes from dynamic regime and testbed. The only solution published for this split task in IEEE'14 by Kimotho [40] is developed based on Particle Filter method with a self-healing factor adapted in each iteration. It tried 5 different state modes in the adaptive particle filtering process and combine them using ensemble technique.

6. Conclusion: To conclude, all competitions in the past 9 years mainly involve the three basic PHM tasks, health assessment estimating the machine or system health condition, diagnostics finding the machine or system failure root cause, and prognostics predicting the machine or system failure time before failure actually occurs. We found that health assessment always supports diagnostics and prognostics. Based on the objectives of PHM Data Challenge competitions, PHM'11, PHM'14 and IEEE'14 are categorized into health assessment problems; PHM'09, PHM'13 and PHM'15 are categorized into fault classification problems; PHM'08, PHM'10, IEEE'12 and IEEE'14 are categorized into RUL prediction problems.

Most RUL prediction problems and fault prediction problem in PHM'15 consider multi-asset and multi-regime issues. A fleet-based analysis, such as similarity-based approach, is effective for RUL prediction with a large number of assets in PHM'08. Initial wear estimation and ensemble technique are necessary in RUL prediction with data scarcity issue. Multi-stage degradation identification is observed to be very effective in RUL prediction problem with unclear end-of-life signature. Domain knowledge would contribute the RUL modeling process only when some special effects have to be considered. A good example of applying domain knowledge in RUL prediction is that a factor which represents the self-healing behavior of fuel cell stacks is specially adapted into the particle filter technique in the solution to RUL prediction in IEEE'14 [40]. A negative example is that common bearing diagnosis technique such as characteristic features does not assist in degradation trend detection in IEEE'12 [39]. After reviewing all solutions, two basic strategies can be summarized in RUL prediction: 1) finding a degradation indicator and predicting the trend until a threshold; 2) direct mapping from extracted features to RUL by regression. According to our review, both two strategies are capable of solving RUL prediction as long as the algorithm of the selected strategy is used for a clear purpose and targeting to the challenges.

Fault classification tasks focus less on multi-asset and multi-regime issues in the PHM Data Challenge competitions. High quality features which require thorough exploration directly impact the method's performance. Prior expertise knowledge and data exploration is critical to classification without training data in PHM'09. Building a multiclass classifier in each time step has been seen as a general strategy for multi-objective task of fault classification and prediction in PHM'15. Data selection and filtering is significant in the situation with heterogeneous and complex dataset especially in inference based classification problems. The benchmark in PHM'13 gives us a lesson that choosing a right tool would be much more effective and efficient than trail-and-error based algorithm selection.

Health assessment tasks in competitions have less concerns on regimes. Three problems including PHM'11, PHM'14 and IEEE'14 in this category represents three levels of health assessment according to their requirements, binary health/unhealthy detection, quantitative health condition estimation using a self-defined health indicator and quantitative health condition estimation using special indicator, respectively. According to our review, all health assessment solutions rely on domain knowledge, e.g. wind profile equation used for handling shear data in PHM'11, expertise related to maintenance policy used to define PM/SM and CM in PHM'14, fuel cell property related domain knowledge used for modeling fuel cell dynamic behavior in EIS spectrum. Since we have already known that health assessment can possibly collaborate with any PHM related tasks, it implies that any other problems, either diagnostics or prognostics, would heavily rely on domain knowledge if health assessment is included.

6.1. Further Work: In this work, we only benchmark the algorithm performance for PHM'13. In the future, similar algorithm benchmark needs to be conducted for other competitions. In addition, other issues such as tradeoff between accuracy and efficiency of PHM solutions and use of ensemble techniques also deserve further analysis. The problem-centered review proposed in this paper could be extended to a wider range of resources, such as publications based on the same datasets, publication with the same objectives or using the same class of techniques, etc.

References:

[1] Accenture and GE. (2015). *Industrial Internet Insights Report. Industrial Insights Report.* Available: https://www.accenture.com/us-en/_acnmedia/Accenture/next-gen/reassembling-industry/pdf/Accenture-Industrial-Internet-Changing-Competitive-Landscape-Industries.pdf

[2] J. Lee, *Industrial Big Data*: Mechanical Industry Press, China, 2015.

[3] N. Millman, " Big data to unlock value from the industrial internet of things.," ed, 2015.

[4] A. Alter, P. Banerjee, P. E.Daugherty, and W. Negm, "Driving Unconventional Growth through the Industrial Internet of Things.," ed, 2014.
[5] *PHM Competitions*. Available: https://www.phmsociety.org/competition
[6] *2009 PHM Challenge Competition Data Set*. Available: https://www.phmsociety.org/references/datasets
[7] K. G. A. Saxena, ""PHM08 Challenge Data Set", NASA Ames Prognostics Data Repository," 2008.
[8] *2011 phm society conference data challenge*. Available: https://www.phmsociety.org/competition/phm/11
[9] (11 March). *2013 PHM Society Conference Data Challenge*. Available: https://www.phmsociety.org/events/conference/phm/13/challenge
[10] F. Ahmadzadeh and J. Lundberg, "Remaining useful life estimation: review," *International Journal of System Assurance Engineering and Management,* vol. 5, pp. 461-474, 2014// 2014.
[11] X.-S. Si, W. Wang, C.-H. Hu, and D.-H. Zhou, "Remaining useful life estimation – A review on the statistical data driven approaches," *European Journal of Operational Research,* vol. 213, pp. 1-14, 8/16/ 2011.
[12] A. K. S. Jardine, D. Lin, and D. Banjevic, "A review on machinery diagnostics and prognostics implementing condition-based maintenance," *Mechanical Systems and Signal Processing,* vol. 20, pp. 1483-1510, 10// 2006.
[13] J. Lee, F. Wu, W. Zhao, M. Ghaffari, L. Liao, and D. Siegel, "Prognostics and health management design for rotary machinery systems—Reviews, methodology and applications," *Mechanical systems and signal processing,* vol. 42, pp. 314-334, 2014.
[14] E. Ramasso and A. Saxena, "Performance Benchmarking and Analysis of Prognostic Methods for CMAPSS Datasets," *International Journal of Prognostics and Health Management,* vol. 5, pp. 1-15, 2014.
[15] G. Zhang and J. Isom, "Gearbox vibration source separation by integration of time synchronous averaged signals," 2011.
[16] P. Boškoski and A. Urevc, "Bearing fault detection with application to PHM Data Challenge," *International Journal of Prognostics and Health Management Volume 2 (color),* p. 32, 2011.
[17] T. Nakagawa, "Periodic and sequential preventive maintenance policies," *Journal of Applied Probability,* vol. 23, pp. 536-542, 1986/006/001 1986.
[18] S. M. Rezvanizaniani, J. Dempsey, and J. Lee, "An effective predictive maintenance approach based on historical maintenance data using a probabilistic risk assessment: PHM14 data challenge," *International Journal of Prognostics and Health Management,* vol. 5, 2014.
[19] D. Siegel and J. Lee, "An auto-associative residual processing and K-means clustering approach for anemometer health assessment," *International Journal of Prognostics and Health Management Volume 2 (color),* p. 117, 2011.
[20] L. Sun, C. Chen, and Q. Cheng, "Feature extraction and pattern identification for anemometer condition diagnosis," *Int. J. Progn. Heal. Manag,* vol. 3, pp. 8-18, 2012.
[21] J. Cassity, C. Aven, and D. Parker, "Applying weibull distribution and discriminant function techniques to predict damaged cup anemometers in the 2011 PHM competition," *Int. J. Progn. Heal. Manag,* vol. 3, pp. 1-7, 2012.

[22] H. Kim, T. Hwang, J. Park, H. Oh, and B. D. Youn, "Risk Prediction of Engineering Assets: An Ensemble of Part Lifespan Calculation and Usage Classification Methods," 2014.
[23] T. Kim, H. Kim, J. Ha, K. Kim, J. Youn, J. Jung, *et al.*, "A degenerated equivalent circuit model and hybrid prediction for state-of-health (SOH) of PEM fuel cell," 2014, pp. 1-7.
[24] W. O. L. Vianna, I. P. d. Medeiros, B. S. Aflalo, L. R. Rodrigues, J. P. P. Mal, x00E, *et al.*, "Proton Exchange Membrane Fuel Cells (PEMFC) impedance estimation using regression analysis," in *Prognostics and Health Management (PHM), 2014 IEEE Conference on*, 2014, pp. 1-8.
[25] F. Wu and J. Lee, "Information Reconstruction Method for Improved Clustering and Diagnosis of Generic Gearbox Signals," *International Journal of Prognostics and Health Management Volume 2 (color),* p. 42, 2011.
[26] H. Al-Atat, D. Siegel, and J. Lee, "A systematic methodology for gearbox health assessment and fault classification," *Int J Prognostics Health Manage Soc,* vol. 2, p. 16, 2011.
[27] W. Xiao, "A Probabilistic Machine Learning Approach to Detect Industrial Plant Faults," *arXiv preprint arXiv:1603.05770,* 2016.
[28] H. Kim, J. M. Ha, J. Park, S. Kim, K. Kim, B. C. Jang, *et al.*, "Fault Log Recovery Using an Incomplete-data-trained FDA Classifier for Failure Diagnosis of Engineered Systems," 2016.
[29] C. Xie, D. Yang, Y. Huang, and D. Sun, "Feature Extraction and Ensemble Decision Tree Classifier in Plant Failure Detection," presented at the Annual Conference of the Prognostics and Health Management Society, 2015.
[30] S. Das, "Maintenance action recommendation using collaborative filtering," *Int. J. Progn. Health Manag,* vol. 4, pp. 7-12, 2013.
[31] G. Linden, B. Smith, and J. York, "Amazon. com recommendations: Item-to-item collaborative filtering," *IEEE Internet computing,* vol. 7, pp. 76-80, 2003.
[32] V. Katsouros, V. Papavassiliou, and C. Emmanouilidis, "A Bayesian Approach for Maintenance Action Recommendation," *International Journal of Prognostics and Health Management, International Journal of Prognostics and Health Management,* vol. 4, p. 034, 2013.
[33] F. O. Heimes, "Recurrent neural networks for remaining useful life estimation," 2008, pp. 1-6.
[34] L. Peel, "Data driven prognostics using a Kalman filter ensemble of neural network models," in *Prognostics and Health Management, 2008. PHM 2008. International Conference on*, 2008, pp. 1-6.
[35] T. Wang, Y. Jianbo, D. Siegel, and J. Lee, "A similarity-based prognostics approach for Remaining Useful Life estimation of engineered systems," in *Prognostics and Health Management, 2008. PHM 2008. International Conference on*, 2008, pp. 1-6.
[36] S. Das, R. Hall, S. Herzog, G. Harrison, M. Bodkin, and L. Martin, "Essential steps in prognostic health management," 2011, pp. 1-9.
[37] H. Chen, "A multiple model prediction algorithm for CNC machine wear PHM," *International Journal of Prognostics and Health Management Volume 2 (color),* p. 129, 2011.

[38] E. Sutrisno, H. Oh, A. S. S. Vasan, and M. Pecht, "Estimation of remaining useful life of ball bearings using data driven methodologies," 2012, pp. 1-7.
[39] T. Wang, "Bearing life prediction based on vibration signals: A case study and lessons learned," 2012, pp. 1-7.
[40] J. K. Kimotho, T. Meyer, and W. Sextro, "PEM fuel cell prognostics using particle filter with model parameter adaptation," in *Prognostics and Health Management (PHM), 2014 IEEE Conference on*, 2014, pp. 1-6.

RISK ASSESSMENT OF TRANSITION TO LEAD-FREE ELECTRONICS ASSEMBLY

Guru Prasad Pandian, Diganta Das, Michael Osterman, and Michael Pecht
Building 89, Room 1103,
University of Maryland,
College Park, MD 20742
Telephone: 301-405-5323
diganta@umd.edu

Abstract: The European Union's Restriction on Hazardous Substances (RoHS) Directive, imposed on electronics manufacturers in 2006, banned the use of certain toxic substances such as lead and cadmium in components and assemblies. Removal of lead which was used in solder, die attaches and surface finishes has introduced reliability risks such as thermo-mechanical fatigue, tin whiskers, tin pest, electro-chemical migration, and corrosion. Due to these concerns, certain reliability-critical industries such as medical, defense, and automotive were either exempted or excluded from these restrictions. Since the commercial electronics sectors have switched to lead-free materials, few suppliers now produce lead-based solders and surface finish boards. Hence there is a growing supply chain pressure on these exempted and excluded industries to switch to lead-free materials. However, they are hesitant to transition to lead-free due to previously said reliability concerns remaining with lead-free assemblies.

This study analyses possible failure modes and mechanisms to assess these reliability risks in this transition. The critical failure mechanisms included thermal and mechanical fatigue and tin whiskers. Simulation was conducted using CalcePWA software to compare the reliability between tin-lead and SAC305 solder under temperature cycling and vibration loading in addition to assessing the risk due to tin whiskers. The discussion is mainly focused on concerns about the changes in manufacturing practices, effects of storage and handling conditions on manufacturing defects, while susceptibility to other failure mechanisms are briefly discussed. The study provides assessment and important factors to be considered and monitored during lead-free transition.

Keywords: Lead-free, ENIG, Reliability, RoHS, Transition

Introduction: In 2003, the European Union (EU) passed the Restriction of Hazardous Substances (RoHS) directive limiting the use of certain hazardous substances in electrical and electronic equipment in EU member states and provided a mechanism for restricting additional substances in the future [1]. The RoHS directive (2002/95/EC) became effective on July 1, 2006, and was applicable to the ten categories of products listed in the WEEE directive, as well as to electric light bulbs and luminaires used in households. In particular, the RoHS 2 directive required medical devices and monitoring and control instruments to comply with current RoHS restrictions by July 2014 and industrial control and monitoring instruments to comply by July 2017. For all other equipment, unless explicitly excluded, compliance is required by July 2019 [2]. The electrical and electronic equipment explicitly excluded were equipment used in military and space applications, large-scale stationary industrial tools, large-scale fixed installations, implantable medical devices, transportation applications (except for electric two-wheel vehicles), non-road mobile machinery, photovoltaic panels designed for permanent use, and equipment designed solely for the purpose of research and development [3].

With consumer electronics driving the trends in technology, the majority of electronic component manufacturers have already transitioned to the use of lead-free materials [4]. On the other hand, electronic manufacturers associated with aerospace, military, and space applications and that are excluded from government-imposed restrictions have generally attempted to maintain lead-based parts and assembly processes due to long-term reliability concerns with lead-free parts and assemblies. However, even some of these manufacturers have been investigating, and in some cases using, lead-free parts because they are the only parts that are affordable and available on the market. For example, it is nearly impossible to purchase high-density BGA packages in leaded versions [4]. As a result, manufacturers using tin-lead solder face the decision of assembling a lead-free BGA with tin-lead solder or replacing the lead-free solder balls on the BGA with tin-lead solder balls. Companies that have been excluded by the environmental legislation have been forced to make last-time buys and store spares or to use re-worked lead-free components. With potential reliability concerns from such re-worked assemblies as well as the risks associated with the inclusion of counterfeit components and the shrinking manufacturing base, these companies are now being compelled to evaluate the transition to lead-free materials.

This paper overviews the key reliability risks in transitioning to lead-free assembly. The intent is to address some of the reliability risks associated with process related to manufacturing, storage and handling of lead-free parts, but not to be an extensive analysis of any particular failure mechanism; the interested reader is referred instead to key references. Potential failure modes and mechanisms were studied to assess these reliability risks in the transition. Simulations were conducted to analyze risks of thermal, mechanical fatigue and tin whiskers in lead-free solders.

Analysis Sample Information: The board model used for simulated reliability assessment and risk assessment was an industrial controller board populated with integrated chips (ICs), resistors, capacitors (such as ceramic, electrolytic) and other standard electronic components used for controlling signals. Table 1 lists the materials that were considered for risk assessment through simulation and literature review:

Table 1 Materials Considered for Risk Assessment

	Lead-based Board	Lead-free Board
Solder Material	Sn63-Pb37	SAC305
Board Material	FR4	FR4

	Lead-based Board	Lead-free Board
Metallization Surface Finish	HASL (Hot Air Solder Leveling)	ENIG (Electroless Nickel Immersion Gold)

Table 2 outlines the potential failure modes and mechanisms of reliability concerns associated with lead-free manufacturing and operational practices. This is by no means the exhaustive list of all the problems with the usage of lead-free materials. But these are some of the most prominent reliability concerns. The failure mode here is broadly categorized as short and open circuits. The mechanisms are not listed based on risk priority as the effect, severity and frequency of occurrence depends on the application and lifecycle conditions the electronic packages are used in.

Table 2 Potential Failure Modes and Mechanisms of Lead-free Process

<table>
<tr><th>Potential Failure Site</th><th>Potential Failure Modes</th><th>Potential Failure Causes</th><th>Potential Failure Mechanisms</th></tr>
<tr><td rowspan="9">Plastic encapsulation of integrated chips (a-A)

Solder bumps (a-B, C, E; b-H)

Soldered leads (a-B, C; b- H, I)

PCB trace surface finish (a-C)

Pad/Board Interface (a-D)

Solder Joints (a-F)

Plated through holes (b-G)</td><td rowspan="6">(a) Open/parameter drift/intermittent</td><td>Reflow process, improper storage conditions</td><td>A. Popcorning</td></tr>
<tr><td>Temperature cycling, power cycling, mechanical vibration</td><td>B. Solder cracks</td></tr>
<tr><td>High current density (>10^4 A/cm^2)/manufacturing defects</td><td>C. Electromigration</td></tr>
<tr><td>Shock or vibration loading/poor handling, thermal cycling</td><td>D. Pad cratering</td></tr>
<tr><td>Higher reflow temperature, poor flux</td><td>E. Solderability</td></tr>
<tr><td>High temperature (>50), salt content</td><td>F. Solder corrosion</td></tr>
<tr><td rowspan="3">(b) Short/parameter drift/intermittent</td><td>Impurities in resin, high reflow temperature, moisture ingression</td><td>G. Conductive anode filament</td></tr>
<tr><td>Humid conditions, voltage bias, contaminants</td><td>H. Dendrites/ Electrochemical migration</td></tr>
<tr><td>Temperature cycling, humidity,</td><td>I. Tin whiskers</td></tr>
</table>

Potential Failure Site	Potential Failure Modes	Potential Failure Causes	Potential Failure Mechanisms
Electrical connectors (b-I)		electroplating process, CTE mismatch b/w conductor and finish	

Reliability Risks: Some of the reliability risks mentioned in the above analysis are evaluated in the following section using literature review and simulation. The evaluation provides a comparison between tin-lead and lead-free packages in terms of possible risks associated with manufacturing, storage and usage of lead-free materials.

Solderability: Currently available lead-free solders have a solidus/liquidus point about 20-50°C higher (depending on the alloy) than the lead-based pastes currently in use [6]. The primary challenge lead-free solders will present electronics assemblers with is higher process temperatures. Minimizing process temperatures limits the thermal stress on boards and components, reducing the potential for manufacturing defects. Higher reflow process thermal cycles expose PCBs to significant amounts of stress on plated through holes and barrels, which can lead to cracking. Higher first pass temperatures on double-sided assemblies expose bottomside finished surfaces to oxidation or interdiffusion, which in turn can lead to solderability problems on the second pass.

Popcorning: Limiting peak temperatures limits intermetallic growth, especially on topside components that are exposed to two passes, and also limits the potential for the popcorning of components with high moisture content. Increasing the maximum reflow temperature is expected to increase the moisture sensitivity of plastic packaged components. Moisture diffusion in epoxy molding compounds (EMCs) is one of the major reliability concerns in plastic encapsulated microcircuits (PEMs), because many failure modes observed in these devices are believed to arise from the diffusion of moisture during manufacturing, storage, or operation [7]-[11]. Moisture desorption in EMCs takes place at reflow process. Moreover, it is important to understand the mechanism of moisture desorption, since during the assembly of PEMs, the packages undergo baking to remove moisture and thus reduce the probability of moisture-induced failures such as popcorn cracking and interfacial delamination. In the most severe case, the stress can result in external package cracks. This is commonly referred to as the "popcorn" phenomenon because the internal stress causes the package to bulge and then crack with an audible "pop" [12]. Plastic-encapsulated components have to be checked for their MSL (moisture sensitivity level) to find the maximum exposure time allowed outside the moisture-resistant bag during storage. Improper storage, handling, or packaging of plastic encapsulated semiconductor devices can allow the introduction of moisture. Moisture trapped inside plastic-encapsulated packages can damage them during soldering, as the moisture vaporizes and tries to expand. This internal vapor pressure can cause separation of the plastic package from the semiconductor chip or lead frame, internal or external cracks, and damage to thin films and wire bonds.

Simulation: The simulation of stress response of these boards under temperature cycling, vibrations, and susceptibility of these materials to tin whiskers were carried out using CalcePWA software. The CalcePWA software consists of a set of simulation tools that use various thermo-mechanical stress and damage models.

The model of the board with components assembled on it, as shown in Figure 1, was used for the simulation.

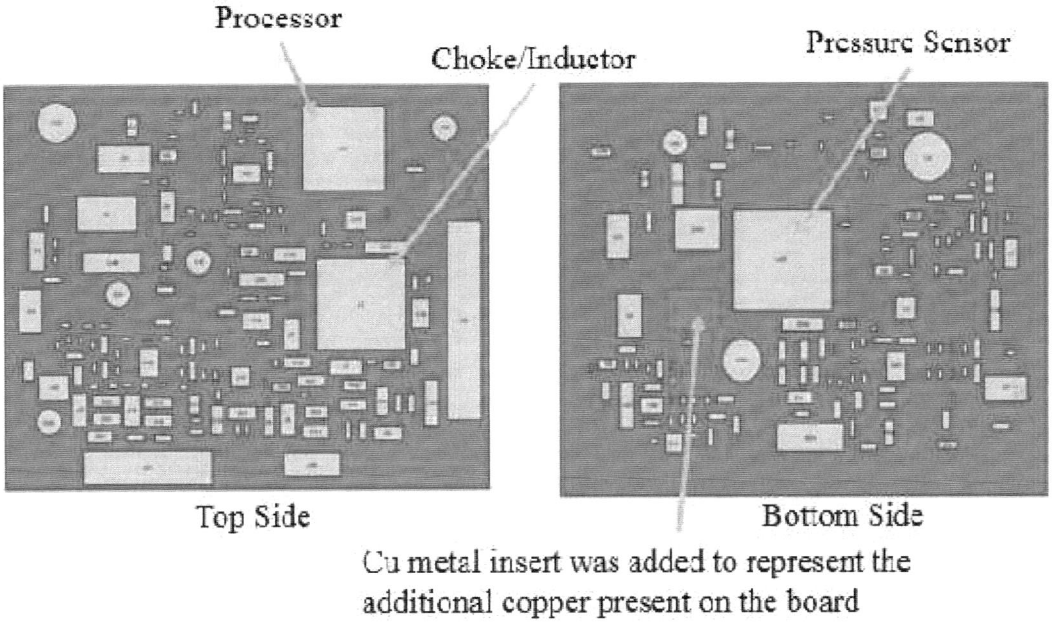

Figure 1 CalcePWA CAD Model of the Board with Components

Thermal analysis was conducted to determine the temperature gradient across the board and the temperature rise of the components under a given temperature. The power dissipation values of components, obtained from typical real-time applications, were taken up for analysis. No particular cooling solution was considered and hence conduction across board and natural vertical convection to the ambience was chosen as the heat transfer model. The thermal analysis was conducted for each extreme temperature that was to be used for the temperature cycling tests (shown in Table 3, Table 4). It was observed that certain ICs (transceivers-U15), fuse (F4), ceramic capacitors, ceramic diodes, and processor exhibited the highest rise in temperature of 17-18 °C (as shown in an example in Figure 2) during each thermal stress while many other components such as electrolytic capacitors, op-amps, and a few other ICs exhibited temperature rise of 3-5 °C.

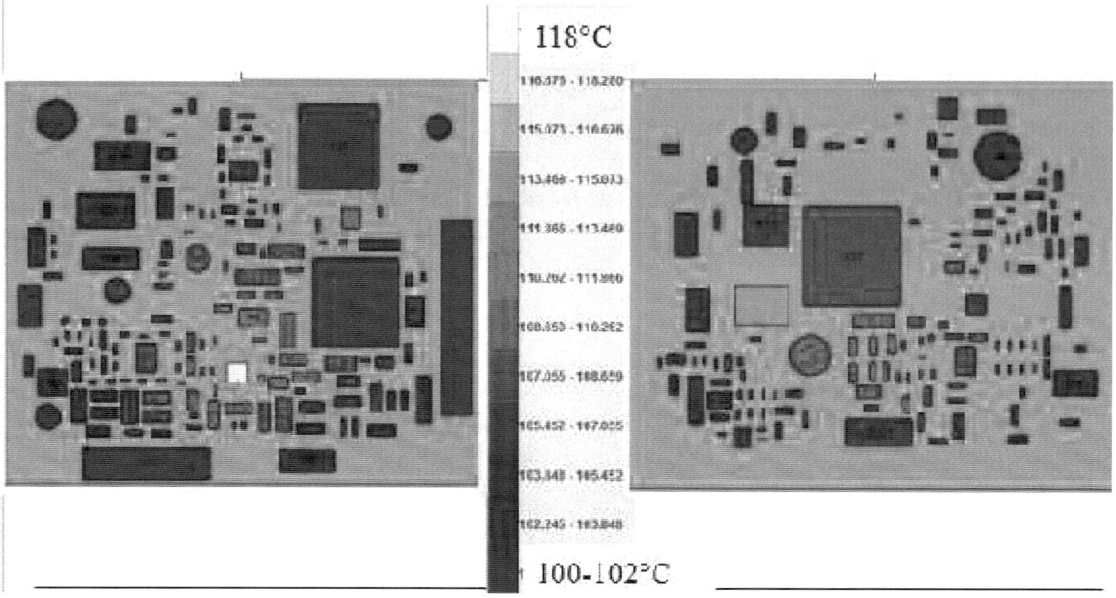

Figure 2 Temperature Distribution at 100 °C Ambient Temperature

Temperature cycling analysis was conducted to assess relative risks of interconnect failures for the components. Simulation test conditions were adapted from standards for accelerated temperature cycling. Temperature cycling was conducted for different conditions:

- Standard accelerated test as per JESD22-A104D [5] for 1000 cycles (Table 3 and Table 4).
- Assumed extreme real application conditions: Death Valley yearly temperature: Each of the temperature range was assumed to undergo 2 cycles per day and for 3 months continuously before the next temperature range starts. This condition was applied for a 10 year operating period.

The thermal analysis results obtained previously were used as inputs for each maximum and minimum temperature. Dwell time and ramp times were provided as shown in the Table 3 and Table 4.

Table 3 Temperature Cycling Condition as per JESD22-A104D

	Temp (°C)	Dwell Time (min.)
Minimum Temp.	0	10
Maximum Temp.	100	10
Ramp Time = 5mins		

Table 4 Temperature Cycling Condition Based on Death Valley Average Yearly Temperature

Temperature range	Dwell Time at max temp. (mins)	Dwell Time at min temp. (mins)	Ramp time (mins)	No. of cycles
-5 to 20 °C	240	240	120	1800 (30 months)
0 to 30 °C	240	240	120	1800 (30 months)
10 to 40 °C	240	240	120	1800 (30 months)
20 to 50 °C	240	240	120	1800 (30 months)

Under the Death Valley condition, which can be considered to be an extreme but realistic application condition, both the lead-free board and the tin-lead seemed to survive the given temperature stress condition over a period of 10 years. Figure 4 shows a part of the list of components with their corresponding damage ratio in descending order. Damage ratio indicates the amount of damage accumulated over the entire period of temperature cycling.

It was observed that in Death Valley conditions, the tin-lead board was more susceptible to thermal fatigue than the lead-free board as shown in Figure 3 (snapshot of results as shown in simulation), while in the accelerated condition, the lead-free board was found to be more susceptible than the tin-lead version as shown in Figure 4. This is due to the higher acceleration factor of the lead-free failure model. This is to say, if at a particular accelerated condition the acceleration factor of tin-lead solder is 10, and if the tin-lead interconnect survives x hours in the accelerated conditions, then it is expected to survive 10x hours in the normal usage conditions. The acceleration factors for lead-free solders are generally higher than the tin-lead solders under the same accelerated conditions. Acceleration factor is a function of testing condition and hence changes based on our testing conditions.

Components	Quantity	Failure Model	No. of cycles to failure	Failure Indicator
D10-solder-open	1	1ST_TF_LL_NEW	> Specified (DR:0.72)	View
D41-solder-open	1	1ST_TF_LL_NEW	> Specified (DR:0.72)	View
D18-solder-open	1	1ST_TF_LL_NEW	> Specified (DR:0.72)	View
D12-solder-open	1	1ST_TF_LL_NEW	> Specified (DR:0.71)	View
D20-solder-open	1	1ST_TF_LL_NEW	> Specified (DR:0.71)	View
D4-solder-open	1	1ST_TF_LL_NEW	> Specified (DR:0.70)	View
D36-solder-open	1	1ST_TF_LL_NEW	> Specified (DR:0.70)	View
D3-solder-open	1	1ST_TF_LL_NEW	> Specified (DR:0.69)	View
D22-solder-open	1		> Specified (DR:0.69)	View
D21-solder-open	1		> Specified (DR:0.68)	View

Tin-lead

Components	Quantity	Failure Model	No. of cycles to failure	Failure Indicator
D41-solder-open	1	1ST_TF_LL_NEW	< Specified (DR:1.04)	View
D10-solder-open	1	1ST_TF_LL_NEW	< Specified (DR:1.04)	View
D18-solder-open	1	1ST_TF_LL_NEW	< Specified (DR:1.04)	View
D12-solder-open	1	1ST_TF_LL_NEW	< Specified (DR:1.01)	View
D20-solder-open	1	1ST_TF_LL_NEW	> Specified (DR:0.99)	View
D4-solder-open	1	1ST_TF_LL_NEW	> Specified (DR:0.98)	View
D36-solder-open	1	1ST_TF_LL_NEW	> Specified (DR:0.97)	View
D3-solder-open	1		> Specified (DR:0.93)	View
D22-solder-open	1		> Specified (DR:0.92)	View

Lead free

Figure 3 A Sample of Damage Ratio of Components Under Temperature Cycling

In all the temperature cycling conditions, the parts that were found to be the most susceptible to failure by solder joint cracking in both lead-free and tin-lead boards were ceramic surface mount parts (voltage suppressors) and the ceramic surface mount capacitors.

Components	Quantity	Failure Model	No. of cycles to failure	Failure Indicator
D10-solder-open	4	1ST_TF_LL_NEW	> Specified (DR:0.54)	View
D41-solder-open	4	1ST_TF_LL_NEW	> Specified (DR:0.54)	View
D18-solder-open	4	1ST_TF_LL_NEW	> Specified (DR:0.54)	View
D12-solder-open	4	1ST_TF_LL_NEW	> Specified (DR:0.53)	View
D20-solder-open	4	1ST_TF_LL_NEW	> Specified (DR:0.53)	View
D4-solder-open	4	1ST_TF_LL_NEW	> Specified (DR:0.52)	View
D36-solder-open	4	1ST_TF_LL_NEW	> Specified (DR:0.52)	View
D3-solder-open	4	1ST_TF_LL_NEW	> Specified (DR:0.51)	View
D22-solder-open	4	1ST_TF_LL_NEW	> Specified (DR:0.51)	View
D11-solder-open	4		> Specified (DR:0.50)	View
D21-solder-open	4		> Specified (DR:0.50)	View

Tin-lead

D41-solder-open	4	1ST_TF_LL_NEW	> Specified (DR:0.21)	View
D10-solder-open	4	1ST_TF_LL_NEW	> Specified (DR:0.21)	View
D18-solder-open	4	1ST_TF_LL_NEW	> Specified (DR:0.21)	View
D12-solder-open	4	1ST_TF_LL_NEW	> Specified (DR:0.20)	View
D20-solder-open	4	1ST_TF_LL_NEW	> Specified (DR:0.20)	View
D4-solder-open	4	1ST_TF_LL_NEW	> Specified (DR:0.19)	View
D36-solder-open	4	1ST_TF_LL_NEW	> Specified (DR:0.19)	View
C55-solder-open	4		> Specified (DR:0.18)	View

Lead-free

Figure 4 A Sample of Damage Ratio of Components Under Temperature Cycling

For ***vibration analysis***, we adopted truck transportation vibration standard MIL-STD-810, method 514.6-annex C [13]. Figure 5 shows the vibration profile for different direction of random vibration with different values of rms value of g. Vertical direction profile with an rms value of 1.04g was selected. Table 5 shows the frequency points and corresponding PSD values for the chosen transverse direction.

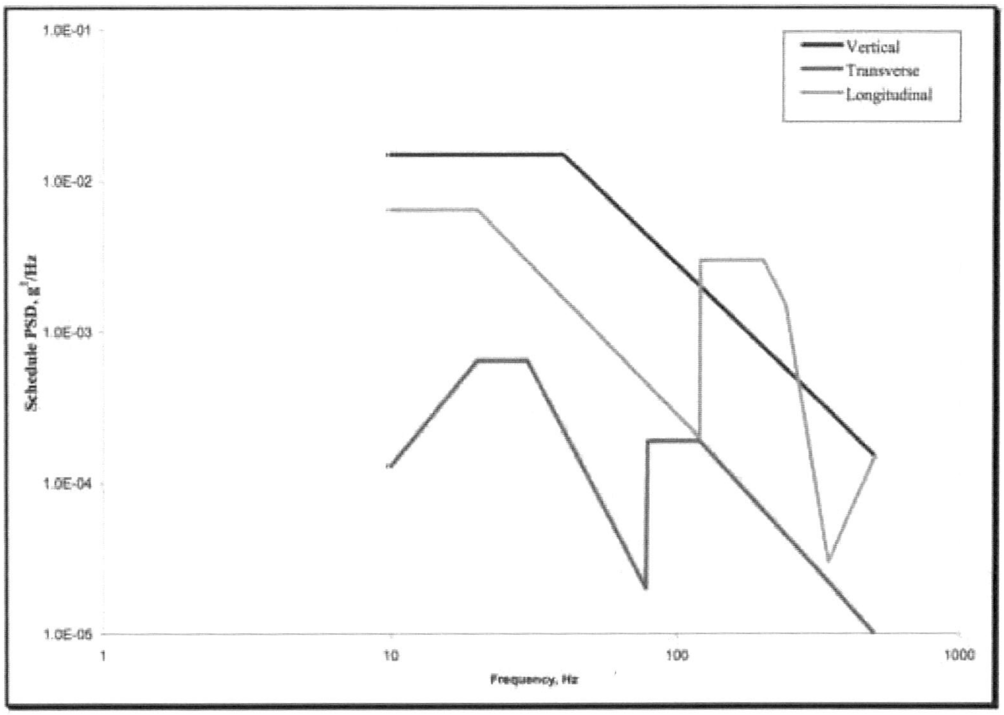

Figure 5 US Highway Truck Vibration Exposure [13]

Table 5 Frequency vs. PSD for Transverse Direction

Frequency (Hz)	PSD (G^2/Hz)
10	0.015
40	0.015
500	0.00015

Figure 6 shows that none of the components experience open failure at interconnects due to the applied vibration. However, it has to be noted that the ceramic components assembled on the board can fail at the component level under vibration even if the interconnects remain intact for high curvature conditions. However, the curvatures estimated for the vibration conditions are not high enough to cause such cracking.

Components	Quantity	Failure Model	Life (DR at 10 years)	Indicator
C42-interconnect-open	1	1ST_VF_RM	> Specified (DR 0.00)	View
Q1-interconnect-open	1	1ST_VF_RM	> Specified (DR 0.00)	View
C41-interconnect-open	1	1ST_VF_RM	> Specified (DR 0.00)	View
Q2-interconnect-open	1	1ST_VF_RM	> Specified (DR 0.00)	View
Q3-interconnect-open	1	1ST_VF_RM	> Specified (DR 0.00)	View
C43-interconnect-open	1	1ST_VF_RM	> Specified (DR 0.00)	View
Q4-interconnect-open	1	1ST_VF_RM	> Specified (DR 0.00)	View
Q5-interconnect-open	1	1ST_VF_RM	> Specified (DR 0.00)	View
Q7-interconnect-open	1	**Tin-lead**	> Specified (DR 0.00)	View
Q8-interconnect-open	1		> Specified (DR 0.00)	View

Components	Quantity	Failure Model	Life (DR at 10 years)	Indicator
C42-interconnect-open	1	1ST_VF_RM	> Specified (DR 0.00)	View
Q1-interconnect-open	1	1ST_VF_RM	> Specified (DR 0.00)	View
C41-interconnect-open	1	1ST_VF_RM	> Specified (DR 0.00)	View
Q2-interconnect-open	1	1ST_VF_RM	> Specified (DR 0.00)	View
Q3-interconnect-open	1	1ST_VF_RM	> Specified (DR 0.00)	View
C43-interconnect-open	1	1ST_VF_RM	> Specified (DR 0.00)	View
Q4-interconnect-open	1	1ST_VF_RM	> Specified (DR 0.00)	View
Q5-interconnect-open	1		> Specified (DR 0.00)	View
Q7-interconnect-open	1	**Lead-free**	> Specified (DR 0.00)	View
Q8-interconnect-open	1		> Specified (DR 0.00)	View

Figure 6 Results of Vibration Analysis

Tin Whiskers: Tin whiskers are filamentary growths that spontaneously grow from electroplated tin surfaces. Whisker growth starts after an incubation period that varies from seconds to years. Whisker diameters range between 0.006 and 10 m [14]. Tin whiskers are spontaneous growths from tin and high tin content alloy finishes. Growths can take on a variety of shapes. Whisker lengths can range from a few microns to over 1 mm. Diameters range from less than 1 m to about 6 m.

Tin whisker risk must be assessed based on product, application, and tolerance to failure. The risk of tin whisker shorts was estimated using CalceTinWhiskerRisk calculator. Tin whisker risk assessment involves identification of potential whisker failure sites, which are adjacent conductor pairs with at least one surface coated with pure tin or a Pb-free tin finish. The other factors to be considered are the characteristics of whiskers itself - whisker density and whisker length. Any mitigation processes employed such as conformal coating, solder dipping and appropriate part selection should also be considered while assessing these risks.

CalceTinWhisker risk calculator assumes: full surface area of a conductor is susceptible to whisker formation; shortest distance between conductors form the whisker bridging; and whisker growth can be extrapolated from measured data. Failure occurs when $l_w \geq l_s$, where l_w is the length of a whisker and l_s is the spacing between the two adjacent conductors. Because whiskers have been observed to change orientation, the growth angle is no longer considered. Based on the CALCE calculator (after 10,000 iterations), the probability for system failure due to tin whisker short was estimated to be 0.03% (refer to attached spreadsheet). Components with the highest probability of tin whisker shorting were found to be surface mount leaded ICs and surface mount electrolytic capacitors (with bent lead).

Other Reliability Concerns: Based on the lifecycle conditions that the electronic assembly is subjected to, there can be other potential failure mechanisms that can lead to system failure. Electrochemical migration can be a serious concern due to the mobility of silver ions under humid conditions [15][16]. Stiffness of SAC solders as compared to tin-lead can cause pad cratering under mechanical vibrations. Some gaseous and sultry environments can lead to solder corrosion. However, lead-free solders have been found to be on par or better than tin-lead solders in corrosion resistance [17].

Conclusion: Temperature cycling stress simulation was conducted for three different condition – accelerated condition, approximate application condition, and board rating condition. There were no significant difference between the damage accumulated in the tin-lead interconnects and lead-free interconnects. Accelerated conditions showed lead-free board accumulating marginally more damage than tin-lead due to the higher acceleration factor in the failure model, while the normal usage conditions showed lead-free board to be marginally safer than tin-lead board. Vibration analysis was conducted based on U.S. truck transportation vibration testing standard MIL-STD-810. It was found that lead-free interconnects and tin-lead interconnects showed similar (no damage accumulation) damage accumulation due to the applied vibration condition. It has to be noted, however, that, although the interconnects themselves might not have failed, brittle material components such as ceramic capacitors and diodes might crack under mechanical vibrations.

Tin whisker risk assessment was conducted for the worst-case scenario: Pure tin coating over copper leads, no conformal coating applied. The CALCE tin whisker risk calculator estimated the probability of failure of the system due to tin whisker shorting over a period of 10 years to be 0.3%. Other reliability concerns such as popcorning, ECM, solder corrosion, and pad cratering related to adopting lead-free materials in electronics assembly were briefly discussed.

Apart from these reliability concerns, there has to be continual monitoring on certain features of products such as lead temperature rating, MSL rating, and part temperature rating. All these are critical while selecting the reflow profile and storage conditions. Any deviation between the rated parameters and followed assembly and storage practice can increase the risks of defects and failures of lead-free systems.

References:

[1] "Directive 2002/95/EC of the European Parliament and of the Council of 27 January 2003 on the restriction of the use of certain hazardous substances in electrical and electronic equipment," Official Journal L 037, pp. 19-23, February 13, 2003, Accessed online: http://eurlex.europa.eu/LexUriServ/LexUriServ.do?uri=CELEX:32002L0095:EN:HTML.

[2] European Commission, "Recast of the RoHS Directive," Accessed online: http://ec.europa.eu/environment/waste/rohs_eee/, Last updated: February 7, 2014.
[3] European Commission, "Environment: Commission proposes revised laws on recycling and use of hazardous substances in electrical and electronic equipment," IP/08/1878, Press Release, December 3, 2008, Accessed online: http://europa.eu/rapid/press-release_IP-08-1878_en.htm.
[4] Mathew, S., M. Osterman, and M. Pecht, "Evaluation of pure tin plated copper alloy substrates for tin whiskers," Circuit World, vol. 35, no. 1, pp. 3-8, 2009.
[5] JEDEC Standard, JESD22-A104D. [Online]. Available: http://www.jedec.org/sites/default/files/docs/22a104d.pdf
[6] Johnson, W., Lugo, R., Sattiraju, S.V., and Jones, G., "Improved Thermal Process Control for Lead-free Assembly", APEX, San Diego, 2001.
[7] Shirangi, M.H., Auersperg, J., Koyuncu, M., Walter, H., Müller, W.H., Michel, B., "Characterization of dual-stage moisture diffusion, residual moisture content and hygroscopic swelling of epoxy molding compounds," Proceedings of the 9th EuroSime2008, Freiburg, Germany, pp. 455–462, 2008.
[8] Shirangi, M.H., Fan, X.J., Michel, B., "Mechanism of moisture diffusion, hygroscopic swelling and adhesion degradation in epoxy molding compounds," Proceedings of the 41st International Symposium on Microelectronics (IMAPS), Providence, USA, pp. 1082–1089, 2008.
[9] Fan, X.J., Zhou, J., Chandra, A., "Package structural integrity analysis considering moisture," Proceedings of Electronic Components and Technology Conference (ECTC), Orlando, USA, pp. 1054–1066, 2008.
[10] Pecht, M.G., Nguyen, L.T., Hakim, E.B., Plastic-Encapsulated Microelectronics. New York, NY: Wiley, 1995.
[11] Ardebili, H., Hillman, C., Natishan, M.A.E., McCluskey, P., Pecht, M.G., Peterson, D., "A comparison of the theory of moisture diffusion in plastic encapsulated microelectronics with moisture sensor chip and weight-gain measurements," IEEE Transaction on Components and Packaging Technologies, 25, 132–139, 2002.
[12] IPC/JEDEC J-STD-020D.1, "Moisture/Reflow Sensitivity Classification for Nonhermetic Solid State Surface Mount Devices". [Online]. Available: http://www.adestotech.com/wp-content/uploads/jstd020d-01.pdf
[13] MIL-STD-810G, Method 514.6 Annex C, Transportation Tailoring Guidance for Vibration Exposure Definition. [Online]. Available: http://www.cascadetek.com/testing-services/white-papers/MIL-STD-810-Transporation-Vibration
[14] H. Leidecker and Brusse, J., "Tin whiskers: A history of documented electrical system failures—A briefing prepared for the Space Shuttle program office 2006," online, available at http://nepp.nasa.gov/whisker/reference/tech_papers/2006-Leidecker-Tin-Whisker-Failures.pdf.
[15] G. Harsanyi, "Electrochemical processes resulting in migrated short failures in microcircuits," IEEE Trans CPMT, Vol. 18, (1995), pp. 602.
[16] Kohman, G. T., H. W. Hermance, and G. H. Downes, 1955, "Silver migration in electrical insulation," Bell Systems Technology Journal, v34, p. 1115.

[17] D. Li, P. P. Conway, C. Liu, "Corrosion characterization of tin–lead and lead free solders in 3.5 wt.% NaCl solution", Corrosion Science, Elsevier, Volume 50, Issue 4, pp. 995-1004, 2008.

Authors Biography:

Guru Prasad Pandian is a second year Ph.D. student at mechanical engineering department at University of Maryland, College park. He has been working as a research assistant with Center for Advanced Life Cycle Engineering at University of Maryland since summer of 2015. His research interests are evaluating reliability of electronic packages and specifically reliability of lead-free solders.

Dr. Diganta Das (Ph.D., Mechanical Engineering, University of Maryland, College Park, B.Tech., Manufacturing Science and Engineering, Indian Institute of Technology) is an Associate Research Scientist at the Center for Advanced Life Cycle Engineering. His expertise is in reliability, environmental and operational ratings of electronic parts, uprating, electronic part reprocessing, counterfeit electronics, technology trends in the electronic parts and parts selection, and management methodologies. In addition, Dr. Das is involved in prognostics-based risk mitigation of electronics. Dr. Das has published more than 75 articles on these subjects, and presented his research at international conferences and workshops. He had been the technical editor for two IEEE standards and is currently vice chair of the standards group of the IEEE Reliability Society. He is a sub-group leader for the SAE G-19 counterfeit detection standards group. He is an Associate Editor of the journal Microelectronics Reliability. He is a Six Sigma Black Belt and a member of IEEE, IMAPS, and SMTA.

Prof. Michael Pecht (M'83–SM'90–F'92) received the M.S. degree in electrical engineering and the M.S. and Ph.D. degrees in engineering mechanics from the University of Wisconsin-Madison, WI, USA. He is the Founder of the Center for Advanced Life Cycle Engineering, University of Maryland, College Park, MD, USA, which is funded by over 150 of the world's leading electronics companies at more than U.S. $6 million/year. He is also the George E. Dieter Professor of Mechanical Engineering and a Professor of Applied Mathematics at the University of Maryland, College Park. He has authored more than 20 books on electronic product development, use, and supply chain management and over 500 technical articles.

ROLLING BEARING FAULT FEATURE EXTRACTION OF CASING VIBRATION SIGNAL

Yizhou Yang and Dongxiang Jiang
State Key Lab of Control and Simulation of Power Systems and Generation Equipment,
Department of Thermal Engineering, Tsinghua University
Beijing 100084, China
Telephone: +86 15201522492
yangyz16@mails.tsinghua.edu.cn

Abstract: The health of rolling bearing plays an important part in the operation of rotating machinery like gas turbine engine. Health monitoring and fault diagnosis of rolling bearings based on vibration signals have been through great development these years. But when sensors are set on the casing instead of the bearing pedestal, and the surrounding structure is very complex, the diagnosis problem becomes much more complicated, which brings more challenges to the signal processing. In this paper, a set of signal processing methods are used to enhance and extract the impact features from casing vibration signals, and to realize the detection of rolling bearing faults. A self-adaptive decomposition method called intrinsic time-scale decomposition (ITD) is applied to decompose the vibration signal into a series of proper rotation components and a monotonic trend, helping to extract dynamic features of the signal. Teager-Kaiser energy operator is a simple algorithm calculating the energy of a signal and is very sensitive to transient impact faults. As the fault feature transmitted to the casing is relatively week, autoregressive model (AR) and minimum entropy deconvolution (MED) are to enhance the non-stationary impact components. Experiments are taken on the rotor-bearing-casing test rig with minor defects in the main shaft bearing. Testing on the casing vibration signal, this fault feature enhancing and extracting method shows its remarkable ability in rolling bearing fault diagnosis.

Key words: Casing vibration; fault diagnosis; feature extraction; rolling bearing.

Introduction: The main shaft rolling bearings in a gas turbine engine is precision components and play vital parts in the safe and efficient operation of the whole machine. Thus the condition monitoring and evaluation of rolling bearings and the detection and maintenance for early bearing faults are necessary. There are plenty of vibration-based fault detection and health monitoring methods [1]. But when dealing with casing signals, some special issues must be taken into consideration. Due to the complex interior structure of a gas turbine engine, displacement sensors sometimes cannot be installed inside the machine [2]. Vibration signals will be collected by the accelerometers installed on the casing. It is difficult to analyze casing signals because of the complex surrounding structure and the long path of transmission from the inside to the outside. The vibrational features will be weaker and much noises will be added, which brings more difficulty in the fault diagnosis of bearing faults.

To detect rolling bearing faults needs to extract fault features from the vibration signals. Once the frequency components concerning bearing faults are found during the signal processing, the occurrence of faults could be substantially confirmed. The vibrational features transmitted to the casing will be much weaker than those of the bearing pedestal, so it is preferred to enhance the features for sensitive diagnosis. Autoregressive model (AR) [3] and Minimum entropy deconvolution (MED) [4] are employed in this paper for feature enhancement. AR is a type of statistics and signal processing method describing certain time-varying processes that can predict the deterministic parts in a signal. MED is a signal deconvolution technique whose result highlight a few large spikes and is therefore suitable for the noise reduction of the impact fault signal of the rotating machinery. Because of multiple components and much noise in the vibration signal, signal decomposition is an alternative preprocessing method. Intrinsic time-scale decomposition (ITD) [5] is a self-adaptive decomposition method that can decompose a signal into a sum of components and a monotonic trend. ITD has many advantages over empirical mode decomposition (EMD) [6], a more famous self-adaptive decomposition technique, like higher calculation speed and no such issues in mode confusion, mislocalization of temporal information and phase shifts and distortion. Teager-Kaiser energy operator (TKEO) [7] is an algorithm calculating the energy of a signal, and is very sensitive to transient impact caused by defects. TKEO is used here as an efficient technique for extracting bearing fault features.

In this paper, a rolling bearing fault feature extraction method combining ITD, TKEO, AR and MED is proposed to apply to the casing vibration signal of a rotor-bearing-casing system. The method is tested on experimental signals collected from the rotor-bearing-casing test rig, where rolling bearing faults can be simulated. The qualities of the methods are shown by comparing the relative size of the character frequency components.

Intrinsic time-scale decomposition: Intrinsic time-scale decomposition (ITD) can decompose a signal into a series of components (so called proper rotation components) and a monotonic trend. L is defined as the baseline extraction operator for the signal X_t. With the baseline component L_t extracted from the signal, the proper rotation component H_t is remained. Thus one step of decomposition is done.

$$X_t = LX_t + (1 - L)X_t = L_t + H_t \tag{1}$$

Supposing $\{\tau_k, k = 1, 2, \cdots\}$ local extrema points, set $\tau_C = 0$. $X(\tau_k)$ and $L(\tau_k)$ are written as X_k and L_k to simplify the expression. Assume that L_t and H_t have definition when $t \in [0, \tau_k]$ as well as X_t having definition when $t \in [0, \tau_{k+2}]$. The piecewise linear baseline extraction operator L is defined on the internal $(\tau_k, \tau_{k+1}]$ between successive extrema as:

$$L = L_k + \left(\frac{L_{k+1} - L_k}{X_{k+1} - X_k}\right)(X_t - X_k) \quad (2)$$

$$L_{k+1} = \alpha\left[X_k + \left(\frac{\tau_{k+1} - \tau_k}{\tau_{k+2} - \tau_k}\right)(X_{k+2} - X_k)\right] + (1-\alpha)X_{k+1} \quad (3)$$

where $\alpha \in [0,1]$ and usually $\alpha = 0.5$. So the proper rotation component is derived:

$$H_t = (1 - L)X_t = X_t - L_t \quad (4)$$

This method of baseline extraction maintains the monotonicity while a high frequency component known as the proper rotation component is remained. The baseline signal is decomposed again using this method until a monotone trend is derived. So the original signal X_t is decomposed into several high frequency components (proper rotation components) and a trend term. The process can be described as:

$$X_t = H_t + L_t = H_t + (HL_t^1 + L_t^1) = \cdots = H\sum_{k=0}^{p-1} L_t^k + L_t^p \quad (5)$$

where HL_t^k is the proper rotation component and L_t^k the baseline signal, after $(k+1)$ times of decomposition.

Like EMD (empirical mode decomposition) and some other self-adaptive decomposition methods, ITD aims at defining a single component signal whose instantaneous frequency has its physical meaning, and then decomposes the original signal into a series of such single component signals and a trend term. But without repeated iterations, ITD has an advantage over EMD in calculation speed. The ITD also overcomes some of the EMD's limitations, like mode confusion, mislocalization of temporal information and phase shifts and distortion.

Teager-Kaiser energy operator: For continuous-time signal $x(t)$ and discrete-time signal x_n, the Teager-Kaiser energy operator [8] is defined as

$$\Psi[x(t)] = [\dot{x}(t)]^2 - x(t)\ddot{x}(t) \quad (6)$$

$$\Psi[x_n] = x_n^2 - x_{n-1}x_{n+1} \quad (7)$$

The simple harmonic system's total energy is the sum of the kinetic energy and the potential energy. Substituting $x(t) = A\cos(\omega \quad + \phi)$ for x, so

$$E = \frac{1}{2}kx^2 + \frac{1}{2}m\dot{x}^2 = \frac{1}{2}m\omega^2 A^2 \propto \omega^2 A^2 \quad (8)$$

And for the sample of the signal of simple harmonic motion $x_n = A\cos(\Omega n + \phi)$, where $\Omega = 2\pi f/f_s$ is the digital frequency, f is the analog frequency and f_s is the sampling frequency of the signal. As the value of Ω is very small, $\sin^2(\Omega) \approx \Omega^2$, so

$$\Psi[x_n] = x_n^2 - x_{n-1}x_{n+1} = A^2\sin^2(\Omega) \approx A^2\Omega^2 \quad (9)$$

Thus the energy operator can track the energy of any single component signal [9]. Only 3 points are needed when calculating the energy at time n in a discrete-time signal. This algorithm is very simple but has a high temporal resolution, which makes this algorithm sensitive to transient components. When localized defects occur on the outer race, the inner race or the rolling elements, the rotation of the main shaft may cause transient impacts, which excite the natural frequencies of the bearing and the surrounding structure. The vibration signal at this time has components with high amplitude and high frequency. The energy operator approximately equal to the square of the product of amplitude and frequency, which highlights the transient impact. Character frequencies of bearing faults can be identified from the Fourier coefficient of the energy operator result. Then whether the bearing has defects could be determined.

Autoregressive model: Autoregressive model is a type of statistics and signal processing method describing certain time-varying processes. It specifies that the output variable depends linearly on its own previous values and a stochastic term. Applying autoregressive (AR) model on the measured vibration signal could predict the deterministic parts in it. By removing the deterministic parts from the signal, the fault impulses will be enhanced and easier to detect. In the linear prediction process of AR model, the predicted value of the signal at time index k is the linear combination of p prior values,

$$x_k = -\sum_{l=1}^{p} a_l x_{k-l} + e_k \quad (10)$$

where a_l are the AR coefficients and e_k is the residual error between the linear prediction and the actual measured value. The residual error is mainly white noise if the signal is stationary. But if the signal is non-stationary, the residual error may also contain impulses or other non-stationary components. There are many kinds of algorithms for estimating the AR model, including least squares method and autocorrelation method.

The selection order of the AR model has a large effect on the output. There are several commonly-used criteria for order selection, such as the Akaike information criterion [11]. But since the main purpose of AR model here is to separate the impulses and the stationary components, the criterion could be maximizing the kurtosis of the residual signal, which is conducive to the extraction of bearing fault features. The appropriate model order p should be less than the spacing between two consecutive impulses to guarantee that the impacts will not be adapted by the model to be part of the deterministic signal but remained in the residual signal.

Minimum entropy deconvolution: Minimum entropy deconvolution (MED) uses the maximum kurtosis value as the iteration termination condition, which makes the deconvolution result highlight a few large spikes and is therefore suitable for the noise reduction of the impact fault signal of the rotating machinery. Thus, it could be a good application in bearing [12] and gearbox [13] fault feature extraction.
Suppose the vibration signal of rolling bearing fault,

$$y(n) = h(n) * x(n) + e(n) \tag{11}$$

where $y(n)$ is the actual measured signal, $h(n)$ is the transfer function, $x(n)$ is the impact sequence of the bearing signal, and $e(n)$ is the noise. After a long transmission path and the effects of noise, the impact characteristics may become no longer significant, which means that the fault features are relatively weaker and the entropy of the signal is increased. The objective of MED is to obtain an inverse filter $w(n)$ recovering the input signal $x(n)$ from the output signal $y(n)$,

$$x(n) = w(n) * y(n) \tag{12}$$

$\hat{w}(n)$ is the estimated value of $w(n)$, and its performance is determined by the result $\hat{x}(n)$ of the deconvolution. The closer the shape of $\hat{x}(n)$ is to $x(n)$, the better the result will be. Since the inverse filter is to help $\hat{x}(n)$ restore the original characteristics or most of the characteristics, that is, minimizing the entropy, so this method is called minimum entropy deconvolution. The entropy of the sequence $\hat{x}(n)$ is measured by its kurtosis, which is chosen as the objective function. The optimal result is achieved by selecting the coefficients of MED to maximizing the kurtosis.

$$O_2^4(w(n)) = \frac{\sum_{n=1}^{N} x^4(n)}{[\sum_{n=1}^{N} x^2(n)]^2} \tag{13}$$

The iteration steps of the MED algorithm can be summarized as follows [14]:

1. Initialize elements of $w^{(0)}$ as 1.
2. Calculate $x(n) = w^{(i-1)}(n) * y(n)$
3. Calculate $b^{(i)}(l) = u\sum_{l=1}^{N} x^3(n)y(n-l)$, where $u = \frac{\sum_{n=1}^{N} x^2(n)}{\sum_{n=1}^{N} x^4(n)}$ and l is the length

of the inverse filter.

4. Calculate $w^{(i)} = A^{-1}b^{(i)}$, where A is the $l \times l$ autocorrelation matrix of the sequence.
5. If $\| w^{(i)} - w^{(i-1)} \|_2^2$ is smaller than the given threshold, stop the iteration. Otherwise $i = i + 1$ and back to step 2.

Experimental system and data: The rotor-bearing-casing coupled vibration experiment system is designed to simulate vibration faults, and to study transmission characteristics of rotor-bearing-casing system in a gas turbine engine. Figure 1 shows the structural drawing and the photo of the test rig. Accelerators are installed on both the bearing pedestal and the casing.

Rolling bearing faults are simulated on this test rig by replacing the main shaft bearing with fault bearings. The main shaft bearing is NF206EM cylindrical roller bearing. Bearing faults are artificially set by grooving on the surface of the outer race, the inner race and one of the rollers. The size of the defects are 0.3mm width and 0.5mm depth, which is relatively small. According to the character frequency formulas of rolling bearing faults, the character frequencies of outer race, inner race and roller are 5.43f, 7.57f and 5.89f, where f is the rotating frequency of the rotor. Except the main shaft bearing, other experimental conditions are the same. The rotating speed will be set to 1800rpm (30Hz) so the character frequencies are 162.8Hz, 227.2Hz and 176.6Hz.

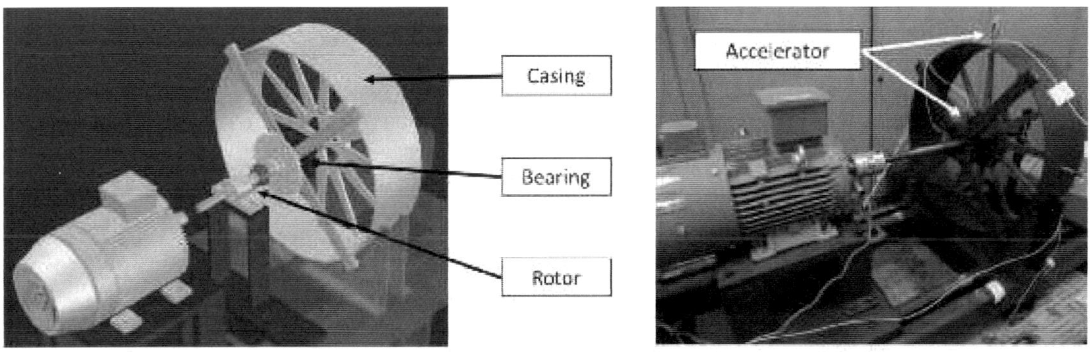

(a) Structural drawing (b) Photo

Figure 1: Rotor-bearing-casing coupled vibration experiment system.

Analyzing the vibration signals of the outer race fault as an example, and comparing with the normal bearing. Although the defect is small, the bearing vibration has obvious changes. The time waveform of the bearing pedestal vibration shows more spikes under the condition of outer race fault than no fault. This could be the preliminary judgment of bearing faults. A bearing fault is usually expressed as the high frequency structural resonance modulated by the periodic impact of the defect. Fault features should be demodulated from the vibration signal when detecting bearing faults. Here the Hilbert transform is employed. The Hilbert envelope spectrum of the bearing pedestal vibration with the normal bearing shows no clear character frequency components. But the components of character frequency and its harmonics and sidebands appear in the envelope spectrum for the outer race fault, which are typical features of rolling bearing faults. However, fault feature extraction of casing vibration is more difficult than bearing pedestal vibration, since the vibrational features will be much weaker when transmitted from the inside to the outside and more structural vibration and noise will be added. There are not many obvious spikes in the time waveform and character frequency components are hard to see in the envelope spectrum. Further analysis is necessary when there are not bearing pedestal signals but only casing signals.

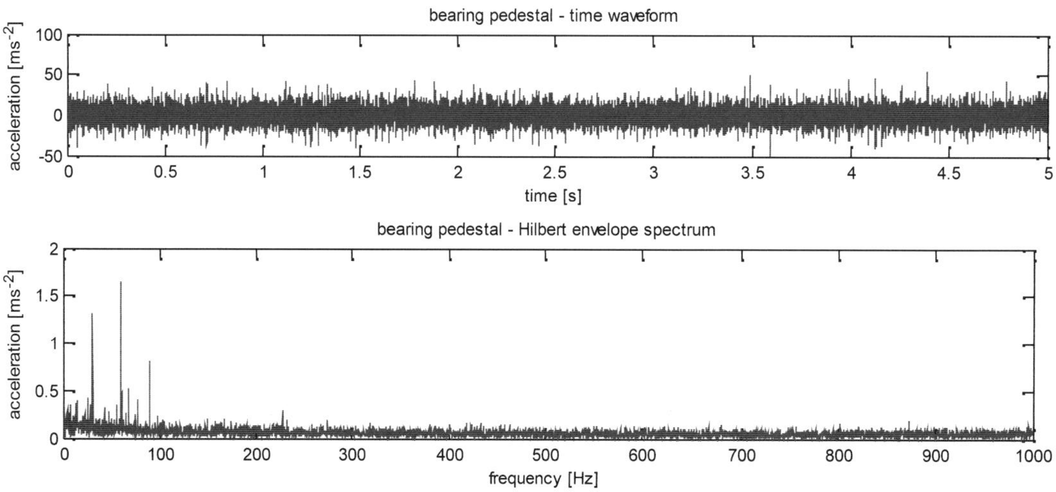

Figure 2: Time waveform and envelope spectrum of bearing signal with no fault.

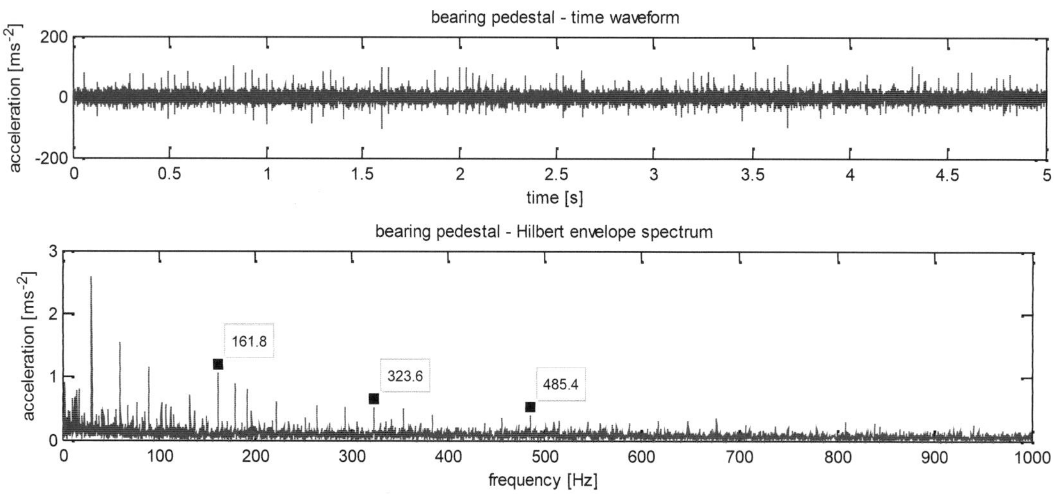

Figure 3: Time waveform and envelope spectrum of bearing signal with outer race fault.

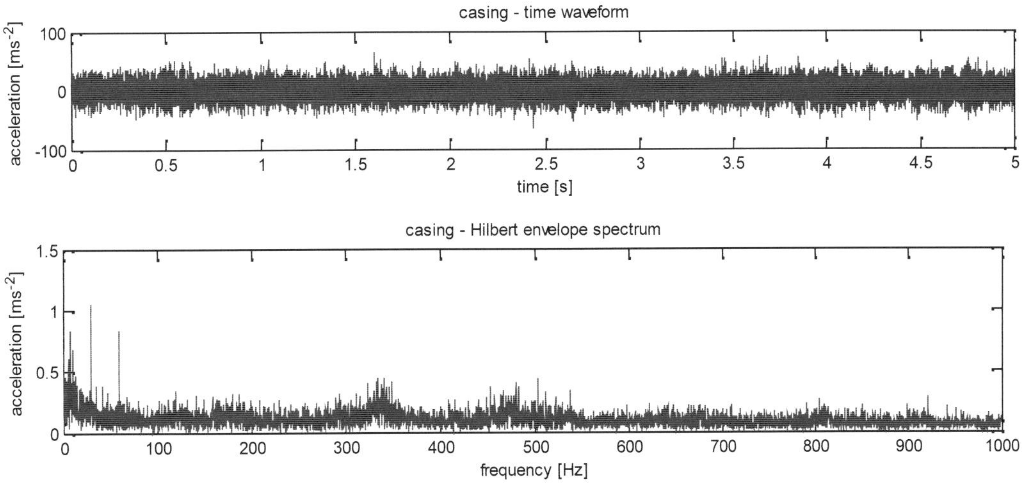

Figure 4: Time waveform and envelope spectrum of casing signal with outer race fault.

Feature extraction based on ITD and TKEO: ITD decomposes the signal into several proper rotation components which are single component signals. So the high-frequency resonance modulated by the fault frequency is more distinct in one of the ITD outputs than in the original signal. Figure 5 displays that the casing signal in decomposed into 11 sub-signals, arranged in descending order of frequency. The modulated features are mostly in the first few sub-signals where the high frequency resonances exist. Figure 6 is Hilbert envelope spectrums of the first 3 ITD sub-signals, where no clear character frequency components can be seen. Instead of envelope spectrum, calculate the Teager-Kaiser energy operator of the sub-signals and take the Fourier transform, and the ITD-TKEO spectrums are in Figure 7. In the first subplot there are 2nd and 3rd harmonic components of character frequency and their side frequencies. Although the features in the ITD-TKEO spectrums are not very prominent, they have been already transmitted to the casing, which show the possibility of bearing fault diagnosis using the casing signal, and prove that TKEO has better capability of highlighting impacts in a signal than Hilbert transform.

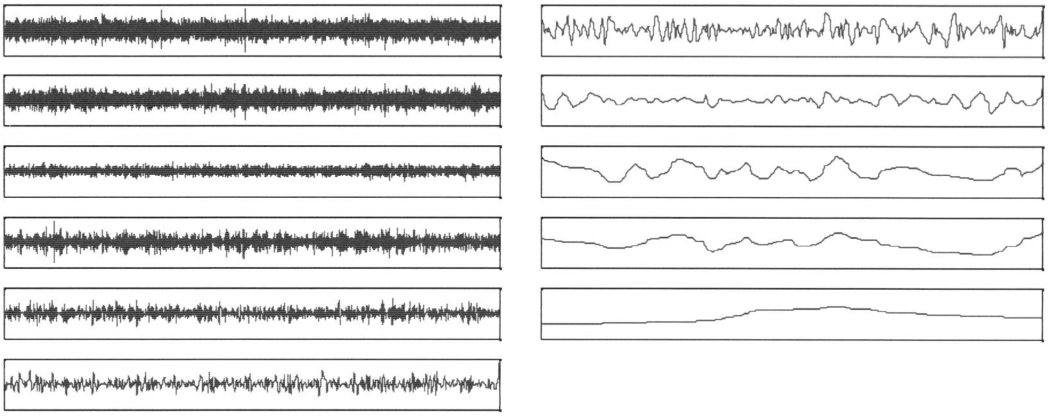

Figure 5: ITD outputs of the casing signal.

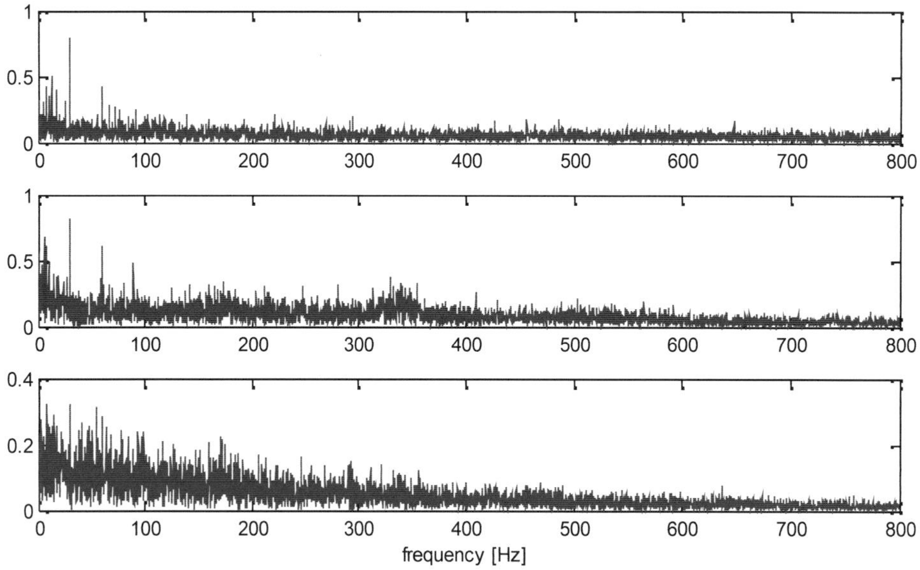

Figure 6: ITD-Hilbert spectrums of the first 3 sub-signals.

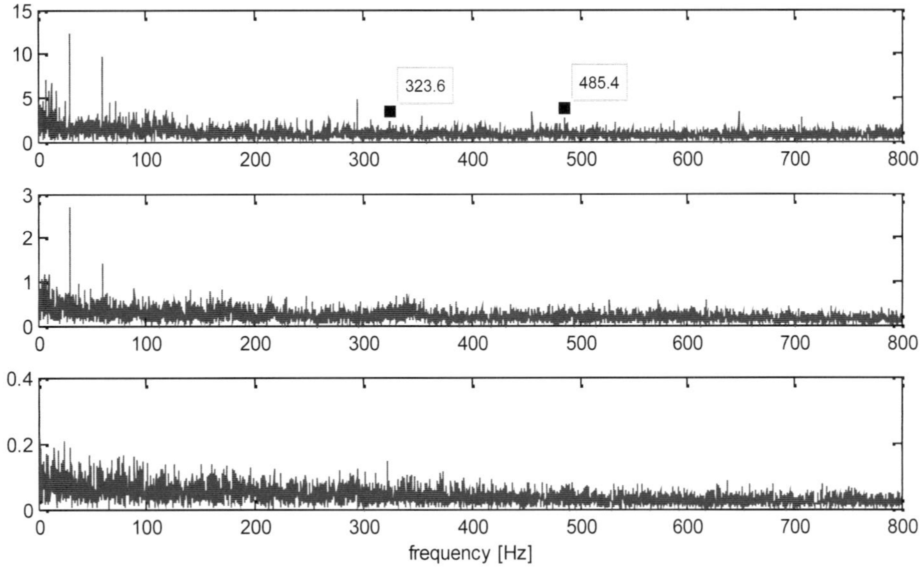

Figure 7: ITD-TKEO spectrums of the first 3 sub-signals.

Feature enhancement based on AR and MED: AR and MED are used to enhance the impulse features in the casing signal. Kurtosis is chosen as an indicator of the impulses, and a bigger value of kurtosis means more significant impulse features. During the optimization for an AR model, kurtosis is also used as an indicator for helping selecting the AR order. Figure 8(a) is the original casing vibration signal whose kurtosis is 3.1174. Figure 8(b) is the residual signal of the AR filter with a kurtosis of 3.9249, which is slightly bigger than the original one. Figure 8(c) is the residual signal of the MED filter with a kurtosis of 7.7049, which has a more significant improvement. And Figure 8(d) is the signal filtered by AR and MED whose kurtosis is 8.4985 and much bigger than the one of the original signal. When the impulses are enhanced and the kurtosis is increased, the spikes are more prominent in the time waveform of a vibration acceleration signal, which is obvious in the figures here.

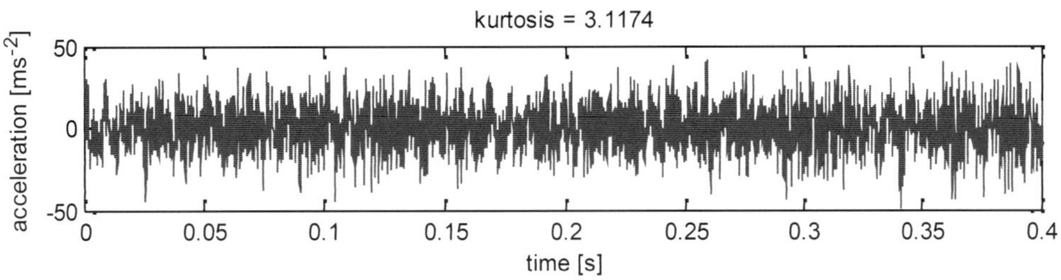

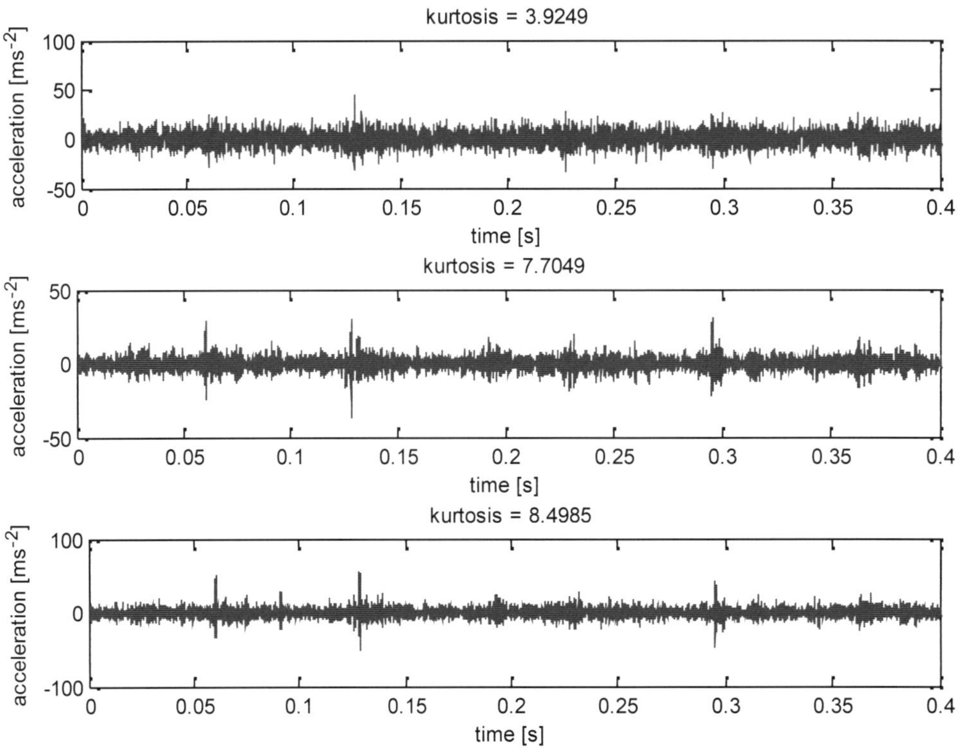

Figure 8: (a) Original casing signal; (b) residual of AR filter; (c) residual of MED filter; (d) residual of AR and MED filter.

Compute the kurtosis for each ITD sub-signal, and the values are in Figure 9. The first 4 high frequency sub-signals have relatively big kurtosis values and could be selected for further analysis since they might carry most of the fault features. Draw an ITD-TKEO spectrum for each of the 4 sub-signals, which are shown in Figure 10. The harmonics of the character frequency and their sideband at the rotating frequency are very clear and much more significant than those components in figures without treatments of AR and MED (Figure 11). That are typical features for the outer race fault of a rolling bearing, and depending on their appearance the existence of a bearing fault could be substantially confirmed.

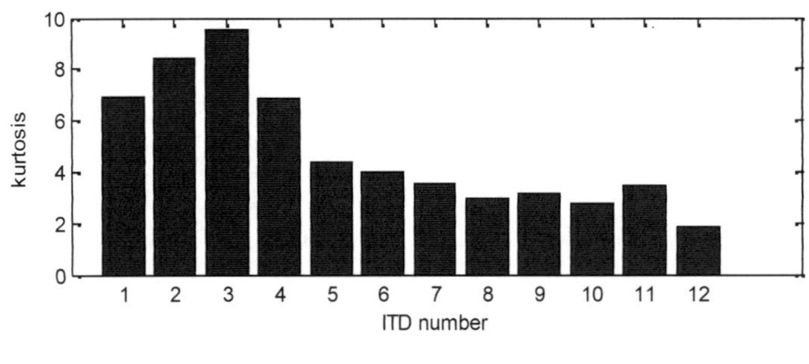

Figure 9: Kurtosis for each ITD sub-signal.

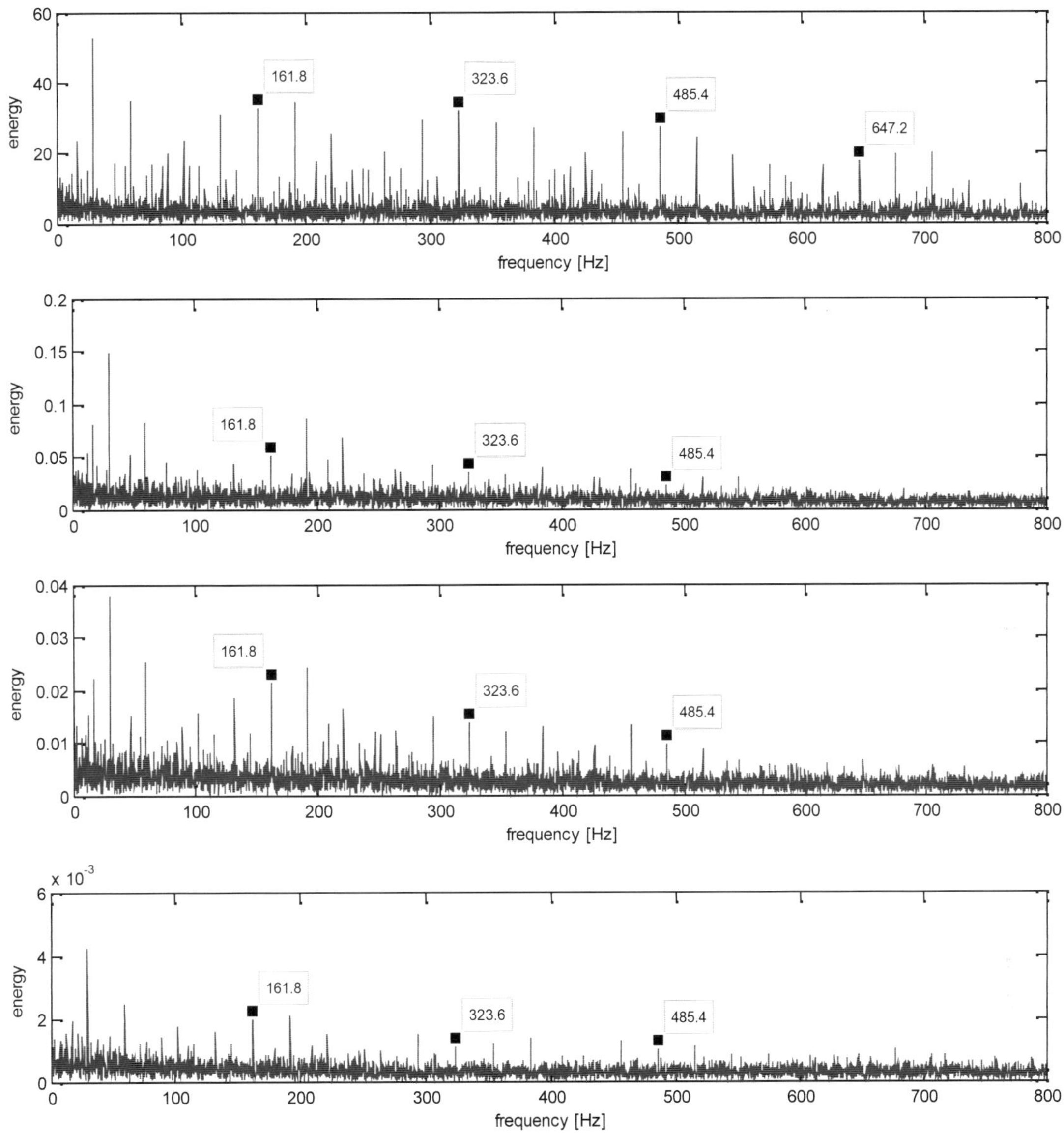

Figure 10: ITD-TKEO spectrums of the first 4 sub-signals with outer race fault.

Draw the ITD-TKEO spectrums of the normal bearing signal as a comparison, which are displayed in Figure. The plots of the first 4 high frequency sub-signals do not show any clear character frequency components, even though they have been processed by the AR and MED method above. This result shows that the deterministic components in a vibration signal will not be enhanced and bring in interference to the diagnosis. Only transient impacts will be enhanced, which will improve the sensitivity of the bearing fault diagnosis method and help to detect minor rolling bearing faults.

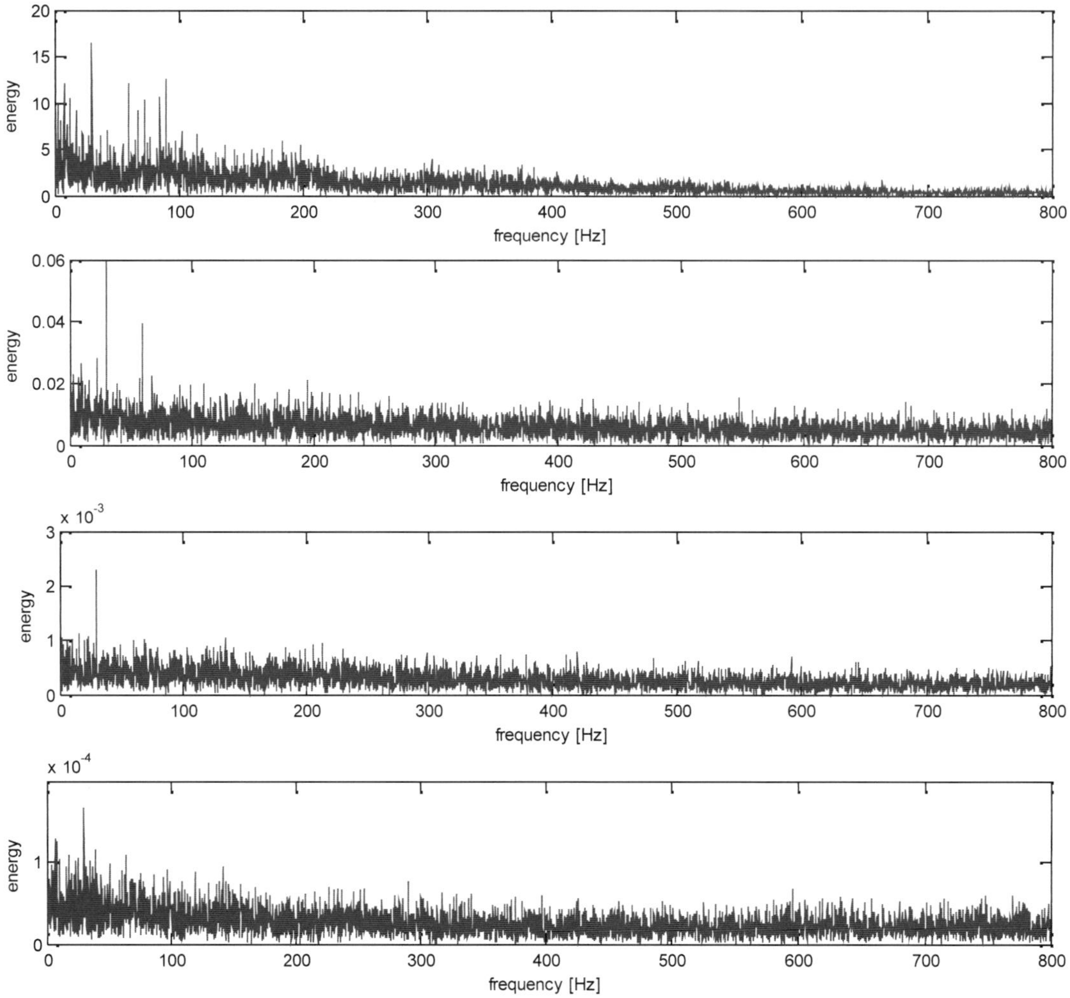

Figure 11: ITD-TKEO spectrums of the first 4 sub-signals with no fault.

Conclusion: This paper presents a feature extraction method for rolling bearing faults in a rotor-bearing-casing system based on the casing vibration signal. The method use AR and MED for feature enhancement and ITD and TKEO for feature extraction. Kurtosis is chosen as indicators for optimization processes since it reflects the strength of the impact components in a signal. Rolling bearing faults are simulated on the rotor-bearing-casing test rig and the experimental signals are used for testing the proposed method. The flow chart of the whole procedure is shown in Figure 12.

This method shows its strong capability in extracting vibrational features of minor bearing faults. The pedestal is close to the bearing and vibration signals collected on it have relatively strong fault features, so the feature extraction is easy as the traditional envelope spectrum based on Hilbert transform could work. But the feature extraction of casing signal is more difficult since the fault features will be weaker and more structural vibration and noise will be added when the vibration is transmitted from the bearing pedestal to the casing. The proposed method has a good performance for this casing signal diagnosis problem, yielding prominent character frequency components.

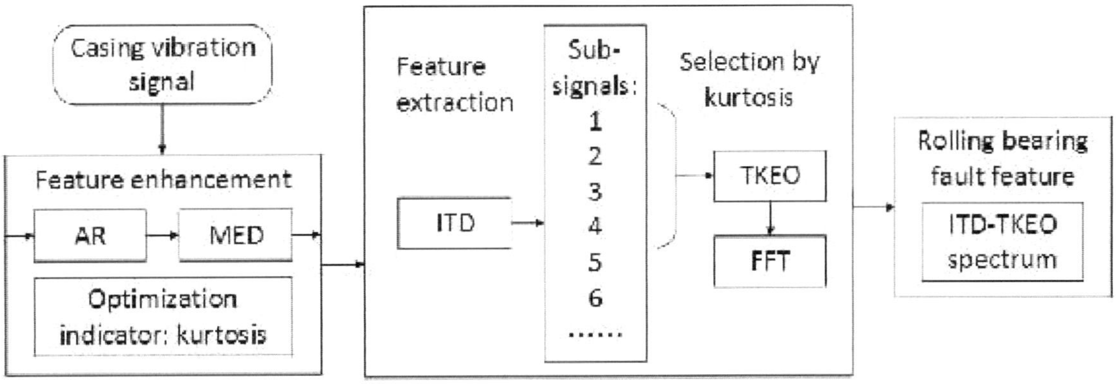

Figure 12: Procedure of the proposed feature extraction method.

References:

[1] El-Thalji, Idriss, and Erkki Jantunen. "A summary of fault modelling and predictive health monitoring of rolling element bearings." Mechanical Systems and Signal Processing 60 (2015): 252-272.

[2] Guo, Chen. "Study on the recognition of aero-engine blade-casing rubbing fault based on the casing vibration acceleration." Measurement 65 (2015): 71-80.

[3] Sawalhi, N., and R. B. Randall. "Spectral kurtosis enhancement using autoregressive models." ACAM Conference, Melbourne, Australia. 2005.

[4] Wiggins, Ralph A. "Minimum entropy deconvolution." Geoexploration 16.1-2 (1978): 21-35.

[5] Frei, Mark G., and Ivan Osorio. "Intrinsic time-scale decomposition: time-frequency-energy analysis and real-time filtering of non-stationary signals." Proceedings of the Royal Society of London Series A 463 (2007): 321-342.

[6] Huang, Norden E., et al. "The empirical mode decomposition and the Hilbert spectrum for nonlinear and non-stationary time series analysis." Proceedings of the Royal Society of London A: Mathematical, Physical and Engineering Sciences. Vol. 454. No. 1971. The Royal Society, 1998.

[7] Kaiser, James F. "On a simple algorithm to calculate the 'energy' of a signal." Acoustics, Speech, and Signal Processing, 1990. ICASSP-90., 1990 International Conference on. IEEE, 1990.

[8] Maragos, Petros, James F. Kaiser, and Thomas F. Quatieri. "On amplitude and frequency demodulation using energy operators." IEEE Transactions on signal processing 41.4 (1993): 1532-1550.

[9] Tianjin Wang, Feng Zhipeng, and Hao Rujiang. "Fault diagnosis of rolling element bearings based on Teager energy operator." Journal of Vibration and Shock 31.2 (2012): 1-5.

[10] An, Xueli, et al. "Application of the intrinsic time-scale decomposition method to fault diagnosis of wind turbine bearing." Journal of Vibration and Control 18.2 (2012): 240-245.
[11] Akaike, Hirotugu. "Fitting autoregressive models for prediction." Annals of the institute of Statistical Mathematics 21.1 (1969): 243-247.
[12] Sawalhi, N., R. B. Randall, and H. Endo. "The enhancement of fault detection and diagnosis in rolling element bearings using minimum entropy deconvolution combined with spectral kurtosis." Mechanical Systems and Signal Processing 21.6 (2007): 2616-2633.
[13] Endo, H., and R. B. Randall. "Enhancement of autoregressive model based gear tooth fault detection technique by the use of minimum entropy deconvolution filter." Mechanical Systems and Signal Processing 21.2 (2007): 906-919.
[14] Guangming, WANG Hongchao CHEN Jin DONG. "Fault Diagnosis Method for Rolling Bearing's Weak Fault Based on Minimum Entropy Deconvolution and Sparse Decomposition." Journal of Mechanical Engineering 1 (2013): 014.

SHELF LIFE EVALUATION METHOD FOR ELECTRONIC AND OTHER COMPONENTS USING A PHYSICS-OF-FAILURE (POF) APPROACH

Nga Man Li, Diganta Das, Michael Pecht
Center for Advanced Life Cycle Engineering (CALCE)
University of Maryland
College Park, MD 20742
Telephone: (301) 405-5323
diganta@umd.edu

Abstract: Shelf lives are not well understood in the electronics industry. Despite the existence of recommended shelf lives for some units from standards or manufacturers' documents, many electronic parts are stored well beyond their recommended shelf lives for different reasons. In many cases, 'expired' parts are found to work fine after many years of extended storage, which gives extra motivation for parties along the supply chain, typically part user companies, to extend storage of their parts and to evaluate the 'actual shelf lives' of their components. The combination of motivations to extend shelf life, inadequacies in recommended shelf lives, and the lack of knowledge and guidelines to shelf life determination often results in arbitrary storage periods and conditions of electronic components in the industry.

In this article, common pitfalls of recommended shelf lives are identified. Then, a physics of failure (PoF) approach to evaluate shelf lives overcoming such pitfalls is proposed. The philosophy we introduce in this approach applies to most storage-induced effects for electronic parts and the approach is described with an electrolytic capacitor in this article.

Key words: Shelf life; Storage

Introduction: There are always motivations and attempts in extending storage period of parts in the electronics industry. One of the reasons is stocking life-time buy parts to minimize the impact of part obsolescence. Responding to the modification of parts through generations due to technology advances and compliances with new restrictions, many companies attempt to minimize the impact of part obsolescence for their future products by purchasing and storing additional parts. Depending on the company's role in the supply chain, this situation often applies to sub-assemblies and even final assemblies. Another reason for part and sub-assembly storage is for customer support with fast response to maintenance needs. Some equipment suppliers store spare parts and sub-assemblies at customer locations to improve system availability and in those cases, the immediate out of storage usability is of prime concern.

Although recommended shelf lives for some parts are specified in standards or manufacturers' documents, some of them are considered to be over-generalized to part types and in some cases, too conservative. In many cases, electronic components are found to work fine long after they are "expired" [1][2]. This collective experience from the industry motivates some companies to extend storage periods of their components. While in other cases, the suggestion of unlimited shelf lives under certain conditions from manufacturers raises doubts from the industry despite the appeal. The combination of motivations to extend shelf life, inadequacies in recommended shelf lives, and the lack of knowledge and guidelines to shelf life determination often results in arbitrary storage periods of electronic parts in the industry.

In this article, common pitfalls of recommended shelf lives are identified. Then, a physics of failure (PoF) approach to evaluate shelf lives overcoming such pitfalls is proposed. Compared to the typical procedure of model-based use life estimation, a fundamental difference in use life and shelf life necessitates additional steps in addition to identifying of the most critical degradation mechanism and applying models. This includes an analysis of the most critical storage-induced effect's impact on usability or reliability of the unit after storage, and the definition of acceptance criteria after storage. The philosophy we introduce in this approach applies to most storage-induced effects for electronic parts and systems and the approach is demonstrated with electrolytic capacitor in this article.

Common Pitfalls of Shelf Life: The definition for shelf life is not uniformly understood or accepted in the electronic industry. Numerical values of shelf life are often provided by part manufacturers and some part user companies may have developed guidelines of storage conditions and shelf lives for different parts. Some examples of shelf life interpretation by electronic component manufacturers and electronic component user companies are listed below. These two parties in the supply chain often refer shelf life to two different timeframes which are "from manufacture to shipment" and "from manufacture to use" respectively while using the same terminology. In addition, these interpretations are ambiguous if one is looking to directly adopt the shelf life numbers given in these documents. For example, recipients for shipments are not specified by most component manufacturers listed below while most electronic components go through multiple transportations in along the supply chain.

Example of shelf life interpretation from electronics component manufacturers:

- Fujitsu [3]: Period from product manufacture to shipment to customers
- NXP [4]: The time between assembly date code of the device and date of shipment
- Texas Instrument [5]: The amount of time from the product was manufactured to the time it is delivered by TI or a TI authorized distributor
- EPCOS [6]: A specified time for which a capacitor can be stored "voltage-free"

Shelf Life Interpretation from an Electronic Component Part User Company:

- GE Power Electronics [7]: The time within which an electronic component or mechanical piece part can be stored, at a given ambient condition, is usable without testing, evaluation or special conditioning.

Apart from the ambiguous timeframe regarding shelf lives, numerical shelf life values are often found to correspond only to one presumed degradation mechanism. These degradation mechanisms are often related to solderability or moisture sensitivity. In the shelf life interpretation example by EPCOS [6], the degradation mechanism being addressed is the thinning of dielectric oxide layer of the capacitor. Yet, if application of reforming voltage can be managed during the course of storage, the stated shelf life would not apply and will depend on other competing degradation mechanisms. Any variations in materials of a part, changes in storage conditions, or different procedures taken during and after storage can invalidate recommended shelf lives of a part.

Difference between Shelf Life and Use Life Estimation: The main difference between shelf life and use life lies in the life cycle phases they represent. Use life is usually assumed to be the final phase of a unit's life cycle. Therefore, after identifying the most critical degradation mechanism, use life is often determined by a PoF, or an empirical model under a certain operating condition until failure of the device, which is defined by predetermined failure criteria, i.e. functionality or parametric failure of the device. In the case of shelf life estimation, storage is not the final phase of a unit's life cycle. Therefore, subsequent procedures and life cycle stresses should be taken into account for the most critical storage-induced effect during storage be selected to ensure the desired usability or reliability of the part after storage. A more generalized term "storage-induced effects" is adopted from this section because some of the "effects" that may take place during storage do not resemble the general interpretation of "degradation"- the loss of relevant properties of the unit which proceeds gradually due to exposure to a certain condition. For example, moisture absorption of a plastic packaged component is not considered a degradation mechanism. However, it is considered to be one of the most critical "storage-induced effect" because it can induce delamination or pop-corning during solder reflow.

Physics of Failure Shelf Life Evaluation Approach: In this study, a shelf life estimation approach is developed and the simple philosophy involved is described in this section. In addition to conventional Failure Mode, Mechanism and Effect Analysis (FMMEA)[8], which is typical for model-based use life estimation, the difference between use life and shelf life necessitates the need for two extra steps in shelf life estimation. The steps takes into account the impacts of storage-induced effects when the most critical storage-induced effect is selected for modeling, and defines an acceptance criteria depending on the impact. A flowchart depicting the shelf life estimation approach is shown in Figure 1.

The left branch of the flowchart is a typical model-based use life estimation procedure. It begins with an FMMEA which requires input information such as materials being used in the part, and environmental conditions under which the life of the part is being estimated. Then, the most critical degradation or failure mechanism has to be identified, this determination is often made based on feedback from field failures of similar products, literature review, or risk and criticality analysis. With the most critical mechanism identified and a failure criteria defined, a model can be applied to predict useful life. For shelf life estimation, a parallel branch is added for determination of the most critical storage-induced effect. This is to take into account the effects' impacts on subsequent procedures or life cycle stresses of the part. Key steps in the approach are discussed in detail in the following sections. This approach can be applied to typical storage-induced effects in electronic parts.

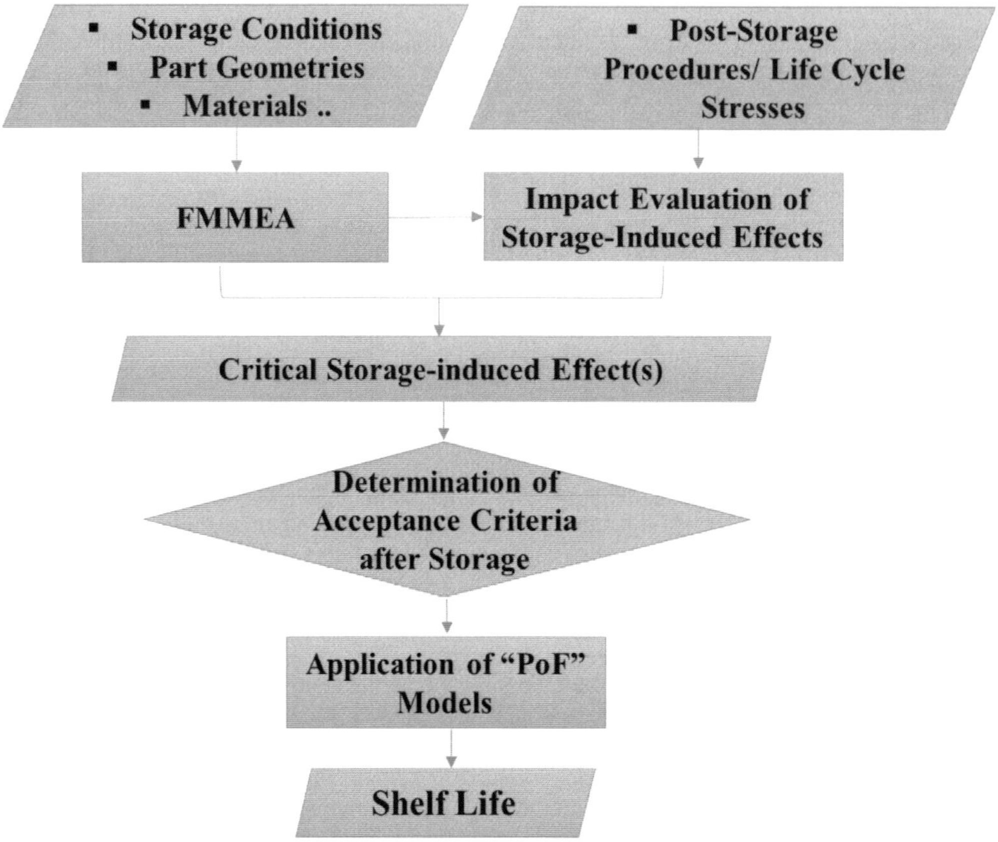

Figure 1. Physics of Failure Shelf Life Estimation Approach

Identify the Most Critical Storage-Induced Effect: As a first step, storage-induced effects during storage have to be identified. As the most common and effective way to identify storage induced effects in electronic parts and systems, FMMEA is adopted for this purpose. FMMEA results depend on the amount of information input. For an accurate and comprehensive study, detail environmental condition, material composition and dimensions are necessary. For material composition, instead of performing individual materials analysis, material declaration management forms IPC-1752A for most electronic parts can be acquired from their distributors or manufacturers. After possible degradation mechanisms are listed in FMMEA, calculation of Risk Priority Number (RPN) [9] can be carried out to identify critical degradation mechanisms out of the list. For the same part, the most critical storage-induced effect can differ depending on the post-storage procedures or use conditions. For example, solderability is a major concern for storage of loose leaded components. However, if the part is already soldered onto an assembly, then the most critical storage-induced effect will shift from lead oxidation to other storage-induced effects. Therefore, an extra step to evaluate impacts of different storage-induced effects is necessary for the selection of the most critical one to be modeled.

Impacts of the Storage Induced Effects: Impacts of storage-induced effects include degradations and environmental effects that can affect reliability or usability of the stored device depending on subsequent life cycle stresses or procedures. For example, reflow, hand-solder and different use conditions. For electronic components, storage-induced effects can be grouped into three categories based on their impacts. Examples of common storage-induced effects under each category are included in Table 1.

> Category 1: Degradation mechanisms that affect reliability of a device by taking place in a continuous manner through use conditions.
>
> Category 2: Degradation mechanisms that can affect usability of the device after storage. These degradation mechanisms may or may not be reversible. This category of degradation mechanisms is critical when the degradation is or becomes irreversible or when reforming procedure cannot be performed.
>
> Category 3: Environmental effects that can initiate degradation mechanisms under storage or subsequent life cycle stresses.

Table 1. Examples of Storage Induced Effects under Each Category

Category 1	**Category 2**	**Category 3**	
Degradation Mechanism	**Degradation Mechanism**	**Environmental Effect**	**Potential Failure Mechanism**
• Electrolyte Evaporation in Electrolytic Capacitors • Lubricant Evaporation in Greased Components	• Reduction in Solderability (e.g., Lead oxidation/ Corrosion)	• Moisture Absorption in Plastic Packaged Parts	• Cracking • Delamination • Pop-corning • Corrosion • Conductive filament formation
• Interconnect Fatigue	• Thinning of Dielectric in Electrolytic Capacitors	• Halogen Infiltration	• Corrosion
• Interconnect Aging (e.g., Excessive Intermetallic Formation, stress relaxation)		• Charge Build-Up	• Electrostatic Discharge

Determination of Acceptance Criteria: After the most critical storage-induced effect is identified, acceptance criteria of the part after storage has to be determined. Acceptance criteria (not failure criteria) is the criteria that should be satisfied for particular requirements or reliability goals after storage, while should be compatible with the metric being modeled for life estimation. The determination of acceptance criteria for each category of storage-induced effects is slightly different.

For Category 1, in which the degradation mechanisms take place in a continuous fashion through use conditions, the acceptance criteria should be determined based on expected use condition and required service time. Take electrolyte evaporation in electrolytic capacitors as an example, electrolyte evaporates at a slower rate during storage (assuming a controlled storage condition) and the evaporation gets faster as the capacitor is put into use condition due to the increase in internal temperature due to self-heating. Therefore, depending on the field condition and required service time, the amount of remaining electrolyte after storage for usage consumption will be the acceptance criteria for this degradation mechanism.

For Category 2, in which the degradation mechanisms can affect usability of the device after storage, acceptance criteria should be determined based on whether the mechanisms are reversible and if the "reworking process" can take place. In cases that the mechanisms are irreversible or is reversible but a 'reworking process' cannot be performed, the acceptance criteria should be determined based on the usability requirements of the part. An example can be a certain percentage loss of parametric measurements of the part. Although mechanisms that are reversible and 'reworking process' can be performed are rarely critical, the acceptance criteria can be the critical point before which the degradation mechanism becomes "unreworkable" or "irreparable. Take lead oxidation or corrosion as an example, the acceptance criteria can be the critical point before which the leads surfaces have degraded significantly that the required solderability cannot be restored using solder dipping.

For Category 3, in which the degradation mechanisms can initiate other failure mechanisms under life cycle stresses. Take moisture absorption in plastic packaged devices as an example, the absorbed moisture can induce delamination or pop-corning during reflow. Moisture can also lead to corrosion of wirebonds which results in permanent or intermittent failures of the device [10][11]. For this type of degradation mechanisms, acceptance criteria should be determined based on the critical accumulation of the effect for the potential failure mechanism to initiate.

After the acceptance criterion of a critical degradation mechanism is defined, physics of failure models or physics based models can be used for shelf life estimation.

Shelf Life Estimation of Electrolytic Capacitor:

In this study, an aluminum electrolytic capacitor is selected for demonstration of the proposed shelf life evaluation approach. As a first step, the materials declaration management form IPC-1752A was acquired from the manufacturer, which provided us materials information of the part. Based on the materials information and a normal warehouse environment, an FMMEA is performed to identify the possible storage induced effects as shown in Table 2. Electrolyte evaporation is considered to be the most critical storage induced effect here because it is one of the most reported degradation mechanism for this type of capacitors and it is not reversible.

Failure Sites	Potential Failure Mode in Use	'Observable Effects' during Storage	Storage Induced Effects	Failure Cause
Electrolyte [12][13]	• Parametric failure • Open	• Weight Loss • Decrease in capacitance • Increase in ESR	**Electrolyte evaporation**	Diffusion or leakage through seal
Electrolyte [13]	• Parametric failure • Open	• Decrease in capacitance	Electrolyte Deterioration	Chemical degradation
Aluminum Electrodes [13]	• Parametric failure • Open	• Increase in leakage current	Corrosion	Infiltration of halogen
Dielectric Oxide Layer [13][14]	• Seal rupture by high leakage current • Short	• Increase in leakage current	Reduction in thickness (recoverable)	Absence of applied voltage
Terminal	Increase in Contact Resistance	Increase in Contact Resistance	Oxidation/ Corrosion	Air contamination, moisture
	Short	Short	Tin whisker	Sn content

Table 2. FMMEA for Storage of Aluminum Electrolytic Capacitor

Gasperi has derived a physics of failure model for electrolyte evaporation for aluminum electrolytic capacitors [12]. In this model, rate of electrolyte loss is related to temperature and vapor pressure of the electrolyte as shown in Equation (1).

$$\frac{dV}{dt} = Kexp(\frac{-A}{T} + B) \qquad \text{Equation (1)}$$

$\frac{dV}{dt}$: Rate of Electrolyte Loss
K: Leak Rate Constant
A, B: Vapor Pressure Constants
T: Temperature (K)

Given that the rate of electrolyte loss is assumed to be independent of time in this model, with V_o as the original volume of the electrolyte in the electrolytic capacitor, and x% as the percentage loss of electrolyte as the **failure** criteria, rearranging Equation (1) and substituting V with $x\%V_o$ gives the relationship of time required to cause the certain loss of electrolyte as shown in Equation (2).

$$Time\ to\ Failure = \frac{x\%V_O}{Kexp\left(\frac{-A}{T}+B\right)} \qquad \text{Equation (2)}$$

The electrolyte used is ethylene glycol as indicated in IPC-1752A for the capacitor. The vapor pressure constants A and B for ethylene glycol were reported to be 7060 and 21.63 respectively in the Handbook of Chemistry and Physics in 1970 [15]. In Gasperi's study, K was determined from experiment and is found to be 0.00031unit/mmHg/100Hr. Failure criteria for electrolyte evaporation typically range from 30% to 40% electrolyte loss [12], [13]. Assuming the same leak rate constant for the capacitor and using 40% electrolyte loss as the failure criteria, capacitor life at different temperatures can be calculated separately with Equation (2). Like other model-based life estimations, the accuracy of the estimated life can be improved by running experiments to determine constants in the available models. This step is not performed in this study as it does not add value to the purpose of demonstrating the proposed shelf life estimation approach. Then, maximum shelf life of a capacitor (considering the required use conditions and service life) can be calculated using a version of "Miner's Rule" of damage accumulation as shown in Equation (3). Simulated results based on this calculation are shown in

Capacitor Temperature during Operation	60°C	65°C
Life at Use Condition	9.5 years	7 years
Application Time	6 years	6 years
Maximum Shelf Life at 25°C	42 years	16.4 years
Maximum Shelf Life at 30°C	28 years	11.1 years

.

$$\frac{Required\ Application\ Time}{Life\ at\ Use\ Condition} + \frac{Maximum\ Shelf\ Life}{Life\ at\ Storage\ Condition} = 1 \qquad \text{Equation (3)}$$

Table 3. Estimated Shelf Life of Electrolytic Capacitor under Different Use Conditions

Capacitor Temperature during Operation	60°C	65°C
Life at Use Condition	9.5 years	7 years
Application Time	6 years	6 years
Maximum Shelf Life at 25°C	42 years	16.4 years
Maximum Shelf Life at 30°C	28 years	11.1 years

Conclusions: In this article, common pitfalls of recommended shelf lives are identified. Then, a physics of failure (PoF) approach to evaluate shelf lives overcoming such pitfalls is proposed for electronic units. Compared to the typical procedure of model-based use life estimation, fundamental difference in use life and shelf life necessitates a modified procedure in addition to identification of the most critical degradation mechanism and application of models. This includes an analysis of the most critical storage-induced effect's impact on usability or reliability of the unit after storage, and the definition of acceptance criteria after storage. The philosophy we introduce in this approach applies to most storage-induced effects for electronic parts and systems and the approach is demonstrated with electrolytic capacitor in this article.

References:

[1] Madsen R., "Component Reliability after Long Term Storage", *Texas Instrument Application Report SLVA304*, 2008

[2] McCluskey F.P., Hakim E.D., Fink J., Fowler A., Pecht M.G., "Reliability Assessment of Electronic Components Exposed to Long-Term Non-Operating Conditions", *IEEE Transactions on Components, Packaging, and Manufacturing Technology-Part A*, vol. 21, no. 2, pp. 352-358, 1998.

[3] Fujitsu Semiconductor Europe, "Product Storage Rules", 2012 [Online]. Available: https://www.fujitsu.com/uk/Images/product-storage-rules.pdf

[4] NXP, "Freescale 5 year Shelf Life.",2014 [Online]. Available: http://www.nxp.com/assets/documents/data/en/reports-or-presentations/FSLSHELFLIFEPOL.pdf

[5] Texas Instrument, "Semiconductor Shelf Life & Search Tool." [Online]. Available: http://www.ti.com/quality/docs/productshelflife.tsp.

[6] EPCOS, "Aluminum Electrolytic Capacitor-General Technical Information", 2016 [Online]. Available: https://en.tdk.eu/download/530704/5f33d2619f a73419e2a4af562122e90c/pdf-generaltechnicalinformation.pdf

[7] Shyamsundar A., Cervone A., "Storage Requirements (Shelf Life) for Components, Materials and PWBs", GE Power Electronics Business, 2012.

[8] M., Pecht and A. Dasgupta, "Physics-of-Failure: An Approach to Reliable Product Development", *Journal of the Institute of Environmental Sciences*, Vol. 38, pp. 30-34, September/October 1995 also in 1995 International Integrated Reliability Workshop Final Report, Lake Tahoe, CA, pp. 1-4, October 22-25, 1995.

[9] Krasker G., *Failure Modes and Effects Analysis: Building Safety Into Everyday Practice*, HCPro, Inc., 2004.

[10] Manoharan, S., Krishnanramaswami, G., Mccluskey, F. P., & Pecht, M. G., "Failure mechanisms in encapsulated copper wire-bonded devices" *IEEE 23rd International Symposium on the Physical and Failure Analysis of Integrated Circuits (IPFA)*, 2016

[11] Bakhshi R., Kunche S., and Pecht M., "Intermittent Failures in Hardware and Software," *Journal of Electronics Packaging*, vol. 136, no. 1, pp. 11014–11014, 2014

[12] Gasperi M., "Life Prediction Model for Aluminum Electrolytic Capacitors", Advanced Technology Labs, Rockwell Automation, 1996
[13] Kulkarni C., Celaya J., Goebel K., Biswas G., "Physics Based Electrolytic Capacitor Degradation Models for Prognostic Studies under Thermal Overstress",*European Conference of the Prognostics and Health Management Society*, 2012.
[14] J. L. Stevens, J. S. Shaffer and J. T. Vandenham, "The service life of large aluminum electrolytic capacitors: effects of construction and application" *IEEE Industry Applications Conference 36th IAS Annual Meeting*, Chicago, IL, USA, 2001, vol. 4. pp. 2493-2499, 2001.
[15] Weast, Robert C., *Handbook of Chemistry and Physics*, CRC Press, Cleveland, OH, pp. D-146, 1970

Bibliography:

[1] **Nga Man Li** received her B.S degree in materials engineering from City University of Hong Kong in 2013. She is currently pursuing her PhD degree in mechanical engineering at the Center for Advanced Life Cycle Engineering (CALCE), University of Maryland, College Park, MD, USA. Her research interests include the effectiveness of testing methods and long term storage of electronic parts and systems.

[2] **Diganta Das** (Ph.D., Mechanical Engineering, University of Maryland, College Park, B.Tech., Manufacturing Science and Engineering, Indian Institute of Technology) is an Associate Research Scientist at the Center for Advanced Life Cycle Engineering. His expertise is in reliability, environmental and operational ratings of electronic parts, uprating, electronic part reprocessing, counterfeit electronics, technology trends in the electronic parts and parts selection, and management methodologies. In addition, Dr. Das is involved in prognostics-based risk mitigation of electronics. Dr. Das has published more than 75 articles on these subjects, and presented his research at international conferences and workshops. He had been the technical editor for two IEEE standards and is currently vice chair of the standards group of the IEEE Reliability Society. He is a sub-group leader for the SAE G-19 counterfeit detection standards group. He is an Associate Editor of the journal *Microelectronics Reliability*. He is a Six Sigma Black Belt and a member of IEEE, IMAPS, and SMTA.

[3] **Michael Pecht** (M'83–SM'90–F'92) received the M.S. degree in electrical engineering and the M.S. and Ph.D. degrees in engineering mechanics from the University of Wisconsin-Madison, WI, USA. He is the Founder of the Center for Advanced Life Cycle Engineering, University of Maryland, College Park, MD, USA, which is funded by over 150 of the world's leading electronics companies at more than U.S. $6 million/year. He is also the George E. Dieter Professor of Mechanical Engineering and a Professor of Applied Mathematics at the University of Maryland, College Park. He has authored more than 20 books on electronic product development, use, and supply chain management and over 500 technical articles.

THE FLINT WATER CRISIS: TIMELINE OF EVENTS, EVALUATION OF TECHNICAL AND HUMAN FACTORS, AND WHY IT MATTERS

Debbie Aliya
Aliya Analytical, Inc.
PO Box 2407
Grand Rapids, MI 49501-2407
Telephone: (616) 475-0059
DaAliya@iotothen.com

Abstract: The lead contamination of the Flint, Michigan drinking water supply had its easiest specific "point source" in the economic downturn, which resulted in financial problems for the city, as well as the individual residents, and the eventual appointment of an Emergency Manager by the Governor of The State of Michigan. Bewilderingly, this seemingly well qualified financial manager acted as if he had no legal obligation to consider scientific facts related to enforcement of the Federal Safe Drinking Water Act, and repeatedly ignored warnings given by those with technical backgrounds. He prohibited the City of Flint from continuing their long-term practice of purchasing Lake Huron water from Detroit, and then refused to approve the $137.00 daily cost for the corrosion prevention chemical to be used in the new water supply, taken from the historically polluted Flint River. The combined result of these major factors was a drinking water supply that, at least once, fell within the Federal range for hazardous waste.

The most significant physical root cause of the lead contamination was the low pH of the untreated water, which attacked the protective layer of lead oxide which builds up in properly maintained lead and leaded brass water lines. These are an unknown fraction of the total services lines that constituted much of the piping system that delivered the water to the individual homes in Flint. At this time, it is unclear to the writer if the low pH was related to the inherent nature of the water drawn from the river, or to the chlorine compounds used to sterilize it, or both, or other factors. The water was so corrosive that General Motors was forced to stop using it for lubrication in their machining operation, due to corrosion of their engine blocks. [1] Members of Governor Rick Snyder's inner circle urged the governor's office to request the Emergency Manager to return the system to the original Detroit source. Yet, the residents of Flint continued to get untreated corrosive water for another year after General Motors switched to a new supply. [2] Almost two years after a local pediatrician started diagnosing multiple infants and children with lead poisoning, two State of Michigan and one City of Flint employee were served with criminal indictments. The City of Flint employee, who, early on, tried to blow the whistle on the situation, has made a plea deal.

Potential sources of the severity of the crisis, including the low-ranking science education in the USA, political divisiveness, and lack of critical thinking, will be explored. The human costs of lead poisoning will be summarized.

Key words: Civil Rights, Corrosion, Crime Rates, Critical Thinking, Department of Environmental Quality, Drinking Water, Educational Outcomes, Environmental Protection Agency, Failure of Democracy, Flint River, Lead Contamination, Lead and Copper Rule, Lead Oxide, Social Justice, Systemic Discrimination, Water Safety,

Introduction: In November of 2011, elected officials in Flint, Michigan were stripped of their powers by the state governor, through use of the Emergency Manager Law. In April of 2014, the Emergency Manager ended the decade's long water purchase agreement with the neighboring city of Detroit, and ordered the Flint Water Department to pump their water from the nearby Flint River. The Emergency Manager (EM) also refused to approve the $137.00 daily expense for the corrosion prevention treatment. The EM was warned by the Supervisor of the Flint Water Department that they had inadequate staff to monitor the water quality. [3] There does not appear to have been a specific warning at that time about the danger of lead poisoning. However, the requirements of the Safe Drinking Water Act are complex, requiring attention to biological and inorganic contaminants, as well as balancing the detrimental health effects of residual traces of the biocides routinely used to minimize the chances of immediate outbreaks of waterborne illnesses. [4] There were multiple obvious problems with the water first pumped from the Flint River, including strong odors, dark coloration, numerous reports of skin rashes from bathing, and a possible outbreak of Legionnaires' Disease.

A recent report by the Michigan Civil Rights Commission has concluded that the "people of Flint have been subjected to unprecedented harm and hardship, much of it caused by structural and systemic discrimination and racism that have corroded your city, your institutions, and your water pipes, for generations." [5]

Why did the Commission come to that conclusion? My research has revealed that systemic racism isn't what I thought it was. I believe that most people from all parts of the political spectrum want all of us to do well, and, aside from a small population of people who voluntarily associate with hate groups, are not racist. How is systemic racism different from regular racism? In short, systematic racism is not a personal characteristic. It's perhaps best defined as an effect, rather than an intentional action. We end up with systemic racism because of decisions made by people who are most likely long gone from the positions where they made the decisions. For the purposes of this paper, I'll define systemic racism an unanticipated effect of multiple complex factors. Feeling guilty, or trying to make others feel guilty, is not helpful to eliminate systemic racism.

I will look at the necessity of using critical thinking skills to address the problems of systemic and structural racism; perhaps even to understand the concepts of structural and systemic discrimination. One of the most important benefits of using critical thinking is the expanded ability to see a broader, more relevant frame of reference, from which to address problems.

This paper is intended to help us focus on the wider background of this monumental failure. Flint is but one city that has lead service lines. In fact, the EPA estimates that there are about 10 million lead service lines / connections with the public water systems of the United States. [6] The New York Times recently ran an article documenting multiple additional lead related water crises in major American cities. [7]

This paper is also intended to remind its likely readers, most of whom have strong technical backgrounds, that lead is a potent neurotoxin, which has been linked to behavioral problems, including ADHD and an increase in violent criminal behavior, trends that affect every human being. [8-10] The health costs of the lead poisoning, much more potent in young children, were not taken into consideration, as may be required by federal law, when the benefits of lower cost water were being taken advantage of in the City of Flint. Perhaps that is related to the fact that it is extremely difficult to determine these costs, human or financial, as the above referenced researchers, and their predecessors and colleagues, have learned.

As economic challenges in the US have arisen since the boom days of World War II, some people have taken to commenting that we are "becoming a third world country." In fact, the water access situation for many Flint residents failed to meet the recommendations of The Sphere Handbook [11], a resource used by non-governmental organizations involved in disaster relief efforts in developing nations. This document suggests 7.5 - 15 liters per person per day for drinking, cooking, and personal hygiene. It also suggests that people should not have to travel more than 500 meters to get their water, nor queue for more than 30 minutes. The actual average indoor water use per household in the US is 210 gallons per day. [12] This is an approximate factor of 41 difference, using the lower value from Sphere, and assuming 2.53 people per household in the US today. [13]

The writer of this document had no personal involvement in the investigation. References used are all publicly available. Several news organizations have made up timelines, still available on the internet, which were helpful in preparing the oral presentations on which this paper is founded, and this introduction. If the topic interests you further, please search for "Flint Water Crisis Timelines" on the internet. The conclusions that the ACLU reached about the civil rights violations have now been entered into the public record.

Some people reacted negatively to a portion of my original presentation, which reported that there was a contribution of systemic racism to the poor water quality. As previously mentioned, the role of systemic racism by US Government officials, in reference to their actions toward communities of low income African-Americans, has now been established. The important thing to understand about systemic racism is that it keeps on doing its damage, even if the current people running the system are not racist. The damaging effects are embedded in the way daily business is carried out. Thus, the accusation of systemic racism does not mean that the current government officials and employees are necessarily racist. There is no need to blame, but our Western cultural heritage might push some to act to minimize the effects of systemic discrimination. When I first started researching the Flint Water Crisis, I assumed, or perhaps heard someone say, that the cause of the problem was a white governor forcing a racist and incompetent white Emergency Manager to oversee African-American's affairs. This, I found out, could not have been further from the truth. Darnell Earley, the EM in charge at the time of the water source switch, was an upstanding African-American citizen, and well qualified in his field. He was a member of the National Forum for Black Public Administrators, and a former president of the International City/County Management Association. [14] See Figure 1.

Figure 1: Former Flint Emergency Manager Darnell Earley
Credit Steve Carmody / Michigan Radio

Many people, on all parts of the political spectrum, do not understand the concept of systemic bias, perhaps a manifestation of the generally bewailed lack of critical thinking. People on the left are more likely to "believe" that systemic discrimination exists, even if they can't give a definition, and certainly would not give highly consistent definitions if they tried to give one. People on the right are more likely to simply not hear the term, or to write off its purported effects as whining by the lazy and willfully ignorant. My own understanding of this term is greater as a result of my research for this paper.

Simple racism comes from both our cultural programming, and our experience. Chris Hayes, a conservative leaning white journalist and author who has investigated the economic divide in America, pointed out on the NPR talk show 1A, that fear of those outside of our own cultural group is justified and natural in certain situations. [15]

It is much easier to understand the concept of simple racism. Its manifestations are well known. Systemic bias, also known as institutional or structural bias, is harder to grasp. It requires knowledge of historical facts, and to truly understand, statistical analysis by specialists is needed to show whether real damage is being done. In this case, the policies of the US federal government at the time of the end of World War II are a factor. Many African-Americans had served in the armed forces, and there were political promises made of civil rights to be enjoyed on arrival back home. The way that the US government kept those promises to help the black Americans was to create segregated housing developments for blacks and whites, [16] in the name of racial harmony. This was before the (now overturned) legal doctrine of "separate but equal" for educational institutions. The persistence of resource inequality in education is thus self-evidently linked to inequality in housing, through the mechanism of local tax bases funding schools. Of course, there is more to school quality than money.

The writer believes that another contributing factor to the Flint Water Crisis, in its full manifestation, is the generally low standard of science education in the US, and accompanying low levels of respect for those working in the field. Engineers in manufacturing routinely bemoan the "bean-counters" getting to make decisions that are technically without merit. This has nothing to do with racism. It is related to the influence given to those with financial backgrounds. The financial manager apparently felt entitled to ignore the technical supervisor. Some statistics on educational achievement will be provided, as will some more uplifting facts regarding public responses to the crisis.

In researching this paper, I have found a lot of recent academic activity relating to the effects of lead poisoning. The complex and troubling situation of health damaging lead levels in the drinking water of significant numbers of residents of the United States of America is certainly going to continue to be a subject of public concern for the foreseeable future, as we strive to replace and maintain our aging water infrastructure.

Background: In December of 2011, Governor Rick Snyder made the first of what would become five successive appointments of Emergency Managers, by April of 2015, to deal with an ongoing financial crisis in the City of Flint. [17] Note that Flint used to be a thriving city with a large tax base, and the infrastructure reflected that. By one measure, the current GM workforce in Flint is 8% of what it was in its heyday. [18] In November of 2012, the voters of Michigan repealed the law that authorized Emergency Manager appointments, but almost immediately, in a lame duck session of the state legislature, a new measure, which was immune to citizen repeal, was put into place. [19] This is at least one foundation of the argument that, at its center, the Flint Water Crisis was due to a failure of democracy. Of course, the US is not a democracy, but a republic, and thus this failure was legal, and in line with our Constitution, which gives enhanced power to our elected officials. Early in the tenure of the Emergency Managers, a decision had been made to get a new water source. However, that source was not ready when Emergency Manager Darnell Earley decided that the water costs had to be reduced right away. Thus, he decided to have the water department temporarily get their water from the Flint River, a historically polluted body, [20] over the strong and repeated objections of the water plant supervisor. The writer is still struggling to understand this latest inappropriate trumping of evidence based technical opinions, and the law, by financial "realities." However, the idea that spendable dollars in a nation are to be limited in the same way that they are in a household is problematic, due to the existence of the system of banks, which are allowed to leverage collateral to a high degree. Money has never represented a fixed amount of goods. It has always represented an amount of goods modified by some function of good will. This is one of the major premises of the fascinating book "Debt: The First 5,000 Years." [21] Nations take the right to print, control and manipulate the money source, for political ends. It causes major problems sometimes, as history has shown, and continues to show. There is a spectrum of attitudes about how much money governments should make available.

Timeline: Note that the following events are documented on Michigan Radio's website, www.michiganradio.org, among other places.

The switch of water supply was made toward the end of April 2014, and by May, the brown, stinking water was called to the attention of the city leadership by concerned citizens who had a sudden increase in skin rashes. To no avail, General Motors tried reverse osmosis purification to prevent corrosion of the engine blocks machined with lubricant diluted with the new water. Elevated coliform bacteria and mutenogenic levels of trihalomethanes were also found in the water. [1, 2] By January of 2015, Detroit's water department offered to reconnect their prior customer, but the Emergency Manager declined.

In February of 2015, resident Lee Anne Walters notified the Environmental Protection Agency (EPA, a Federal Agency), that 104 ppb of lead had been detected in the water coming out of her faucet, far in excess of the (allowed but unsafe) EPA "Action Level" of 15 ppb. By this time, Miguel Del Toral, a water expert at the EPA, had reported that the testing method was likely to be understating the lead levels. EPA regulations require the water to be tested after the water system has remained out of use for the night. Instead, residents were instructed to flush the pipes before collecting the samples. A second round of testing at Ms. Walters' home found 397 ppb, while the third round showed over 13,000 ppb, almost three times the level that would be classified as hazardous waste. How was this happening? Despite the historical pollution of the Flint River, it was not polluted with lead. It took until July of 2015 for a brave EPA employee to leak a memo with the information about the high lead levels to the ACLU, who publicized it, with the facts of the high lead levels.

Immediately, the EPA stated that no action was required. Maybe they just did not believe that the lead levels could be so high, as widespread knowledge about the effect of the low pH water did not yet exist among politicians and the general public. However, in August 2015, the (Michigan) Department of Environmental Quality (DEQ), acting on a documented level of 11 ppb between January and June 2015, instructed the City of Flint to "optimize corrosion control." By this time, it appears that someone in the government was aware that the lead was coming from service lines leading off the mains (rather than the main lines themselves), to and within the homes, as a result of the low pH, possibly related to the high chlorine needed to kill the coliform bacteria [1, 2] and other corrosion enhancing factors, that were breaking down the previously passive and protective film of lead oxide that builds up in lead pipes over time.

Also in August, a citizens group hired some researchers from Virginia Tech to test the water, which resulted in confirmation of the high lead levels. However, the Michigan DEQ denied the accuracy of the Virginia Tech tests. Again, denial from government authorities was the response in September, when pediatrician Dr. Mona Hannah-Attisha reported that she had found a tripling of blood lead levels in some Flint neighborhoods, along with signs of lead poisoning in multiple children. The Michigan authorities denounced her work. A week later, they finally publicly admitted that she was correct, and by October, the water supply was switched back to Detroit. However, the damage to the passive lead oxide film could not be immediately repaired, and lead levels remained sporadically high. Figure 2 (color key in 2A) shows the results of lead tests from January of 2016. From April of 2014, until October of 2015, there was, as far as the writer can determine from the sources described, no effective corrosion control. It appears that the Director of the Department of Environmental Quality for the State of Michigan ruled that the water analysis tests themselves constituted "corrosion control." Only in October of 2015, did the Director of the DEQ admit to any mistakes.

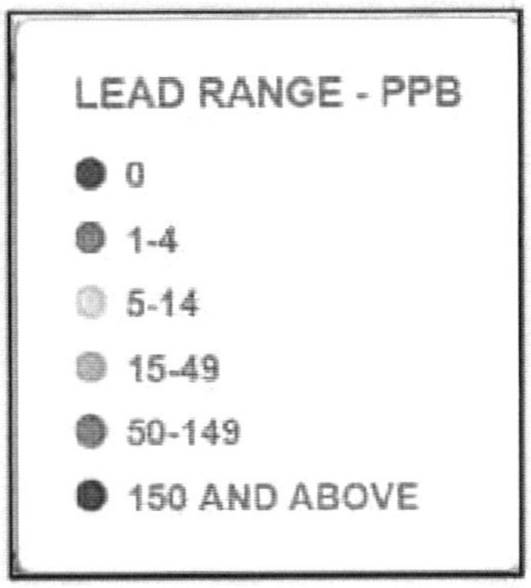

Figure 2: Key to Color Code from Figure 1, same attribution

By October, the complexity of the technical issues was starting to become apparent. Records could not be located to determine the location of the service lines, and their materials of construction were unknown. Furthermore, some attempts at replacing old lines started to reveal that it was steel lines with leaded brass fittings that were sometimes the "culprits" in the higher lead water supplies. It was starting to become apparent that the high costs of providing bottled water, water filters, medical help and behavioral training to affected children, were only a small portion of the costs to be incurred in replacing a system that was now essentially untrustworthy.

In December of 2015, Flint added additional chemical corrosion controls, based on the new understanding that was emerging about the corrosion mechanisms. December 2015 also saw the first Emergency Declaration (aside from the financial emergency that started the whole problem), which set the stage for help from the state. A task force held Michigan DEQ Director Wyant responsible for the situation, and he resigned before the end of the year.

On January 5, 2016, Governor Rick Snyder declared a State of Emergency for Genesee County, where Flint, Michigan, is located. Governor Snyder followed his declaration with a request for help from the Federal Emergency Management Administration. This

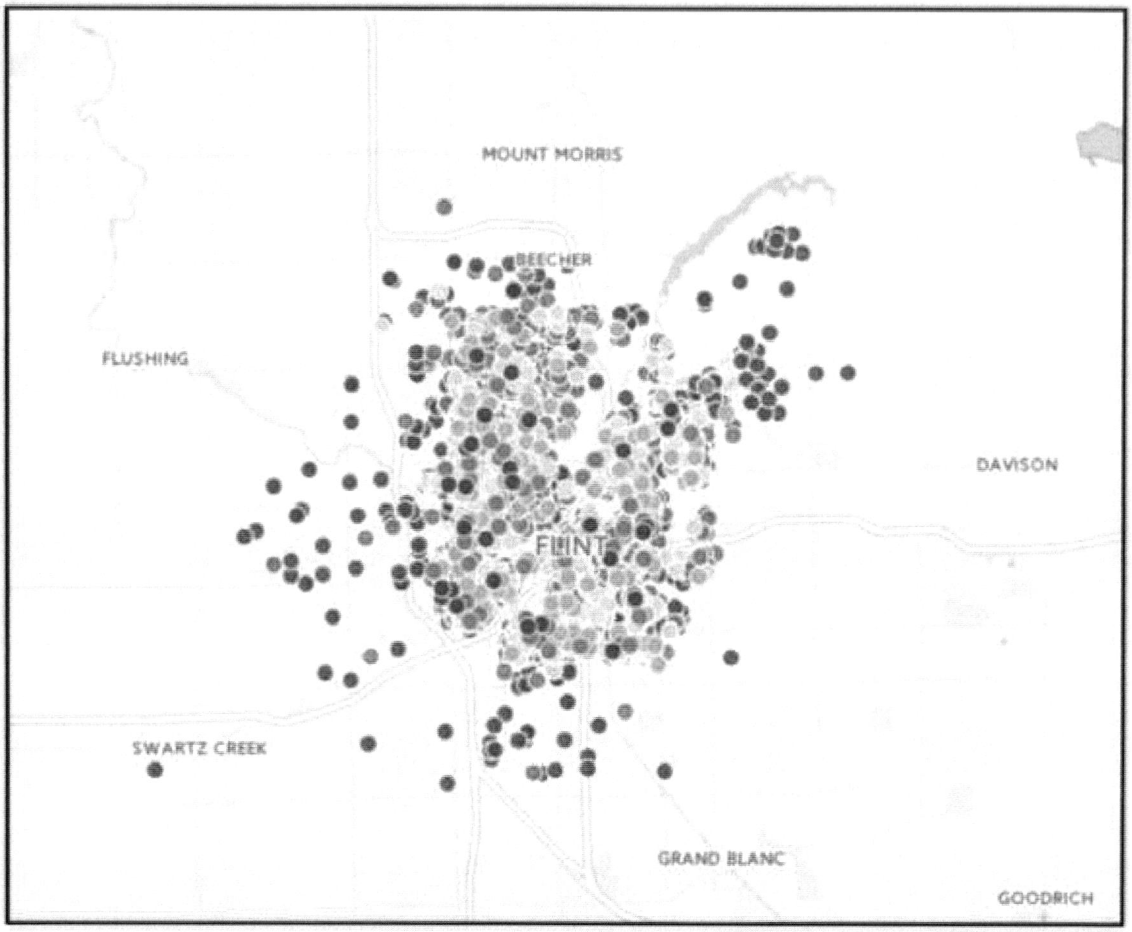

Figure 2A: A screenshot of a map generated by Mark Brush, with the help of Elias Brush, Cass Adair, and SmartyStreets, found at slide 2 of 3 on http://michiganradio.org/post/when-will-we-know-water-flint-safe-drink

eventually resulted in $5 million in Federal Aid, but the US Senate declined the $600 million aid request from the Michigan delegation. The Michigan State legislature approved $28 million in aid for bottled water, filters, medical treatment for lead poisoning, services for those children who have developed or are at risk to develop life-long behavioral problems, and an infrastructure integrity study. By early February of 2016, the EPA Regional Administrator had resigned. In April of 2016, the first criminal charges were filed.

At some point, subsequent to the switch to the Flint River, the Michigan Department of Environmental Quality ruled that the monitoring of the water itself constituted the required corrosion control. This was perhaps, to me, the most egregious human failure. It certainly shows a lack of critical thinking, and a willful lack of understanding of the purpose of the law, which is to protect the health of the members of the public. Despite this ruling (that the testing constituted corrosion control), someone in the chain of command thought it best to falsify the data, by eliminating the high outlier values of lead from the tests record.

Calling the testing corrosion control is using logic perhaps similar in some ways to the inaccurate thinking that lets the health care industry call screening tests "preventive care." Change of life-style, medications, and meditation are all preventive care. Blood tests and mammograms, in and of themselves, do nothing to improve our health or prevent disease. Yet I have never heard anyone of the media, or being interviewed by the media, make this point. It is, in my opinion, a sad state that people cannot distinguish monitoring from prevention. This situation is representative of the intellectual level of our national discussion on questions of interest to the political body. Even people who have a science background are not immune from confusion regarding monitoring and control.

Mike Glasgow, the Flint Water Department Supervisor, who had tried to keep the state officials from making the switch to the Flint River, based on concerns of lack of personnel trained to monitor the water quality, was ultimately charged with evidence tampering. Citizen action resulted in his being able to make a plea deal, as he was seen as having tried to do the right thing.

Governor Snyder eventually activated the Michigan National Guard to distribute bottled water and filters. Note that water filters cannot be used with hot water, so all bathing water had to be filtered and then separately heated. The unfiltered tap water was even unsafe to wash vegetables.

By June of 2016, the new corrosion control measures had partially "healed" the lead oxide film, and the EPA announced that children and even pregnant women could safely drink the tap water, as long as it was filtered. In November of 2016, the situation was unchanged.

Lead and Its Effect on Humans

The first two times I presented on the topic of "The Flint Water Crisis," I was surprised that listener response was more appreciative of the material regarding mechanisms of the corrosion than the human factors. Perhaps the listeners, one audience of which were 100% Michigan residents, were unaware of the severity of the consequences of even small amounts of lead ingestion. Maybe, due to the success of environmental regulation, we are too far from the days of the 1970's, when we had to worry about lead in the air from leaded gasoline. The fact is that health researchers and the Environmental Protection Agency say that there is no safe level of lead in drinking water, especially for children. [22]

The paper of Feigenbaum and Muller [9], who obtained data from the early 1900's, a time when lead exposure was due to use of the latest new-fangled trend: city-wide, tap-water delivery systems, is interesting. The authors showed that lead related behavioral problems can be disentangled from sociological and economic factors. Rich white people were just as likely to become murderers as poor black people, if they were hooked up to the same low pH water that was delivered through lead pipes. (This is a simplification, interpretation and selection from the data they display. Please read the paper.) While the original leaded water problem had random victims, today it is a social justice problem that our democracy has been unable to solve. Interestingly, the paper was written with help from the NSF-IGERT Multidisciplinary Program in Inequality & Social Policy at Harvard University (Grant No. 0333403) and the Robert Wood Johnson Foundation Health & Society Scholars Program. There are 18 named authors, and additional groups that contributed to this paper. The well documented correlation between lead and homicides is only one facet to the lead problem. See Figure 3. The World Health Organization (WHO) Lead Fact Sheet provides the information shown in Figure 4. [23]

The History of Water Treatment: Water treatment goes back to the Bronze Age [24] in Egypt, where aluminum sulfate was used to clarify drinking water in 1500 BCE. There was not much progress made until the 1700's of the Common Era, when Europeans started using sand to filter out particulate. Not until the 1800's did humans understand the link between epidemics and water. In the 1900's, Americans started to use chlorine for disinfection, while the Europeans used ozone. In 1914, the first American standards limited bacteria in drinking water. Major revisions were made to the US Public Health Service standards in 1925 and 1946. By the 1962 revision, there were 28 regulated substances in drinking water. However, it was not until 1975 that the Safe Drinking Water Act was passed, which was the first time America had enacted enforceable standards for drinking water. Even then, the standards did not go into effect until 1977. The 1996 version of the regulation was the first time that the (long term) health risks from the disinfectants had to be balanced against immediate risks from bacteria borne illness.

[Leave one blank line between title and paragraph]

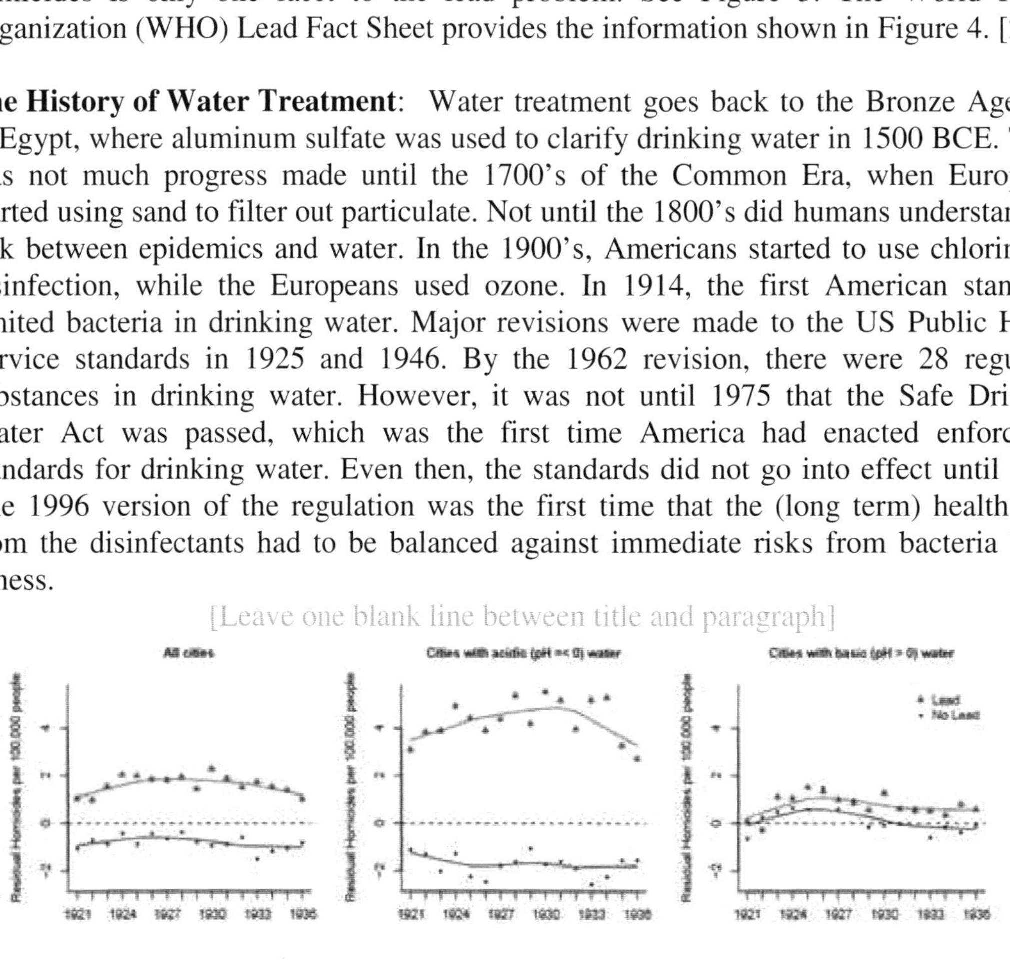

Figure 5: Residual homicide rate in US cities, 1921-1936. Homicide rates were consistently higher in cities using lead pipes than in cities using non-lead pipes. The gap between the lead and the non-lead cities is much wider in the sample of cities whose water was acidic and negligible in the sample of cities whose water was basic. The trend lines are constructed by capturing the residuals after regressing the homicide rate on all covariates other than the lead pipe indicator. The homicide data are drawn from cities reporting data to the *Mortality Statistics* between 1921 and 1936. The population data are interpolated using decennial census data. Cities with only lead pipes and cities with a mix of lead and iron pipes are included in the "lead" category. Cities using either galvanized iron or wrought iron are included in the "no lead" category.

Figure 3: From Reference 9, Used by permission. Their pH value of "0" being neutral is a standardized value equivalent to a regular pH value of 7 (page 64 of the paper)
Note: Left graph is all data, center graph is low pH, right graph is high pH.

Water Quality Monitoring: The concerns of the Supervisor of the Flint Water Department, that they had inadequate staff to monitor the water quality, might seem unjustified. The Flint Water plant had been open for years. However, there's a difference between monitoring drinking water taken from a source that has been used for decades, and a new source with a long history of industrial contamination. There are currently seven micro-organisms, four disinfection by-products, three disinfectants, 16 inorganic chemicals, 53 organic chemicals, and four radio-nucleotides that are regulated by the Environmental Protection Agency. [4] The Nalco Water Handbook [24] gives the Equation 1 to calculate the time water needs to be in contact with the disinfectant, chlorine in this case, based on Giardia, a hard-shelled parasite, harder to kill than viruses.

$$\text{Contact Time (CT)} = 0.2828\,(\text{pH}^{2.69})(\text{FRC}^{0.05})(0.933^{(T-5)})(L) \qquad (1)$$

Where
FRC= Free Residual Chlorine, T= Temperature, Degrees C, L= Log removal

Proper use of this equation obviously requires someone who has a working knowledge of chemistry, as well as high school math.

The State of Science Education in the USA: As of 2015, the Programme for International Student Assessment, an effort of the Organization for Economic Co-operation and Development, showed the United States at 25th place among 70 nations whose 15-year-olds took a standardized science test. [25] See Figure 5. Figure 6 shows the comparison between the three top and bottom ranked countries and the USA, between performance of girls and boys. Note that in the very bottom ranked countries, where the averages are dismal, the girls are doing better than the boys in two of those countries. Overall, students in the US are slightly more likely to have to attend at least one science class per week than the average of all 70 countries. Only 6.4% of US students do not have to attend even one science class per week. Compare that to the number 1 country, Singapore, where only 1.3% of students attend no science class. [26]

How did this happen? When I first decided to look into the social causes of the Flint Water Crisis, and started to understand that corrosion was nominally "the physical root cause," I wanted to "blame" the influence of the evangelical religious movement, for confusing the natures of belief in revealed truth (through the "Creation Science" movement) and the fruits of the scientific process of learning about the world through our senses and intellect. Many people, since the time of World War II and the Cold War, also struggle with the difficulties brought along with the benefits of technology (overpopulation, pollution, threat of nuclear war, etc.), and don't want to have anything to do with science. The existential angst of the postwar period has been documented elsewhere. [27]

Social Justice and Educational Opportunity: I have already provided references which document the scourge of systemic racial discrimination in the American housing market. Its consequences in the justice system, a recent topic of frequent discussion in the news media and talk shows since the events that precipitated the Black Lives Matter movement, is a disproportionately small fraction of two parent households in minority communities. Raising kids is hard enough when there are two parents, and one parent families have greater struggles. This is one factor in the decline of the stable, nuclear family.

This decline of the stable, nuclear family, and its consequences in all aspects of education, influence the amount of learning that takes place in our country. Finally, I thought about the fact that systemic discrimination against minorities, who are making up an increasingly large fraction of our country, will have a larger and larger effect on the overall average educational achievement. When a large sub-group does poorly, the whole group does worse than it would have if the large sub-group had done well. It's simple algebra. The lower educational achievement of poorer Americans has been well documented in the news for decades.

In concluding this section on the effects of social injustice, I would like to encourage the reader to do a web-search on the term "social justice." It's not a new-fangled idea. It's the foundation of the moral imperatives of charity in the Judeo-Christian tradition. I did find one reference on-line that addresses the reason that part of our political spectrum stops listening when they hear the term "social justice."

R. Dozier Gray, on the web-page of the National Leadership Network of Conservative African-Americans, writes "In one sense, social justice is the basis for a sound and civil society. The struggle for social justice is, in its purest form, the struggle for equality of opportunity over outcome. That's not necessarily a problem." But, Gray goes on to say, "In this new interpretation, social justice can more appropriately be considered 'collective retribution' or 'restorative justice.' The lingering question, however, is to restore what to whom and at what cost. It opens up a Pandora's Box of unsettling possibilities." [28]

I thank R. Dozier Gray for his insightful piece. I'd also like to thank David Levy [29] for articulating an important concept of critical thinking, which is that some questions have either-or answers, while most issues have a spectrum along the black-white continuum. I believe that we can't be forced into an either-or position on the subject of social justice. It has been said that the major contribution at the time of the founding of the Jewish religion was an emphasis on the balance of the responsibility of the community members to first, preserve the community, and second, to preserve the freedom and independence of the individual. Social justice, or its lack, in a society that has systemic racial discrimination, is always going to come down to the issue of whether pure, real time justice is "fair," indeed, is really justice, or whether restorative justice is a beneficial policy to follow. A national dialogue about how much restorative justice Americans want to have, and how we want to implement it, would be more productive than the one we're currently having, where progressives say they want social justice, and conservatives close their ears.

Lead poisoning and health

Key facts: Lead is a cumulative toxicant that affects multiple body systems and is particularly harmful to young children.

- Lead in the body is distributed to the brain, liver, kidney and bones. It is stored in the teeth and bones, where it accumulates over time. Human exposure is usually assessed through the measurement of lead in blood.
- Lead in bone is released into blood during pregnancy and becomes a source of exposure to the developing fetus.
- There is no known level of lead exposure that is considered safe.
- Lead poisoning is entirely preventable.

Lead is a naturally occurring toxic metal found in the Earth's crust. Its widespread use has resulted in extensive environmental contamination, human exposure and significant public health problems in many parts of the world.

Important sources of environmental contamination include mining, smelting, manufacturing and recycling activities, and, in some countries, the continued use of leaded paint, leaded gasoline, and leaded aviation fuel. More than three quarters of global lead consumption is for the manufacture of lead-acid batteries for motor vehicles. Lead is, however, also used in many other products, for example pigments, paints, solder, stained glass, lead crystal glassware, ammunition, ceramic glazes, jewelry, toys and in some cosmetics and traditional medicines. Drinking water delivered through lead pipes or pipes joined with lead solder may contain lead. Much of the lead in global commerce is now obtained from recycling.

Young children are particularly vulnerable to the toxic effects of lead and can suffer profound and permanent adverse health effects, particularly affecting the development of the brain and nervous system. Lead also causes long-term harm in adults, including increased risk of high blood pressure and kidney damage. Exposure of pregnant women to high levels of lead can cause miscarriage, stillbirth, premature birth and low birth weight, as well as minor malformations.

Figure 4: WHO Fact Sheet: Lead Poisoning and Health (Aliya highlights)

The Hebrew Bible [30] provides the description of the "perfect" human judge: "and he shall not judge after the sight of his eyes, neither decide after the hearing of his ears; But with righteousness shall he judge the poor, and decide with equity for the meek of the land." When I first read this, I really wondered how the judge who ignored the plain physical evidence was going to do a good job. Trying to put myself into the mindset of an ancient prophet, I realized that Isaiah must be predicting a person of divine knowledge, perfect intuition, an interpretation that Christians have long assumed to be obvious. As Niels Bohr, the great physicist said, "The opposite of a great truth is another great truth."

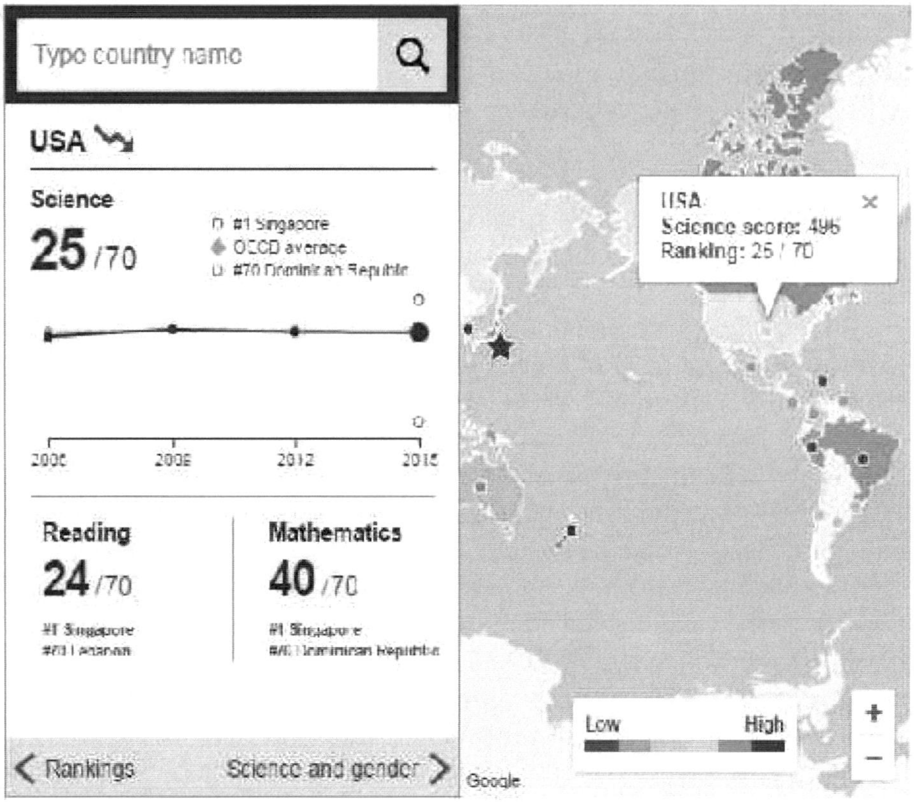

Figure 5: PISA Data from Interactive BBC Website

Science rankings

#		Girls Score	Boys Score
01	Singapore	552	559
02	Finland	541	545
03	Estonia	533	536
28	**USA**	**493**	**500**
68	Kosovo	383	374
69	Algeria	383	369
70	Dominican Republic	331	333

Figure 6: PISA Data from Interactive BBC Website

As noted by journalist and political writer Chris Hayes, in an interview with Joshua Johnson on the NPR program 1A, on March 22 (podcast available at http://www.npr.org/podcasts/510316/1a), the difficulty in eliminating systemic injustice is related to the idea of a zero sum game, the belief that if the poor have more, others must have less. The New Testament articulates and addresses this fear: "To everyone who has something, more will be given, and he'll have more than enough. But from the person who has nothing, even what he has will be taken away from him." [31]

As Gray points out in his essay, life is unfair, and his piece seems to indicate that "equity," social justice, or just plain "justice," (there should only be one kind, according to him) will leave an unfair world unfair, and that unfairness is better than the alternative: unfairly restoring justice, thereby adding new layers of unfairness. That is the human condition, and we're fated to experience it.

I reminded myself as I was writing this paper, that since the days of the ancient Israelites (Bronze Age) and Classical Greeks (6th to 4th Centuries BCE), humans have been struggling to understand exactly what justice is. Sophocles (c496-406) [32] addressed this in his classical tragedy, Antigone, where the heroine invokes her moral requirement to disobey the king, because he has prohibited her from carrying out a religious obligation (burying her brother), and she gets the death penalty. The King was using his political power to carry out his duty to preserve order in Thebes, depriving a person unrelated to the crime committed, of carrying out her duty to the highest truth.

The reason, I believe, that we still have human judges in this age of Artificial Intelligence is that it is often a difficult balancing act to provide both individual and community justice. In other words, the judge is obligated to be fair to the individual while avoiding setting precedents that would be bad for the group. This is the true job of judges, and the reason that the idea of "Originalism" in regards to Constitutional law is problematic. Specific human laws are always approximate recipes for good behavior, as no legal system can anticipate every potential frame of reference. Going back to Sophocles, it wasn't Antigone's fault that her brother took the actions that he did, but she was the one who ultimately suffered. King Kreon perceived no way to achieve both justice and fairness, if he even cared about fairness. Sophocles paints him as fairly heartless. A new interpretation of the Hebrew Bible's story of the Tower of Babel indicates that the "sin" of the residents of Babel was the leaders' forcing the individuals of that city into the excessive degree of conformity in their private lives, in order to achieve the public goal of building the tower. [33] In my view, the regular citizens of Babel were also at fault, for tolerating the bad leadership. Otherwise, the leaders could have been punished rather than the whole community.

The Safe Drinking Water Act: Title XIV of the Public Health Service Act (also known as US Code Title 42 Public Health and Welfare Chapter 6A Subchapter XII) is called "Safety of Public Water Systems" or the Safe Drinking Water Act. [34] It is very complicated and difficult to follow and understand, for anyone who is not a water treatment specialist, which the writer of this paper is not. A few highlights of recent changes in the law will be presented.

Until very recently, "plumbing fixtures" could have up to 11% lead. Only since Jan 4, 2014, three months before the switch to the Flint River water, were new systems require to be "lead free." The maximum weighted average of lead in a "lead free" system is 0.25%. Calculations are based on the maximum allowable lead for the alloy and the wetted surface area. It is not specified whether the minimum or maximum value of inside surfaces of the pipes is to be used. Nowhere in this 80-page document are maximum contaminant levels given. But the rule does require systems to monitor drinking water. The monitoring must be done at customer taps. If lead concentrations exceed an action limit of 15 ppb or copper concentrations exceed an action limit of 1.3 ppm in more than 10% of customer taps sampled, the system must undertake additional actions to control corrosion. If the action level for lead is exceeded, the system must also inform the public about steps they (the members of the public) should take to protect their health. This may include advising them to replace lead service lines under their control. Note the law only protects "the herd" (the totality) of citizens. No actions are required if the number of homes with excess contaminants is below 10%. This is another example of how our society manages responsibility to the individual (none in this case) versus the community. Advice, education, and even pressure to vaccinate children follows the same "protection of the herd" logic.

At the time of the water crisis, the lead "action limit" for both the US and Michigan was 15 ug/L. By April 15, 2016, Governor Snyder was calling the federal lead standard "dumb and dangerous." His new proposal calls for lowering Michigan's lead action level standard to 10 parts per billion by 2020. His plan also calls for removing all lead service lines in Michigan within 10 years. The governor's cabinet director says it will take time just to identify where all the lead service lines are. The proposal prioritizes removal of lead service lines "based on vulnerable populations, high blood levels (and) high test results." [35]

Conclusion: The roots of the Flint Water Crisis are complex, and understanding its origin requires many forays into the most basic ideas about what it means to be a human, a member of a complex, modern society, and to enjoy the exercise of free will. Dealing with the consequences of lead exposure in the Flint, MI, population, and elsewhere, will take time and money and thought. New academic approaches to understanding and documenting injustice, in all its forms, are being brought forward in a divisive political environment. Those who wish to set an example of using critical thinking techniques to address these problems have their work cut out, as implied by the artwork shown in Figure 7.

The sword of Justice is rusted.
The eyes of the Lady are crusted
with the Dust of Ages.
The pans of her balance droop
under the weight of the years,
while softly Mercy calls us
through the masks of our fears.

The sword of Justice is rusted.
Her daughter's echoes faintly reach our ears.

Figure 7: Sculpture "Simple Justice," [36] created by Vivienne Thaul Wechter, near Fordham School of Law, in New York City. (Aliya photo credit)

Acknowledgements: Thanks to Michigan Radio for their coverage of the Flint Water Crisis, and permission to use images and charts from their website, and to the ACLU for their support of Flint residents who were working for environmental justice. Thanks also to my colleague Patrick Doyle, Doyle & Associates, Edwardsburg, MI, for his help in understanding the detailed statistical analysis in [9], and his suggestions to improve the clarity of the document, particularly how to emphasize the difference between regular and systemic racism. Thanks to my colleagues in the ASM International Failure Analysis Society, who accepted my proposal to speak on this topic at the MS&T conference in Salt Lake City in October of 2016, and my local West Michigan ASM and SAE International Chapter members who listened to and critiqued the first two versions of the talk. Thanks to Peter Ried, Ried And Associates, Portage, MI, James Maynard, Ionia, MI, and Steve Batzer, PhD, of Batzer Engineering, Fife Lake, MI, colleagues and friends with more conservative leanings than I have, who gave me feedback on how to communicate with conservatives. Thanks also are due to my local writing group, The Mid Michigan Word Gatherers, who listened to a long section of a technical paper to critique my presentation, and Karen Bota, the group's founder, for help editing. This document contains more than the usual amount of opinion for a technical paper. Any errors or fact or interpretation are strictly my own responsibility. Finally, I thank all who have read this far. When I started studying the Flint Water Crisis, I had no idea of how time consuming it would be to document all the "things I knew," that created the foundation for my viewpoint, some of which I have since found out were incompletely understood, if not misunderstood. That is the purpose of critical thinking, to see more clearly.

References:

[1] Livengood, Chad, Emails: Flint Water Warnings Reached Gov's Inner Circle, The Detroit News, Feb. 26, 2016, http://www.detroitnews.com/story/news/michigan/flint-water-crisis/2016/02/26/snyder-aides-urged-switching-flints-water-oct/80967048/
[2] Colias, Mike, How GM Saved Itself from the Flint Water Crisis: Rusting Engine Blocks Flagged Big Problem, Automotive News, January 31, 2016, http://www.autonews.com/article/20160131/OEM01/302019964/how-gm-saved-itself-from-flint-water-crisis
[3] Fonger, Ron, Flint Water Supervisor Warned State of Problems before Switch to River, mlive.com/news/flint/index.ssf/2016/02/flint_water_supervisor_warned.html
[4] Safe Drinking Water Act, US Code 42, Office of the Law Revision Counsel of the U.S. House of Representatives, 1974 (Amended 1984, 1996)
[5] Demashkieh, Rasha and Reyes Kopack, Laura, JD, et al, The Flint Water Crisis: Systematic Racism Through the Lens of Flint, Michigan Civil Rights Commission, 2017
[6] Flynn, Daniel J., Ed, Nalco Water Handbook, Third Edition, McGraw Hill, 2009, p 41.10
[7] Wines, Michael and Schwartz, John, Unsafe Lead Levels in Tap Water Not Limited to Flint, The New York Times, Feb. 8, 2016
[8] Cook, Philip J. and Laub, John H., After the Epidemic: Recent Trends in Youth Violence in the United States, Crime and Justice, 2002, Vol. 29, pp. 1-37
[9] Feigenbaum, James J. and Muller, Christopher, et al, Lead Exposure and Violent Crime in the Early Twentieth Century, 2016, Scholars at Harvard (scholar.harvard.edu), Cambridge, MA.
[10] Nevin, Rick, How Lead Exposure Relates to Temporal Changes in IQ, Violent Crime, and Unwed Pregnancy, US National Library of Medicine, National Institutes of Health (PubMed.gov), 2000, 83(1):1-22.
[11] The Sphere Handbook, The Sphere Project, 1997, www.sphereproject .org
[12] The Environmental Protection Agency, updated Jan 26, 2017 https://www3.epa.gov/watersense/our_water/water_use_today.html
[13] https://www.statista.com/statistics/183648/average-size-of-households-in-the-us/ (viewed March 3, 2017)
[14] https://en.wikipedia.org/wiki/Darnell_Earley
[15] Hayes, Chris, A Colony in a Nation, W.W. Norton & Co., 2017
[16] Massey, Douglas and Denton, Nancy, American Apartheid: Segregation and the Making of the Underclass, Harvard University Press, 1993, referenced in The Future of Fair Housing: Report of the National Commission on Fair Housing and Equal Opportunity at http://www.civilrights.org/publications/reports/fairhousing/historical.html#_ednref25
[17] https://en.wikipedia.org/wiki/Financial_emergency_in_Michigan
[18] Hakala, Josh, Michigan Radio, 2016, http://michiganradio.org/post/how-did-we-get-here-look-back-michigans-emergency-manager-law
[19] Lee, Trymaine, The Rust Belt: Once Mighty Cities in Decline- An Auto Giant's exit brings the region to its knees, MSNBC, Part 3 of 4 of A Geography of Poverty, Paragraph 10 in http://www.msnbc.com/interactives/geography-of-poverty/ne.html

[20] Report on Water Pollution in the Lake Huron Basin: Flint River, United States Department of the Interior, Federal Water Pollution Control Administration, Great Lakes Region, December 1966, pp 69-75

[21] Graeber, David, Debt: The First 5,000 Years, Melville House, 2011

[22] Environmental Protection Agency, https://www.epa.gov/ground-water-and-drinking-water/basic-information-about-lead-drinking-water, first sentence under the title.

[23] World Health Organization, Lead Poisoning and Health: Fact Sheet, reviewed September 2016, http://www.who.int/mediacentre/factsheets/fs379/en/

[24] Flynn, Daniel, Editor, The Nalco Water Handbook, Third Edition, 2009, Section 41

[25] The Organization for Economic Cooperation and Development, https://www.oecd.org/pisa/pisa-2015-results-in-focus.pdf

[26] The British Broadcasting Company, http://www.bbc.com/news/education-38212070

[27] Fromm, Erich, Escape from Freedom, Holt Paperbacks, 1994

[28] Gray, R. Dozier, New Visions Commentary - Social Justice: Not what it Used to Be, The National Center for Public Policy Research, http://www.nationalcenter.org/P21NVGrayFreedom90410.html

[29] Levy, David, Tools for Critical Thinking: Metathoughts for Psychology, 2nd Edition, Waveland Press, 2010, pp 25-29

[30] Isaiah, Mishneh Torah (The Hebrew Bible), Mechon Mamre, http://www.mechon-mamre.org/p/pt/pt1011.htm

[31] Mathew 25:29, International Standard Version of the New Testament, http://biblehub.com/matthew/25-29.htm

[32] Sophocles, tr. by Young, George, Dover Publications, 1993

[33] Klitsner, Judy, Subversive Sequels in the Bible, Koren Publishers, Jerusalem, 2011

[34] US Government Printing Office, 42 USC 6A – Public Health Service, 2010

[35] Smith, Lindsey, Michigan Radio, Sept. 20, 2016, http://michiganradio.org/post/state-lawmaker-flint-says-delay-strengthening-lead-rules-beyond-ridiculous

[36] Thaul Wechter, Vivienne, Simple Justice Sculpture, Fordham School of Law, an artwork in New York City, honoring the writer, Richard Kluger, and his book about a landmark Supreme Court ruling. The book's full title is "Simple Justice: The History of Brown v. Board of Education and Black America's Struggle for Equality."

THERMAL DEGRADATION OF POLYIMIDE INSULATION AND ITS EFFECT ON ELECTROMAGNETIC COIL IMPEDANCE

N. Jordan Jameson, Michael H. Azarian, and Michael Pecht
Center for Advanced Life Cycle Engineering (CALCE)
Building 89, Room 1100C
University of Maryland, College Park, MD 20742
Telephone: (301) 405-8038
jjameson@umd.edu

Abstract: The failure of insulation in electromagnetic coils is a significant cause of coil failure and can have severe implications for the system in which the coil is used. Impedance monitoring of coils has emerged as a promising avenue for evaluating the insulation health of electromagnetic coils in-situ. Due to its excellent mechanical properties and ability to endure high temperatures, polyimide is widely used as an insulator in the electromagnetic coil manufacturing industry. However, little information is known about the electrical behavior of polyimide insulation when subjected to the variety of stresses experienced when used as electromagnetic coil insulation, casting uncertainty on the use of impedance monitoring when monitoring polyimide insulation health. This paper presents an experimental analysis of how the insulation electrical parameters evolve over time, and their consequent effect on the coil impedance spectrum, thus providing useful empirical evidence for impedance monitoring for electromagnetic coil insulation health monitoring.

Keywords: Condition monitoring, electromagnetic coil insulation, insulation health monitoring, impedance measurement, insulation capacitance, insulation resistance, polyimide thermal degradation

Introduction: Electromagnetic coils are used ubiquitously in a variety of industries and industrial applications, including electric motors, solenoid valves, and transformers. Studies have shown that between one-quarter [1], [2] and one-third [3] of motor failures are due to failures in the insulation. A study of U.S. nuclear power plants [4] showed that over 50% of solenoid valve failures originated with failures in the electromagnetic coil (e.g., coil opens or coil shorts).

There has been much work addressing the problem of insulation health monitoring in machinery. In 2004, Stone *et al.* [5] released a book that is a standard reference for insulation used in rotating machinery. In it, the available methods of diagnostics for insulation were discussed. For in-situ insulation health monitoring, the available methods are partial discharge testing or particulate measurement (i.e., smoke detector) [6], [7]. However, many machines, such as low-voltage motors or solenoid-operated valves, operate at voltages that do not induce partial discharge (or induce such low levels as to be unmeasurable). Offline methods include the measurement of the dissipation factor, insulation capacitance, polarization index, insulation resistance, insulation power factor, and offline partial discharge. A common theme for the offline methods discussed is direct access to the insulation. In other words, in order to measure the insulation capacitance, for example, one must have the ability to place electrodes outside the insulation, in effect using the insulation as a dielectric between the inner conductor and the outer electrodes. This is not practical for most coils, since measurement of these parameters requires disassembly of the system in which the coil is used. Hence, it is necessary to find a method that can detect insulation degradation and does not rely on these traditional methods.

Early efforts at using coil terminal impedance as an insulation health monitoring feature were outlined in patents from Kendig and Rogovin [8], [9]. The authors did not disclose any quantitative analyses of their proposal, nor did they publish any technical papers; it was an idea without a technical foundation. In 2006, Werynski *et al.* [10] asserted that the impedance resonant frequency could be used as a feature for coil insulation health monitoring. Their work was followed in 2007 and 2009 by Perisse *et al.* [11], [12], and in 2013 by Savin *et al.* [13]. These studies showed that changes in the insulation capacitance cause increases in impedance resonant frequency. The drawbacks of their work are as follows. First, the aging processes were based upon a uniform temperature in the coil since the coils were aged in a high temperature oven. Second, insulation resistance was not studied, but Younsi *et al.* [14] showed that insulation resistance can play a significant role in understanding the aging condition of electromagnetic coil insulation.

Polyimide, a common insulation material used in electromagnetic coils, is a polymer constructed of imide monomers, which consist of two acyl groups bound to a nitrogen. The chemical structure of “Kapton”, a classic polyimide, is shown in Figure 1. The high resistance of polyimide to thermal degradation and its high mechanical strength leads to diverse applications including high temperature fuel cells, liquid crystal alignments, and magnet wire insulation for use in electromagnetic coils [15], [16]. Polyimide insulation is given a 240 °C temperature rating, which means that the insulation is expected to survive for 20,000 hours (about 2.3 years) at 240 °C before experiencing electrical breakdown, according to ASTM D2307 [17]. Previous studies of polyimide aging [18]–[20] only examined films (1.5 < thickness < 125 μm) of polyimide under various degradation mechanisms (thermal, radiation, saline exposure). The results from these studies are mixed and inconsistent. Further, none of these studies examined the mechanical stresses that would be present when polyimide is applied as an electromagnetic coil insulator (due to Joule heating and the resulting expansion of the conductor). Moreover, the ASTM standard does not address the mechanical stress that the magnet wire will experience when applied in electromagnetic coils.

In this paper, dual-wound coils constructed using polyimide insulation are aged using Joule heating effects from applied power, which is a more realistic aging condition than the uniform temperature aging provided by an oven. Further, the coil impedance, the insulation impedance, and the insulation resistance are measured to provide insight into the aging process of polyimide insulation. The experiments provide data and features that can be incorporated into electromagnetic coil insulation health monitoring schemes.

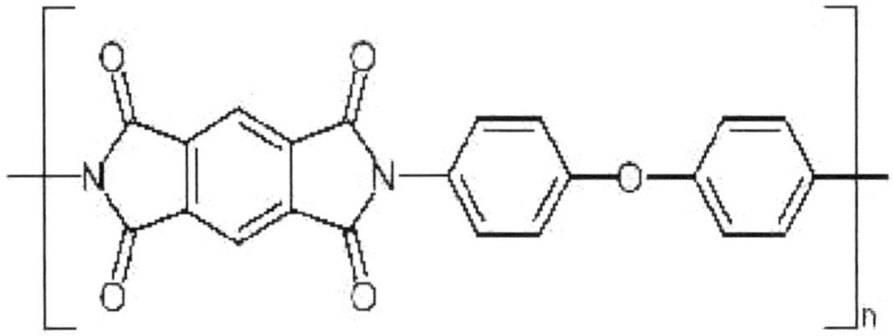

Figure 1: Chemical structure of "Kapton" polyimide

Experimental Setup: Dual-wound electromagnetic coils were manufactured by Magnecomp, Inc. (South Carolina, USA), as shown in Figure 2 (a). The parallel magnet wires were wound on a rectangular Pyroglass bobbin. An example of this winding arrangement (though not a replica of the arrangement of the experimental coils) is illustrated in Figure 2 (b), where the base numbers indicate the turn number and the superscript numbers indicate the two parallel windings of magnet wire. Each winding had 230 turns of AWG 27 (conductor diameter of 361 μm ± 3 μm) of Heavy Allex® polyimide insulation. This produced 7 layers of parallel windings, and each winding had a DC resistance of about 2.44 Ω.

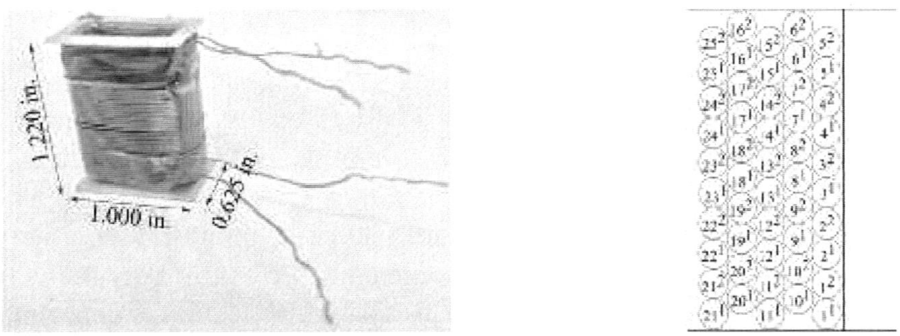

(a) (b)

Figure 2: Dual-wound coil (a) received units; and (b) winding arrangement illustration.

In order to age the insulation in a manner closely resembling the stresses that would be likely in an application condition, one of the parallel windings was powered to produce an average internal temperature of 200 °C. This temperature was determined using the temperature-resistance relationship as shown in (1), where $\alpha = 0.00393/°C$ for copper, and R_C and T_C are reference resistance and temperature, respectively. The final average temperature of the coil is given by $T(R_T)$ and the final resistance is given by R_T. Generally, the required voltage was between 7.7 and 7.8 V, and after the transient behavior settled, the current was between 1.8 and 1.85 A.

$$T(R_T) = \frac{1}{\alpha}\left(\frac{R_T}{R_C} + \alpha T_C - 1\right) \tag{1}$$

The coils were removed from the power periodically and three measurements were performed. First, the terminal impedance of one of the windings was measured using an

Agilent E4980A LCR meter at 501 distinct frequencies over the frequency range $f \in [20, 2e6]$ Hz with a 500 mV rms signal. This is illustrated for a single winding in Figure 3 (a). Second, the impedance of the insulation was measured using an Agilent E4980A LCR meter at 201 distinct frequencies in the frequency range $f \in [20, 2e6]$ Hz with a 1 V rms signal. This measurement was performed by shorting the terminals of each winding as illustrated in Figure 3 (b), which allowed the electric field to be directed through the insulation only. Third, the response of the insulation to a step application of 100 V was measured, allowing for measurement of DC resistance characteristics of the insulation. This measurement was performed with the coil windings in the same setup as for the insulation impedance. For this measurement, a 100 VDC potential was placed across the insulation and held for 20 minutes, while the current was sampled once per second.

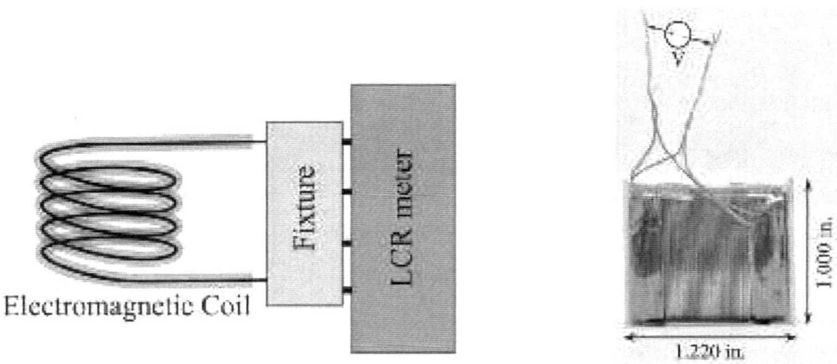

(a) (b)

Figure 3: Measurement setup for (a) winding terminal impedance; and (b) insulation impedance and resistance.

Experimental Results: Impedance is a complex-valued variable, which means it can be represented in polar form, $Z = |Z|e^{j}$, or Cartesian form, $Z = \Re\{Z\} + j\Im\{Z\} = R + j$, where $j = \sqrt{-1}$. In Cartesian form, the real part is referred to as resistance (not to be confused with DC resistance), and the imaginary part is referred to as reactance. Reactance, the imaginary part of impedance, can be further broken into inductive reactance, $X_L = \omega$, and capacitive reactance, $X_C = -(\omega\ )^{-1}$, such that $X = \Im\{Z\} = X_L + X_C$, where $\omega = 2\pi$ is angular frequency in rad/s (and f is frequency in Hz). Since any given reactance measurement is either positive or negative, when $X > 0$, the total reactance is said to be inductive, and when $X < 0$, the total reactance is said to be capacitive. The impedance spectra, shown in Cartesian form, for the one of the windings over 1103 hours of aging is shown in Figure 4. The resulting increase in the resonant frequency (the frequency where reactance crosses zero) was reproduced in the other samples.

The insulation impedance measurements can also be split into real and imaginary parts. In these measurements, reactance was negative (capacitive) over the entire frequency range, $f \in [20, 2e6]$ Hz. Therefore, each reactance measurement was converted into capacitance using the relationship $C = -(\omega X_C)^{-1}$. As can be seen in Figure 5, the insulation capacitance (at 100 kHz) decreased over the aging period. This behavior was replicated in all three samples. It must be noted that the samples aging period was limited by the life of the terminals. Since the terminals are made of magnet wire, the mechanical stresses introduced by bending caused the terminals to fail prior to the polyimide insulation. Nevertheless, the aging trend of the insulation is clear and repeatable over the three samples shown. In one of the samples, there is an increase in the capacitance between about 450 hours and 620 hours. This is a partial recovery after the coil was removed from thermal stress for 1 week. Hence, the insulation degradation process, as reflected in the electrical parameter measurement, is at least partially reversible. However, after being reconnected to power, the capacitance not only returned to its previous value, but the trend continued downward.

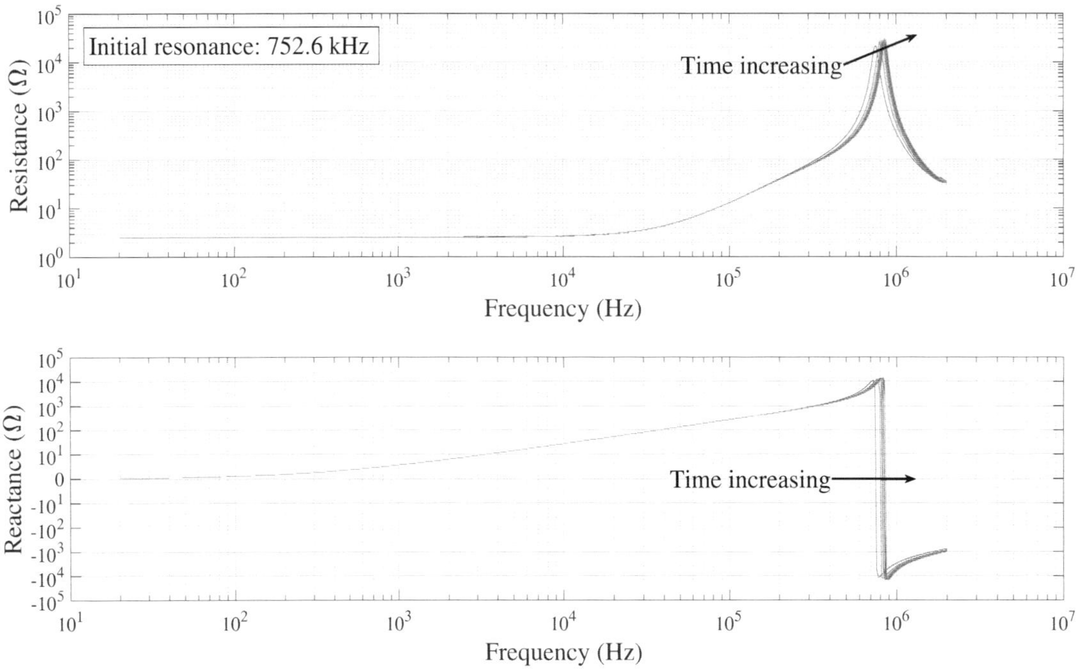

Figure 4: Resistance (above) and reactance (below) spectra of one coil winding over the course of an experiment (1172 hours).

The analysis of the insulation current was performed as follows. First, the insulation current of healthy insulation was measured over an extended period of time, in order to allow the current to settle. This was done by placing 100 VDC across the insulation, and measuring the resulting current. The initial insulation current measurements for two coils are shown in Figure 6. One coil was measured for about 8.3 hours with current measurements taken every 3 seconds and the other for 13.8 hours with current measurements taken every 5 seconds. Then, in order to measure the effect of aging on the insulation resistance, the current was sampled every 1 second for 20 minutes. These results are shown in Figure 7 and Figure 8. (For the insulation current measurements, only two samples were available due to experimental complications.)

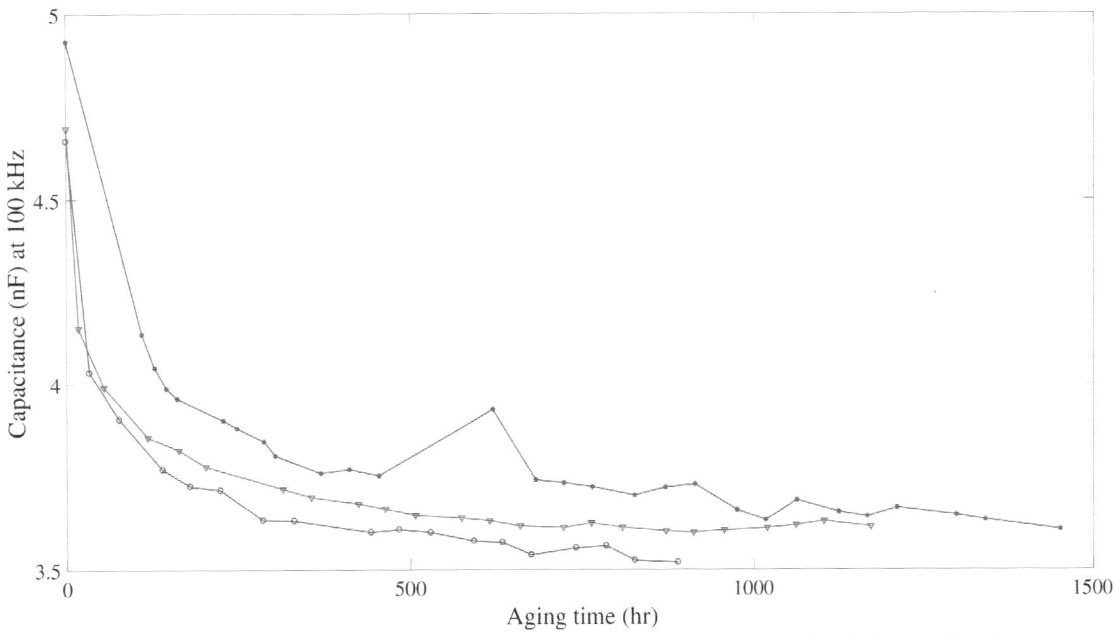

Figure 5: Change in polyimide insulation capacitance as a result of thermal aging at an average coil temperature of 200 °C. The line colors and markers correspond to three different coils.

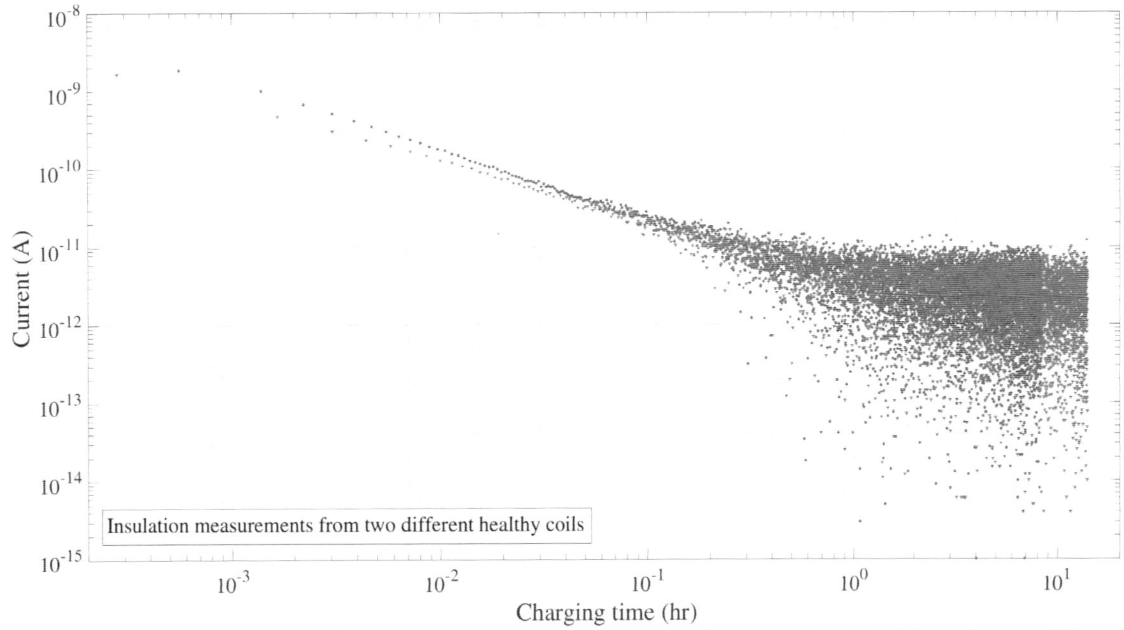

Figure 6: Polyimide insulation current measurements for new (healthy) insulation. The two colors and markers represent two different coils.

The relationship between the polarization current and the charging time can be expressed using the Curie-von Schweidler "universal law" for dielectrics [21]–[25] as shown in (2), where $i_p(t_c)$ is the polarization current at charging time, t_c, and $0 < n < 1$ is the power law exponent.

$$i_p(t_c) \propto t_c^{-n} \tag{2}$$

The curves shown in Figure 6 clearly show this behavior, with the current settling to a constant value such that the total current is as expressed in (3).

$$i_p(t_c) = at_c^b + c \tag{3}$$

In (3), a is the coefficient of proportionality, c is the steady-state (or conduction) current, and $-1 < b < 0$ is the power law decay of the polarization current. The curve (3) can be fit to the data using a robust non-linear least squares algorithm. The robustness in the algorithm ensures that the adjustments to the function are less sensitive to outliers in the data. In this work, the least absolute residual (LAR) method for robust regression was used, which minimizes the absolute value of the residual instead of minimizing the square of the residual. Therefore, the extreme values have a lesser influence on the fit. Some insulation current measurements are shown in Figure 7. The measurement procedure was altered after the first two measurements to begin the current measurements at 1 second as opposed to 3 seconds. This is the reason for the difference in the starting times for the data.

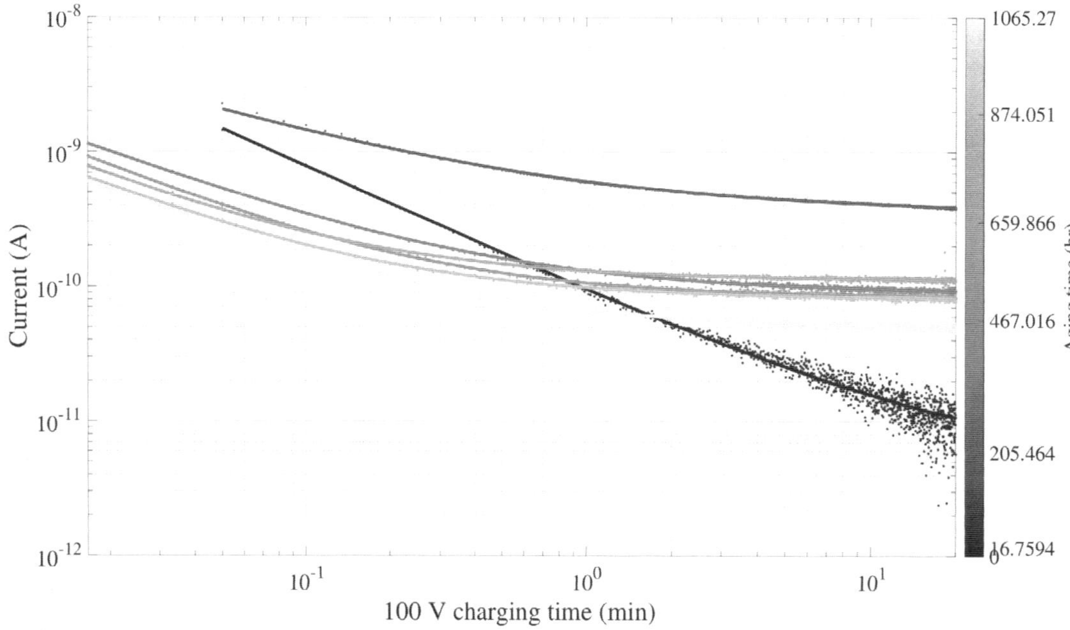

Figure 7: Measurements of insulation current versus charging time at 100 VDC as the insulation aged at an average coil temperature of 200 °C. The lines are drawn over the data and are shown in the same color as the raw data.

It is clear that as the insulation is aged, the insulation resistance changes. In fact, from the first measurement to 16.76 hours of aging, there is a significant decrease in the insulation resistance. In order to illustrate how the insulation resistance changed for two samples as the insulation aged, the regression fit parameters are shown in Figure 8. The data appears to be agree between the two samples for the parameters $a(t)$ and $c(t)$, whereas there are large differences in the parameter $b(t)$ between the two samples. The values of $c(t)$ show that the conduction current of the insulation increased over the aging period, or equivalently, the resistance of the insulation decreased as the insulation aged.

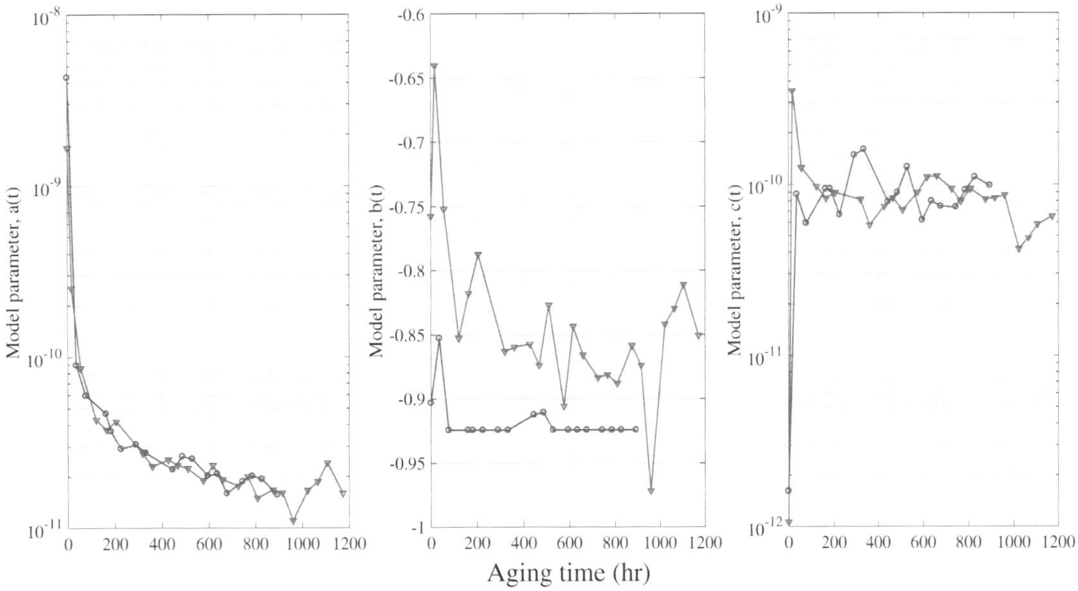

Figure 8: Evolution of regression fit of the polarization current as the insulation aged at an average coil temperature of 200 °C. The marker styles and colors correspond to two different coils.

Terminal impedance data analysis: Previous work [26]–[28] showed that not only do terminal impedance measurements capture changes in the insulation, but there are frequencies within the impedance spectrum that capture the insulation degradation information better than others. To explore this effect in the coils studied in this paper, a Spearman correlation spectrum was constructed. The Spearman correlation coefficient for n samples, X_i, Y_i is computed by first converting the raw data into ranked variables x_i and y_i, then forming the distance measure $d_i = y_i - x_i$.

$$\rho_S = 1 - \frac{6}{n(n^2-1)} \sum_{i=1}^{n} d_i^2 \tag{4}$$

The Spearman correlation coefficient is a measure of linear and nonlinear monotonic correlation [29], and hence, is more general that Pearson correlation, which only measures linear correlation. The Spearman correlation coefficient was computed using the vector of time entries when the coil was removed from power and measured, and the time series of each impedance measurements at each frequency. This resulted in a vector that is the same length as the number of frequencies, $N = 501$, where each entry is a Spearman correlation coefficient quantifying the degree of monotonicity of any given impedance measurement at a particular frequency over the degradation time period. Thus, the impedance at each frequency is assessed as a potential health indicator independently of the other frequencies.

The Spearman correlation spectra show the differences in behavior of resistance ($\Re\{Z\}$) and reactance ($\Im\{Z\}$) as the coils aged. In general, there is significant agreement among the reactance Spearman spectra, with root mean square (RMS) difference values of 0.0905, 0.1052, and 0.0941 between the three curves. The resistance Spearman spectra have much larger RMS difference values of 0.6898, 0.3589, and 0.7331 between the three curves.

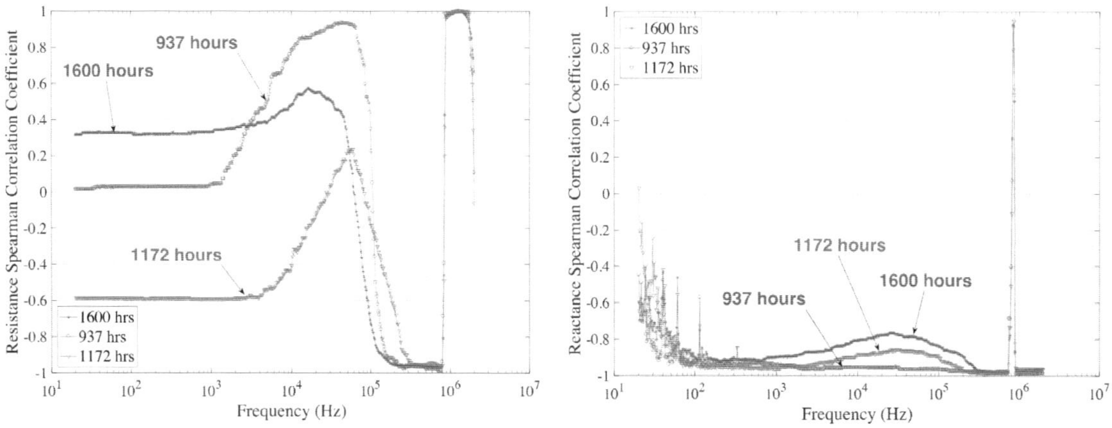

Figure 9: Resistance (left) and reactance (right) Spearman correlation spectrum for 3 coil samples aged at an average coil temperature of 200 °C. The line colors and markers correspond to three different coils aged for different lengths of time.

Consider the time series of reactance at $f = 604$ kHz (one of the minimum points in the three reactance Spearman spectra) and $f = 853.2$ kHz (the maxima of the three reactance Spearman spectra) as shown in Figure 10. The "recovery" in one of the coils (as mentioned with the capacitance measurements) is visible between 450 and 620 hours. Similarly, consider the resistance (i.e., $\Re\{Z\}$) at $f = 778.1$ kHz and $f = 1.262$ MHz as shown in Figure 11.

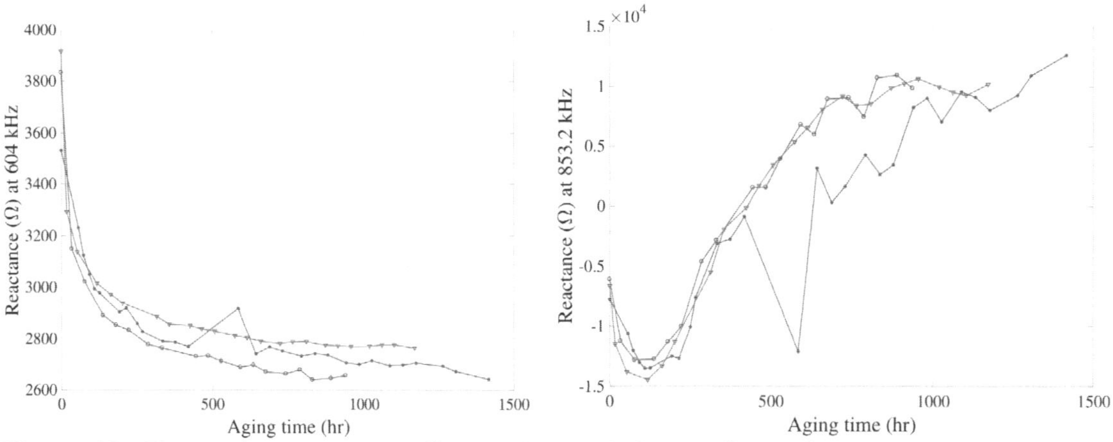

Figure 10: Change in reactance at $f = 604$ kHz (left) and $f = 853.2$ kHz (right) over the aging period for three coils. The line colors and markers correspond to three different coils.

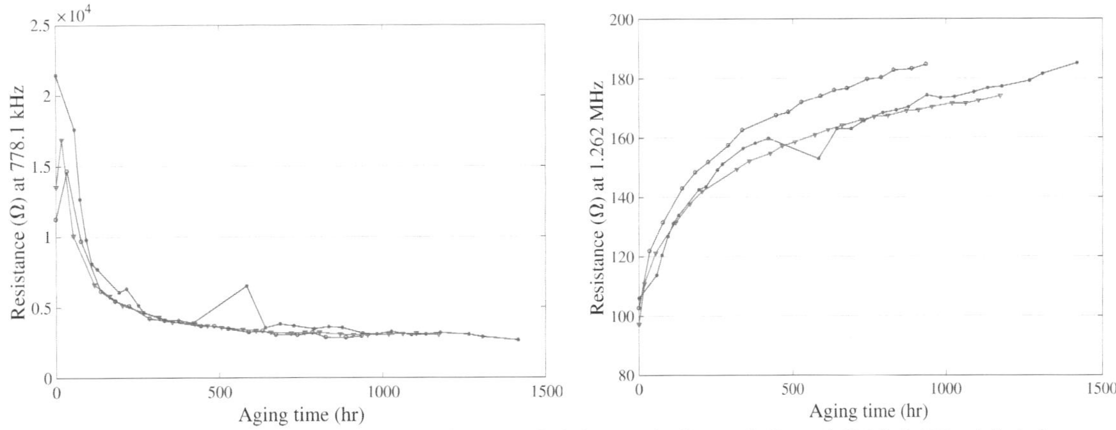

Figure 11: Change in reactance at f = 778.1 kHz (left) and f = 1.262 MHz (right) over the aging period for three coils. The line colors and markers correspond to three different coils.

Table 1: Percent changes in relevant frequencies for resistance and reactance, and the changes in the resonant frequency for all three coils.

	Percent change: % change = $\lvert (x_C - x_f)/x_C \rvert \times 100\%$		
Frequency	Resistance, $R = \Re\{Z\}$	Reactance, $X = \Im\{Z\}$	Resonant frequency
778.1 kHz	73.65, 77.20, 85.59	*	*
1.262 MHz	79.93, 76.57, 66.75	*	*
604 kHz	*	30.69, 29.43, 23.60	*
853.2 kHz	*	263.1, 254.3, 217.7	*
*	*	*	15.60, 14.89, 13.30

The data shown in Table 1 demonstrates the relative sensitivity of frequencies below and above the coil resonant frequency to changes in the electrical properties of the insulation. Further, the trends are repeatable across three samples. Not all coils (with various geometries and insulation materials) experience the same amount of migration of the resonant frequency with aging. It is therefore necessary to observe the most sensitive feature, which is shown not to be the resonant frequency, contrasting with suggestions in [10]–[12].

Conclusion: The degradation of insulation in electromagnetic coils can result in failure of the coil, which in turn can result in the failure of the system using the coil. Past studies have shown insulation failure to be a significant cause of failure in electric motors and solenoid valves.

This paper investigated the changes in the electrical behavior of polyimide insulation while being aged at an average coil temperature of 200 °C. Polyimide is a widely-used insulation with high resistance to thermal degradation (rated for 240 °C) and high mechanical strength. The study presented herein showed that as polyimide insulation aged, the coil resonant frequency increased, the capacitance dropped, and the steady-state DC resistance of the insulation decreased. This data can be used to understand the thermal degradation nature of polyimide and to design impedance-based health monitoring systems for electromagnetic coils that use polyimide insulation.

The terminal impedance data was analyzed showing that within the impedance spectrum, there are frequencies that are very sensitive to changes in the electrical properties of the insulation. Four frequencies were given that had percent changes between 2 and 16 times larger than that experienced by the resonant frequency. This provides further experimental support to the hypothesis presented in [26] and [27]. Moreover, a Spearman correlation spectrum provided a holistic view of the changes in the impedance spectra as the coils aged, which can be related to the changes in the insulation electrical parameters. In other words, the decreasing insulation capacitance and insulation resistance resulted in the observed changes in the impedance spectrum as the coils aged. In knowing the relationship between these variables, a suitable threshold can be established, allowing for condition-based maintenance to be properly implemented.

Acknowledgement: The authors would like to thank Magnecomp, Inc. for generously donating the coils used in the presented experiments. Further thanks go to the more than 100 companies that sponsor research activities at the Center for Advanced Life Cycle Engineering (CALCE) at the University of Maryland annually.

References:

[1] Motor Reliability Working Group, "Report of large motor reliability survey of industrial and commercial installations, Part I," *IEEE Trans. Ind. Appl.*, vol. IA-21, no. 4, pp. 853–864, Jul. 1985.

[2] Motor Reliability Working Group, "Report of large motor reliability survey of industrial and commercial installations, Part II," *IEEE Trans. Ind. Appl.*, vol. IA-21, no. 4, pp. 865–872, Jul. 1985.

[3] O. V. Thorsen and M. Dalva, "A survey of faults on induction motors in offshore oil industry, petrochemical industry, gas terminals, and oil refineries," *IEEE Trans. Ind. Appl.*, vol. 31, no. 5, pp. 1186–1196, Sep. 1995.

[4] V. P. Bacanskas, G. C. Roberts, and G. J. Toman, "Aging and service wear of solenoid-operated valves used in safety systems of nuclear power plants. Volume 1: Operating experience and failure identification," Oak Ridge National Laboratory, Mar. 1987.

[5] G. C. Stone, E. A. Boulter, I. Culbert, and H. Dhirani, *Electrical Insulation for Rotating Machines: Design, Evaluation, Aging, Testing, and Repair*. John Wiley, 2004.

[6] S. Grubic, J. M. Aller, B. Lu, and T. G. Habetler, "A survey on testing and monitoring methods for stator insulation systems of low-voltage induction machines focusing on turn insulation problems," *IEEE Trans. Ind. Electron.*, vol. 55, no. 12, pp. 4127–4136, Dec. 2008.

[7] P. J. Tavner, "Review of condition monitoring of rotating electrical machines," *IET Electr. Power Appl.*, vol. 2, no. 4, pp. 215–247, 2008.

[8] M. W. Kendig and D. N. Rogovin, "Method of conducting broadband impedance response tests to predict stator winding failure," U.S. Patent 6 323 658 B1, 27-Nov-2001.
[9] M. W. Kendig and D. N. Rogovin, "Method of conducting broadband impedance response tests to predict stator winding failure," U.S. Patent 6 483 319 B1, 19-Nov-2002.
[10] P. Werynski, D. Roger, R. Corton, and J.-F. Brudny, "Proposition of a new method for in-service monitoring of the aging of stator winding insulation in AC motors," *IEEE Trans. Energy Convers.*, vol. 21, no. 3, pp. 673–681, Sep. 2006.
[11] F. Perisse, D. Mercier, E. Lefevre, and D. Roger, "Robust diagnostics of stator insulation based on high frequency resonances measurements," *IEEE Trans. Dielectr. Electr. Insul.*, vol. 16, no. 5, pp. 1496–1502, Oct. 2009.
[12] F. Perisse, P. Werynski, and D. Roger, "A new method for AC machine turn insulation diagnostic based on high frequency resonances," *IEEE Trans. Dielectr. Electr. Insul.*, vol. 14, no. 5, pp. 1308–1315, Oct. 2007.
[13] S. Savin, S. Ait-Amar, and D. Roger, "Organic enameled wire aging monitoring based on impedance spectrum analysis," in *2012 Annual Report Conference on Electrical Insulation and Dielectric Phenomena (CEIDP)*, 2012, pp. 874–877.
[14] K. Younsi, P. Neti, M. Shah, J. Y. Zhou, J. Krahn, K. Weeber, *et al.*, "On-line capacitance and dissipation factor monitoring of AC stator insulation," *IEEE Trans. Dielectr. Electr. Insul.*, vol. 17, no. 5, pp. 1441–1452, Oct. 2010.
[15] D.-J. Liaw, K.-L. Wang, Y.-C. Huang, K.-R. Lee, J.-Y. Lai, and C.-S. Ha, "Advanced polyimide materials: Syntheses, physical properties and applications," *Prog. Polym. Sci.*, vol. 37, no. 7, pp. 907–974, Jul. 2012.
[16] A. Georgiev, D. Dimov, E. Spassova, J. Assa, P. Dineff, and G. Danev, "Chemical and Physical Properties of Polyimides: Biomedical and Engineering Applications," in *High Performance Polymers - Polyimides Based - From Chemistry to Applications*, M. J. M. Abadie, Ed. InTech, 2012.
[17] "D2307 – 07a: Standard Test Method for Thermal Endurance of Film-Insulated Round Magnet Wire." ASTM International, 2013.
[18] L. Li, N. Bowler, P. R. Hondred, and M. R. Kessler, "Dielectric response of polyimide to thermal and saline degradation," in *2010 Annual Report Conference on Electrical Insulation and Dielectric Phenomena (CEIDP)*, 2010, pp. 1–4.
[19] S. Diaham, M. L. Locatelli, and T. Lebey, "Improvement of Polyimide Electrical Properties During Short-Term of Thermal Aging," in *2008 Annual Report Conference on Electrical Insulation and Dielectric Phenomena*, 2008, pp. 79–82.
[20] R. Khazaka, M. L. Locatelli, S. Diaham, and P. Bidan, "Effects of mechanical stresses, thickness and atmosphere on aging of polyimide thin films at high temperature," *Polym. Degrad. Stab.*, vol. 98, no. 1, pp. 361–367, Jan. 2013.
[21] J. Curie, "Recherches sur le pouvoir inducteur specifique et la conductibilite des corps cristallises," *Ann. Chim. Phys.*, vol. 17, pp. 385–434.
[22] E. R. von Schweidler, "Studien über die Anomalien im Verhalten der Dielektrika (Studies on the anomalous behaviour of dielectrics)," *Ann. Phys.*, vol. 329, no. 14, pp. 711–770, Jan. 1907.
[23] A. K. Jonscher, *Dielectric Relaxation in Solids*. Chelsea Dielectrics Press Limited, 1983.

[24] A. K. Jonscher, "Dielectric relaxation in solids," *J. Phys. Appl. Phys.*, vol. 32, no. 14, p. R57, 1999.
[25] "IEEE Recommended Practice for Testing Insulation Resistance of Electric Machinery," *IEEE Std 43-2013 Revis. IEEE Std 43-2000*, pp. 1–37, Mar. 2014.
[26] N. Jordan Jameson, Michael H. Azarian, and Michael Pecht, "Impedance-based condition monitoring for insulation systems used in low-voltage electromagnetic coils," *IEEE Trans. Ind. Electron.*, vol. PP, no. 99, pp. 1–10, 2017.
[27] N. Jordan Jameson, Michael H. Azarian, Michael Pecht, Carlos Morillo, and Kai Wang, "Health monitoring of solenoid valve electromagnetic coil insulation under thermal deterioration," in Proceedings of MFPT 2016/ISA's 62nd IIS, Dayton, OH, 2016.
[28] N. Jordan Jameson, Michael H. Azarian, and Michael Pecht, "Impedance-based health monitoring of electromagnetic coil insulation subjected to corrosive deterioration," in *Proceedings of the Annual Conference of the Prognostics and Health Management Society 2016*, Denver, CO, 2016.
[29] J. L. Myers, A. Well, and R. F. Lorch, *Research Design and Statistical Analysis*. Routledge, 2010.

VIBRATION SENSOR WITH TWO-WIRE INTERFACE AND BIAS USED FOR MEASURE TEMPERATURE

George Zusman
Vibration Measurement Solutions, Inc.
Houston, TX 77077
832-277-307-
zusger@gmail.com

Abstract: A sensor with integrated mechanical transducing and temperature monitoring capability is described. The sensor includes housing containing a transducer, a temperature sensor with associated bias, a summing circuit, and a two-wire cable connector. The transducer is operable to output a dynamic transducer waveform that corresponds to dynamic mechanical perturbations sensed by the transducer. The temperature sensor is operable to output a quasi-static temperature waveform that corresponds to temperatures sensed by the temperature sensor. The summing circuit is operable to combine the transducer waveform and the temperature waveform into a composite modulated voltage bias output signal or modulated current bias output signal. The two-wire cable standard connector is accessible on an outside of the housing and is connectable to a two wire cable that delivers power to the sensor from a power source and delivers the composite output signal from the sensor to a remote data acquisition circuit.

Key words: Vibration, temperature, sensors, two wire interface

Introduction: The present paper describes sensors with vibration transducing capability, in particular the piezoelectric or MEM transducer with an integrated preamplifier and a two-wire interface connected via a cable to a constant current IEPE/ICP® (ICP is an abbreviation of "integrated circuit piezoelectric", and is a registered trademark (no. 1,603,466 in the USA) of PCB Group, Inc. There is a similar non-proprietary standard called IEPE, which stands for "integrated electronics piezoelectric.) [1] or constant voltage CLD (Current Line Drive) [2] power source for power input, and to a data acquisition circuit for signal output.
Such piezoelectric (or MEMs) based transducers used for detection of vibrations and other dynamic physical parameters are often installed at locations that are remote from monitoring equipment that evaluates the sensor output. To overcome the noise and interference inherent in such environments, line drive piezoelectric sensors have been developed that combine a piezoelectric transducer and a built-in preamplifier that transforms the high impedance charge mode output of the transducer into a low impedance voltage or current output signal. A two-wire interface and associated two-wire cable (such as a coaxial cable or two wire cable) connects the sensor to a remote constant current or constant voltage power source for power input, and to a remote data acquisition circuit for signal output. The two-wire cable may be as long as several hundred meters. In some piezoelectric sensor designs, the sensor is powered by a constant current power source and the sensor's output is a modulated bias

voltage signal. In other piezoelectric sensor designs, the sensor is powered by a constant voltage power source and the sensor's output is a modulated bias current signal.
Line drive piezoelectric sensors as described above are designed to measure only a single parameter, such as vibration. If an additional parameter needs to be measured, such as temperature in situations where both parameters are need to be monitored at the same measurement location, a separate output and lengthy cable connection to a remote unit are required. This redundancy adds design complexity and expense.
Bellow the integrated sensor is proposed that combines mechanical transducing with temperature monitoring [3].

Description: The presented sensor includes housing containing a transducer, a temperature sensor, a summing circuit, and a two-wire cable connector. The transducer is operable to output a dynamic transducer waveform that corresponds to dynamic mechanical perturbations sensed by the transducer. The temperature sensor is operable to output a quasi-static temperature waveform that corresponds to temperatures sensed by the temperature sensor. The summing circuit is operable to combine the transducer waveform and the temperature waveform to generate a composite bias-modulating output signal. The two-wire cable connector is accessible on an outside of the housing and is connectable to a two wire cable that delivers power to the sensor from a remote power source and delivers the composite output signal from the sensor to a remote data acquisition circuit.
The Figure 1 shows a schematic diagram of the sensor with combination of a constant current power source and a data acquisition circuit.
The Figure 2 shows a schematic diagram of the sensor with combination of a constant voltage power source and a data acquisition circuit.

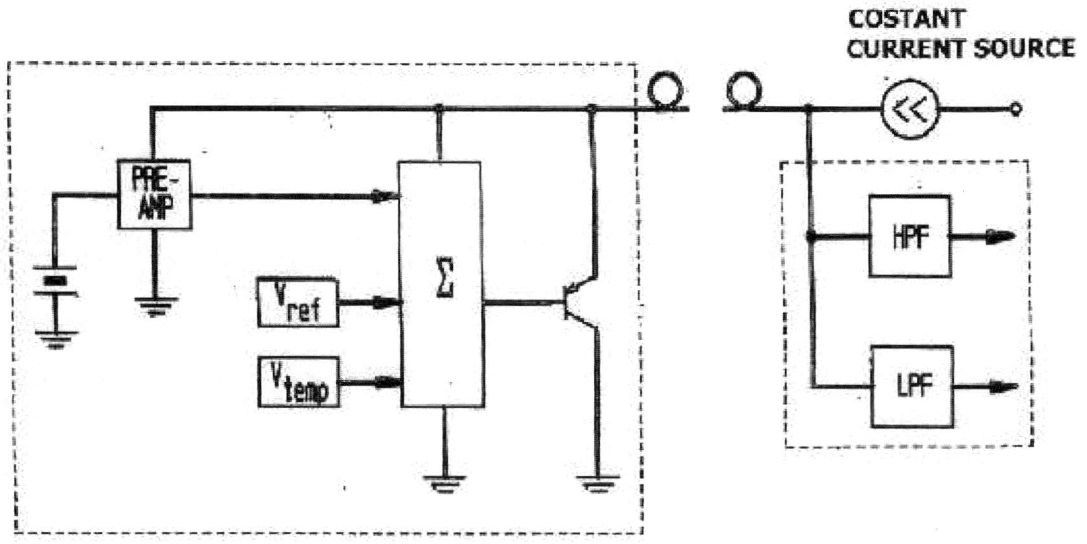

Figure 1 IEPE vibration sensor included charge amplifier, reference voltage source, temperature sensor, summarize circuit and voltage repeater, where DC part of bias voltage consist of constant voltage and voltage proportional to temperature, and AC part of voltage proportional to vibration signal.

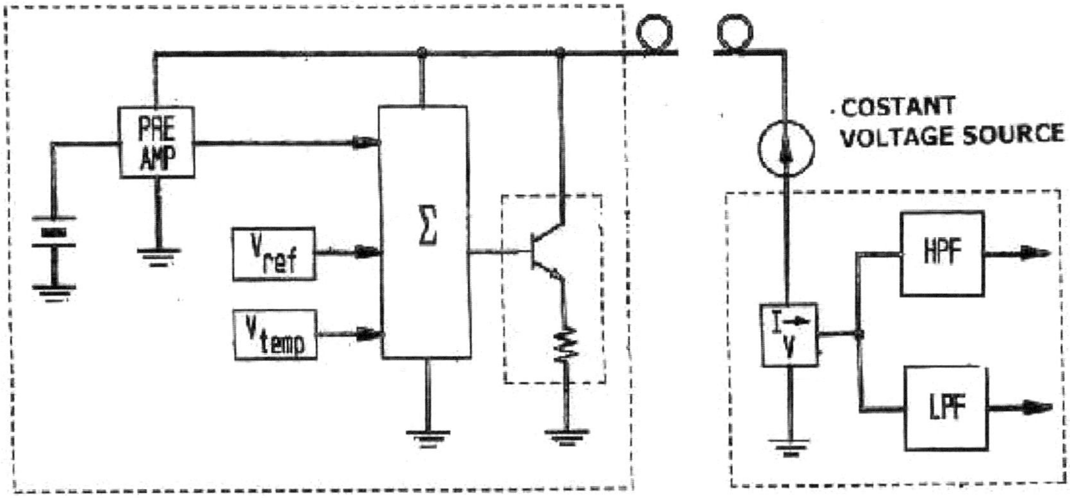

Figure 2 CLD vibration sensor included charge amplifier, reference voltage source, temperature sensor, summarize circuit and voltage to current transmitter, where DC part of bias current consist of constant current and current proportional to temperature, and AC part of current proportional to vibration signal.

The summing amplifier outputs a time-varying voltage output embodying a composite waveform that represents a summation of the transducer waveform and the temperature waveform, with direct current bias provided by the reference voltage from the voltage reference source. Unlike the dynamic transducer waveform, which tends to vary relatively rapidly, the temperature waveform output by the temperature sensor will normally be quasi-static in nature. The temperature waveform will be time-varying, but of relatively low frequency compared to the frequency of the transducer waveform.

Testing results: On the Figure 3 and Figure 4 the testing results of the described sensor are provided. The sensor typical specification is placed in the Table 1.

Table 1. Sensor technical specification

Name	Value
Nominal sensitivity, mV/g	100
Frequency range with +/-3 dB , Hz	2...10000
Dynamic rang, g pk	0...50
Sensitivity tolerance, %	<+/-5
Cross exes sensitivity, %	<3
Non-linearity, %	<1
Nominal bias voltage at 20 Degree C, V	12
Bias voltage tolerance at 20 Degree C, V	<+/-0,1
Nominal bias voltage temperature coefficient, mV/Degree C	25
Temperature measuring range, Degree C	-40...+120
Temperature measurement accuracy, Degree C	<+/-4
Working temperature range, Degree C	-55...+125

Interface	2-wire IEPE
Nominal power voltage, V	24 +/-10%
Constant current range, мA	2...8

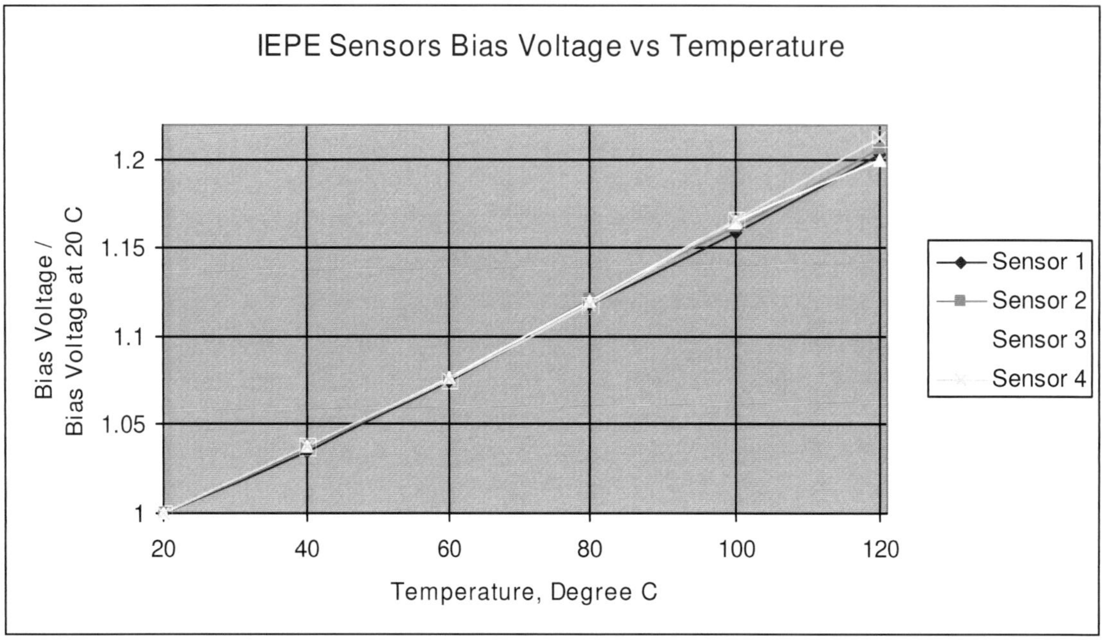

Figure 3. Designed IEPE sensor Relative Bias voltage vs Temperature

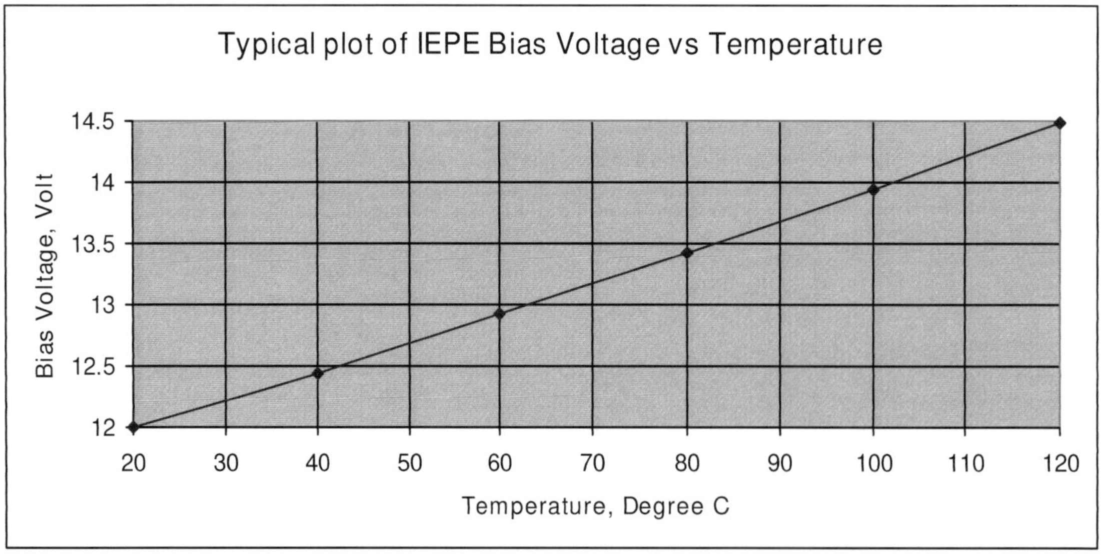

Figure 4. Typical IEPE designed sensor Absolute Bias Voltage vs Temperature

Conclusion:

This paper introduced a vibration sensor with two wire output and possibility to measure acceleration and temperature without requiring any additional third wire. It is done by making bias of the sensor accurately linear increasing follow to the temperature. The sensor could be used by usual way with two wire caballing and to measure the temperature additionally to acceleration no any changes required in the installation procedures. We presented the sensor testing results and typical specification. These sensors practically would be not expensive than common accelerometers and it is might be important then user have a possibility to decide of measuring temperature at any time after installation and just for particular channels without any additional equipment or action.

References:

[1]IEPE Standard http://www.mmf.de/iepe_standard.htm

[2]V. Petrovich, G. Zusman "Apparatus for vibration measurement", Pat. USSR 1545090, Jan 20, 1987, published Feb. 23, 1990.

[3]G. Zusman "Two wire electronics interface sensor with integrated mechanical transdusing and temperature monitoring", USA Pat. Application 14/821,727. Published Feb. 9, 2017

AUTHOR INDEX